METAL FORMING

Interrelation Between Theory and Practice

The Metallurgical Society of AIME Proceedings
published by Plenum Press

1968–Refractory Metal Alloys: Metallurgy and Technology
Edited by I. Machlin, R. T. Begley, and E. D. Weisert

1969–Research in Dental and Medical Materials
Edited by Edward Korostoff

1969–Developments in the Structural Chemistry of Alloy Phases
Edited by B. C. Giessen

1971–Metal Forming: Interrelation Between Theory and Practice
Edited by A. L. Hoffmanner

A Publication of the Metallurgical Society of AIME

METAL FORMING

Interrelation Between Theory and Practice

Proceedings of a symposium on the Relation Between Theory
and Practice of Metal Forming, held in Cleveland, Ohio, in October, 1970

Edited by
A. L. Hoffmanner
Principal Engineer
Materials Technology
Equipment Group of TRW, Inc.
Cleveland, Ohio

SPRINGER SCIENCE+BUSINESS MEDIA, LLC

Originally printed by Plenum Press, New York in 1971
Softcover reprint of the hardcover 1st edition 1971

Library of Congress Catalog Card Number 70-171698
ISBN 978-1-4613-5708-7 ISBN 978-1-4615-1757-3 (eBook)
DOI 10.1007/978-1-4615-1757-3

METAL FORMING: THE INTERRELATION BETWEEN THEORY AND PRACTICE

A Memorial Volume Dedicated to the Memory of Professor Horace Polakowski

It is infrequent that a man rises in the technical community to successfully contribute to both basic science and commercial technology. Horace Polakowski was one of those rare individuals destined by both ability and desire to significantly contribute to both.

This man not only attained eminence in the world of science as well as the world of technology, but literally spanned the globe in his work experience. He was fluent in English, German, Polish, and Russian and had a working knowledge of French and Italian.

He was born on December 4, 1914, in Poland, where in 1938, he obtained a diploma in engineering (mechanical engineering) from the Technical University of Lvov, Poland. His first technical experience was in industrial operations in Poland, a Mannesmann steel tube mill, a stamping and pressing shop, an alloy foundry, large military vehicle repair shops, and in a gray iron foundry. In 1939 he was awarded a fellowship to study at the University of Swansea but was unable to join the Metallurgical Engineering Department until after World War II. In 1948, he joined the University of Swansea and in 1952 was granted a Ph.D. in metallurgy. In 1950 he was awarded the Andrew Carnegie Silver Medal by the Iron and Steel Institute in London.

As a stateless person, Dr. Polakowski came to the United States to join the Armzen Company as Development Manager in 1953. Subsequently he joined La Salle Steel Company where he worked prior to becoming a member of the staff of Illinois Institute of Technology. He was appointed a full Professor at IIT in 1958. He published prolifically in many technical areas. In 1965 the University of Wales recognized his important technical contributions and conferred upon Dr. Polakowski a Doctor of Science.

In recent years his interest turned more and more to commercial applications of his extensive background in metalworking. As he so aptly put it he became "a consultant to the world." Equipment designed according to his patents have found worldwide acceptance in the roller leveling of strip.

Truly, Dr. Natalis Horace Polakowski made unique contributions to metalworking science and technology. A recitation of his accomplishments unfortunately cannot truly do justice to the flavor of this man who could galvanize an audience, convulse his listeners with laughter, and at the same time bring to bear an incisive and inquiring mind on important technical problems.

This volume is indeed a fitting memorial. Horace Polakowski may have been a stateless person when he landed in the United States on December 28, 1953, but he was an illustrious citizen of the world when he died while returning after delivering a lecture series in Argentina in 1970.

Elliot S. Nachtman

January, 1971

FOREWORD

On October 21 and 22, 1970, the Shaping and Forming Committee, Institute of Metals Division, The Metallurgical Society of AIME, held a Conference on "The Relation Between Theory and Practice in Metal Forming" at the Sheraton-Cleveland Hotel during the Fall Meeting of AIME in Cleveland, Ohio. This conference was devoted to recent applications of theory to metal forming to establish a milestone in the current ability to predict phenomena during deformation processing and, thereby, demonstrate the utility of theory for process design. The papers were selected by first requesting presentations of relevant recent work from 68 recognized authorities in metal forming which resulted in 17 papers. A subsequent call for papers resulted in the submission of 19 abstracts from which 4 papers were selected. The selection criteria required that the paper coupled theory with practice, and that the work was recent, unpublished and worthy of permanent record. The selection was performed by the Conference Chairman.

The papers in this volume have been organized in accordance with the following subjects:

Extrusion
Drawing and Sheet Metal Forming
Forming Loads and Friction
Workability

These papers appear to assess the salient recent applications of mechanics to the deformation processing of alloys at the present time, i.e., circa 1970.

A. L. Hoffmanner
Conference Chairman

May, 1970

LIST OF CONTRIBUTORS

Taylan Altan, Metalworking Division, Columbus Laboratories, Battelle Memorial Institute, Columbus, Ohio

W. A. Anderson, Physical Metallurgy Division, Alcoa Research Laboratories, New Kensington, Pennsylvania

Betzalel Avitzur, Professor of Metallurgy and Materials Science, Lehigh University, Bethlehem, Pennsylvania

C. Baker, Reynolds Metals Company, Richmond, Virginia

John T. Berry, Department of Mechanical Engineering, University of Vermont, Burlington, Vermont

J. H. Brophy, Paul D. Merica Research Laboratory, The International Nickel Company, Incorporated

T. Chandra, Department of Metallurgical Engineering, McGill University, Montreal, Canada

H. Darlington, Homer Research Laboratories, Bethlehem Steel Corporation, Bethlehem, Pennsylvania

Vincent DePierre, Air Force Materials Laboratory, MAMN, Wright Patterson Air Force Base, Ohio

M. L. Devenpeck, Edgar C. Bain Laboratory for Fundamental Research, United States Steel Corporation, Monroeville, Pennsylvania

R. W. Dunlap, Carnegie Mellon University, Pittsburgh, Pennsylvania

R. C. Gibson, Paul D. Merica Research Laboratory, The International Nickel Company, Incorporated

H. W. Hayden, Paul D. Merica Research Laboratory, The International Nickel Company, Incorporated

D. O. Hobson, Metals and Ceramics Division, Oak Ridge National Laboratory, Oak Ridge, Tennessee

A. L. Hoffmanner, T R W Incorporated, Cleveland, Ohio

J. J. Jonas, Department of Metallurgical Engineering, McGill University, Montreal, Canada

L. J. Kashar, United States Steel Corporation, Monroeville, Pennsylvania

Shiro Kobayashi, Department of Mechanical Engineering, Division of Mechanical Design, University of California, Berkeley, California

E. H. Kottcamp, Jr., Homer Research Laboratories, Bethlehem Steel Corporation, Bethlehem, Pennsylvania

S. A. Levy, Reynolds Metals Company, Richmond, Virginia

Alan T. Male, Westinghouse Astronuclear Laboratories, Pittsburgh, Pennsylvania

H. J. McQueen, Mechanical Engineering, Sir George Williams University, Montreal, Canada

James A. Mullendore, Chemical and Metallurgical Division, Sylvania Electric Products, Incorporated, Towanda, Pennsylvania

T. E. O'Connell, Carnegie Mellon University, Pittsburgh, Pennsylvania

Malcolm H. Pope, Department of Mechanical Engineering, University of Vermont, Burlington, Vermont

H. C. Rogers, Department of Metallurgical Engineering, Drexel University, Philadelphia, Pennsylvania

R. W. Rogers, Jr., Physical Metallurgy Division, Alcoa Research Laboratories, New Kensington, Pennsylvania

George Saul, Air Force Materials Laboratory, MAMN, Wright Patterson Air Force Base, Ohio

John A. Schey, Department of Materials Engineering, University of Illinois, Chicago Circle, Chicago, Illinois

Aly H. Shabaik, Assistant Professor of Engineering, University of California, Los Angeles, California

Oleg D. Sherby, Department of Materials Science, Stanford University, Stanford, California

Conrad M. Young, Department of Materials Science, Stanford University, Stanford, California

Z. Zimerman, Homer Research Laboratories, Bethlehem Steel Corporation, Bethlehem, Pennsylvania

CONTENTS

STUDY OF FLOW THROUGH CONICAL CONVERGING DIES*

Betzalel Avitzur

Professor of Metallurgy and Materials Science

Lehigh University - Bethlehem, Pennsylvania

ABSTRACT

Experimental study of metal flow encounters inherent limitations in the absence of adequate theoretical support. Such support is provided in this presentation of an analytical approach to the study of drawing or extrusion through conical converging dies. For such flow the analysis is relatively complete and its practical applicability has been proven. This approach -- limit analysis -- can be applied to rolling, forging, and other metal forming processes.

Limit analysis is based on the principle of minimum energy and leads through stated assumptions to approximate solutions which place upper and lower bounds on drawing or extrusion force as a function of the semicone angle of the die, reduction, friction, and, later, a material property variable. The analysis brackets the conceptual exact solution between these upper and lower bound solutions. Comparison of outcomes shows reasonably good agreement with the data of Wistreich's classic experimental study. Explicit criteria, in which friction is a predominant factor, are derived for fracture and other flow patterns.

Strict procedures which establish a solution as an upper bound solution are demonstrated: proposal of an appropriate velocity field to describe each mode of flow studied, computation of the associated internal power, shear, and friction losses; and determination of the range of parameters in which total stress is minimized.

*This work was supported by the National Science Foundation Grant No. GK-10916. Editorial work by Mr. J. D. Leith is greatly appreciated.

Noteworthy is the fact that sound flow, shaving, and central burst (i.e., phenomena which belong traditionally to the theory of plasticity, the field of metal cutting, and fracture mechanics respectively) have here been brought under a single uniform treatment by the upper bound approach using the principle of minimum energy and identical analytical tools.

NOMENCLATURE

A	area
cr	critical (subscript)
F	force
f	final (subscript)
m	friction
0	origin of coordinate system
o	original (subscript)
opt	optimum (subscript)
R	radius
r	radial distance in spherical coordinates
r%	percent reduction in area
$\dot{U}$	velocity
$\dot{W}_i$	internal power of deformation
$\dot{W}_f$	power associated with friction
$\dot{W}_s$	redundant power associated with shear
$\dot{W}_t$	total power (work per unit time)
α	semicone angle of die
β	inclination on true stress-true strain curve
Γ	boundary of velocity discontinuity
Δv	velocity difference, velocity discontinuity
μ	Coulomb's coefficient of friction
θ	angular position in spherical coordinates
σ_o	effective flow stress
σ_{xb}	back push stress
σ_{xf}	front pull stress
τ	friction stress

INTRODUCTION

The Subject

The study of metal flow through conical converging dies covers such processes as wire drawing, open die extrusion, hydrostatic extrusion, and extrusion through a confined chamber. The approach used in this presentation can be applied to rolling, forging, and other metal forming processes. The present analysis for conical dies was chosen because of its completeness and because its practical applicability has been proven.

Fig. 1 represents a billet and die. The billet is a single material of constant strength; consideration will be given later to variable material properties. In form, the billet is a cylindrical rod of radius R_o; the rod is reduced to radius R_f by forcing it to pass through the conical converging die. Reduction is measured from the cross section area of the billet at the entrance to the die (A_o) to that at the exit (A_f).

Three variables involved in the reduction process are noted at once: (1) The radius ratio R_o/R_f or one of the related expressions, area ratio $[(R_o/R_f)^2]$ or relative reduction $(A_o-A_f)/A_o \equiv 1-A_f/A_o \equiv [1 - (R_f/R_o)^2]$. (2) The semicone angle of the die, i.e., α, half the die angle; in wire drawing α is relatively small, possibly 6° to 12°; open die extrusion may employ $\alpha=60^o$; for extrusion through a closed chamber the die angle may be straight, with $\alpha=90^o$. The third variable is friction between the die and the rod.

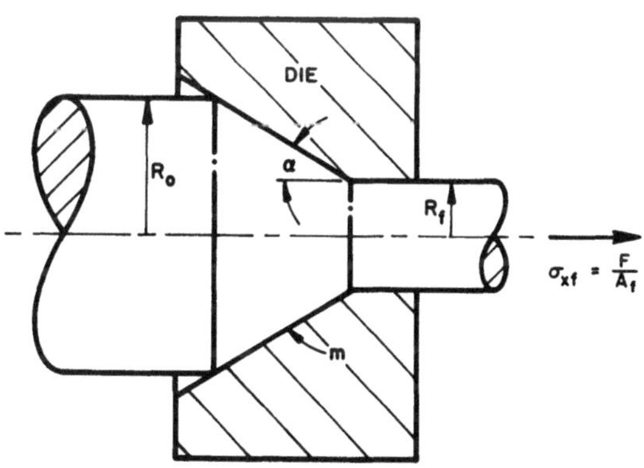

$\sigma_{xf} = f(\frac{R_o}{R_f}, \alpha, \text{ AND } m)$

FIG. 1 DRAWING FORCE

Friction

Whenever there is a relative motion between two surfaces there is resistance to this motion, and this resistance is called friction. The mechanics of friction are complex. Although the fundamentals of the phenomenon have been given much study, yet very little that is known would facilitate formulation of the exact functional relationship between friction and the other process variables. The most common simplifying assumptions made with regard to friction stress (τ) are the following:

(a) Coulomb friction. It is assumed that the shear stress τ is proportional to the pressure p between workpiece and die. Then $\tau=\mu p$, where the proportionality factor μ is called the Coulomb coefficient of friction, assumed constant for a given die, workpiece, and lubricant.

(b) Constant friction. It is assumed that the shear stress is proportional to the strength of the workpiece material. Then $\tau= m\sigma_0/\sqrt{3}$, where the proportionality factor m is called the shear factor, with $0 \leq m \leq 1$ assumed constant for a given die, workpiece, and lubricant.

(c) Hydrodynamic lubrication prevailing. When a lubricant film separates the workpiece from contact with the die, then hydrodynamic lubrication prevails together with its special laws of shear within the lubricant medium.

These three process variables -- reduction, cone angle, and friction -- are independent in that the process planner may exercise a degree of freedom in choosing their values. The severity of friction, for instance, is controlled, within limits, by choices of lubricant, die material and finish, speed, etc.

The Dependent Variable

The force required for drawing or extrusion can now be characterized in related terms. In Fig. 1 the drawing force F (or drawing stress $\sigma_{xf}=F/A_f$) is obviously a function of reduction (larger reduction requires higher force), of cone angle, and of friction, and similarly for extrusion force F (or extrusion stress $\sigma_{xb}=F/A_o$). In short, the force or stress involved in drawing or extrusion is a dependent variable which is a function of reduction, cone angle, and friction.

Description of the drawing force (say) as a function of these three independent variables may be undertaken by either an experimental approach or an analytical approach. Each approach can be aided by the other. Both approaches will be reported here and the results will be compared.

THE EXPERIMENTAL APPROACH

The Equipment

A drawbench is used to measure the force required in wire drawing. In Fig. 2 the conical die is viewed from the front. The workpiece rod passes through the die from the left, its forward end being held by grips. The grips are pulled by an hydraulic system -- a cylinder with piston -- or otherwise. The force applied is transmitted through a load cell (Fig. 3); in this cell strain gages are applied to the tensile rod to produce a Wheatstone bridge in which the voltage differential, properly calibrated, provides a measure of the drawing force.

FIG. 2 WIRE DRAWING

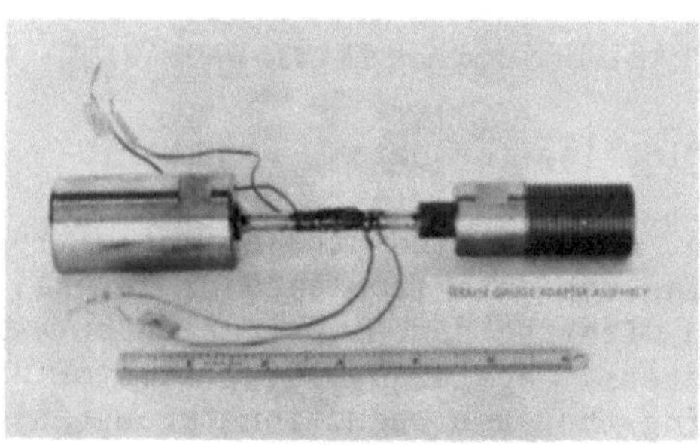

FIG. 3 LOAD CELL

Recording of the Readings

The measure of drawing force is produced continuously on the recorder (Fig. 2) as a function of grip displacement, i.e., length of wire drawn. Fig. 4 represents a single run of a wire through a die. With the application of motion to the hydraulic piston the load on the wire rises: the steep slope of the curve, due to the elastic stretching of the system with increasing force, occurs before any wire is drawn through the die. The peak on the curve is due to inertia forces: the sudden application of force accelerates the wire from standstill to full speed in a short time. The inertia force subsides and the drawing force drops to its steady state constant value for the constant speed drawing. This measure of the steady state drawing force is ordinarily taken as the force reading. When the drawing force is divided by the cross section area of the wire as it emerges from the die, the result is the drawing stress for the specific combination of reduction, cone angle, and friction.

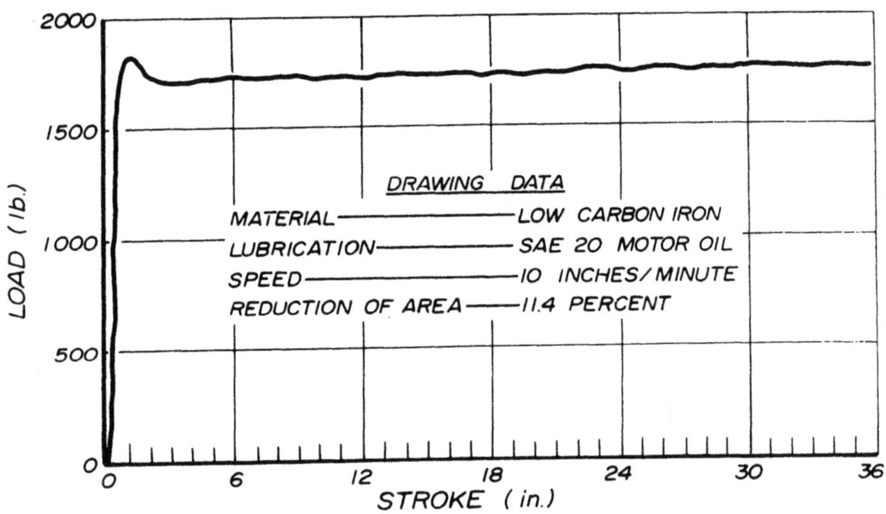

FIG. 4 PEAK PHENOMENON IN WIRE DRAWING

Plotting the Data

Experimentally, then, one may study the effect of reduction on drawing force by recording the data from each of several runs through the same die and with the same lubrication but with incoming rod of several diameters. Again, by using dies of different cone angles, but effecting the same reduction through all dies, one may study the effect of cone angle on drawing force.

The wire drawing process is limited to small ranges over both reduction and cone angle. Hydrostatic extrusion, however, in which the rod is pushed through the die by way of a pressurized liquid (Fig. 5), does not suffer the same limitations: the reductions possible are much larger and die angles may vary from very small to straight, i.e., to $\alpha=90^0$. With pressure of the liquid measured by gage, the effects on pressure required by variations in reduction, cone angle, and friction can be recorded over wide ranges.

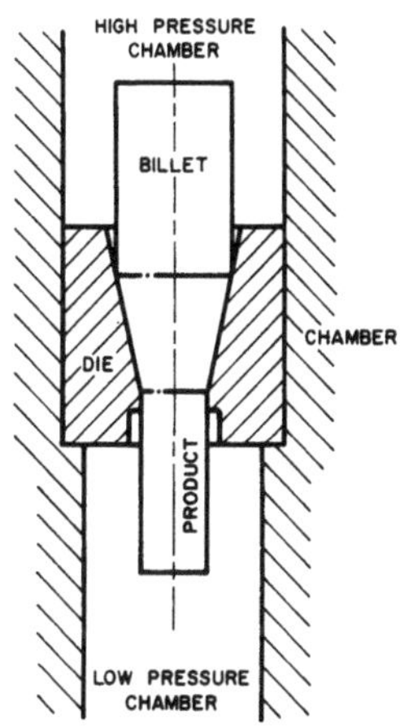

FIG. 5 HYDROSTATIC EXTRUSION

Another variable needed to account for, is the properties of the material processed. Fig. 6, for example, is a record of extrusion pressure for lead as a function of extruded rod position. The extrusion rate at the beginning was slow and this required low pressure. When the extrusion rate was increased by forcing the ram to enter the extrusion chamber more quickly, the extrusion pressure rose immediately because plastic flow stress in lead rises with a rise in the rate of straining: lead is a strain rate sensitive material. Lead may creep at very slow speed with very light load, but if one tries to impose higher strain rates on this material, its resistance to deformation increases. The force required for drawing or extrusion is thus a function not only of the initial independent variables -- reduction, cone angle, and friction -- but also of a fourth variable: the properties of the material itself. For the present, further consideration of this material variable will be deferred.

The experimental method, clearly, can become intolerably tedious in any effort to cover representative sets of reduction ratios, cone angles, and friction values in the combinations which may be relevant in studies of drawing or extrusion forces as functions of no more than three independent variables; and especially so when, beyond this, one is aware that whatever picture is so obtained may be clouded in unknown ways by the unrecorded presence of material property variations.

One can only conclude that experimental work in this area must be of limited value in the absence of adequate theoretical support.

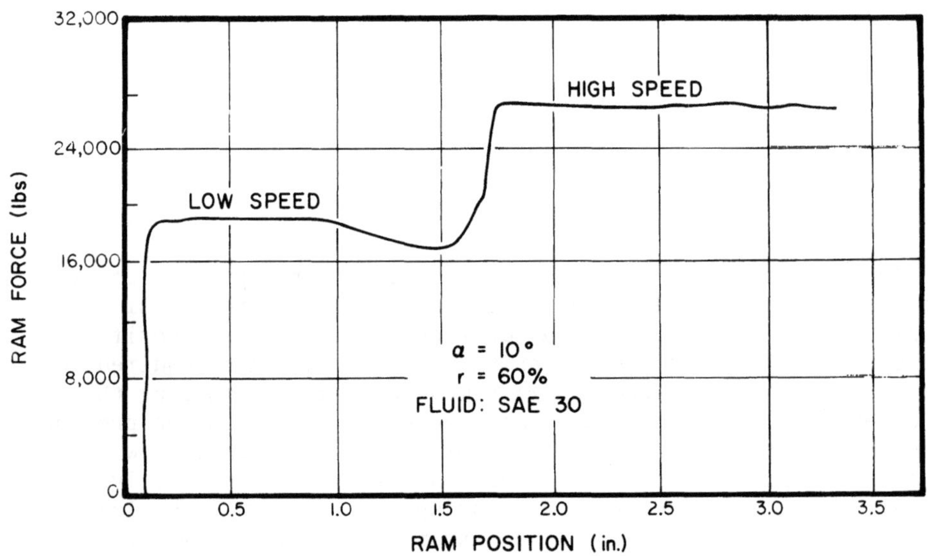

FIG. 6 SPEED EFFECT ON EXTRUSION PRESSURE

THE ANALYTICAL APPROACH

Exact Solutions vs. Limit Analysis

Exact solutions are not available for such problems in metal forming as flow through conical converging dies. Approximations and simplifying assumptions are inevitable and many approaches -- slug equilibrium, slip line techniques and others -- have been partially successful.

Limit analysis is a promising approach which is being used with increasing frequency. In this approach, as applied to the study of drawing or extrusion force, two approximate solutions are developed. One, the upper bound solution, provides a value which is known to be higher than or equal to the actual force; the other, the lower bound solution, provides a value which is known to be equal to or lower than the actual force; the actual force thus lies between the two solutions. For example, in Fig. 7 with drawing stress as ordinate and the semicone angle of the die as abscissa, upper and lower bound solutions are plotted for several reductions together with corresponding measured values of actual stress. Even when experimental results are not available, it is expected that the actual stress and the exact solution, if these were available, would lie between the upper and lower bounds as obtained analytically. Thus, by limit analysis, an approximate solution is given with an estimate of the maximum possible error. The gap between upper and lower bound solutions may be narrowed by providing several upper bounds, choosing the lowest upper bound, and by providing several lower bounds, choosing the highest lower bound. Upper and lower bound solutions are obtained by following strict rules (including requirement of proper description of friction behavior and material characteristics) and thereby making the solutions upper and lower bounds.

The rules and procedures for developing an upper bound solution will be demonstrated in what follows, keeping in mind that several upper bound solutions may be obtained for any specific process. The first step is to assume a certain pattern of deformation.

Flow Patterns

The assumed pattern of deformation is based on one's conception of the actual flow, which the assumed pattern resembles but to which it will not, in general, conform perfectly.

It is in the observation of actual flow that one finds the characteristics of flow; but even when sound flow prevails, with no fracture and no failure, flow patterns will change with variations in reduction, cone angle, and friction.

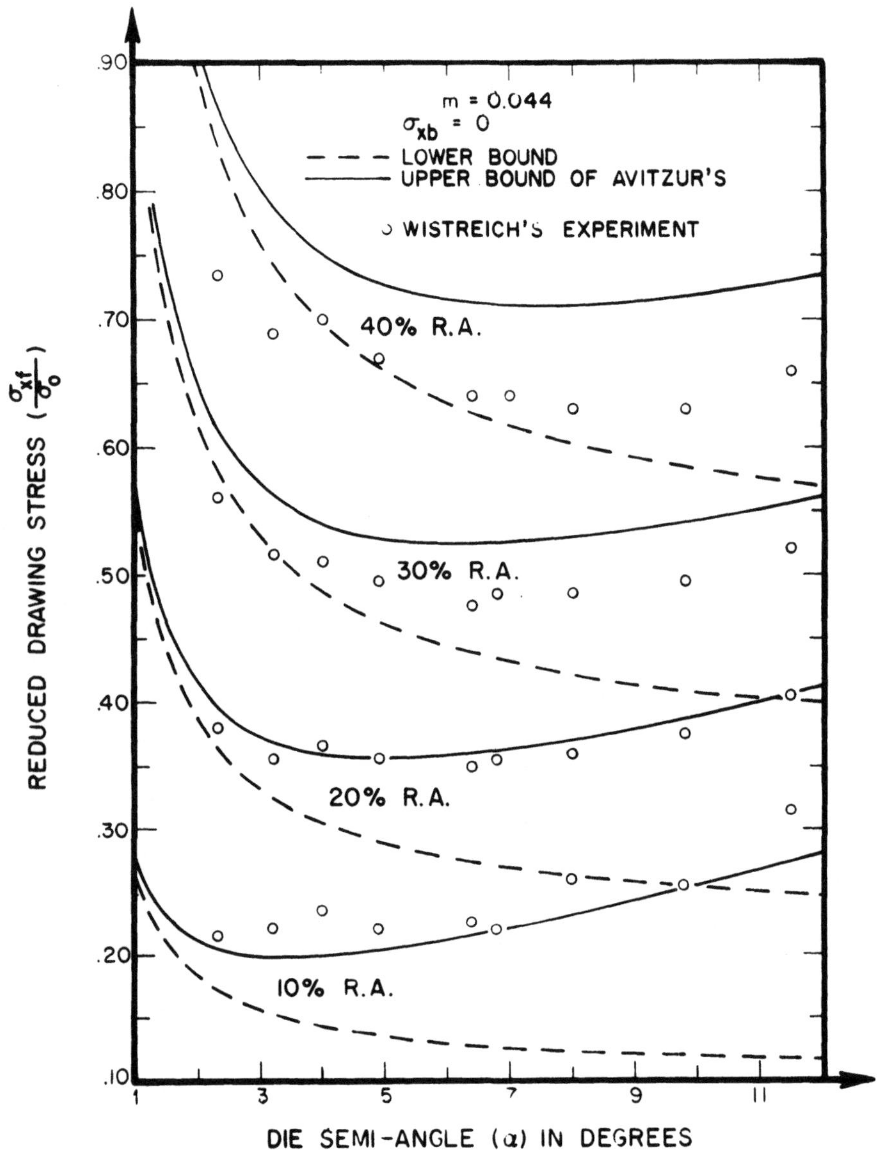

FIG. 7 REDUCED DRAWING STRESS VS. DIE SEMI-ANGLE AND PERCENT REDUCTION IN AREA

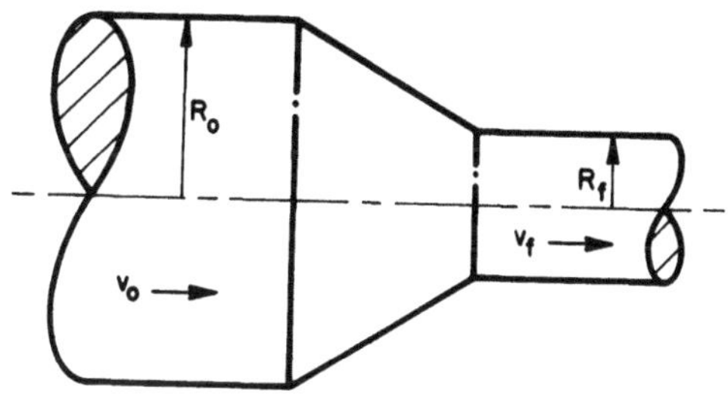

$$\frac{v_o}{v_f} = \left(\frac{R_f}{R_o}\right)^2$$

FIG. 8 VELOCITY AT ENTRANCE AND EXIT

In all observations far away from the die, both at entrance and at exit, one notices that the wire moves as a rigid body in the axial direction. If the wire enters the die at velocity v_o and leaves it at velocity v_f (Fig. 8), then, by volume constancy (i.e., the requirement that the volume of the material worked must remain unchanged), the ratio of these velocities, for sound flow with no fracture, is $v_o/v_f=(R_f/R_o)^2$.

The plastic deformations occur under the influence of the die and in the vicinity of the die. From overall observation one notices that on occasion (Fig. 9) the wire entering the die bulges before coming in contact with the die; on other occasions the wire converges before entering the die; it may even bulge and converge in sequence. Flowing through the die the wire moves in a general converging flow; its cross section reduces as the wire elongates; velocity increases as the wire approaches the exit. At the exit the wire may or may not separate from the die. All of these variations are explained by saying, in the 'als ob' sense, that under the imposed conditions the wire chooses that pattern which accommodates easy flow.

To obtain a picture of the flow many methods are available. One of these, visioplasticity, has been described by Thomsen[1] et al. (Fig. 10). Some investigators, when concerned with flow patterns for a specific metal, prefer to test the metal itself; others prefer to deform materials which are easier to work with, like lead or even plasticine. Whatever the method chosen, one must assume a certain, if approximate, picture of the flow in order to proceed to an upper bound solution.

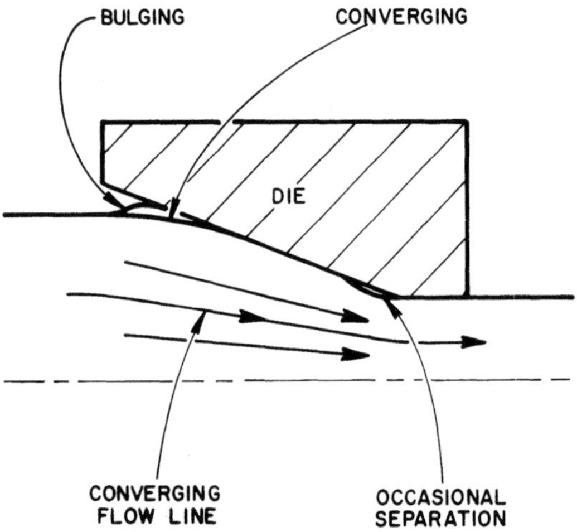

FIG. 9 BULGING AND CONVERGING

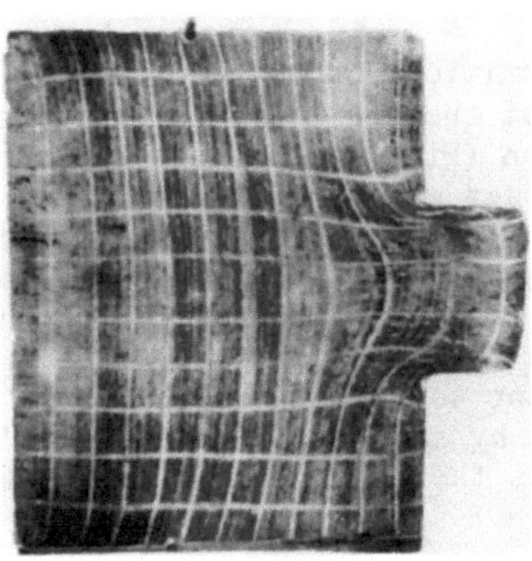

FIG. 10 VISIOPLASTICITY THROUGH GRID LINE TECHNIQUE

The Spherical Velocity Field

For the present solution assumption was made (Fig. 11) that at any point within the zone (II) of plastic deformation the material is moving toward the apex of the die. This assumption may be based on visioplasticity studies or on simple inspection of Fig. 11. Any material point on the surface of the entering wire reaches the surface of the die. It must then move parallel to the surface of the die: it cannot get into the die but it does not separate from the die. So moving, within the zone of plastic deformation, it is moving into the apex. Any material point in the center line is at the axis of symmetry and must therefore move into the apex. If it is then said that any material point in an intermediate position, similarly, is moving into the apex, a pattern of flow has been described from which very little deviation is to be expected for sound flow with no fracture.

It is next in order to note that the flow direction at the entrance side of the die (zone I) is not parallel to the direction of flow in the plastic zone (II), nor is the latter parallel to that on the exit side (III), i.e., there are two transient regions where the direction of flow is changing. These regions may be wide and undergoing gradual change which cannot be described mathematically, but it has been found[2] that precise mathematical description is not an absolute necessity at this point. What can be done, simply, at entrance and exit, is to continue one direction up to the intersection with the other: in each case the aggregate of changes in direction constitute a locus of velocity discontinuity occurring along a spherical surface with center at the apex of the die.

For this study, then, drastic change in direction is assumed at each of two spherical surfaces of velocity discontinuity, Γ_2 and Γ_1. At entrance, the motion is that of a rigid body, with every point moving parallel to the axis of symmetry. As soon as each material point reaches the first of the assumed surfaces, it undergoes drastic change in direction, i.e., it starts to move toward the apex of the die. This material point moves faster and faster until it reaches the spherical surface at the exit, where again it undergoes drastic change in direction, restoring motion parallel to the axis of symmetry.

In the velocity field which has been described a material point on the surface of the rod is moving somewhat slower than material points increasingly closer to the center. The velocity of a material point entering the plastic zone through Γ_2 is the product of the original velocity v_0 and $\cos\theta$, where θ is the angular position of the point. The angle θ varies from zero on the center line (and then the velocity is v_0) to the value α on the surface (and then the velocity is $v_0\cos\alpha$). If the semicone angle α is small, $\cos\alpha$ is close to unity, but if α is close to 60° (say), then the velocity at the surface is about half the velocity at the center. The reason for this slowing lies in the requirement of volume constancy.

As the material point moves into the deformation zone it moves along a radial line and its velocity increases. Leaving the surface Γ_2 its velocity is $v_o\cos\theta$ at distance r_o from the apex of the die. As the material point moves along a radial line its angular position θ does not change but its radial distance r decreases; as r decreases, velocity increases, by volume constancy, because the cross section area of the flow channel is narrowing (Fig. 11).

The proposed flow pattern has been described by the present author elsewhere (Ref. 2) as follows:

> The assumption is made that the die is a rigid body of the geometry shown. A kinematically admissible velocity field is described. (Fig. 11) The wire is divided into three regions in which the velocity field is continuous. In zones I and III the velocity is uniform and has an axial component only. In zone I the velocity is v_o and in zone III the velocity is v_f. Because of volume constancy $v_o=v_f(R_f/R_o)^2$. In zone I deformation has not yet begun. It includes the incoming rod, which is separated from the deforming zone II by the surface Γ_2. Surface Γ_2 is spherical, of radius r_o with the origin at the apex O of the cone of the die. Zone II is the zone of deformation bounded by the surface of the die, with a cone of an included angle 2α, and two concentric spherical surfaces Γ_1 and Γ_2. The surface Γ_2 is the previously mentioned spherical boundary between zones I and II. The spherical surface Γ_1 of radius r_f, with the origin at the apex O of the cone, separates zone II from the emerging product of zone III. In zone II the velocity is directed toward the apex O of the cone, with cylindrical symmetry.

Distorted Grid Pattern

Before proceeding to power demand and forces requirements, prior coverage of distortions resulting from the velocity field assumption is appropriate. Analytically, the velocity field represented by Fig. 11 is given by the equations $\dot{U}_r=v= -v_f r_f^2\cos\theta/r^2$ and $\dot{U}_\theta=\dot{U}_p=0$.

In Fig. 12 any straight line parallel to the axis of symmetry at the distance R_1 in zone I will end up in the product as a straight line at the distance R such that $R/R_f=R_1/R_o$. A cylinder of original radius R_1 changes to a cylinder of radius R.

A straight line which passes through the axis of symmetry (Fig.13) and is initially perpendicular to it will distort during the deformation. Consider any point A on this straight line: the closer the

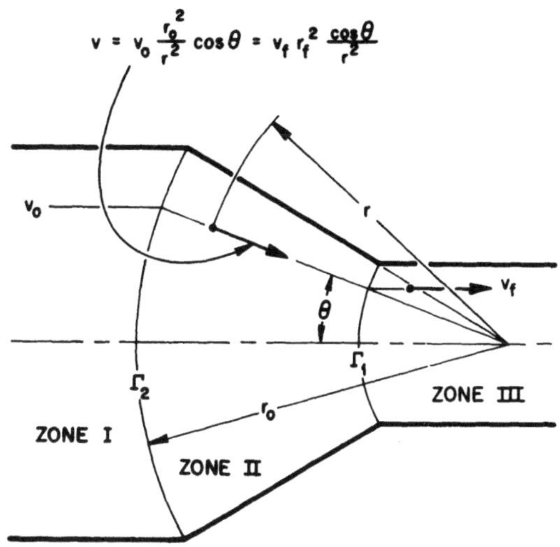

FIG. 11 FLOW LINE

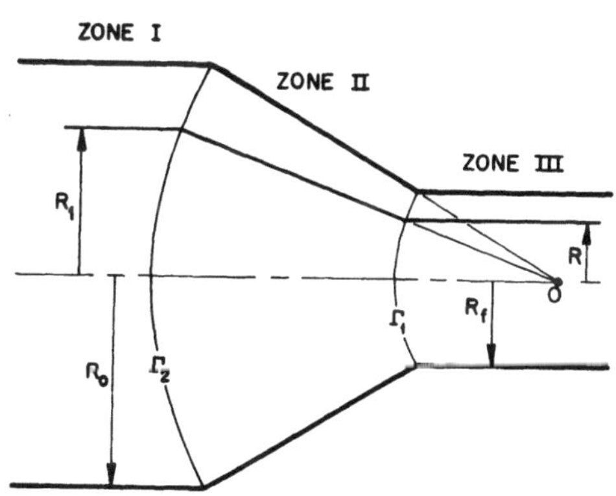

$$\frac{R}{R_f} = \frac{R_1}{R_0}$$

FIG. 12 DISPLACEMENT OF AXIAL LINES

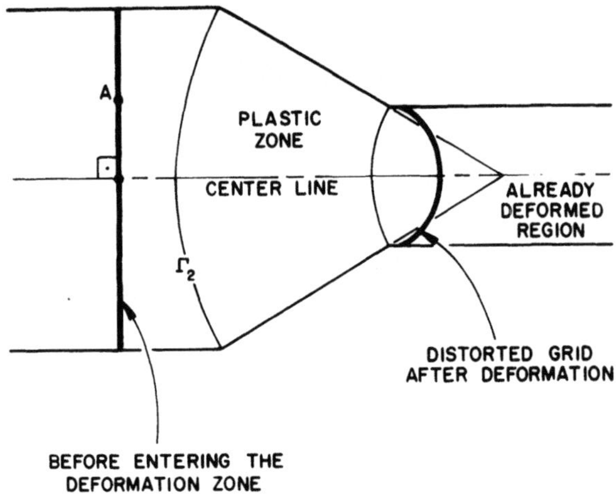

FIG. 13 A PERPENDICULAR GRID LINE

point is to the center line, the sooner it will reach the plastic zone and the faster it will move toward the exit once it is in the plastic zone. Meanwhile, points closer to the outer surface of the wire enter the plastic zone later, move more slowly while they are in that zone, and thus exit last to the already deformed region. The result is distortion of an originally rectangular grid pattern.

The distortion to be expected from the presently assumed velocity field is represented by Fig. 14, in which the abscissa is the relative distance from the center line, with only half of the billet shown, the distortion being given by the ordinate on a scale so weighted that all reductions can be presented in a single graph. From Fig. 14, as well as from experience, the following observations can be made:

(1) The larger the cone angle and reduction, the more distorted is the grid;
(2) The angle between a grid line originally normal to the axis of symmetry and one originally parallel to that axis remains 90° at the center but grows increasingly acute as the surface is approached;
(3) Since the actual cone angle for flow may be that of a die itself or that of a formed dead zone, the possibility of dead zone formation (which will be discussed below) should not be overlooked when comparisons of predicted and experimental grid data are made.

If the process is interrupted at constant intervals and the instantaneous shape of the grid is superimposed on the half-billet sections, distortion patterns which develop are represented in the

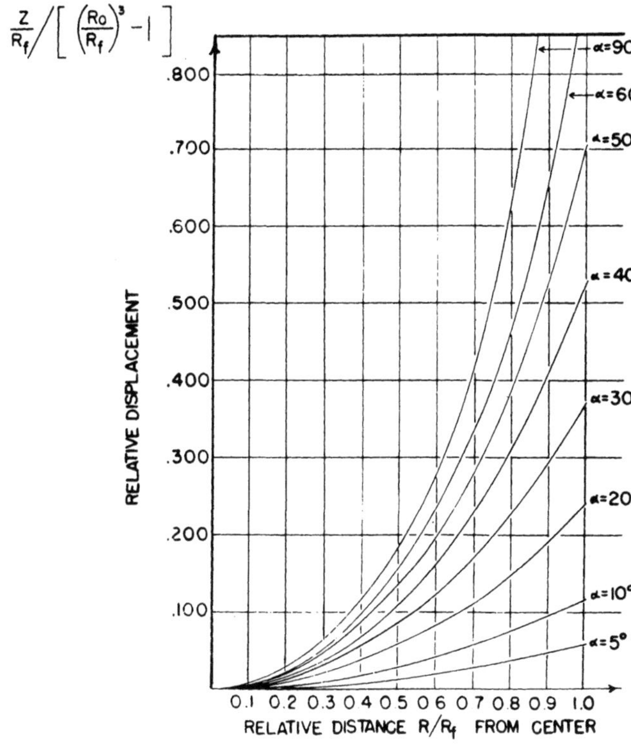

FIG. 14 RELATIVE POSITION OF DISTORTED GRID VS. DISTANCE FROM CENTER AND CONE ANGLES

collective Fig. 15: Fig. 15a is an initial view for cone angle $\alpha=15^o$ and reduction $R_o/R_f=1.5$; distortion increases with increasing reduction (Fig. 15b) and with increasing die angle (Fig. 15c). For very large reductions and large cone angles (Fig. 15d), and moving toward combinations possible only during extrusion through a confined chamber (Fig. 15e) or during hydrostatic extrusion, the line which was originally perpendicular to the axial direction (Fig. 15f) becomes almost parallel to the radial direction. Lines which were mutually perpendicular originally (Fig. 15g) become almost parallel close to the surface of the wire (Fig. 15h). This amounts to much shear or distortion, although it is noted that shear or distortion is not a reliable measure of damage. For a study of the distorted grid pattern, see Refs. 2 and 3.

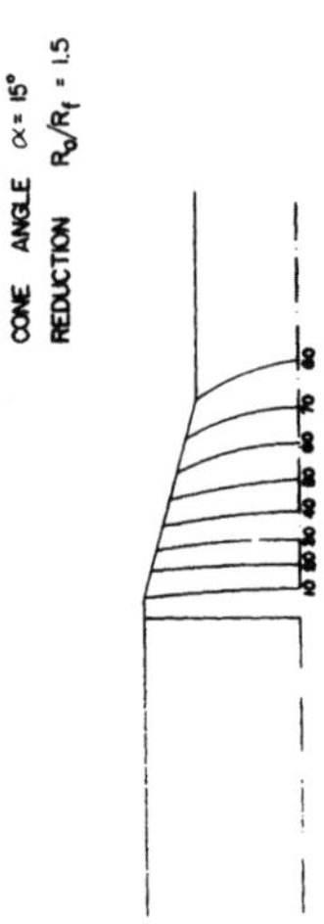

FIG. 15a

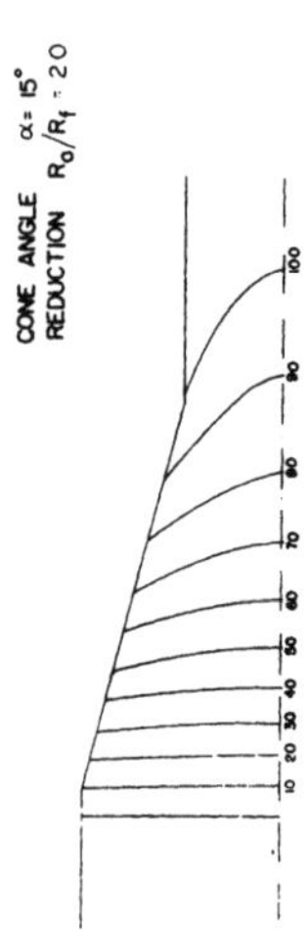

FIG. 15b

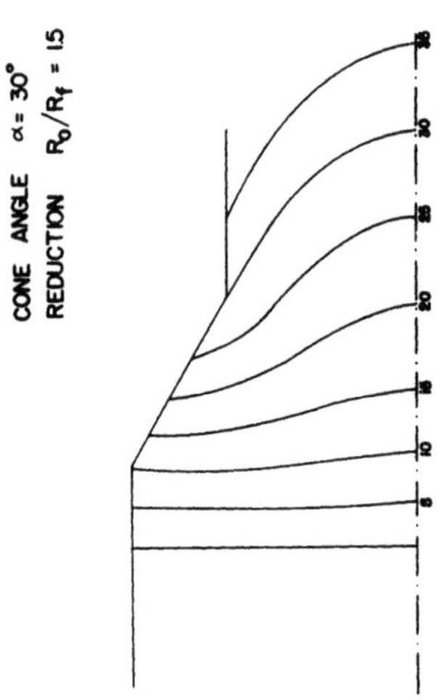

FIG. 15c

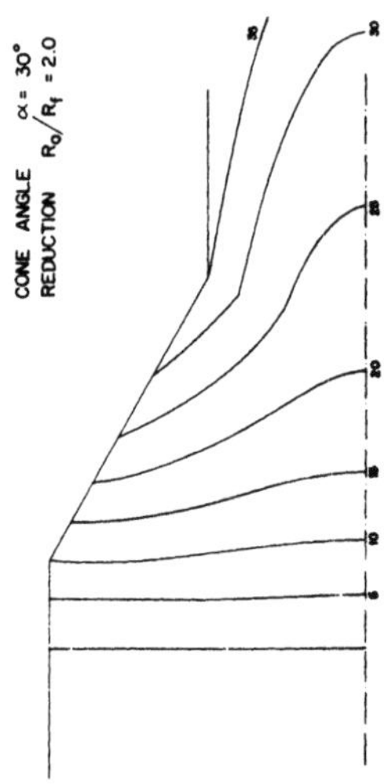

FIG. 15d

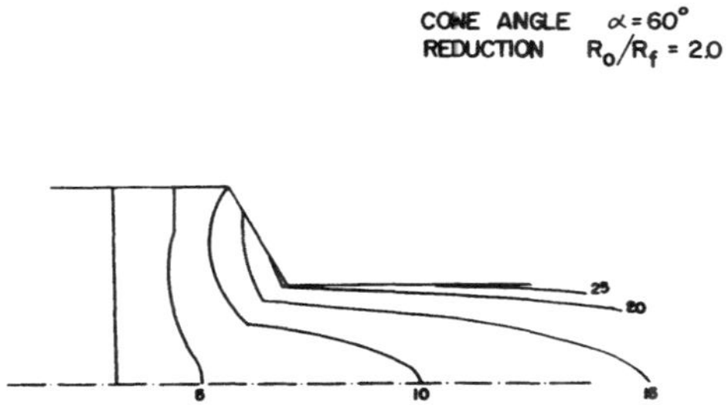

FIG. 15e

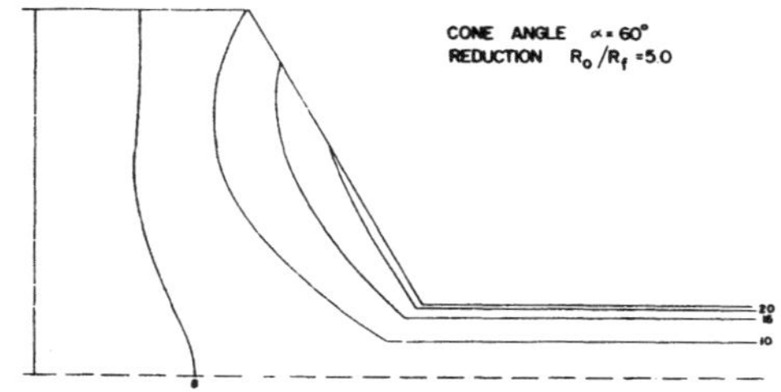

FIG. 15f

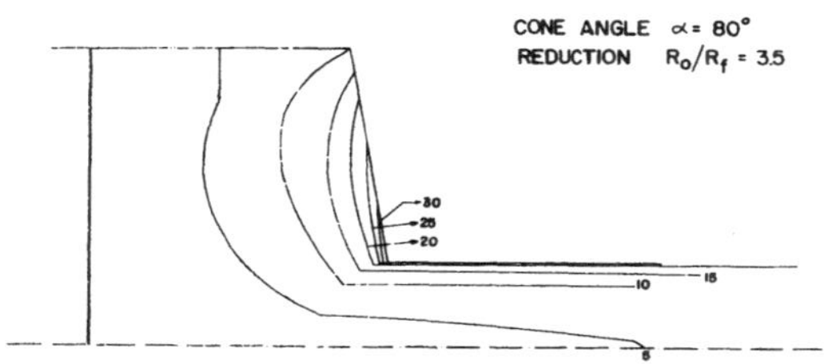

FIG. 15g

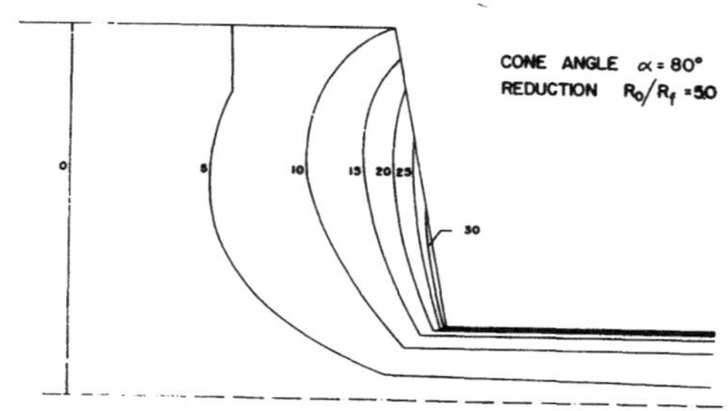

FIG. 15h

FIG. 15 INTERMEDIATE GRID DEFORMATION PATTERN

The Powers

Choice of a velocity field having been made, procedures will be described for computing the force needed to require the material to follow any specific pattern of deformation; the force associated with any other flow pattern will be different. The material can choose ('als ob') to flow in one or another of the geometrically possible patterns: the material does choose that pattern of deformation which requires the least amount of energy. The process minimizes the entropy.

Given the velocity field, it is a simple matter of mathematics to find the strain rate field; from the strain rate field one determines the amount of energy required for deformation from initial diameter to final diameter. The entire change in diameter takes place in the plastic region, zone II of Fig. 12. Deformations do not occur in zones I and III and no energy is spent in the rigid body motions in these zones. The power or force required to overcome resistance to deformation is thus computed through the velocity and strain rates field for zone II alone. (See Refs. 2,4, and 5.)

Another place where power is spent is at the two surfaces of velocity discontinuity, Γ_1 and Γ_2 . In order to bring about change in direction, even if it is not drastic, energy must be spent. The measure of velocity discontinuity associated with each of the spherical surfaces is a function of the angle θ: where Δv is the measure of change in direction of flow, $\Delta v = v_f \sin\theta$ on Γ_1; on Γ_2, $\Delta v = v_o \sin\theta = v_f (R_o/R_f)^2 \sin\theta$.

With the velocity discontinuity known, and with shear stress on the surfaces Γ_1 and Γ_2 assumed to be the maximum resistance of the material to shear, i.e., $\tau = \sigma_o/\sqrt{3}$, one can compute the power required to cause this shear.

The velocity discontinuity is the main source of the distortion studied through distorted grid patterns. The larger the die angle α, the higher the values of θ which are reached. Thus the distortion and shear are related to the power spent because of the velocity discontinuity. This distortion is not contributing to any change in the size of the wire: it is a redundant distortion, over and above the change in diameter, or redundant work, and the power associated with the velocity discontinuity is therefore called redundant power.

The distortion and its related redundant power are negligible for small die angles: if dies of zero cone angle could be used the distortion and redundant power would be zero. With increasing cone angle both distortion and redundant power increase. As the cone angle becomes very large, the distortion is much more pronounced and the energy required to bring about the distortion becomes greater and greater.

Following computation of the internal power of deformation and of the redundant power of deformation, one must determine a measure of the resistance of the die to the motion of the wire over its surface, i.e., friction resistance.

Over an incremental portion of the die-wire interface friction resistance is equal to the product of friction stress τ and the area. The velocity of the wire, relative to the die, is found from the equation describing the velocity field in zone II (Fig. 11) when $\theta=\alpha$; on the surface, $\Delta v=v_f r_f^2 \cos\alpha/r^2$. Friction is usually assumed to obey one of the three rules previously stated: Coulomb, constant shear, or hydrodynamic rules when full separation of die and wire is maintained by the lubricant. Friction losses are then computed by integrating the product of shear stress, local velocity, and incremental area over the entire area of contact.

Total Power and Force

The drawing force, in any solution, is built up from these three components: the internal power of deformation, the redundant power of deformation, and the power needed to overcome friction losses. Fig. 16 illustrates characteristics of the resulting summation. With too small a cone angle, friction is predominant and makes the force excessive. As the cone angle increases, friction drops very drastically, so that in spite of increase in redundant work there is a drop in the total power requirement. An optimal value is reached for the power when further increase in cone angle causes the redundant work to offset what has been gained on friction, and thereafter redundant work is the predominant factor, not friction. Further increase in die angle produces further increase in total power.

Each of the three components of total power must be determined independently. Internal power of deformation is the ideal portion, i.e., it is the power required for reduction from one diameter to the other in an ideal situation. But the ideal pattern of deformation can never be accomplished because this forming must be accomplished through dies, and when one uses a die, one gets distortion. This redundant distortion power must be computed independently of the internal power because with larger cone angles one has higher values of redundant power than with small cone angles. One must be able similarly to account for friction: a solution which assumes zero friction may not be a good solution for very small cone angles but might be very good for large cone angles.

Use of the upper bound approach permits accounting for ideal power, redundant power, and friction, each independently of the others.

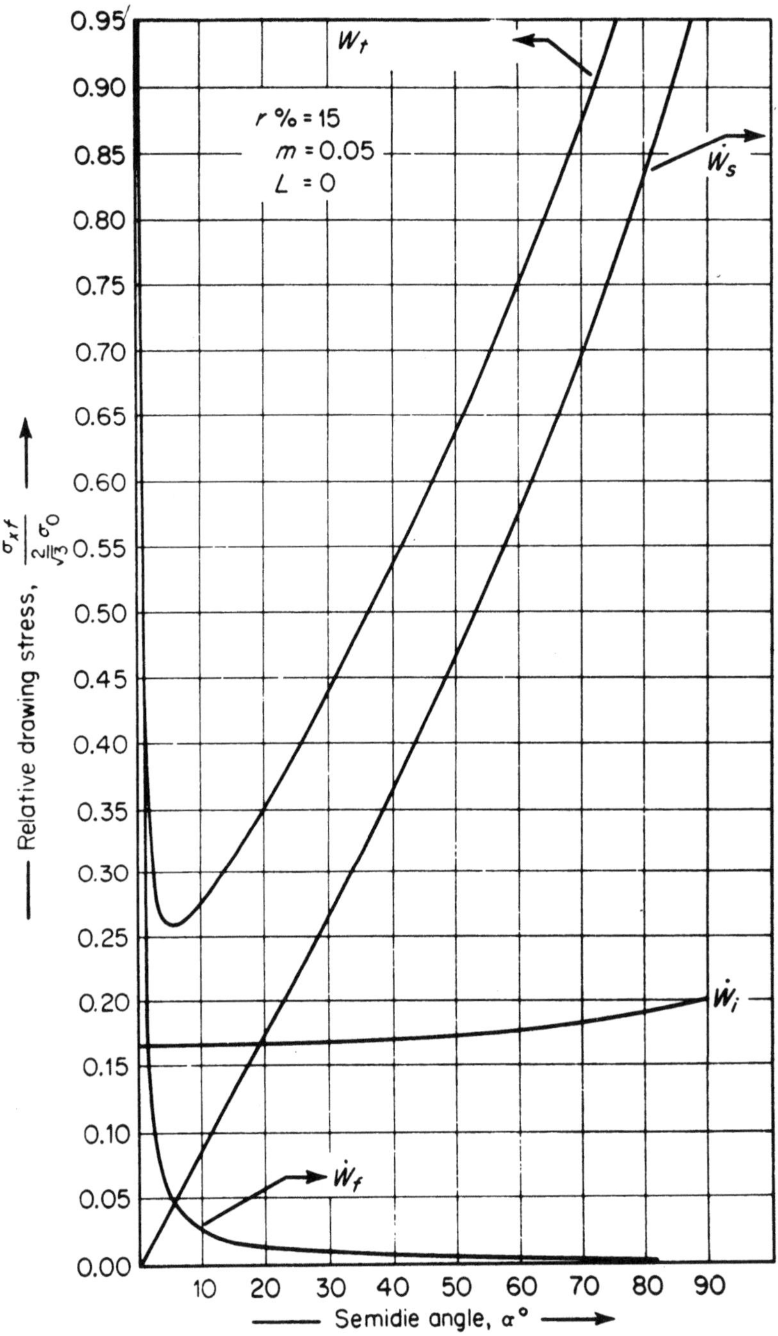

FIG. 16 CONTRIBUTION FROM DIFFERENT POWERS TO THE TOTAL DRAWING STRESS

COMPARISON BETWEEN ANALYTICAL AND EXPERIMENTAL DATA

Many experimental studies verify the results developed by the preceding analysis. One of the first of these was reported by Wistreich[6] in 1955: this work has become a classic example because of its completeness and precision.

Electrolytic copper was drawn through a series of dies of varying cone angle. For each die angle reductions were effected from 5 percent to 45 percent in increments of 5 percent. The drawing force was recorded for each run. The values of friction and the strength of the wire were measured independently for each reduction. In Fig. 17 the observed drawing stress values have been plotted against die angles for reduction values at the even 10 percent intervals, with curves representing corresponding analytical results superimposed on the experimental data. All of the independent variables are reported by Wistreich -- reduction, cone angle, friction, and flow strength of the material -- together with the dependent drawing stress values. The analytical solution has no 'fudge factor' at all. Fig. 17 shows a reasonably good agreement between the experimental results and the analytical upper bound solution.

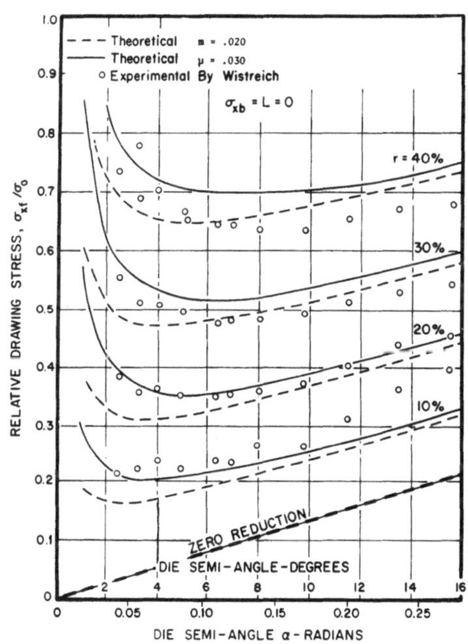

FIG. 17 VARIATION OF RELATIVE DRAWING STRESS WITH DIE ANGLE, FROM WISTREICH'S PAPER

SOUND FLOW VS. OTHER PATTERNS OF FLOW

Possible Patterns

From the preceding comparison of analytical and experimental data it may be concluded that the analysis involved in the upper bound solution for sound flow is capable of predicting fairly well the required drawing or extrusion forces as well as the optimal die angle which minimizes these forces. Such prediction is helpful in design of equipment or process. Beyond this use, however, is the considerable value of the analysis when insight into the factors governing metal flow is sought.

Sound flow occurs only within a limited domain among all possible combinations of die angle, reduction, and friction values. In Fig. 18 the upper level represents sound flow diagramatically along with two other common modes of flow: dead zone formation and shaving; any of these patterns can occur when wire drawing or extrusion is undertaken, as well as others yet to be considered. Dead zone formation occurs when a material, whether drawn or extruded through dies of increasingly high cone angle, shears within itself to develop a dead metal zone which no longer takes part in the flow but adheres to the die, forming a die-like channel through which the billet passes in a still-converging kind of flow. Shaving develops when the dead zone material does not adhere to the die but starts to move backward, peeling off as in a metal cutting operation; the core of the billet no longer deforms at all, but moves through the die with essentially no change in diameter and with exit velocity the same as entrance velocity.

The production planner or plant foreman is very much interested in the factors which create or prevent dead zone formation, shaving, and other possible flow patterns. Knowledge of these factors and establishment of criteria for creation or prevention of any specific kind of flow will enable process design and choice of proper reductions, die angles, and lubricants to be governed on a more soundly scientific basis.

The Concept of Minimum Energy

How the analysis can help in development of criteria for sound and for unsound flow patterns becomes clearer when the drawing stress (whether measured experimentally or computed analytically) is plotted as a function of die angle as on the lower level of Fig. 18 in appropriate relationship to the inserts which identify the flow patterns presently under view: sound converging flow, converging flow through a dead zone, and shaving.

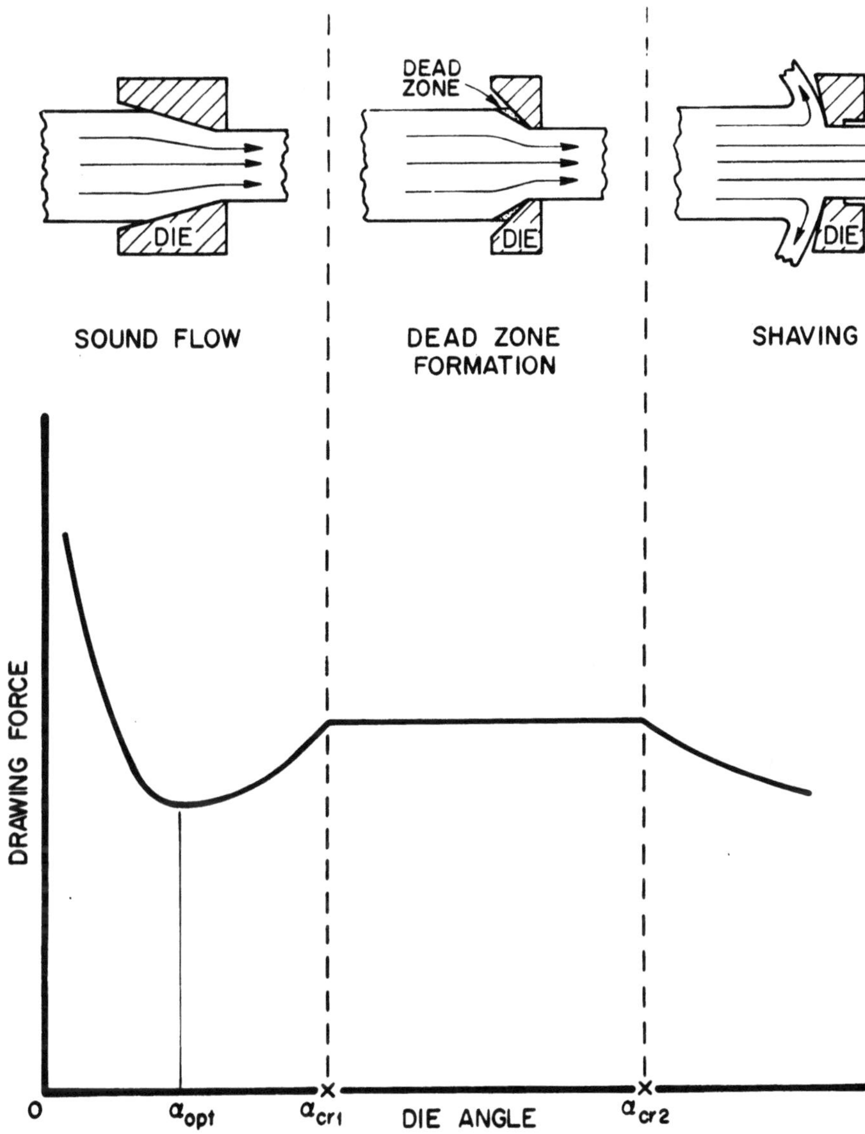

FIG. 18 SCHEMATIC ILLUSTRATION OF THE EFFECT OF TOOL ANGLE AND MODE OF FLOW ON DRAWING FORCE

In the range of sound converging flow the drawing stress is high for very small cone angles because of high friction losses. With increasing cone angle, friction losses reduce and so does drawing stress until an optimal angle is reached; beyond this, further increase in die angle causes increasing drawing stress because of increasing shear or redundant power of deformation due to excessive distortion. This gradual rise in drawing stress continues until a first critical angle (α_{cr_1}) is reached, at which point ordinary sound flow changes into flow through a dead zone formation. With the change in flow pattern the slope of the characteristic curve describing the drawing stress undergoes a discontinuity: the drawing stress thereafter continues unchanged with increasing die angle up to a second critical angle (α_{cr_1}) at which shaving begins.

If sound flow could have been continued for cone angles larger than α_{cr_1}, the consequent excessive distortion and redundant power would have caused excessive drawing force. The angle of the dead zone, however, is smaller than that of the die: the material has found a way to eliminate excessive distortion and to preserve total process energy.

At this point the concept of minimum energy reappears. The material has a choice of several or many patterns of deformation. All that one imposes on the wire is forced motion on the exit side in the axial direction and confinement by the walls of the die. Within these 'boundary conditions' the material can choose any flow pattern which is geometrically possible and which obeys the requirement for volume constancy. With changing conditions (e.g., changing die angle) the flow pattern may change (with small cone angles dead zone will not form, with large cone angles dead zone will form).

Dead Zone Formation

The analytical solution for the range within which converging flow takes place through a dead zone calls for a new velocity field. The dead zone material forms a new conical surface. Continuing this surface to its apex, two spherical surfaces of velocity discontinuity can be described, as before, with common center at the apex and with radii r_o and r_f. Fig. 19 shows that the resulting geometry is identical with that of the velocity field assumed initially for sound flow except for the single difference that the cone angle is now determined by the material and not by the process planner through his choice of die. The new cone angle is determined in such fashion that the required power is minimized. The power required for this new velocity field, as a function of reduction, cone angle, and friction is then computed to provide a solution for the drawing force.

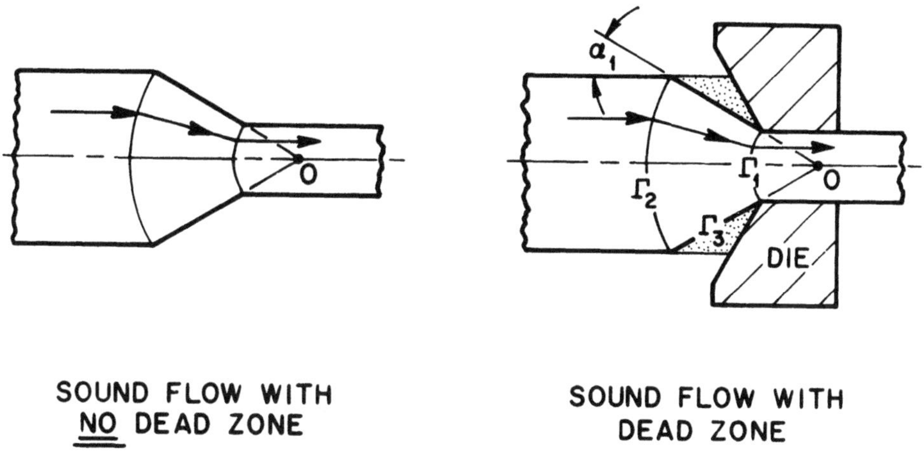

FIG. 19 IDENTICAL VELOCITY FIELDS FOR SOUND FLOW WITH AND WITHOUT DEAD ZONE FORMATION

When a dead zone has formed, the value of its related angle does not change with increasing die angle. The angle α_1 is that angle which by compromise between the combined shear over Γ_1 and Γ_2 and the shear over Γ_3 minimizes the drawing force. Once the die angle is above its first critical value, changes in die angle do not affect the angle of the formed dead zone, nor does the die angle affect the drawing force which remains constant as long as dead zone exists.

Proceeding analytically, two characteristic equations are obtained by computing the relative drawing stress for sound flow with no dead zone formation and the relative drawing stress associated with dead zone formation. Fig. 18 represents the characteristics graphically for (say) 20 percent reduction with a constant specific friction value. The intersection of these curves is at the critical value of the cone angle for the specific reduction and friction combination involved. The characteristic curves are described mathematically by equations of the form $\sigma_{xf}/\sigma_o = f_1(r\%,\alpha,m)$ for sound flow and $\sigma_{xf}/\sigma_o = f_2(r\%)$ for dead zone formation (See Ref. 2, chapter 8.) For these equations the predicted drawing stress values are the same at the intersection point. Equating the right members of the equations one may then solve for the critical semicone angle. This critical angle is exhibited in Fig. 20 as a function of the reduction with the parameter friction ranging from m=0 to maximum possible friction. For any value of m the region above the correspond-

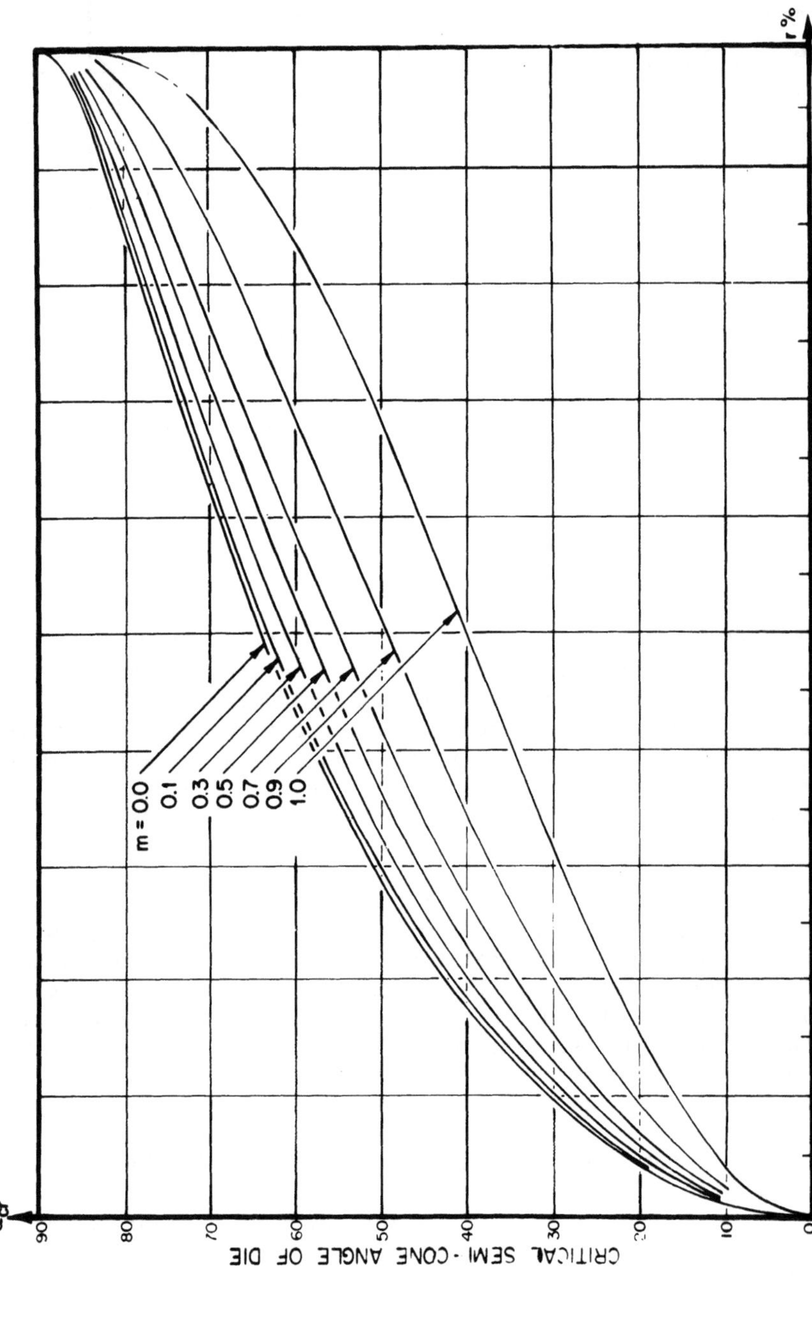

FIG. 20 CRITICAL SEMI-CONE ANGLE VS. REDUCTION AND CONSTANT SHEAR FACTOR

ing curve means dead zone formation; below that value dead zone is not expected.

Shaving

When the semicone angle increases beyond a second critical value (α_{cr_2}), the dead zone material ceases to adhere to the die and starts moving backward. Gradually the backward motion from the dead zone reaches such speed that all the material from the surface layer of the original wire is shaved off; the core moves through the die at equal entrance and exit velocities without deforming plastically. When the relative drawing stress is measured or computed (see Ref. 7), one notices a second change in the slope of the characteristic curve (Fig. 18) and in the range where shaving occurs the drawing stress required by the characteristics for shaving is lower than that required by the characteristics for sound flow with or without dead zone formation. A schematic of the flow associated with shaving is shown in Fig. 21. Shaving may occur either with or without the formation of a built-up edge, just as in metal cutting. While the flow in drawing is axisymmetric, unlike the orthogonal flow in plane strain, shaving, nevertheless, is a metal cutting operation.

Fig. 22 describes a velocity field to accommodate a flow which ends up in shaving. On this basis one can compute the internal power dissipated in the chip, the shear or redundant power lost along surfaces of velocity discontinuity, and the friction losses. Summing these powers gives the relative drawing stress or relative extrusion pressure for shaving flow computed with and without built-up edge. The results are shown in Fig. 23. Equating the forces computed for shaving with the previously computed forces for sound flow with or without a dead zone produces the solution for the second critical angle. With dies of larger angle, shaving (with or without a built-up edge) is expected; with dies of smaller angle, sound flow (with or without dead zone formation) is expected. The characteristic curve for shaving may intersect that for dead zone formation to indicate the die angle beyond which the dead zone material changes into a chip; or it may intersect the characteristic curve for sound flow with no dead zone formation at a point below the characteristic line for dead zone formation, in which case the sound flow will change directly to shaving, with increasing die angle; in this case intermediate dead zone formation is bypassed entirely: when sound flow ceases, shaving commences.

In Fig. 24 the resulting criteria for maintaining or preventing shaving, both for drawing and for extrusion, are described in terms of reduction in area as a function of the semicone angle. The graph exhibits the zones in which one expects sound product, shaving with built-up edge, or shaving with no built-up edge.

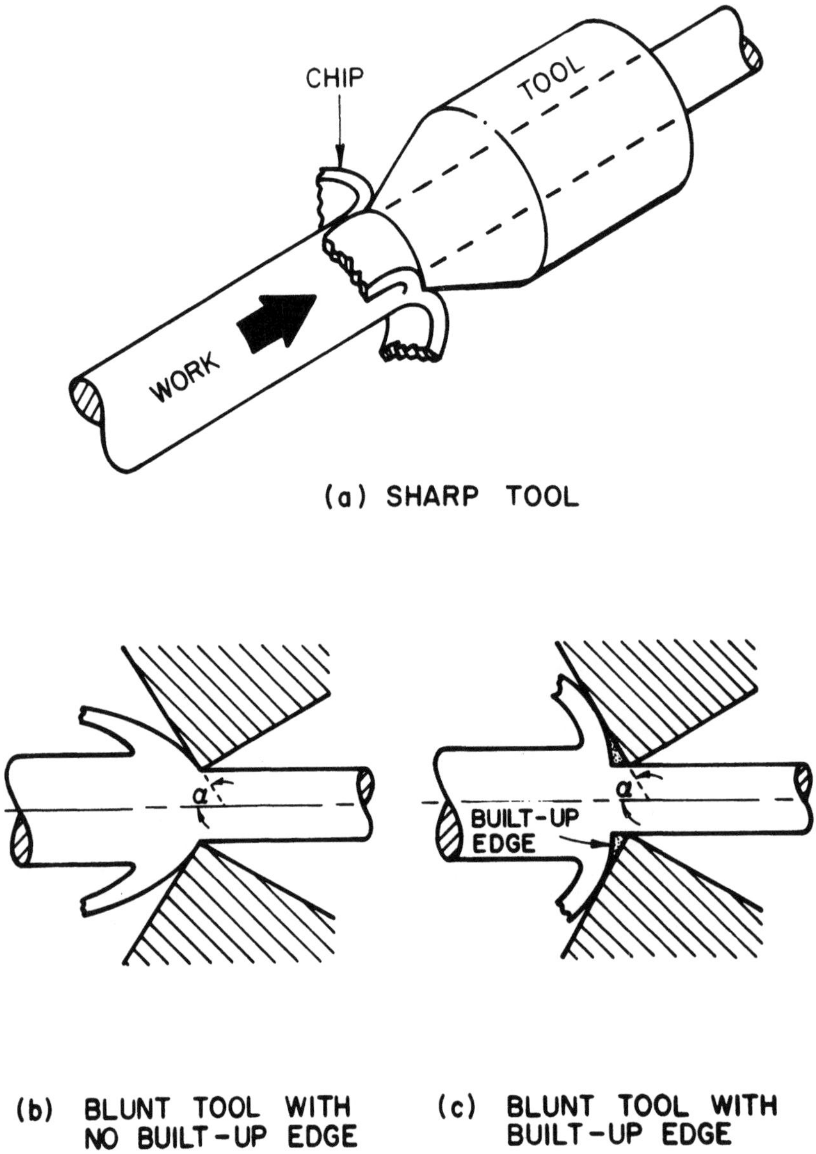

FIG. 21 THE SHAVING PROCESS

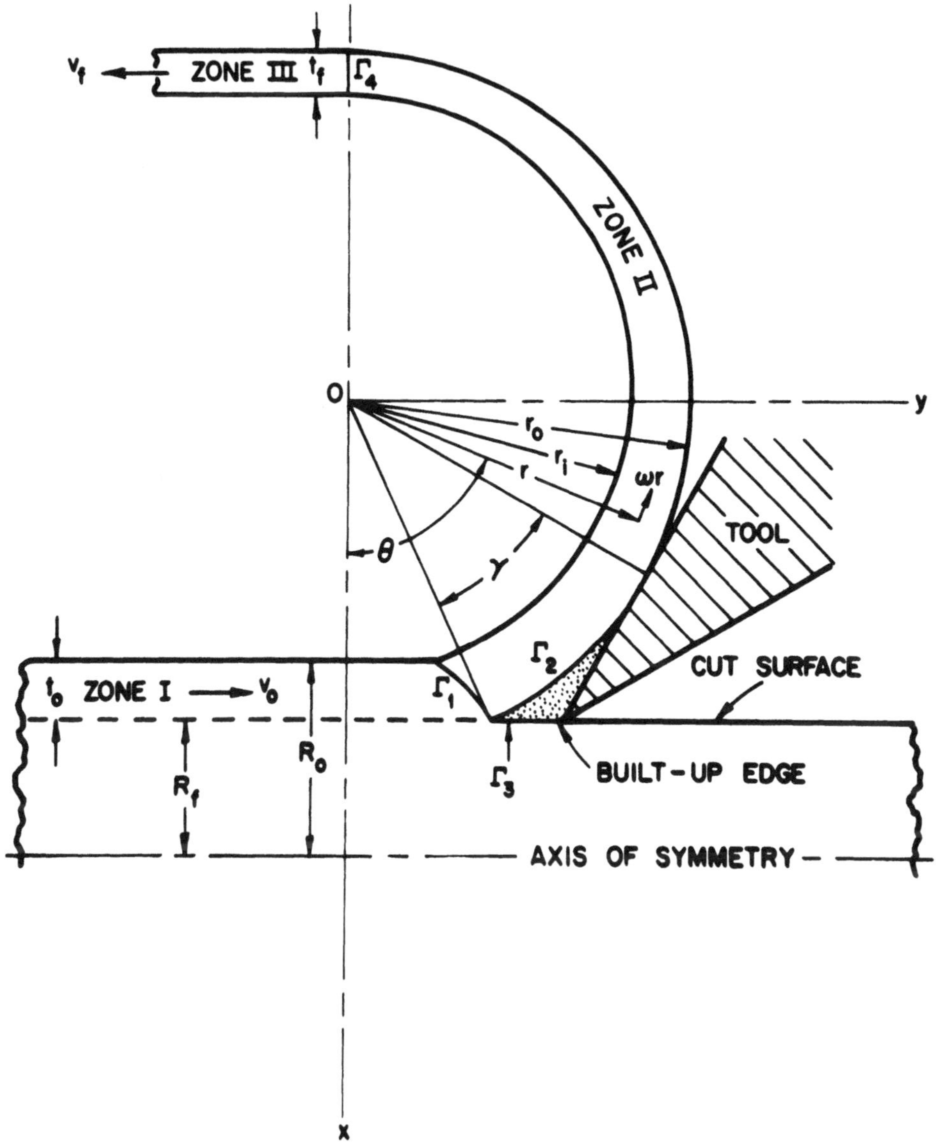

FIG. 22 ASSUMED VELOCITY FIELD FOR SHAVING

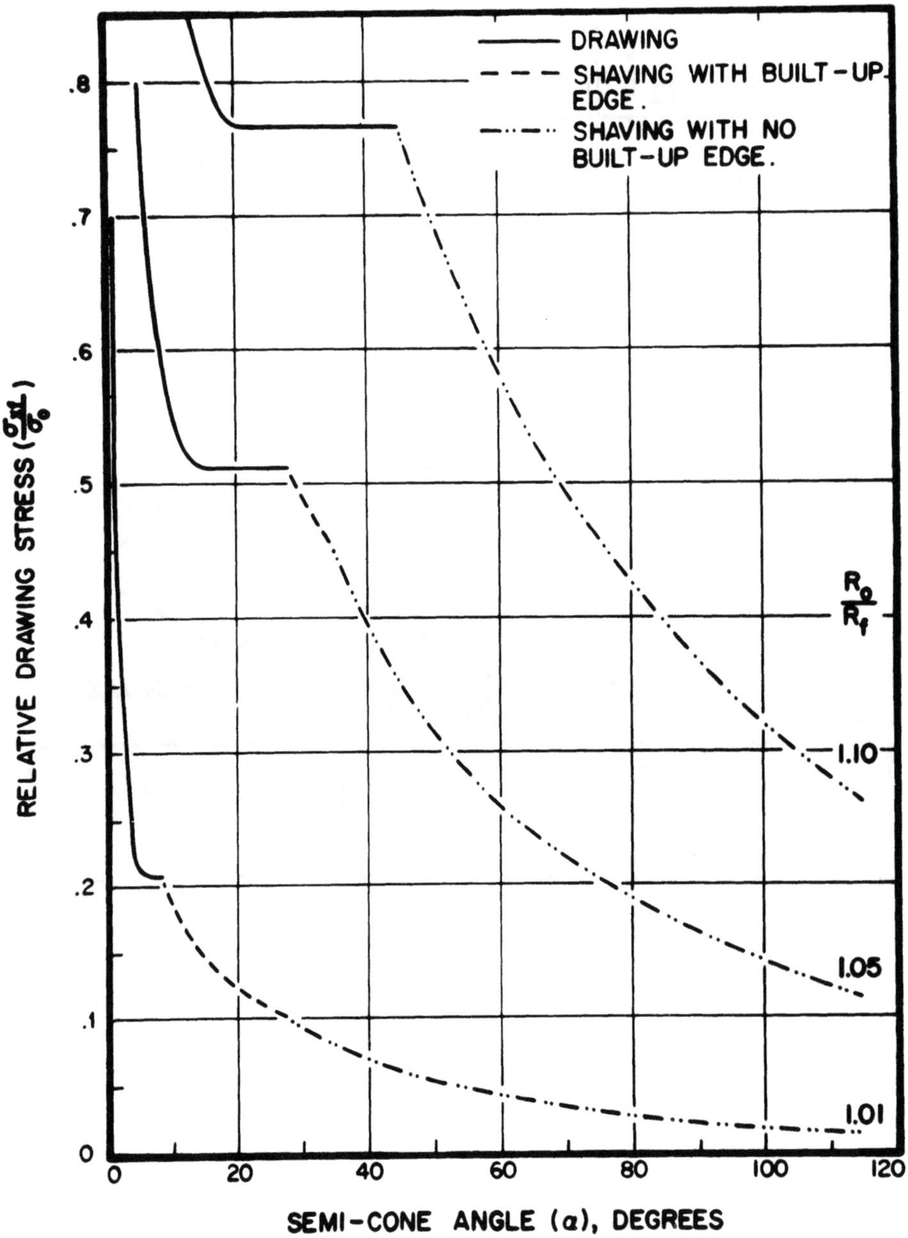

FIG. 23 RELATIVE DRAWING STRESS FOR SHAVING

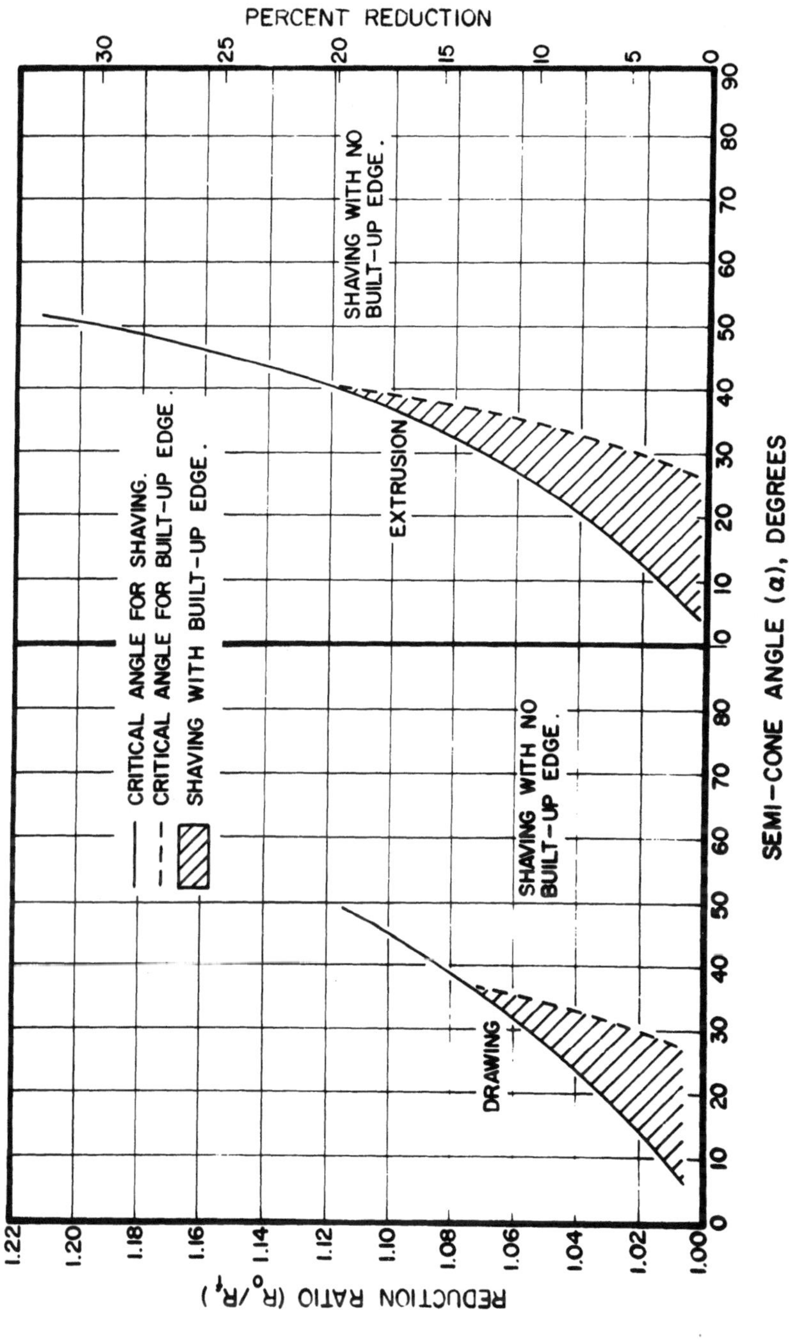

FIG. 24 CRITICAL ANGLES FOR SHAVING AND BUILT-UP EDGE

Central Burst

Many other patterns of flow exist in wire drawing or extrusion. A typical defect, although one of relatively infrequent occurrence, is the central burst defect (called, also, chevroning or the cup and cone fracture). While central burst does not occur often, it is of great concern because it occurs internally: the holes are in the interior of the product where they may not be detected readily but may then cause unexpected failures in service. When these defects were noted in the early 1930's it was proposed by Jennison[8] that improper die design was responsible and that by proper die design the defect could be prevented. In more recent years, while metallurgy and the study of metallography prospered, the early advice of Jennison was forgotten.

Presently, continuing faithful to our principle of minimum energy, it will be shown that the processing factors govern the criterion for the prevention of central burst, and even material characteristics will be accounted for.

The little that was known initially was that if central burst occurred it occurred with relatively small reductions, relatively large die angles, and subsequent to previous severe cold working of the billet. Further, on a mass production basis, central burst occurred in only a limited number of specimens, not in all of them.

Before any analysis was made it was postulated (Refs. 9 and 10) for the range in which central burst actually occurs that the relative drawing or extrusion stress associated with this fracture must be lower than that associated with any of the three other flow patterns already studied. Fig. 25 exhibits a possible domain where central burst may occur and characteristics of the drawing force associated with this flow; Fig. 26 shows the locus of possible central burst when drawing stress is plotted as a function of reduction with friction and cone angle held constant.

Understanding of the phenomenon of central burst grows from further consideration of the spherical velocity field of Fig. 27, which, previously, was found to represent the actual flow pattern rather well for any reduction when the optimal die angle is used. As the die angle is increased, the two surfaces of velocity discontinuity tend to deviate from the spherical shape and to approach one another (see Fig. 28a) . For some reductions and friction values the surfaces actually touch (see Fig. 28b). Since the exit velocity is higher than that at the entrance, when the rigid body associated with the exit touches the rigid body associated with the entrance it follows that a fracture must initiate at the point of contact between Γ_1 and Γ_2. The plastic zone II becomes an annulus. The entire wire continues its motion through the die; while it is moving forward the small fracture grows until it exits from the die. The

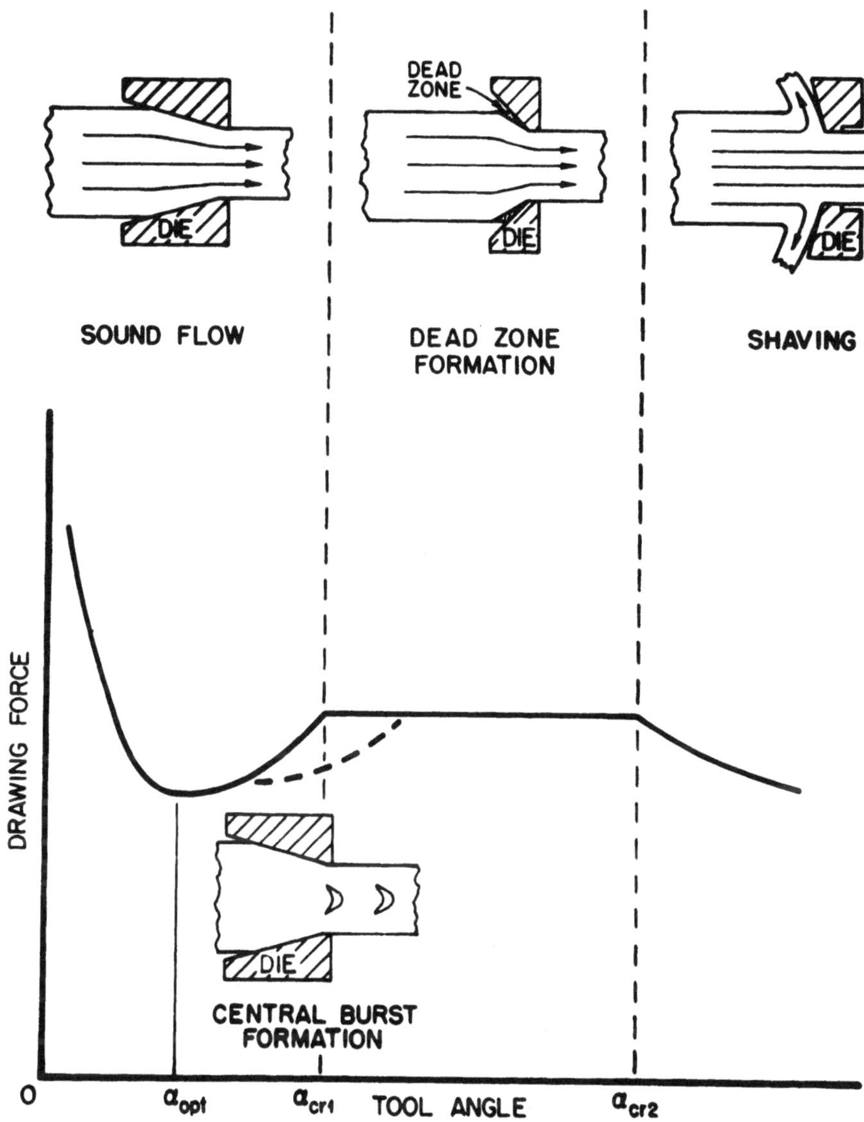

FIG. 25 SCHEMATIC ILLUSTRATION OF THE EFFECT OF TOOL ANGLE AND MODE OF FLOW ON DRAWING FORCE

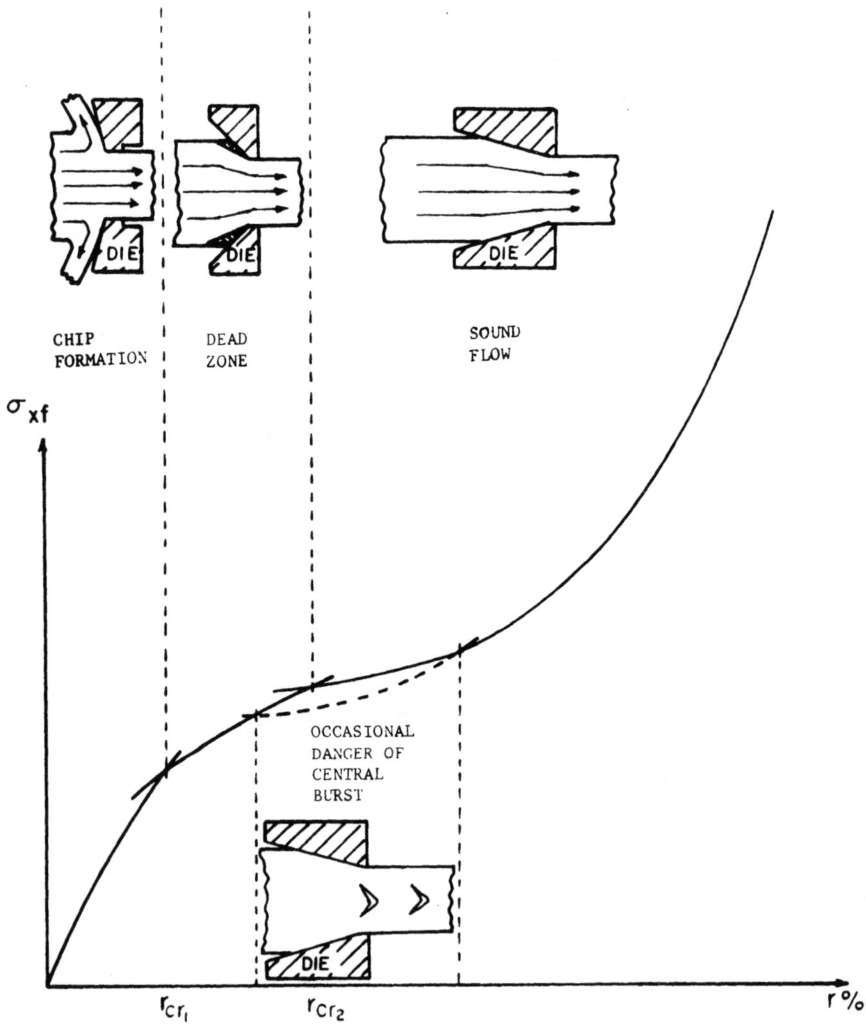

FIG. 26 DRAWING STRESS AS A FUNCTION OF REDUCTION

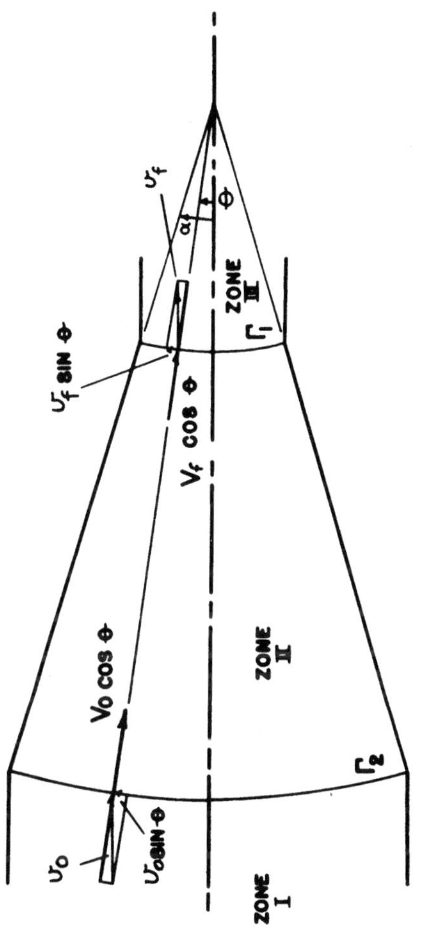

FIG. 27 VELOCITY FIELD

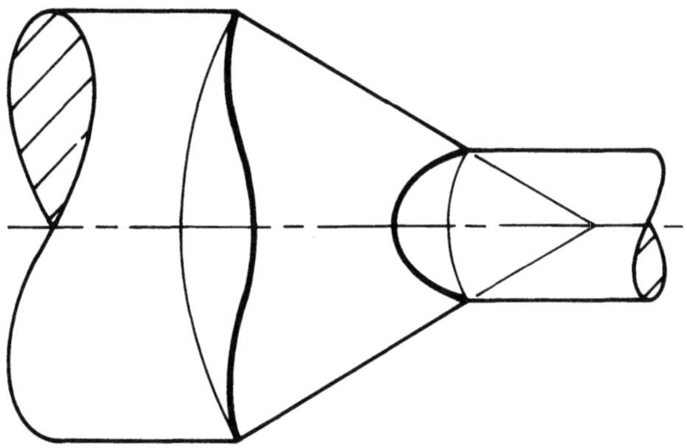

FIG. 28a NON-SPHERICAL SURFACES OF VELOCITY DISCONTINUITY

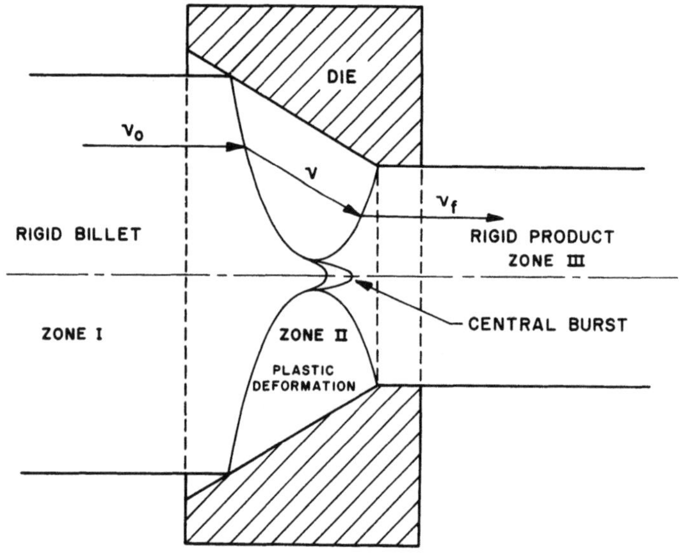

FIG. 28b IMPROVED VELOCITY FIELD WITH CENTRAL BURST

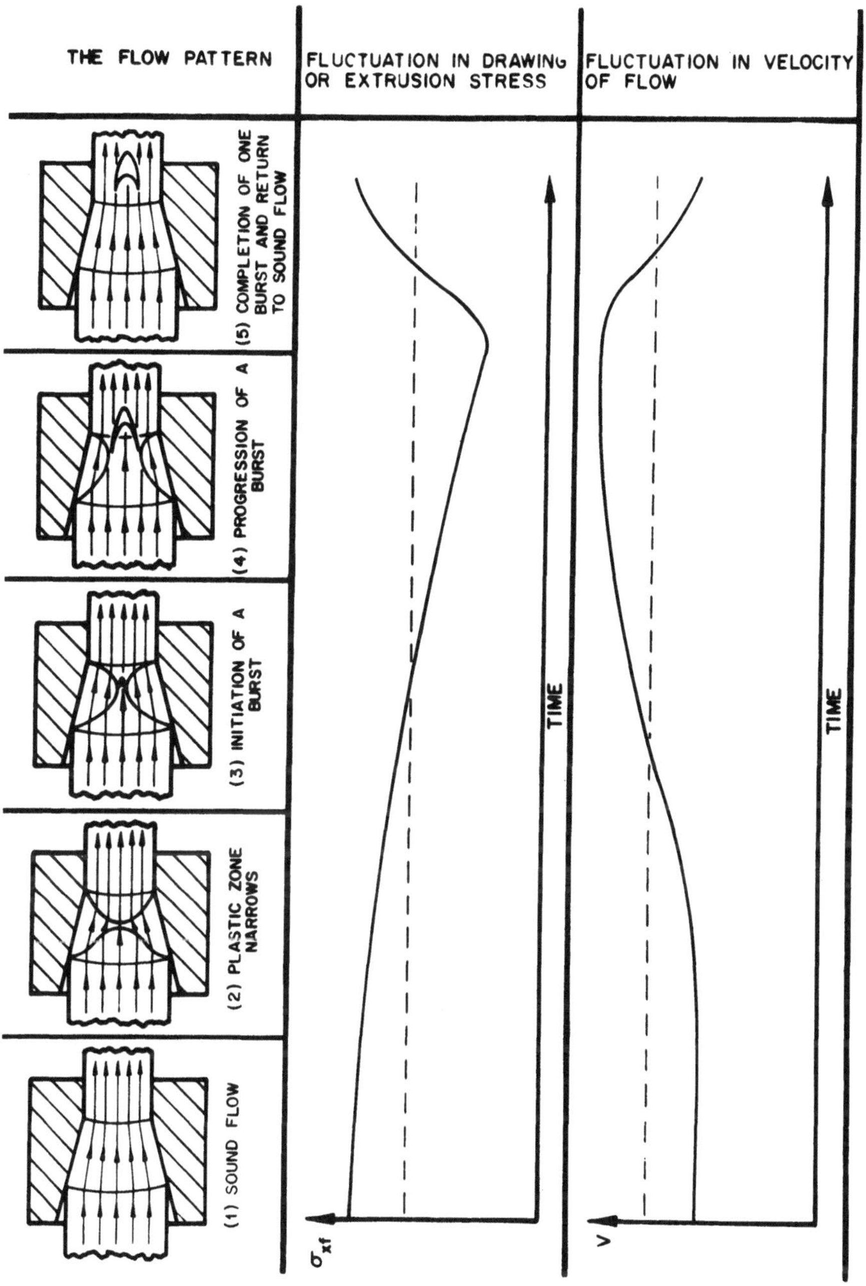

FIG. 29 INSTABILITY OR PERIODICITY ASSOCIATED WITH THE CENTRAL BURST DEFECT

sequence shown in Fig. 29 -- initiation, growth, and departure of a single fracture -- is a periodic event. The flow becomes a non-steady state flow. A velocity field for a non-steady state flow is hard to study.

The velocity field which was actually proposed and studied in Refs. 9 and 10 included a fracture of infinitesimal size only, since only the initiation of central burst was of concern under the assumption that if initial fracture can be prevented no growth is possible. Once again total relative drawing or extrusion stress was computed. The limited range within which the new computed force was lower than the forces computed through the other fields was then taken as the range over which central burst is expected, as represented in Fig. 25, in which drawing force is shown as a function of die angle.

Equating the forces at the intersections of the central burst curve with the characteristic curve for sound flow yields the relationship between the reduction ratio R_o/R_f and the critical semi-cone angle, and this is exhibited graphically in Fig. 30 as the central burst criterion, with friction as parameter; this is the 'Indian feather' criterion. Above and to the left of the appropriate friction curve in each case no central burst is expected; below and to the right one expects that central burst will occur.

In practice, whenever central burst did occur in a reasonably homogeneous material, it was always found that the die angle/reduction combination was to the right of the relevant criterion line; when one of the variables was changed so that the combination fell in the safe zone, central burst ceased to occur. The reverse, however, did not hold: many drawings or extrusions performed in the danger zone occur with no central burst at all. The criterion as presented is always safe although it may be unnecessarily restrictive.

In later investigations it was reasoned that a most important factor in the difference between sound flow and the cup and cone fracture might be the ability of the material to strain harden. This property is measured by the angle β, the inclination at any point on the true stress-true strain curve, so that the corresponding slope diminishes as the straining or cold working becomes more and more severe.

Strain hardening ability was thereupon introduced as a material property in the analysis and on this basis the power consumption was computed again. The resulting new criterion is represented in Fig. 31. The Indian feather of Fig. 30 is still the criterion if $\beta=0$, i.e., if the material is a non-strain hardening material. The danger zone becomes smaller as the coefficient of strain hardening becomes larger. For sufficiently high values of β there exists no

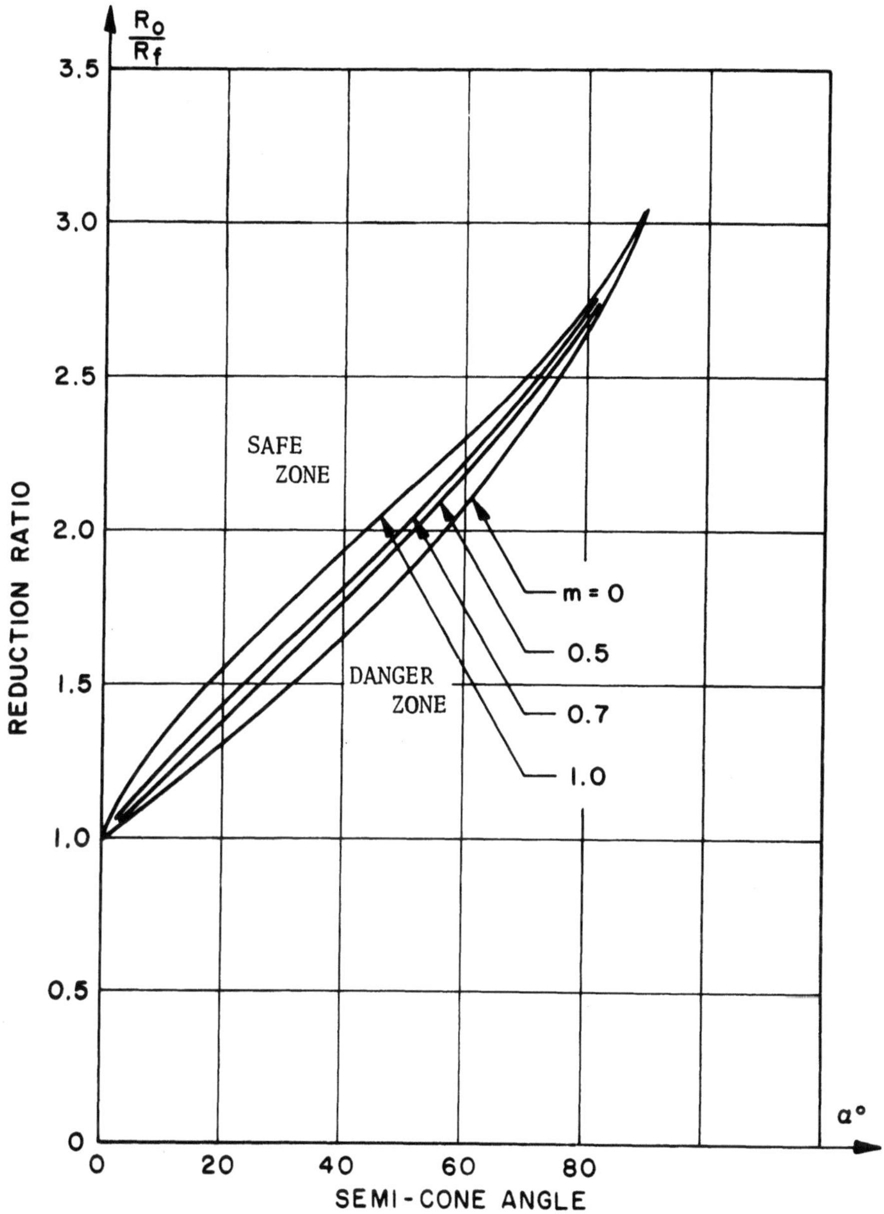

FIG. 30 CRITERION FOR CENTRAL BURST

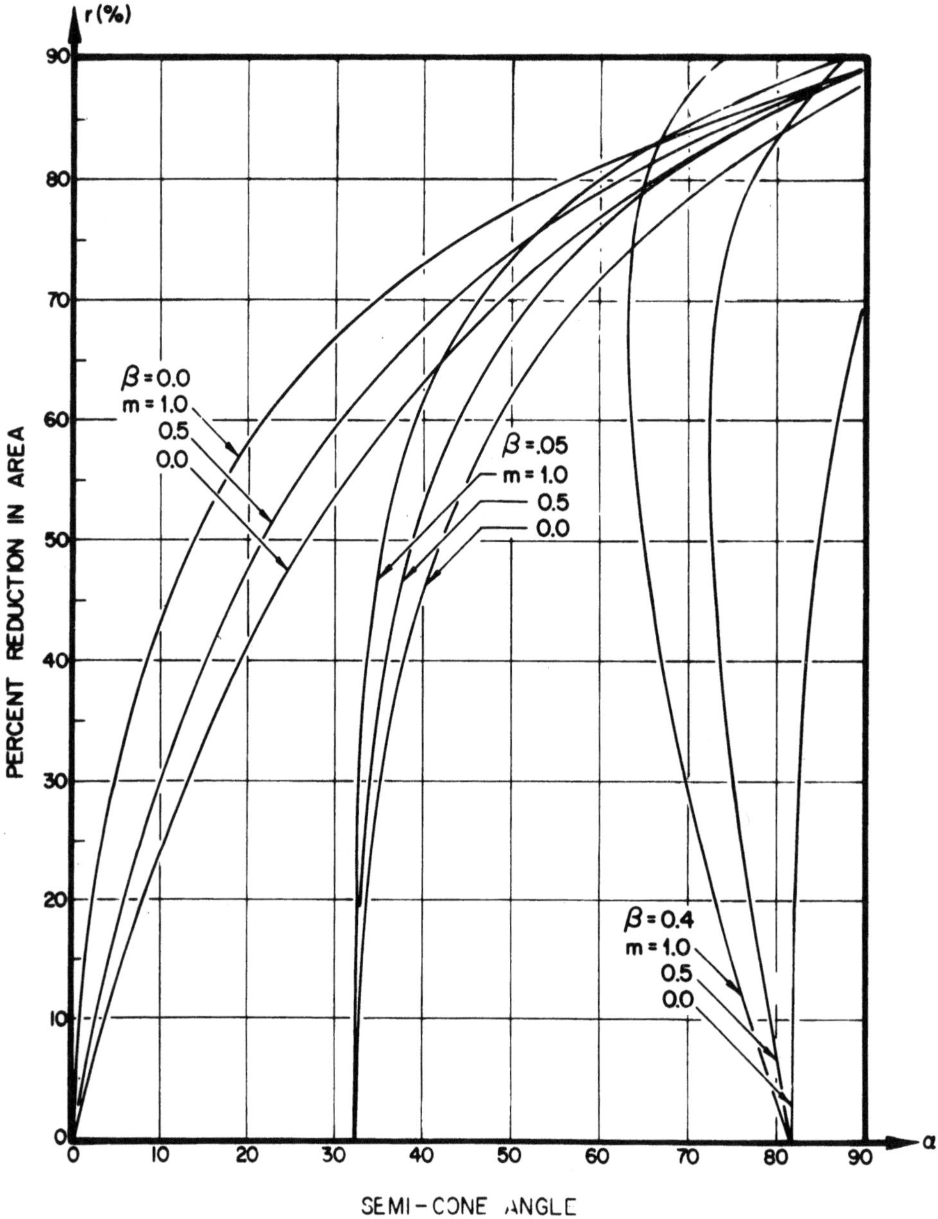

FIG. 31 CRITERION TO PREVENT CENTRAL BURST IN MATERIALS THAT STRAIN HARDEN

danger of central burst defect for any practical die angle or reduction. When this criterion is properly used in die and process design the problem of central burst is eliminated.

Summary

A collective display of results from this analysis is given in Fig. 32. The basic graph shows relative drawing stress as a function of the semicone angle of the die. Inserts represent the various modes of deformation and the several associated criteria.

This display emphasizes the broad scope of possibilities afforded by application to metal forming of the principle of minimum energy. Computation of the required forces and determination of the optimal die angle for maximum possible reduction without tear of the rod at the exit is an impressive achievement by itself; steps beyond that were taken when criteria for fracture and other flow patterns were developed.

Noteworthy is the fact that while sound flow is considered to belong to the theory of plasticity, and shaving to the field of metal cutting, with central burst traditionally in the realm of fracture mechanics, here in this presentation all phenomena have been treated uniformly by the upper bound approach under the principle of minimum energy and with the use of identical analytical tools. Fracture, for example, has not been accounted for by a separate metal property. One does not need a separate property. Indeed, let it be proposed that there is no such property. Fracture is the outcome of processing and the appropriate stress-strain relations and yield criteria alone. Yield criteria indicate when plastic flow will commence. A material can undergo any amount of deformation that one can put to it: it is important only that one puts it in the right fashion, without exhausting the ability of the material to undergo further and further deformation. For all these studies, whether fracture, metal cutting, or plastic flow, a single characteristic approach yields the answers to all the questions: the upper bound approach with the associated concept of minimum energy.

Other phenomena can be studied, likewise. Suppose that one observes a new mode of fracture which has not yet been studied; let Fig. 33 serve as example: no criterion for prevention of this fir tree (or fish skin or snake skin) fracture is yet available. All that one has to do -- and it may require some doing -- is to propose a velocity field which leads to this defect, to compute the internal power, shear, and friction losses associated with the flaw, and to determine the range of parameters in which the total drawing or extrusion stress required is least among the power requirements of all previously known flow patterns. There are many more which occur and which may be studied in this manner.

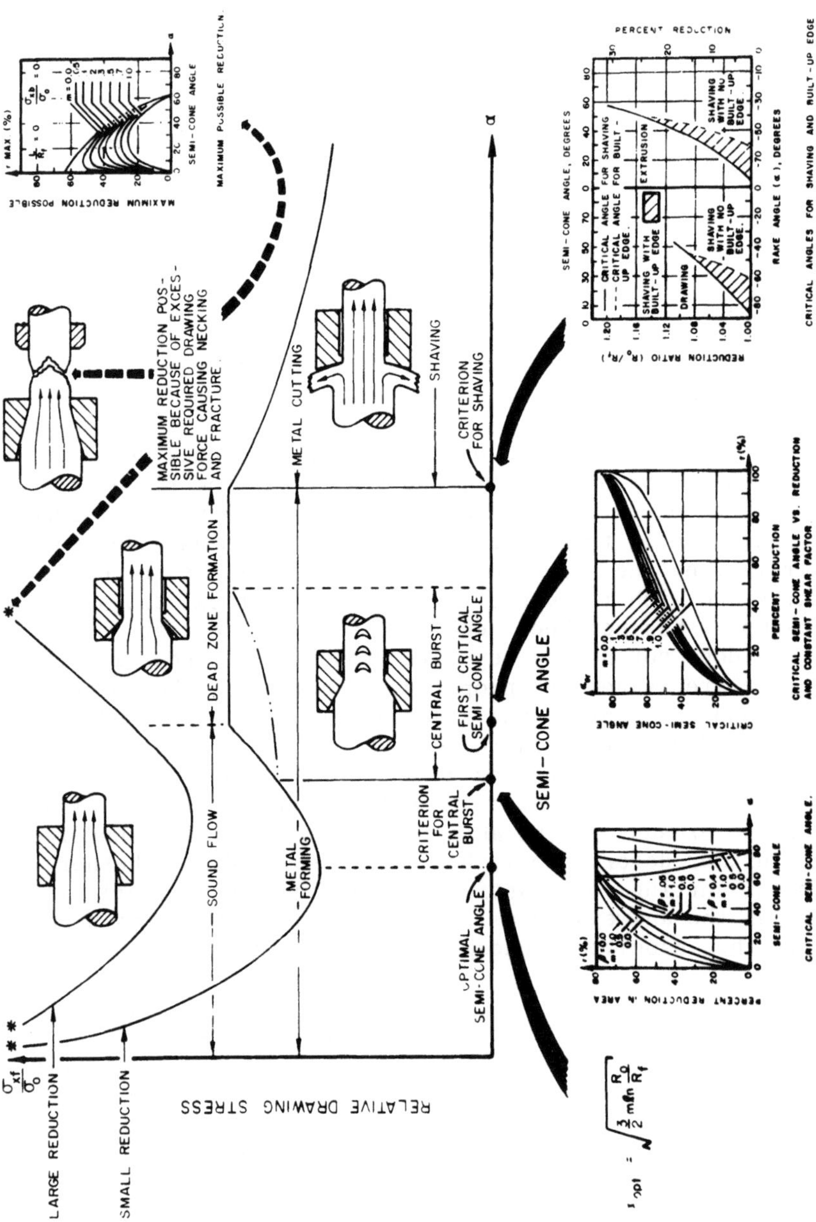

FIG. 32 EFFECT OF CONE ANGLE ON METAL FLOW

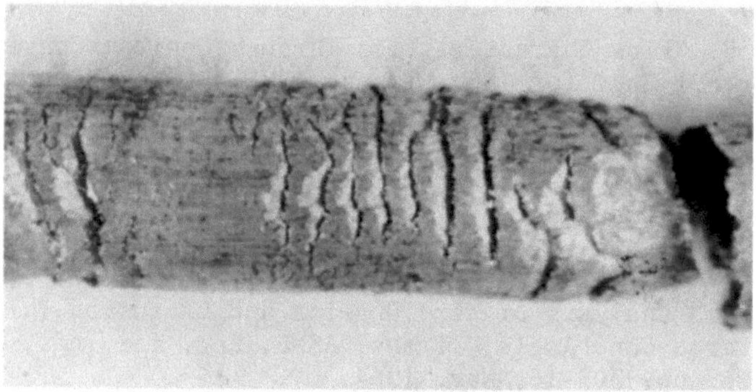

FIG. 33 FIR TREE DEFECT

Friction is a predominant factor in the criteria developed. Quantitative evaluation of friction existing during forming is a necessity. The analysis performed provides also a means to measure friction (see Ref. 11) whether Coulomb, constant, or that of hydrodynamic lubrication.

In Fig. 26 the parameters were reversed: with cone angle and friction held constant, drawing force is exhibited as a function of reduction. When too small a reduction is attempted, shaving occurs;increasing the reduction, one may encounter dead zone formation or occasional central burst defect. To eliminate these possibilities one must take larger reductions and thereby obtain sound flow. Most of the defects in wire drawing occur where very little distortion is expected but where all of these characteristic kinds of failure do occur. If one changes to extrusion, taking a ratio of 40:1 in aluminum, one rarely finds fracture of the product: what one does find is that the product is distorted. The grid line which was straight and perpendicular to the axis of symmetry will be sheared in very high degree: this does not disturb the material nor make it inferior. Distortion, as was said earlier, is no measure for the likelihood of failure, nor is the amount of deformation: the mode of deformation is the main source of many failures.

REFERENCES

1 Thomsen, E. G., Yang, C. T. and Bierbower, T. B., "An Experimental Investigation of the Mechanics of Plastic Deformation of Metals," Univ. Calif. (Berkeley), Publ. Eng., vol. 5 , no. 4, pp. 89-144, 1959.

2 Avitzur, B. "Metal Forming: Processes and Analysis," McGraw-Hill Book Co. , New York, 1968.

3 Avitzur, B. "Flow Characteristics Through Conical Converging Dies," Jr. of Eng. for Ind., Trans. ASME, Series B, vol. 88, pp. 410-20, Nov. 1966.

4 Avitzur, B. "Analysis of Wire Drawing and Extrusion Through Dies of Small Cone Angle," Trans. ASME, Eng. for Ind., Series B, vol. 85, pp. 89-96, Feb. 1963.

5 Avitzur, B. "Analysis of Wire Drawing and Extrusion Through Dies of Large Cone Angle," Trans. ASME, Eng. for Ind., Series B, vol. 86, pp. 305-16, Nov. 1964.

6 Wistreich, J. G., "Investigation of the Mechanics of Wire Drawing," Proc. Inst. Mech. Engrs. (London), vol. 169, pp. 654-65, 1955.

7 Feilbach, W. H., Jr. , and Avitzur, B., "Analysis of Rod Shaving and Orthogonal Cutting," Jr. of Eng. for Ind., Trans. ASME Series B, vol. 90, no. 2, pp. 393-403, May 1968.

8 Jennison, H. C.,"Certain Types of Defects in Copper Wire Caused by Improper Dies and Drawing Practice," American Institute of Mining and Metallurgical Engineers, Technical Publication No. 285, Class E, Institute of Metals No. 97, 1930.

9 Avitzur, B., "Analysis of Center Bursting Defects in Drawing and Extrusion," Jr. of Eng. for Ind., Trans. ASME Series B, vol. 90, no. 1, pp. 79-91, Feb. 1968.

10 Zimerman, Z., and Avitzur, B., "Analysis of the Effect of Strain Hardening on Central Bursting Defects in Drawing and Extrusion," Transactions ASME Series B, vol 92, no. 1, pp. 135-45, Feb. 1970 . Lehigh Publication Oct. 1967.

11 Evans, W.,and Avitzur, B., "Measurement of Friction in Drawing, Extrusion, and Rolling," Jr. of Lub. Tech., Trans. ASME Series F, vol. 90, no. 1, pp. 72-80, Jan. 1968.

SELECTION OF OPERATING PARAMETERS TO PREVENT CENTRAL BURSTING DEFECTS DURING COLD EXTRUSION

Z. Zimerman, H. Darlington, and E. H. Kottcamp, Jr.*

Homer Research Laboratories, Bethlehem Steel Corporation, Bethlehem, Pennsylvania

INTRODUCTION

Central bursts, or chevrons, are the internal arrow-shaped defects occasionally encountered in cold extrusions or drawn wire (Figure 1). In cold extrusion the defect is usually observed in the final, light step of a multistep extrusion. When central bursting occurs, the load-carrying capacity of the part is drastically reduced. The possibility of central burst occurrence has required manufacturers of critical parts, e.g., axle shafts and steering gear components, to install and maintain ultrasonic equipment for 100% inspection of the formed parts. Although inspection can prevent the use of components containing central bursts, the rejection of extruded items increases production costs because both material and time are consumed in the manufacture of the unacceptable parts. Thus, it is desirable to completely prevent the occurrence of central bursts.

When central bursts are detected, the quality of the raw material is often examined. However, as long ago as 1930 a study of wire drawing by Jennison showed that the geometrical features of the deformation process are a major cause for central bursting [1]**. Since then many others, including Tanaka [2], Russell [3], Johnson and Kudo [4], and Pugh and Watkins [5], have

*Dr. Z. Zimerman is an Engineer, Mr. H. Darlington is a Supervisor, and Dr. E. H. Kottcamp, Jr., is Assistant to the Vice President and Director of Research.

**The numbers in brackets designate references listed at the end of the text.

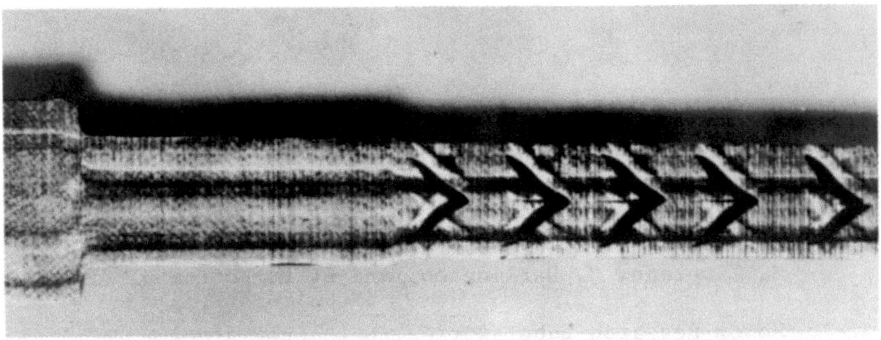

FIG. 1. CENTRAL BURSTING IN A COLD EXTRUDED SHAFT.

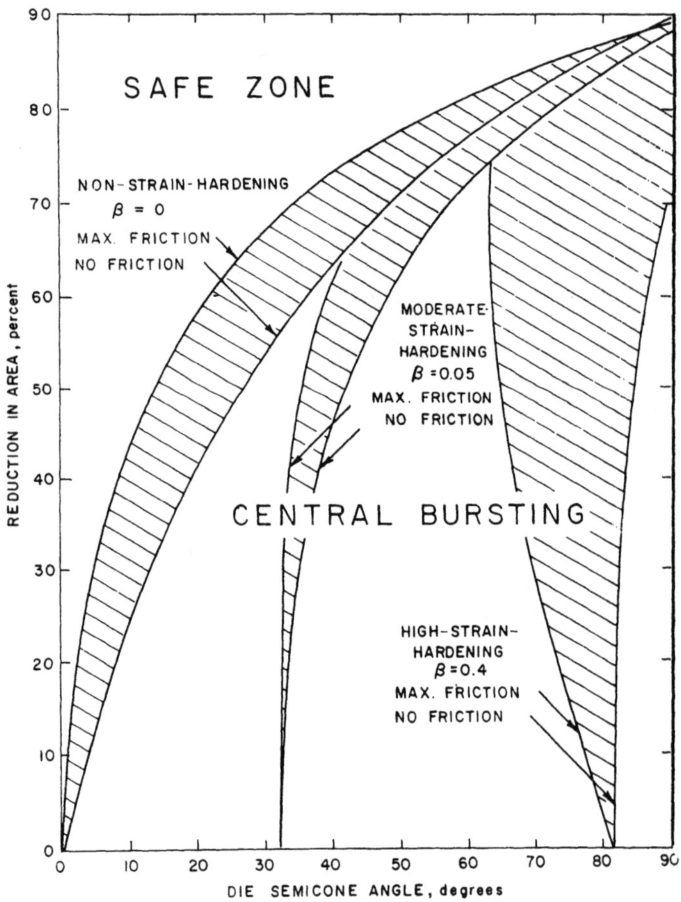

FIG. 2. CRITERION FOR PREVENTION OF CENTRAL BURSTING.

reported qualitative methods to eliminate central bursts. Bethlehem Steel Corporation also recognized the importance of developing a method for the prevention of the defect and supported an analytical investigation of the mechanics of central burst formation at Lehigh University. On the basis of analytical-mathematical studies, this investigation resulted in a criterion for the prevention of central bursting that defines permissible combinations of the two process parameters of reduction of area and die cone angle; the criterion was verified by laboratory extrusion and wire drawing [6].

In the study described in the present paper, we utilized this criterion as a guide for process design under industrial extrusion conditions. In addition, we examined correlations between the occurrence of the defect and the mechanical and metallurgical properties of the tested steels. Finally, examples of commercial central bursting were examined.

CRITERION FOR THE PREVENTION OF CENTRAL BURSTING

Before describing the results obtained in the current industrial investigations, we will review the criterion and also outline the method employed in this work to evaluate the strain hardening capacity.

The Criterion

The criterion, which defines combinations of die semicone angles and reductions in area that will prevent central bursting, is summarized in Figure 2. The criterion takes into consideration the effect of the strain hardening capacity of the deforming material, β, as well as that of friction in the forming operation, m. Any forming operation for which the combination of die semicone angle and reduction in area lies to the left of the appropriate lines in Figure 2 will be in the "safe zone", i.e., will be free of central bursting. In contrast, if the combinations are to the right of the lines, central bursting may occur. The effect of the strain hardening capacity on the criterion is presented in terms of three general groupings of material: non-strain-hardening, moderate-strain-hardening, and high-strain-hardening.

The following examples will demonstrate how the criterion works. A 30% reduction in area when extruding a non-strain-hardening material, i.e., one in which the strain hardening capacity has been exhausted, can cause central bursting if a die of 15-degrees semicone angle is used; but a die of about 5-degrees semicone angle will prevent the defect. Similarly, increasing the reduction in area to 55% while maintaining a

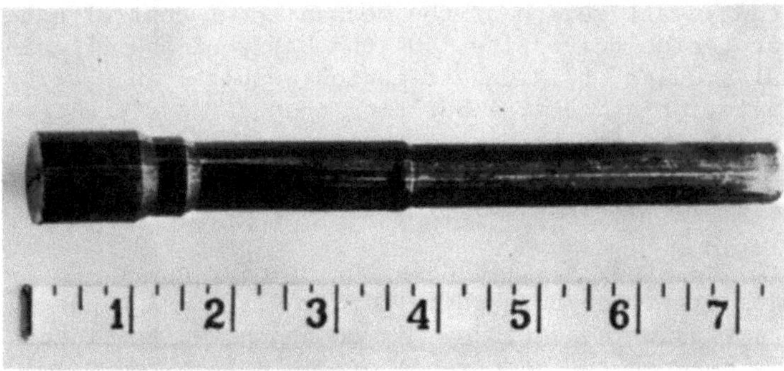

FIG. 3. COLD EXTRUDED SHAFT OF 1024 STEEL.

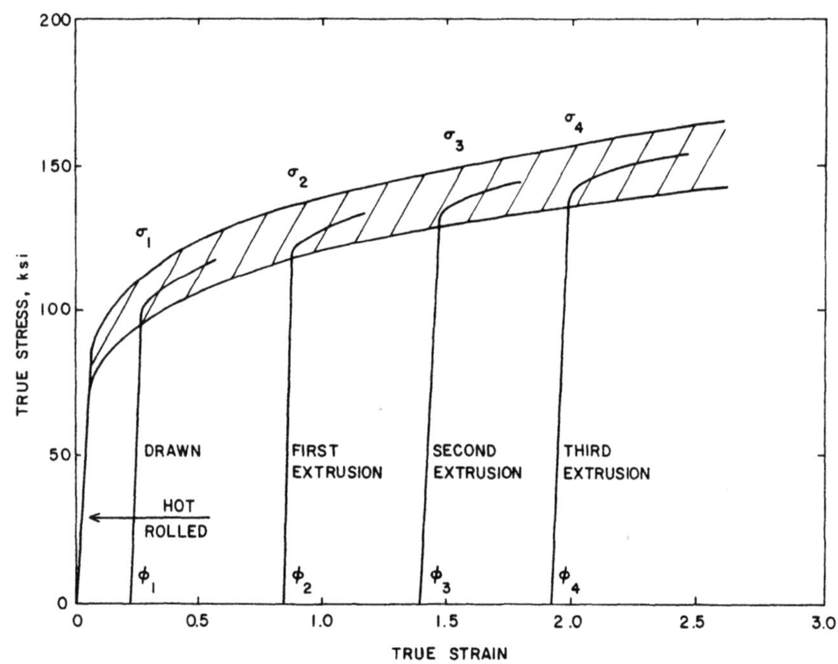

FIG. 4. EXTENDED STRESS-STRAIN CURVE, 1024 STEEL.

15-degree semicone angle die will prevent the defect. An increase in friction between the material and the die will have an adverse effect on the prevention of central bursting. For example, if the friction between the material and die is small, as in the case of phosphate-coated and well lubricated steel in a carbide die, a reduction in area of 30% in non-strain-hardening material extruded through a 9-degree semicone angle die will produce sound extrusions. With the same process geometry, an increase of friction can result in central bursting.

An increase in the strain hardening capacity increases the safe zone. Thus, although a 30% reduction in area through a die of 15-degrees semicone angle can cause central bursting in a non-strain-hardening material; even a moderate strain hardening capacity will permit the same geometry to produce sound extrusions.

The Extended Stress-Strain Curve

The strain hardening capacity of a material can be determined from a tension test by measuring the slope of the true stress/true strain curve. However, the strains introduced by a cold forming operation are usually larger than the total strain in a tension test. The following method, which has been employed in this study to evaluate the true stress/true strain relationship of the deforming material, takes this problem into account.

We will demonstrate the method using, as an example, the cold-extruded shaft illustrated in Figure 3. The shaft was manufactured from hot-rolled 1024 steel by one step of drawing and three of extrusion. To determine the true stress/true strain relationship of the material, we proceed as follows:

Samples from the hot-rolled bar and each forming step are tension-tested and the true stress/true strain relationship is determined for each sample. The results are plotted in the manner shown in Figure 4. The origin of the true stress/true strain curve after each successive forming step is established by the total true strain introduced by the forming operations. This total true strain includes the ideal strain, the strain resulting from the friction, and the strain resulting from distortion due to the die angle, and is determined using an equation developed by Avitzur to calculate the relative forming stress for drawing or forward extrusion [7]. It is this single curve, termed the "extended stress-strain curve", that describes the true stress/true strain relationship of the material over the entire range of the forming strains.

Evaluation of the Strain Hardening Capacity

For convenience in the derivation of the criterion, the true stress/true strain relationship was defined by the equation:

$$\sigma_t = S\ (1 + \beta\ \phi)$$

where S is the initial true stress, ϕ is the true strain, σ_t is the true stress at true strain ϕ, and β is the linear strain hardening coefficient. Since materials do not exhibit linear strain hardening over a large strain range, the linear strain hardening coefficient is a function of the true strain. The linear strain hardening coefficient for a particular forming step can be calculated from the extended stress-strain curve if one knows the initial true strain in the material and the true strain introduced by that particular forming step. For the shaft shown in Figure 3 with the true stress/true strain relationship described in Figure 4, the linear strain hardening coefficient of the steel as it undergoes the second extrusion step would be determined by:

$$\beta = \frac{\sigma_3 - \sigma_2}{\sigma_2 (\phi_3 - \phi_2)}$$

where the subscripts are those noted in Figure 4.

The scatter in the results for the experimental extended stress-strain curve for the 1024 steel in Figure 4 is indicated by the width of the band of the true stress/true strain curve. Although the experimental results show only small variations in the strength of the material, these variations cause large changes in the linear strain hardening coefficient. Because the criterion for the prevention of central bursting is strongly dependent on the strain hardening coefficient (Figure 2), tooling design must take into account the lowest strain hardening coefficient that can be expected. If the value of this coefficient is not certain, the criterion for non-strain-hardening materials should be used.

VERIFICATION OF THE CRITERION IN AN INDUSTRIAL PROCESS

Extrusion

The verification of the criterion in an industrial process was based on the production of the shaft shown in Figure 3. Hot-rolled 7/8-inch-diameter 1024 steel bars were drawn and then extruded in three steps on a five-station 3/4-inch boltmaker. The cold forming steps were:

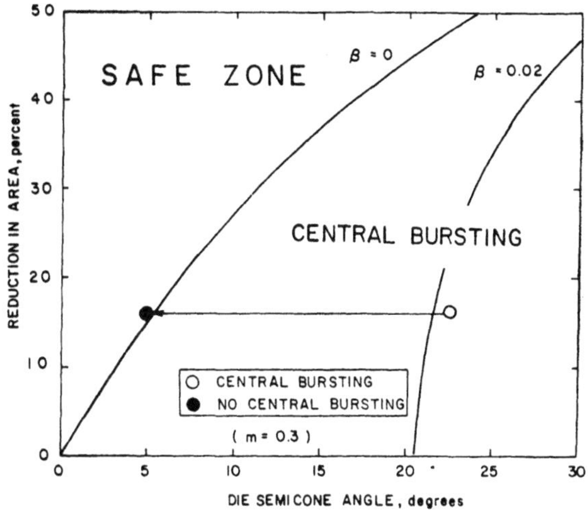

FIG. 5. PREVENTION OF CENTRAL BURSTING IN THE INDUSTRIAL SHAFTS ACCORDING TO THE CRITERION. (FOR COLD EXTRUSION OF PHOSPHATED AND LUBRICATED STEEL m = 0.1 - 0.2).

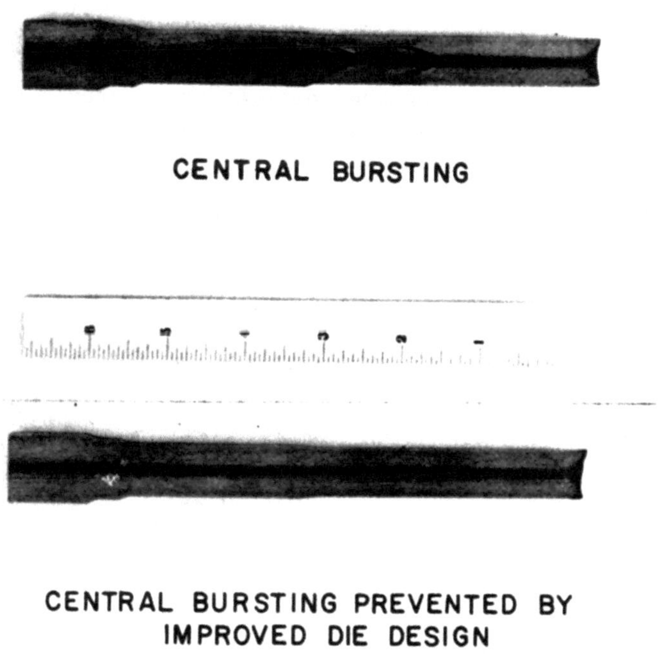

FIG. 6. SECTIONS OF COLD EXTRUDED SHAFTS OF 1024 STEEL, NITAL ETCHED.

	Reduction	Die Semicone Angle
Drawing	8%	9.0 degrees
Extrusion	22%	22.5 degrees
Extrusion	23%	22.5 degrees
Extrusion	16%	through a variety of angles

For the final extrusion step we selected angles that would, according to the criterion, either produce, or prevent, central bursting.

Since central bursting occurs only occasionally and in isolated heats of steel, four different heats of 1024 steel were examined.

The wt % compositions determined from twelve or more tests for each of the four heats were:

Heat No.	C	Mn	P	S	Si
1	0.22	1.52	0.013	0.033	0.18
2	0.21	1.33	0.015	0.025	0.17
3	0.27	1.45	0.019	0.027	0.26
4	0.22	1.44	0.018	0.021	0.22

Initially, about 1000 shafts were produced from each of the four heats using for the last extrusion step a die semicone angle of 22.5 degrees, which is in the "central bursting zone" for non-strain-hardening materials (see Figure 5). All the shafts were examined ultrasonically. Central bursting was detected in about 4.5% of the shafts produced from one of the heats (No. 1), while no bursting was found in any of the shafts from the other three heats.

From the extended stress-strain curve for each of the four heats the linear strain hardening coefficients were determined for the final extrusion step, as follows:

Heat No.	β
1	0.02 - 0.14
2	0.05 - 0.13
3	0.06 - 0.15
4	0.07 - 0.36

Only in part of heat No. 1 was the strain hardening coefficient low enough to extend the central bursting zone to include the extrusion configuration (see Figure 5). In all other cases the linear strain hardening coefficient was 0.05 or larger and the extrusion configuration was in the safe zone, where no central bursting is expected.

Since central bursts had been found only in heat No. 1, a second coil from the same billet of that heat was extruded on the boltmaker. Five hundred shafts were produced with the original 22.5-degree semicone angle, and another five hundred with an improved die design, one in which the die had a semicone angle of 5 degrees.

Ultrasonic inspection detected no central bursting in the shafts produced with the improved die design. Of the shafts extruded through the 22.5-degree semicone angle die, 1.2% contained central bursts. The linear strain hardening coefficients were 0.01-0.18 for the steel and extrusion configurations in this trial.

Samples from the other three heats and from another billet of heat No. 1, were extruded through larger cone angles, but no central bursting was found.

Metallurgical Examination

Examination of the average chemical composition of the steel (see in-text tabulation above) does not reveal differences that could account for the observed central bursting in heat No. 1. We have found in other studies that small differences in chemistry, such as the slightly higher manganese and sulfur contents in heat No. 1, do not result in major changes in mechanical behavior. In this regard, we did not detect any central bursting in heat No. 3 which has the highest carbon, phosphorus, and silicon contents and indeed the highest total alloy content of the four heats.

Etched sections from shafts of heat No. 1 showed a dark marking along the centerline (Figure 6). According to metallographic examination, these dark streaks were associated with a higher carbon content at the center than at the surface of the shafts. Steel with this centerline condition is not typical of the normal product shipped to the cold extrusion industry. However, even with the condition, central bursting was prevented when a die geometry in the safe zone (Figure 2) was used.

The yield strength, tensile strength, percent elongation, and reduction in area in a tension test for the four heats of steel are shown in Figures 7, 8, 9, and 10 as functions of the total true strain introduced in the various extrusion steps. As in the case of the extended stress-strain curve, the

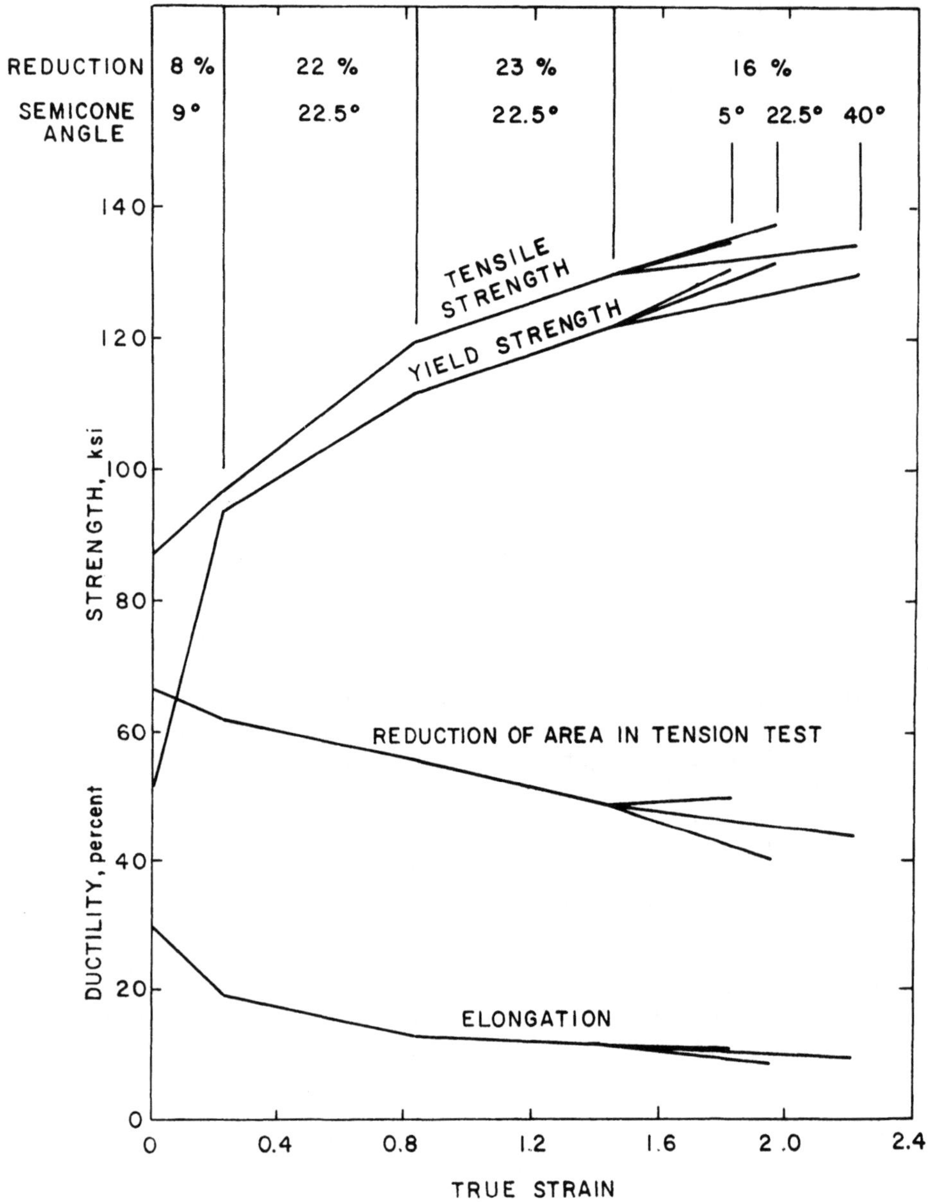

FIG. 7. MECHANICAL PROPERTIES OF COLD EXTRUDED SHAFTS, HEAT No. 1.

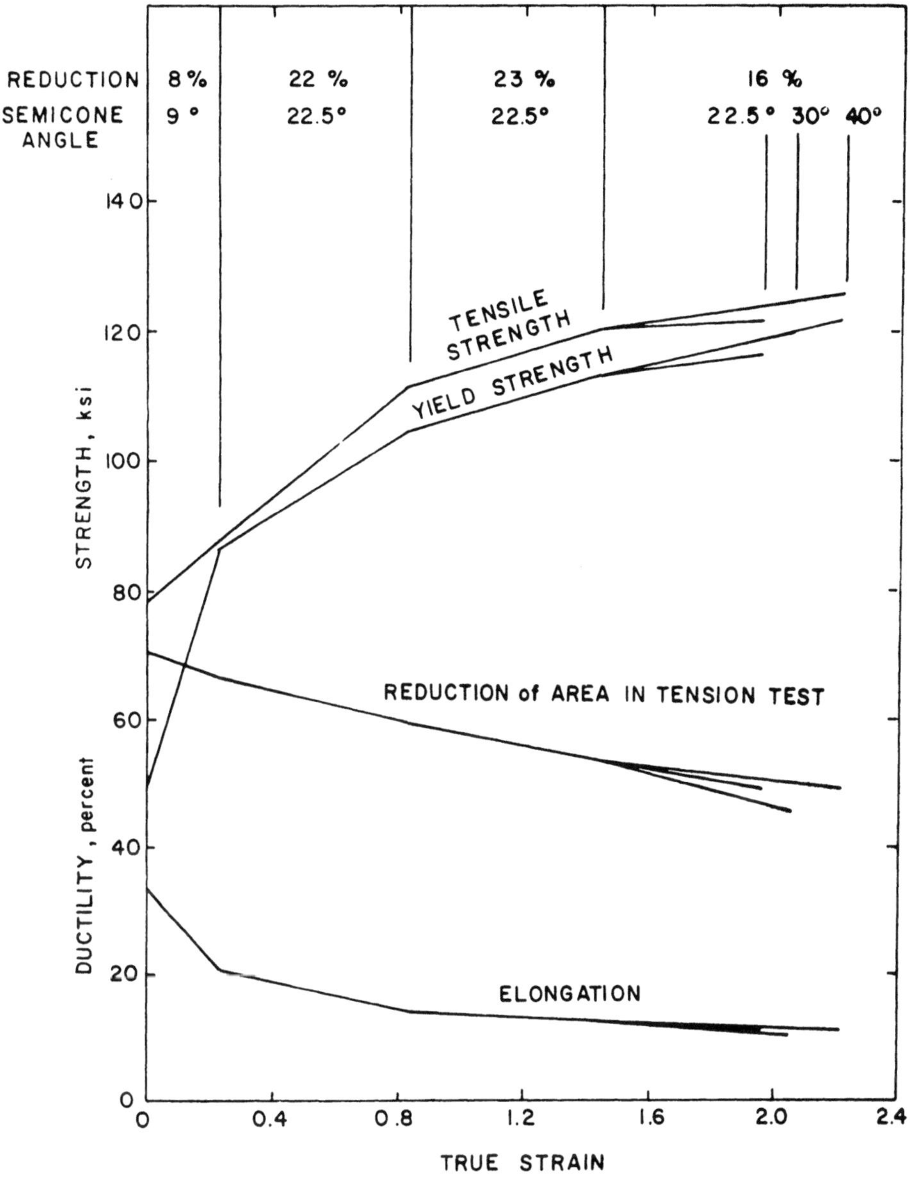

FIG. 8. MECHANICAL PROPERTIES OF COLD EXTRUDED SHAFTS, HEAT No. 2.

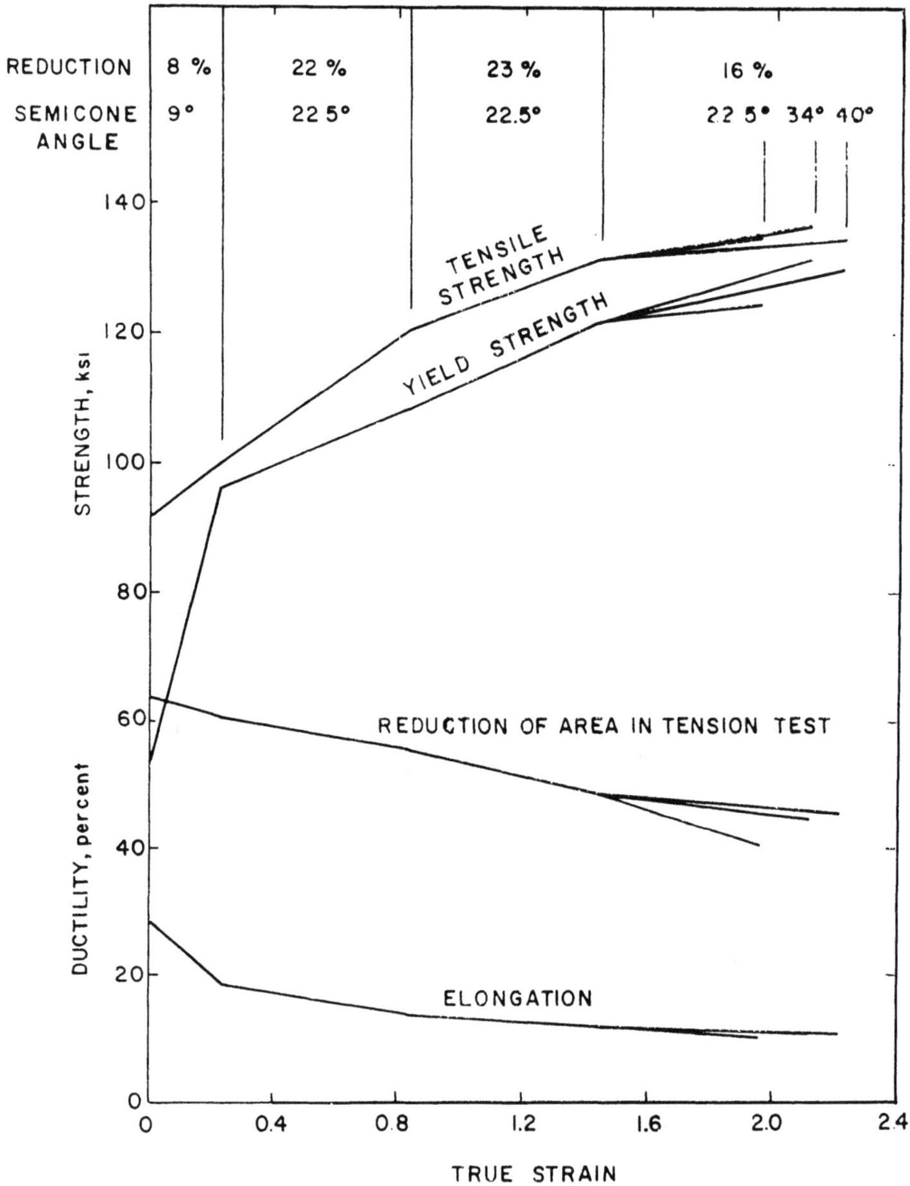

FIG. 9. MECHANICAL PROPERTIES OF COLD EXTRUDED SHAFTS, HEAT No. 3.

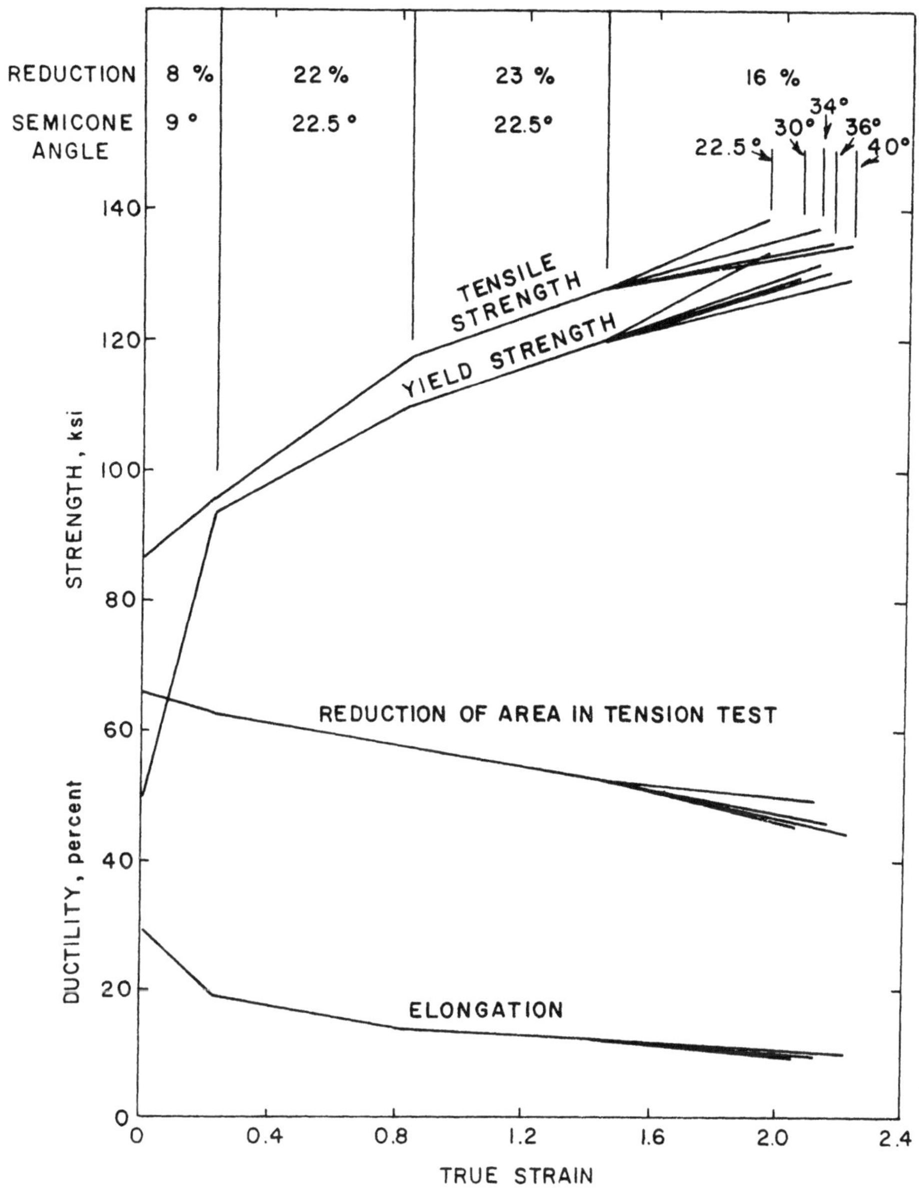

FIG. 10. MECHANICAL PROPERTIES OF COLD EXTRUDED SHAFTS, HEAT No. 4.

Material	Die Semicone Angle, degrees	Reduction in Area, percent	Condition Before Reduction Causing Central Bursts
1038 M Steel	18	9.8	Cold worked 33%
1041 Steel	15	5.4	Hot rolled, cold worked 2%
1041 Steel	10	15.9	Spheroidize annealed, cold worked 57%
1050 Steel	9 4	14.7 14.7	Cold worked 45%
4032 H Steel	16	28.5	Hot rolled, cold worked
4032 H Steel	25	30.3	Hot rolled, cold worked
8620 H Steel	27	20	Cold worked 60%
8625 H Steel	24	27.8	Hot rolled, cold worked 26.5%
Cast 1050 Steel	9 4	14.7 14.7	Cast as bar, cold worked 45%
Zirconium	7.5 7.5	15 28.5	Cold worked wire

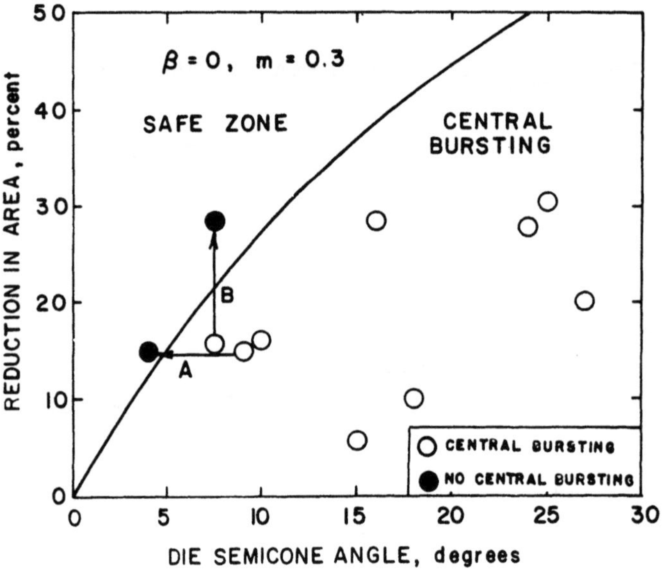

FIG. 11. EFFECT OF DIE GEOMETRY ON THE OCCURRENCE OF CENTRAL BURSTING.

total true strain includes not only the ideal strain but also the strain from the die friction and from the distortion due to the die angle. The ratio between the diameter of the tension test specimen and the diameter of the hot-rolled bar or cold-formed section was kept constant for all tests. The plotted values are averages of two or more tests. These data do not provide a correlation with central bursting occurrence.

EXAMPLES OF COMMERCIAL CENTRAL BURSTING

In addition to the industrial cold extrusion experiments to verify the criterion, we conducted a survey of central bursting examples encountered in the cold extrusion industry which covered a wide variety of parts extruded from many steels and one case of drawn zirconium wire. Figure 11 shows the criterion for the prevention of central bursting and tooling configurations that resulted in central bursting. Each case of central bursting lies in the zone for which the criterion predicts this defect. For two of these cases, the occurrence of central bursting was subsequently eliminated when the manufacturers improved their tooling design in accordance with the criterion. The improved geometry for each case is indicated by a solid circle in Figure 11, and the change in the geometry is shown by an arrow. In case A the die semicone angle was reduced so that the geometry fell to the left of the criterion line, and in case B the percent reduction of area was increased to bring the forming geometry into the safe zone.

CONCLUSION

An experimental investigation on cold extrusion of four heats of 1024 steel and a survey of central bursting occurrence in commerical cold extrusion and wire drawing operations showed that central bursting defects can be prevented in industrial processes by selecting the proper combinations of percent reduction of area and die semicone angle on the basis of the criterion derived by Zimerman and Avitzur [6]. Forming sequences which will prevent central burst formation can be designed on the basis of that criterion.

ACKNOWLEDGMENT

The authors appreciate the support received from their associates at Bethlehem Steel Corporation; in particular, P. S. Villa for his help in the industrial extrusion phase, P. E. Nemchik and R. L. Snyder for their work in the laboratory, J. F. Clark for his assistance in the survey, and B. S. Mikofsky for his editorial expertise.

REFERENCES

1. Jennison, H. C., "Certain Types of Defects in Copper Wire Caused by Improper Dies and Drawing Practice", Trans AIME, 1930, p 121.

2. Tanaka, H., "On the Causes of Cuppy Defects in a Drawn Material", J. Japanese Inst. of Metals, V. 16, 1952.

3. Russell, J. V., "Steels for Cold Forming", Metals Engineering Quarterly, Feb. 1962.

4. Johnson, W. and Kudo, H., The Mechanics of Metal Extrusion, Manchester University Press, 1962.

5. Pugh, H. Ll. D. and Watkins, M. T., "Experimental Investigation of the Extrusion of Metals", Production Engineering, V. 40, No. 4, London, April 1961, p 256.

6. Zimerman, Z., and Avitzur, B., "Analysis of the Effect of Strain Hardening on Central Bursting Defects in Drawing and Extrusion", Journal of Eng. for Industry, Trans. ASME, Series B, Vol. 92, No. 1, Feb. 1970, p 135.

7. Avitzur, B., Metal Forming - Processes and Analysis, McGraw-Hill, New York, 1968, Chapter 8.

THE EFFECT OF MATERIAL PROPERTIES ON TENSION ZONE AND BOUNDARY SHEAR-STRESS IN EXTRUSION*

Aly H. Shabaik

Assistant Professor of Engineering

University of California, Los Angeles, Calif 90024

ABSTRACT

The complete solution of stress and strain were obtained for commercially pure aluminum and superplastic alloy of the eutectic of Pb-Sn in an axisymmetric extrusion process of extrusion ratio 4 and a half cone angle of 45°. The extrusion speed was 0.1"/min for aluminum and .003 in/min for superplastic Pb-Sn. The stress components along and perpendicular to the flow lines were calculated for different values of workhardening and strain-rate exponents. The shear stress along the boundary was determined. The size of the tension zone was compared for different material properties.

INTRODUCTION

The limitations imposed by high strength and other space-age materials, when conventional methods are used to form them, have demanded an evaluation of the applicability of these shaping processes. An understanding of the mechanics of metal deformation is essential before improvement of the current methods of forming can be achieved. To determine the effect of process variables and material properties on the mechanics of forming processes, it is necessary to find the strain, strain rate, and stress distribution in the course of metal deformation. A number of approximate methods have been developed and used for analyzing metal forming problems. Most of these methods are either the outcome of a simplified analysis or offer only a partial solution to the problem. The Visioplasticity method developed by Thomsen et al.[1,2] is the only one

*Submitted to the AIME Symposium on the relation between theory and practice in metal forming. October 21-22, 1970.

that gives a realistic solution since the velocity field is obtained from a series of photographs of the instantaneous grid pattern during an actual forming process. The strain rate, the strain, and the stress fields can then be obtained from the consideration of equilibrium and plasticity equations. The method has had limited application to forming problems and in many cases has been used to examine metal flow rather than to obtain the complete solution, essentially because the calculations involved are too time consuming as a result of a long process of graphical differentiations and integrations. The extrusion problem, of which a major part of the process is considered to be in a steady-state condition, is the only one so far in which this technique has been applied to obtain the complete solution of stress and strain. Shabaik et al.[3,4] have developed a computer program to calculate the complete solution in axisymmetric and plane-strain extrusion from a single photograph of the steady-state flow lines using the concept of the flow function first introduced in the solution of metal forming problems by Shabaik[5]. In this paper, complete analysis of stress and strain in an axisymmetric extrusion of commercially pure aluminum and superplastic alloy of the eutectic lead-tin through conical die of extrusion ratio 4 and half-cone angle of 45° was obtained using the computer program developed in Reference[4]. Aluminum and superplastic alloy were chosen to examine the effects of a workhardening material and a non-workhardening one on the state of tension zone and boundary shear stress in extrusion.

Complete Analysis of Axisymmetric Extrusion

The complete solution of stress and strain in axisymmetric extrusion can be obtained using the method given in Reference[4]. From the experimentally determined steady-state flow-pattern, the radial and axial velocity components (u, v) can be calculated from the known values of the flow function ψ as follows:

$$u = \frac{1}{2\pi r}\frac{\partial \psi}{\partial z},$$

$$v = -\frac{1}{2\pi r}\frac{\partial \psi}{\partial r} \qquad (1)$$

It is readily seen, by direct substitution, that the velocity components given by Eq. (1) satisfy the continuity condition.

When the velocity components are known at all points in the deformation zone, the strain-rate components ($\dot{\varepsilon}_r$, $\dot{\varepsilon}_z$, $\dot{\varepsilon}_\theta$, $\dot{\gamma}_{rz}$) the total effective strain rate ($\dot{\bar{\varepsilon}}$) can be determined:

$$\dot{\varepsilon}_r = \frac{\partial u}{\partial r},$$

$$\dot{\varepsilon}_z = \frac{\partial v}{\partial z},$$

$$\dot{\varepsilon}_\theta = \frac{u}{r} \; ,$$

$$\dot{\gamma}_{rz} = \frac{\partial v}{\partial r} + \frac{\partial u}{\partial z} \; ,$$

$$\dot{\bar{\varepsilon}} = \sqrt{\frac{2}{3}} \left\{ \dot{\varepsilon}_r^{\,2} + \dot{\varepsilon}_\theta^{\,2} + \dot{\varepsilon}_z^{\,2} + \frac{1}{2} \dot{\gamma}_{rz}^{\,2} \right\}^{1/2}$$

The total effective strain ($\bar{\varepsilon}$) can be calculated from the integration of $\dot{\bar{\varepsilon}}$ w.r.t time along the flow lines.

The stress components at any point in the deformation zone can be evaluated by considering the equilibrium equations and the rules of plastic flow.

$$\sigma_z(r,z) = \int_z^a \left[\frac{\partial}{\partial r} \left(\frac{\dot{\gamma}_{rz}}{2\dot{\lambda}} \right) + \frac{\dot{\gamma}_{rz}}{2r\dot{\lambda}} \right] dz$$

$$+ \int_0^r \left[\frac{\partial}{\partial r} \left(\frac{\dot{\varepsilon}_z - \dot{\varepsilon}_r}{\dot{\lambda}} \right) - \frac{\partial}{\partial z} \left(\frac{\dot{\gamma}_{rz}}{2\dot{\lambda}} \right) - \frac{\dot{\varepsilon}_r - \dot{\varepsilon}_\theta}{r\dot{\lambda}} \right]_{z=a} dr + \sigma_z(o,a) \qquad (2)$$

$$\sigma_r(r,z) = \sigma_z(r,z) + \frac{\dot{\varepsilon}_r - \dot{\varepsilon}_z}{\dot{\lambda}} \qquad (3)$$

$$\sigma_\theta(r,z) = \sigma_z(r,z) + \frac{\dot{\varepsilon}_\theta - \dot{\varepsilon}_z}{\dot{\lambda}} \qquad (4)$$

$$\tau_{rz} = \frac{1}{2} \frac{\dot{\gamma}_{rz}}{\dot{\lambda}} \qquad (5)$$

where $\sigma_z(o,a)$ is the axial stress at a reference point (o,a), $\dot{\lambda} = \frac{3\dot{\bar{\varepsilon}}}{2\bar{\sigma}}$ and $\bar{\sigma}$ is the flow stress of the material.

The stress components can be calculated if the mechanical properties of the material are known. Therefore, information on the flow stress at different temperatures, strain rates, and strains is of considerable value for the analysis of the extrusion process.

From the known values of the stress components in the (r,z) coordinate system, the stresses along the coordinate system given by the tangent to the flow line and perpendicular to it at any point, can be determined using the stress transformation equations.

$$\sigma_n = \sigma_z \,\mathrm{Sin}^2\alpha + \sigma_r \,\mathrm{Cos}^2\alpha - 2\,\tau_{rz} \mathrm{Sin}\alpha \,\mathrm{Cos}\alpha \; ,$$

$$\sigma_s = \sigma_z \,\mathrm{Cos}^2\alpha + \sigma_r \,\mathrm{Sin}^2\alpha + 2\tau_{rz} \mathrm{Sin}\alpha \,\mathrm{Cos}\alpha \; ,$$

$$\tau_{ns} = \tau_{rz} \,(\mathrm{Cos}^2\alpha - \mathrm{Sin}^2\alpha) + (\sigma_r - \sigma_z) \,\mathrm{Sin}\alpha \,\mathrm{Cos}\alpha \qquad (6)$$

where: n is the normal to the flow line,
s is the tangent to the flow line,

α is the angle between the tangent to the flow line and z axis.

Results and Discussion

Complete Analysis. The steady state flow lines in the axi-symmetric extrusion process of commercially pure aluminum and super-plastic alloy of the eutectic of lead and tin through conical die of a half-cone angle of 45° and an extrusion ratio of 4 were ob-tained using the visioplasticity technique. Fig. (1) shows a photo-graph of the grid pattern of aluminum after extrusion at a ram speed of 0.1"/min with Fluorocarbon as a lubricant and a photograph of the grid pattern of the superplastic lead and tin extruded at a ram speed of 0.003"/min with Molykote as a lubricant. The super-plastic Pb-Sn was prepared according to the procedures outlined in Ref.[6]. Split specimens of 1" dia and 3" long were used; and grid lines of .002" thick and 0.020" spacing were printed using Kodak photo resist method. Fig. (2) shows the flow lines at equally spaced sections (k) in the axial direction as plotted by a computer. The complete analysis of stress and strain was obtained using the computer program[4] following the method previously outlined.

1) Velocity Field. The velocity components (u,v) were calcu-lated using Eq. (1), and the velocity magnitude along the flow lines was evaluated and plotted as shown in Fig. (3). From this figure, it can be noted that for both aluminum and superplastic Pb-Sn the velocity along the boundary goes to a minimum at the die corner. Furthermore, for the superplastic Pb-Sn, the velocity gradient is somewhat less than that of aluminum and the results of the super-plastic material are closer to uniform deformation than those of aluminum.

From the known velocity components, the strain rates, total effective strain rate $\dot{\bar{\varepsilon}}$ and total effective strain $\bar{\varepsilon}$ were evaluated. Figs. (4) and (5) show the values of $\dot{\bar{\varepsilon}}$ and $\bar{\varepsilon}$ as a function of r for constant k, respectively. In Fig. (5) it can be noted that the trend of $\bar{\varepsilon}$ is similar for both aluminum and superplastic Pb-Sn and that the two results differ only slightly in magnitude. The total effective strain at the center point of the exit section approaches a value equal to that calculated from uniform deformation.

It should be noted that the values of the velocity magnitude and the effective strain rate must be multiplied by the ram velo-city, which is 0.1 in/min. in the case of aluminum and 0.003 in/min. for the case of superplastic Pb-Sn.

2) Stress Field. For the stress calculation, the flow stress of the material must be known in terms of the strain rate, strain and temperature. Quantitative data of this kind can be obtained from simple tests. Fig. (6) shows the stress-strain curves of commercially pure aluminum obtained at constant values of strain rates and for different temperatures in a plane strain compression test by Bailey and Singer[7]. Attempts were then made to fit the empirical formula $\sigma = c\,\dot{\bar{\varepsilon}}^m$ to the experimental results. The

*Figures and Tables for this paper may be found on page 70ff.

constant c the strain rate exponent m were evaluated at different temperatures and the values are given in Table (1). From Table (1), it may be noted that the strain rate exponent of aluminum at room temperature is approximately zero.

The flat curves of stress for strain values >0.8 and at higher temperatures indicate that aluminum changes from a workhardening material at room temperature to a non-workhardening material at higher temperatures.

It should be noted that all stress and strain values given in Fig. (6) are those applying to plane strain compression. Thus, effective stress, strain and strain rate can be obtained by multiplying by $\sqrt{3}/2$, $2/\sqrt{3}$ and $2/\sqrt{3}$ respectively.

From the above results the stress-strain relationship of aluminum can be written as follows:

a) $\bar{\sigma} = c\,f(\bar{\varepsilon})$ at room temperature,

b) $\bar{\sigma} = c_1\,\dot{\bar{\varepsilon}}^{\,m}$ at higher temperature.

$f(\bar{\varepsilon})$ is commonly expressed as a power function of the strain and accordingly Eq. a)becomes $\bar{\sigma} = c\,\bar{\varepsilon}^{\,n}$. Eqs. a) and b) can then be combined in a general form as follows:

$$\bar{\sigma} = c\,\bar{\varepsilon}^{\,n}\,\dot{\bar{\varepsilon}}^{\,m} \tag{7}$$

where n is the workhardening exponent and m is the strain rate exponent. Therefore, for cold working m = o; and, for hot working n = o and $m \approx 0.25$ for the range of temperatures commonly used in hot forming. For warm working, both m and n are not zero.

For the superplastic Pb-Sn, the flow stress is independent of strain,and the stress-strain rate results obtained from tension and compression tests[6] are shown in Fig. (7). Fig. (8) shows the strain rate exponent m as a function of the strain rate. The strain rate exponent m was evaluated in this case from the results between two consecutive points on the stress-strain rate curve. The exponent m serves as a measure of the superplasticity of the material, with values of m = 0.4 and higher indicating that the material has a relatively high stretching characteristic. The following values of m and n were used in Eq. (7) in order to cover a wide range of materials and material properties.

a) m = 0.45, n = o (superplastic)
b) m = 0.25, n = o (hot working)
c) m = o, n = o (ideal plastic mat.)
d) m = 0.15, n = 0.15 (warm working)
e) m = o, n = 0.25 (cold working)
f) m = o, n = 0.45 (highly workhardening material)

For the stress calculations, values of $\dot{\lambda}$ were obtained from

$$\dot{\lambda} = \frac{3\dot{\bar{\varepsilon}}}{2\bar{\sigma}} = \frac{3}{2c\bar{\varepsilon}^{\,n}\dot{\bar{\varepsilon}}^{\,(m-1)}}$$

where the values of $\dot{\bar{\varepsilon}}$ and $\bar{\varepsilon}$ are given in Figs. (4) and (5).

The stress components ($\sigma_r, \sigma_z, \sigma_\theta, \tau_{rz}$) can be calculated from Eqs. (2), (3), (4) and (5). For the stress components along and perpendicular to the flow lines, the transformation Eq. (6) can be used. Fig. (9) shows the values of $\frac{\sigma_n}{c}$, $\frac{\sigma_s}{c}$ and $\frac{\tau_{sn}}{c}$ as plotted by a computer.

Effect of Material Property

a) On Shear Stress Along the Boundary. The shear stress along the boundary for the above values of m and n for both aluminum and superplastic lead-tin are shown in Fig. (10). The trend is the same for both materials and the shear stress reaches a peak value at the die corner. It can be noted from both Figs. 10a and 10b that the shear stress increases in magnitude as the material properties change from those of superplastic to those of ideal plastic material. The values of the shear stress then decrease as the material properties change from those of ideal plastic to workhardening.

The values of the shear stress corresponding to values of m in the superplastic range, case (a), are lower than the values of shear stress corresponding to other cases of m and n. This means that to obtain the same flow-characteristic of a workhardening material under certain friction condition when it is in the superplastic state, a much better lubricant must be used. In other words, this means that the requirements on lubricants in hot extrusion are much more than in the cold extrusion.

a) On Tension Zone. From the known values of the stress components, the line of zero axial stress can be determined. Fig. (11) shows the boundary of the tension zone for all cases of m and n considered and for both materials. It can be noted from this figure that the tension zone is almost the same for all cases of m and n except for m = o, n = o. It must be noted that values of axial stress in the tension zone are different for different m and n values since the values of c, in Eq. (7), are different.

CONCLUSION

The visioplasticity method was used to obtain the steady state flow lines in an axisymmetric forward-extrusion process of commercially pure aluminum and an eutectic alloy of lead-tin in the superplastic condition. An extrusion ratio of 4 and a half-cone angle of 45° were used to determine the flow pattern at a ram speed of 0.1 in/min (for aluminum) and 0.003 in/min (for Pb-Sn eutectic). The coordinates of 20-25 points on each flow line were used as input data to the computer program, and the complete analysis of velocity, strain rates, strain and stresses was obtained. It was found that, for both workhardening and non-workhardening materials, the velocity along the tool-billet interface goes through a minimum in the vicinity of the intersections of die and cylinder, and the shear stress changes from a positive shear along the entry portion of the die to a negative shear along the exit portion. The normal pressure reaches a peak value in the vicinity of the point where shear reversal takes place.

The shear stress along the die corner decreases as the material property changes from that of a workhardening to a strain rate dependent material, assuming the flow lines remain the same. The values of the shear stress along the boundary corresponding to values of m in the superplastic range are lower than values of shear stress corresponding to other cases of m and n.

The size of the tension zone was found to vary slightly with varying values of m and n, except for m = o, n = o for both aluminum and superplastic Pb-Sn.

REFERENCES

1. Thomsen, E. G., Yang, C. T. and Bierbower, J. B., "An Experimental Investigation of the Mechanics of Plastic Deformation of Metals," pp. 89-144, University of California Press (1954).

2. Thomsen, E. G., "Visioplasticity," CIRP Annalen, Bd. XII, Heft 3, p. 127.

3. Shabaik, A. H., Lee, C. H., Kobayashi, S., "Application of the Visioplasticity Method to Extrusion Through a Conical Die," Proceedings of the 7th MTDR Conf., Birmingham, England, September 1966.

4. Shabaik, A. H., Kobayashi, S., "Computer Application to the Visioplasticity Method," ASME J. Eng. for Ind., May 1967, pp. 339-346.

5. Shabaik, A. H., "Theoretical Methods for Analysis of Metal Deformation Problems," Ph.D. Dissertation, University of California, 1966.

6. Jovane, F., Shabaik, A. H., Thomsen, E. G., "Some Extrusion Studies of the Eutectic Alloy of Pb and Sn," ASME J. Eng. for Ind., Aug 1969, pp. 680-686.

7. Bailey, J. A., Singer, A.R.E., "Effect of Strain Rate and Temperature on the Resistance to Deformation of Aluminum, Two Aluminum Alloy, and Lead," J. Institute of Metals, 92, 1963-64, 404-408.

Temp., °C	m				σ_0, lb-force/in^2	
	$\varepsilon = 0{\cdot}1$	$\varepsilon = 0{\cdot}25$	$\varepsilon = 0{\cdot}5$	$\varepsilon = 2{\cdot}5$	$\varepsilon = 0{\cdot}25$	$\varepsilon = 2{\cdot}5$
22	0·055	0·06	0·035	0·025	12,300	22,400
100	0·070	0·075	0·040	0·035	8,100	15,800
200	0·100	0·110	0·045	0·050	6,600	10,000
300	0·120	0·120	0·080	0·095	5,000	5,700
400	0·140	0·140	0·120	0·125	3,200	3,200
500	0·150	0·155	0·170	0·175	1,800	1,800
550	0·160	0·165	0·210	0·200	990	890
600	0·205	0·230	0·220	0·215	720	700

TABLE I Ref. (7)

Variation of n and σ_0 with Temperature and Strain (ε) for Aluminium

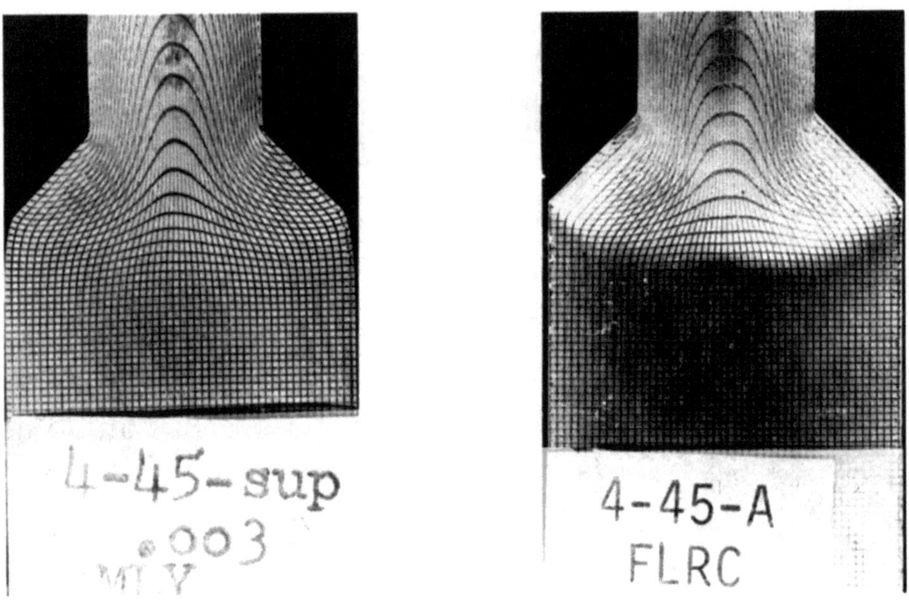

A B

Fig. 1. Distorted Grid Pattern in Axisymmetric Extrusion. Ext. Ratio = 4.0, Half Cone Angle = 45°. (A) Superplastic Lead-Tin, Speed = 0.003"/min, Molykote as Lubricant. (B) Aluminum, Speed = 0.1"/min, Fluorocarbon as Lubricant.

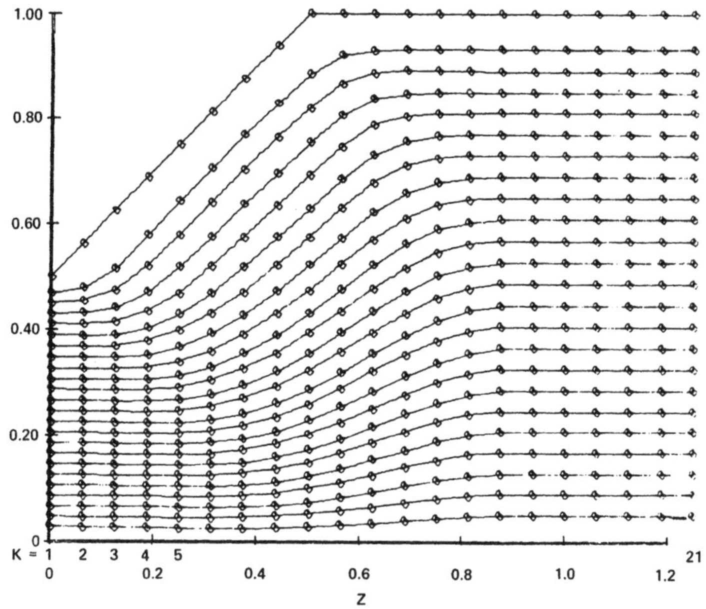

Fig. 2a. Steady State Flow Lines of Aluminum Obtained from Fig. 1b.

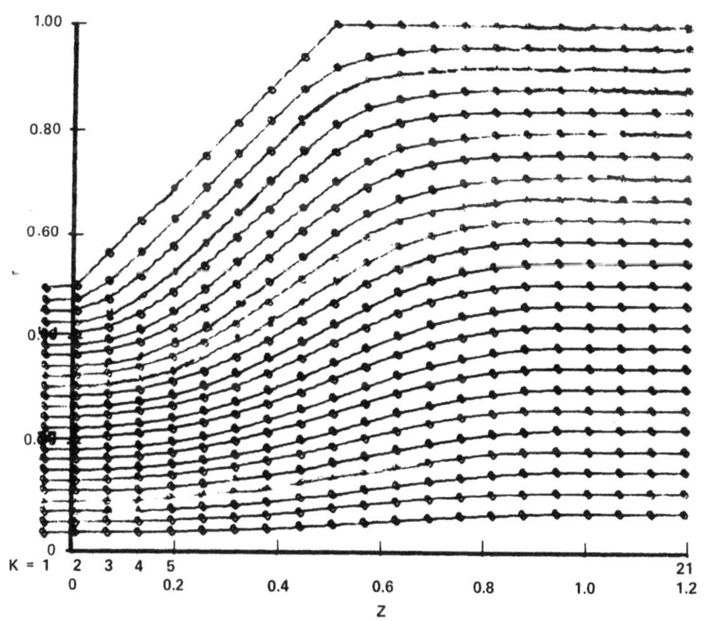

Fig. 2b. Steady State Flow Lines of Superplastic Pb-Sn Obtained from Fig. 1a.

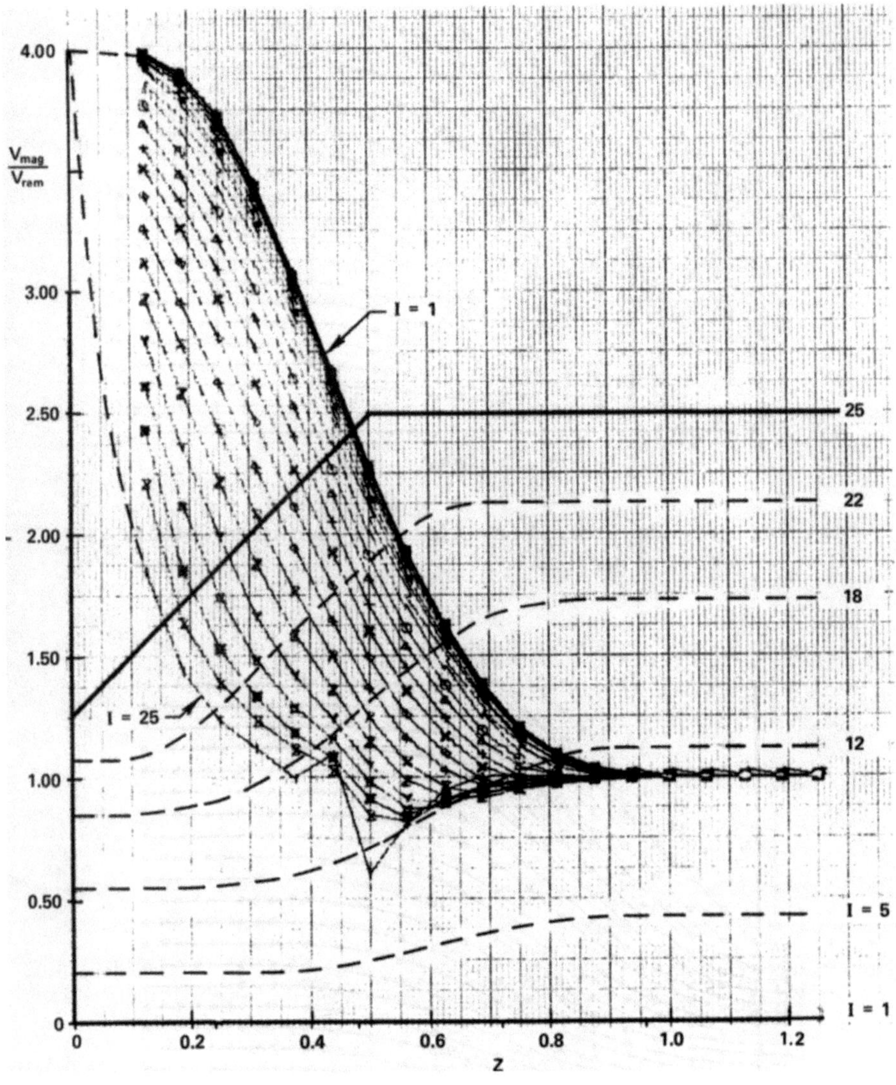

Fig. 3a Velocity Magnitude Along the Flow Lines in Axisymmetric Extrusion of Aluminum Shown in Fig. 1b

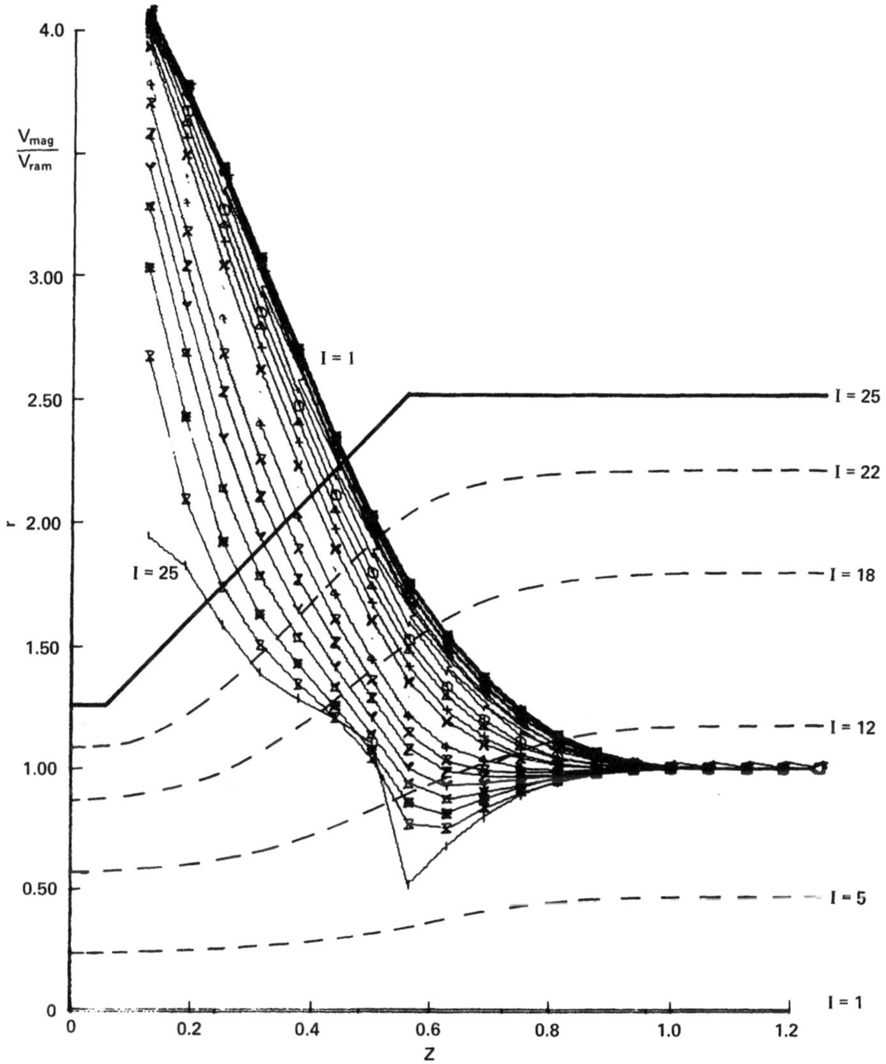

Fig. 3b Velocity Magnitude Along the Flow Lines in Axisymmetric Extrusion of Superplastic Pb-Sn Shown in Fig. 1a

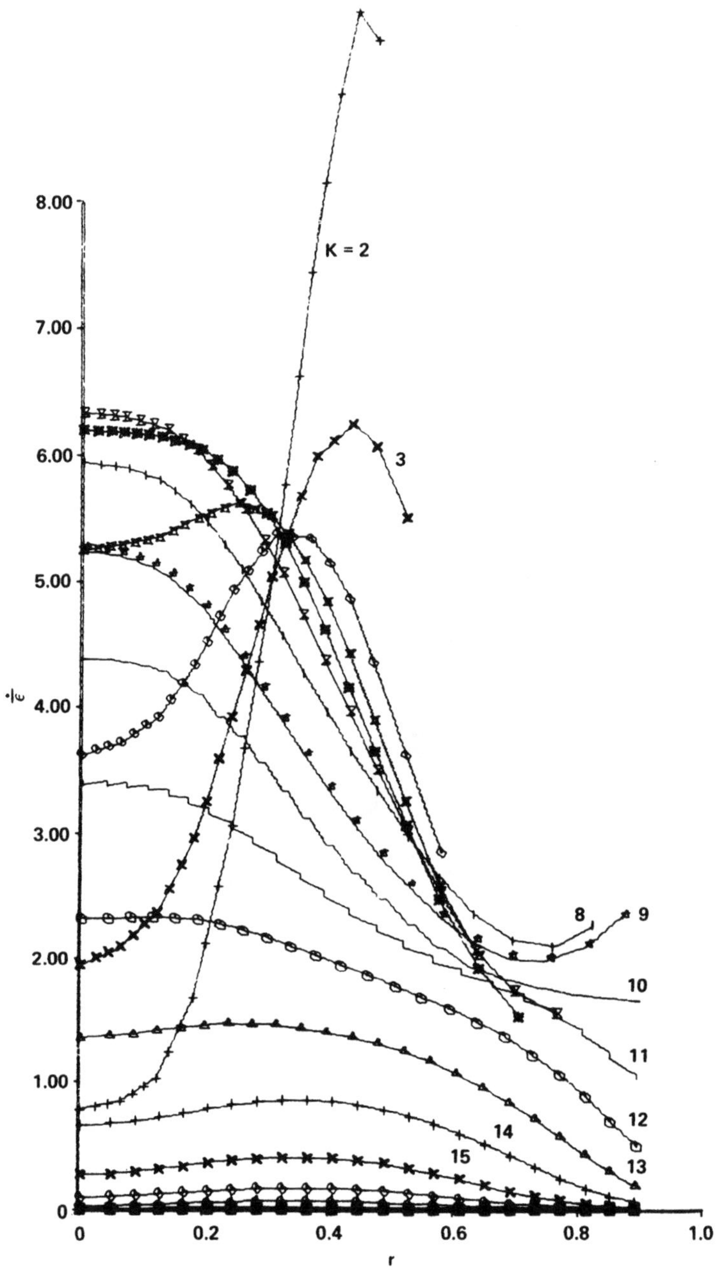

Fig. 4 Total Effective Strain Rate Distribution in Axisymmetric Extrusion of Aluminum Shown in Fig. 1b

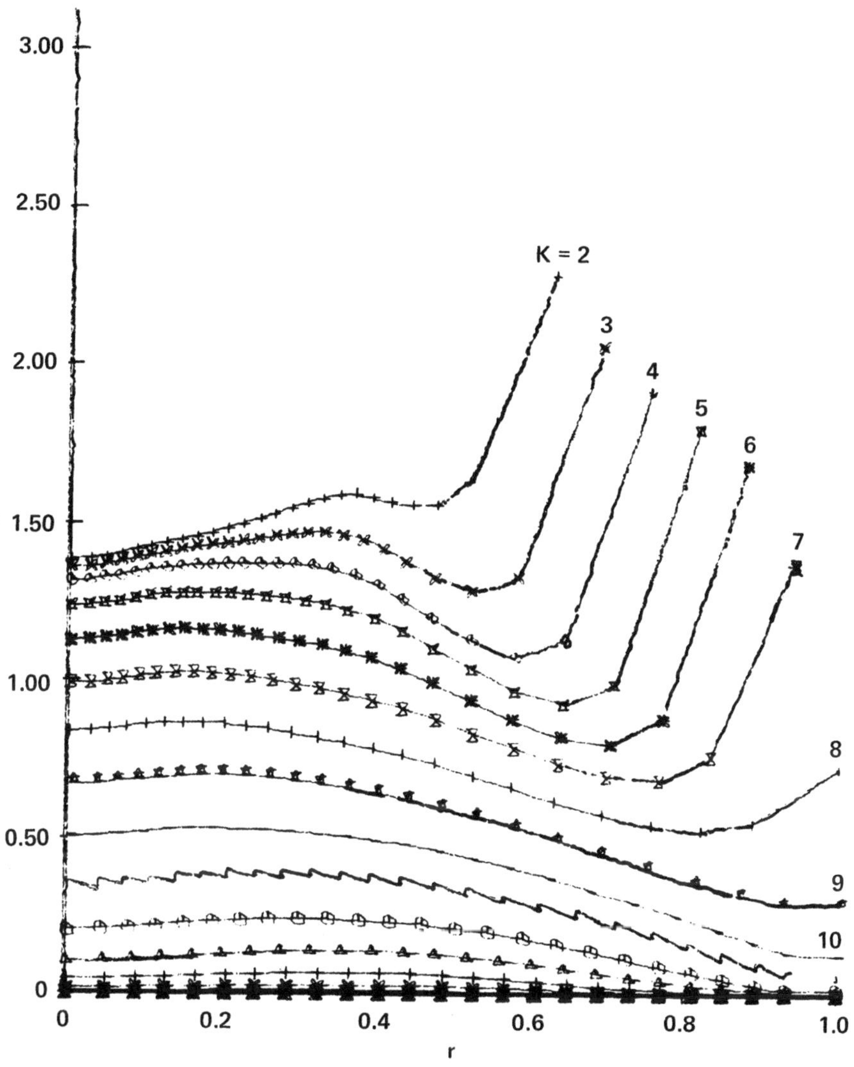

Fig. 5a Total Effective Strain Distribution in Axisymmetric Extrusion of Aluminum Shown in Fig. 1b

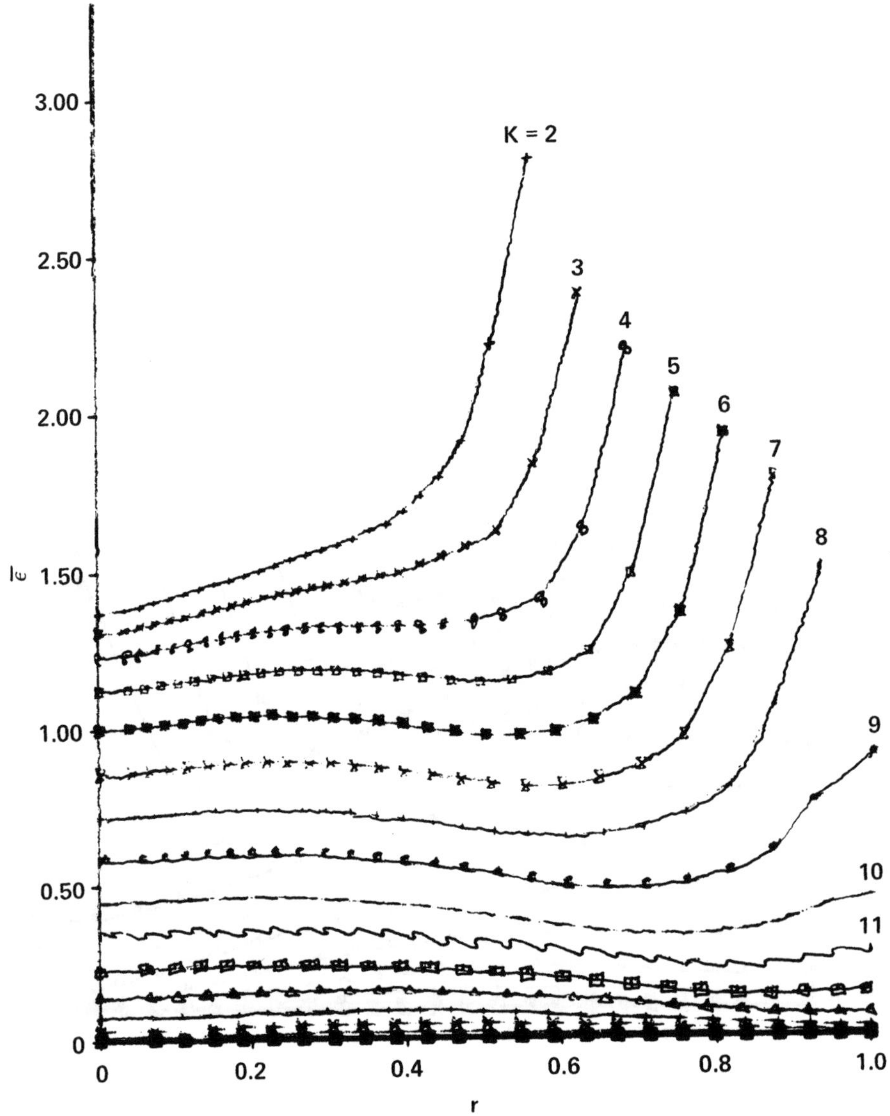

Fig. 5b Total Effective Strain Distribution in Axisymmetric Extrusion of Superplastic Pb-Sn Shown in Fig. 1a

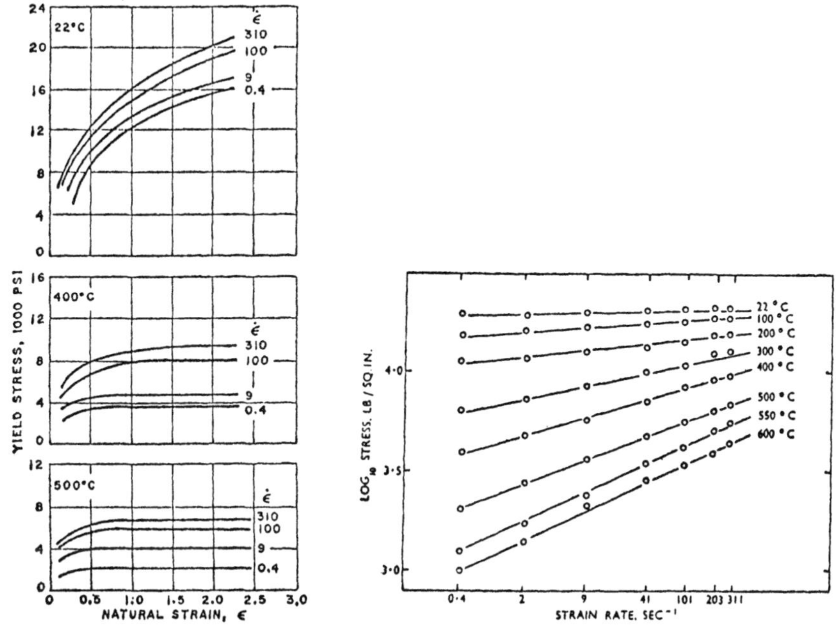

Fig. 6 Stress-Strain Results of Aluminum at Different Temperature and Strain Rate Reference [7]

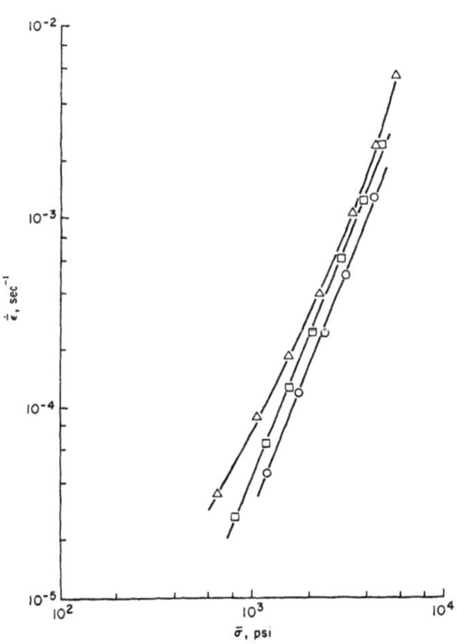

Figure 7. Test Results of Compression, Reference 6 .

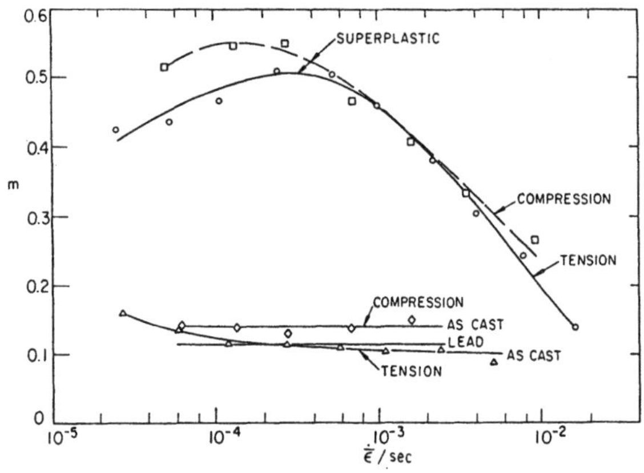

FIG. 8 EXPONENT m AS FUNCTION OF STRAIN RATES FOR EUTECTIC OF LEAD AND Sn

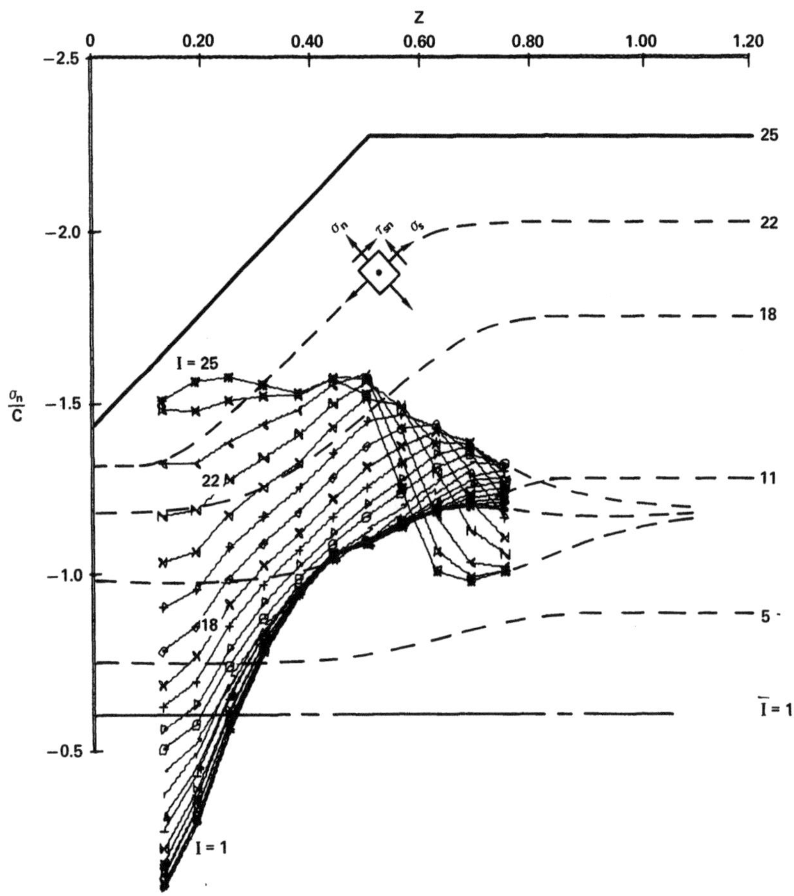

Fig. 9a. Normal Stress Distribution Along the Flow Lines in Axi-symmetric Extrusion of Aluminum Shown in Fig. 1b.

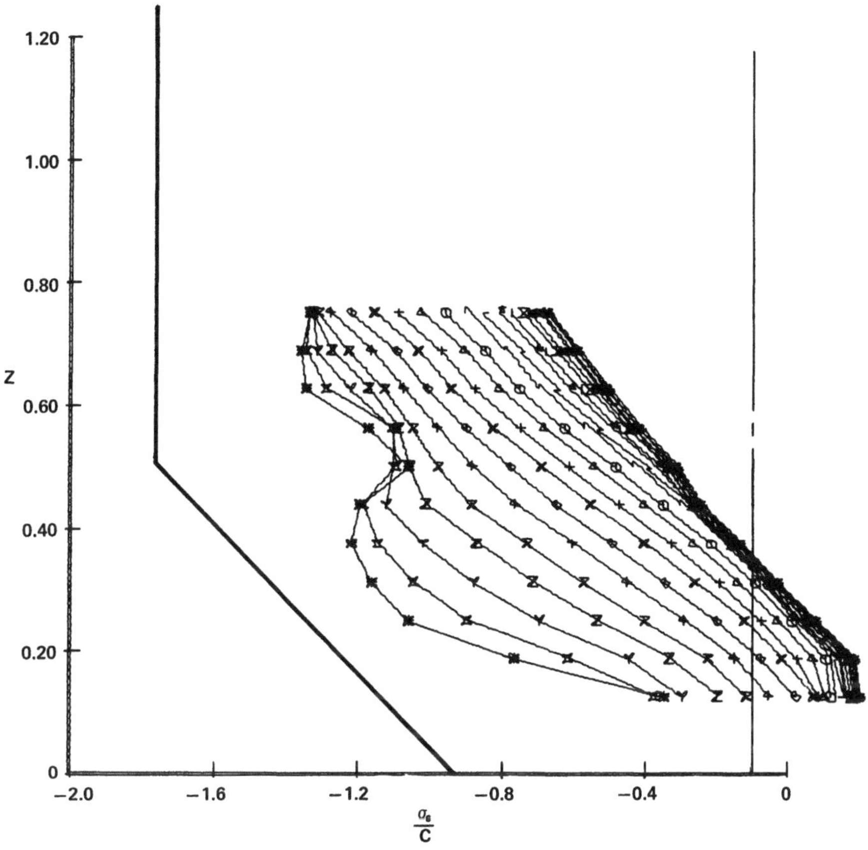

Fig. 9b Tangential Stress Distribution Along the Flow Lines in Axisymmetric Extrusion of Aluminum Shown in Fig. 1b

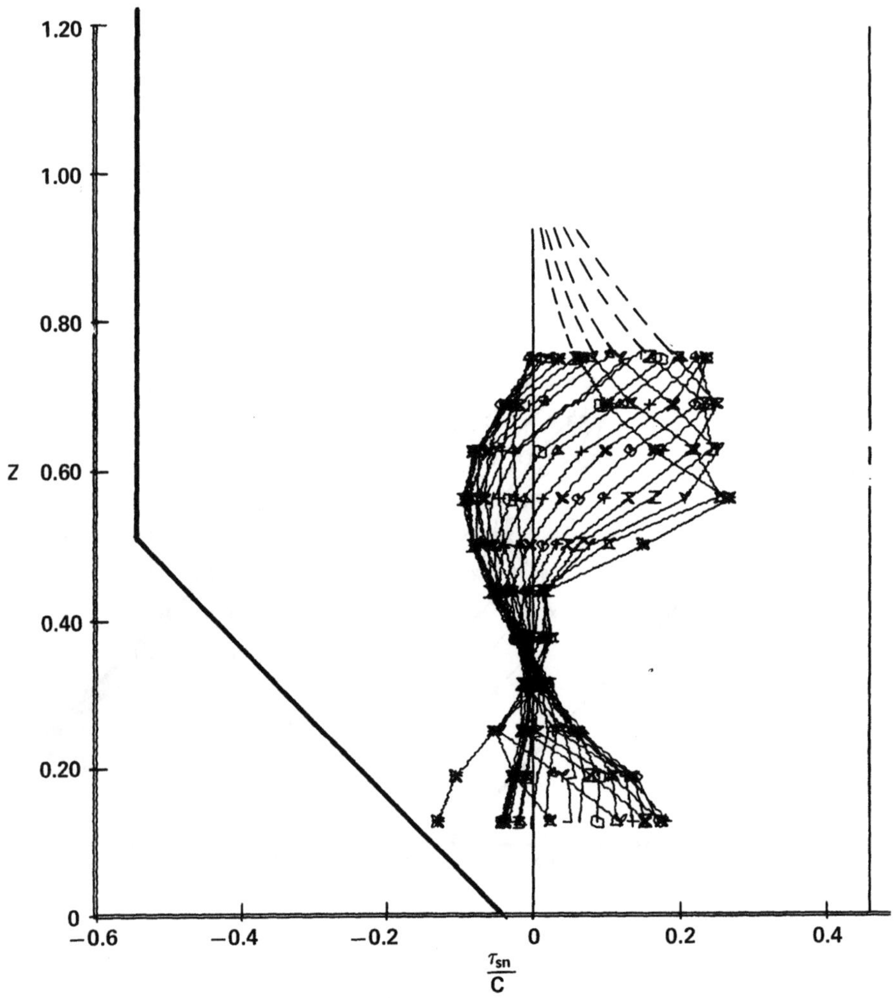

Fig. 9c Shear Stress Distribution Along the Flow Lines in Axisymmetric Extrusion of Aluminum Shown in Fig. 1b

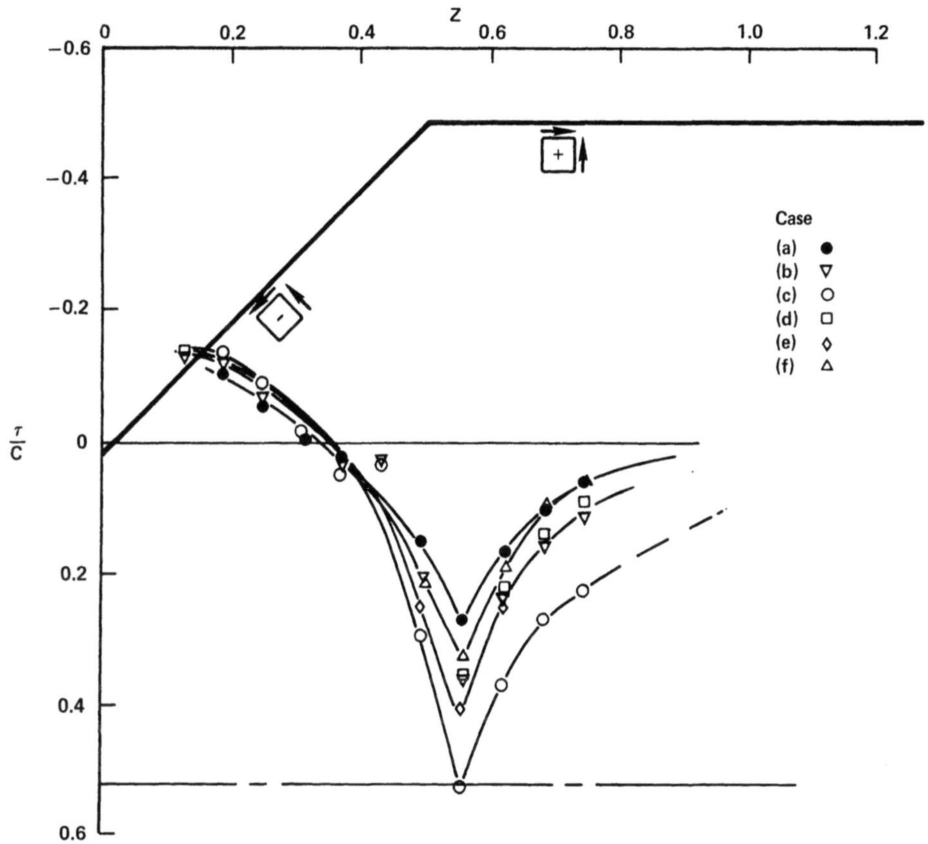

Fig. 10a Shear Stress Distribution Along the Boundary in Axisymmetric Extrusion of Aluminum

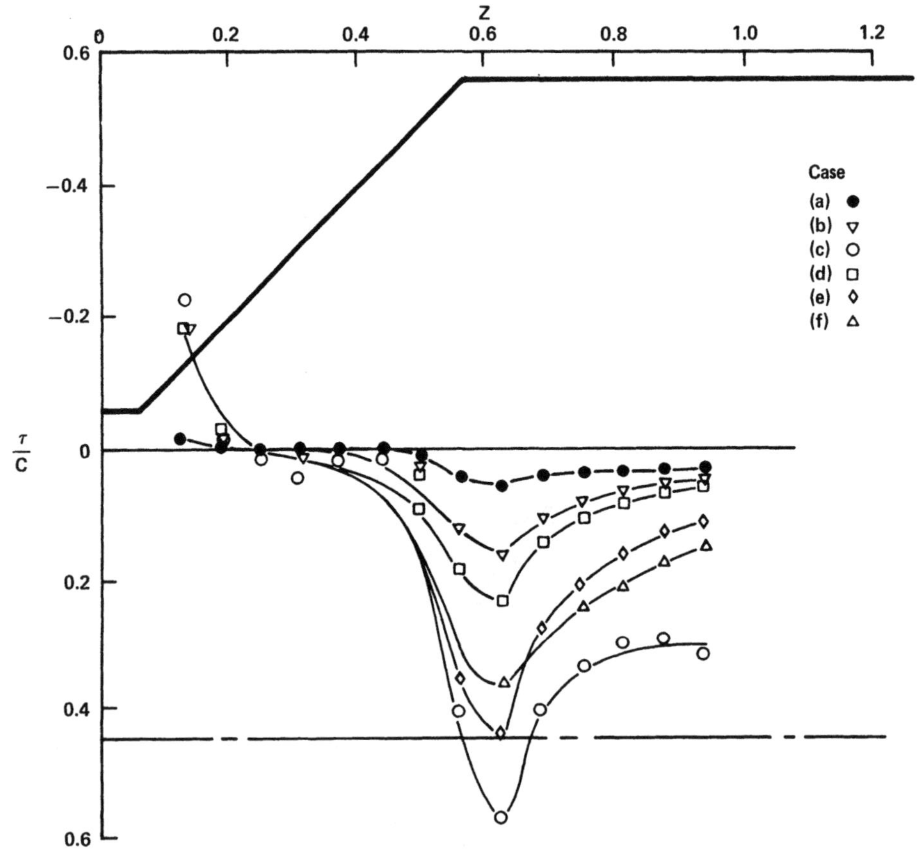

Fig. 10b Shear Stress Distribution Along the Boundary in Axisymmetric Extrusion of Superplastic Pb-Sn

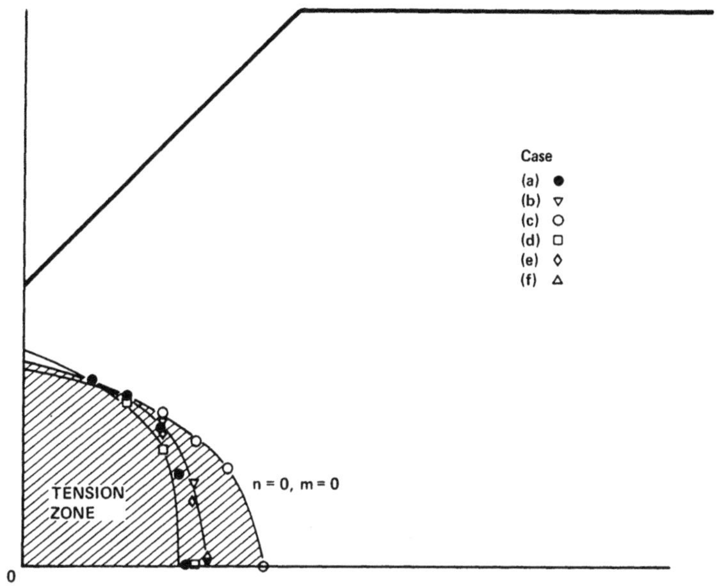

Fig. 11a. Tension Zone in Axisymmetric Extrusion of Aluminum.

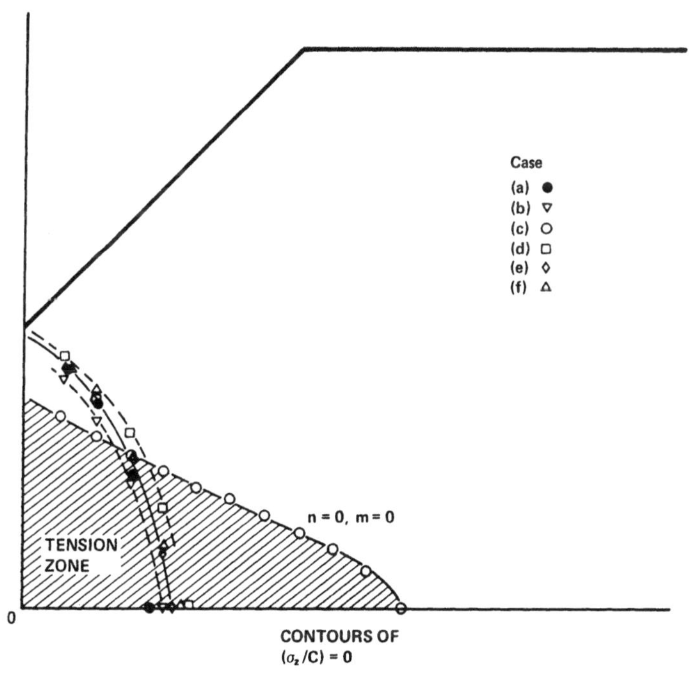

Fig. 11b. Tension Zone in Axisymmetric Extrusion of Superplastic Pb-Sn.

APPLICATION OF VISIOPLASTICITY TECHNIQUES TO AXISYMMETRIC EXTRUSIONS

R. Medrano, P. Gillis, C. Hinesley and H. Conrad

Department of Metallurgical Engineering and Materials Science, University of Kentucky, Lexington, Kentucky

ABSTRACT

The visioplasticity techniques employed to evaluate the strain and strain rate fields throughout the deforming portion of an axisymmetric extrusion are reviewed. Examples are given of the application of these techniques to extrusions exhibiting the usual single maximum flow pattern and to others exhibiting various degrees of an uncommon wavy double maxima pattern. It is shown that there is good agreement for all flow patterns between the positions of the transverse grid lines, final normal strains and final angle of intersection of the horizontal and transverse grid lines calculated from the flow function analysis and those measured directly from the deformed grids. Of significance is that the geometry of the deformation zone varies with the type and degree of flow pattern.

INTRODUCTION

To compare theories of extrusion with experiments usually only a few measureable parameters are available, for example, the die and ram forces[1] during extrusion. Additional experimental data can be obtained by visioplasticity[2-4]. In this technique the cylindrical billet is split longitudinally in half prior to extrusion and a grid is applied to one of the split faces. This grid comprises of a set of lines parallel to the billet axis, called flow lines, and a set initially perpendicular to the first, called transverse lines. The two halves are put back together, the billet is partially extruded, removed from the die, separated along the same plane and the grid line deformations observed.

The present paper deals with the extraction of kinematical information from visioplasticity data taken from an axisymmetric extrusion. Of particular interest is the determination of the strain and strain rate fields throughout the deforming portion of the extrusion. While this may seem to be a relatively trivial problem in comparison to the construction of a complete theory of extrusion, it is one which must be done carefully and correctly if any complete theory is ever to have many points of comparison with reality. Furthermore, the kinematical analysis alone generates much information that is directly useful in both the development of theories and the design of dies.

A simple method for obtaining strains is to measure directly the deformed grids and to compare each measurement with the original grid dimensions. Although sophisticated techniques are now being developed for applying this method[5], we have used instead the flow function method of analysis[2,3]. Flow function theory and the associated numerical techniques of grid line analysis have been discussed in detail in two earlier papers[6,7], but in the spirit of the present symposium the theory is reviewed here and the exact numerical procedure is specified by means of the computer program included in Appendix A.

THEORY

The flow function Φ is defined with reference to Fig. 1 as

$$\Phi = 2\pi \int_0^r v\, r\, dr \tag{1}$$

Here v is the axial component of particle velocity and r is the radial coordinate in ordinary cylindrical coordinates r, θ, z. According to this definition Φ is the axial volume-flow rate. That is, Φ measures the volume rate of flow through a circle of radius r centered on the axis of symmetry. Obviously, Φ will depend on both r and z.

For isochoric flow Φ is proportional to the mass rate of flow, or the rate of material flow through the circle of radius r. The material flow can then be described in terms of the size of the circle a given amount of material passes through at various sta-

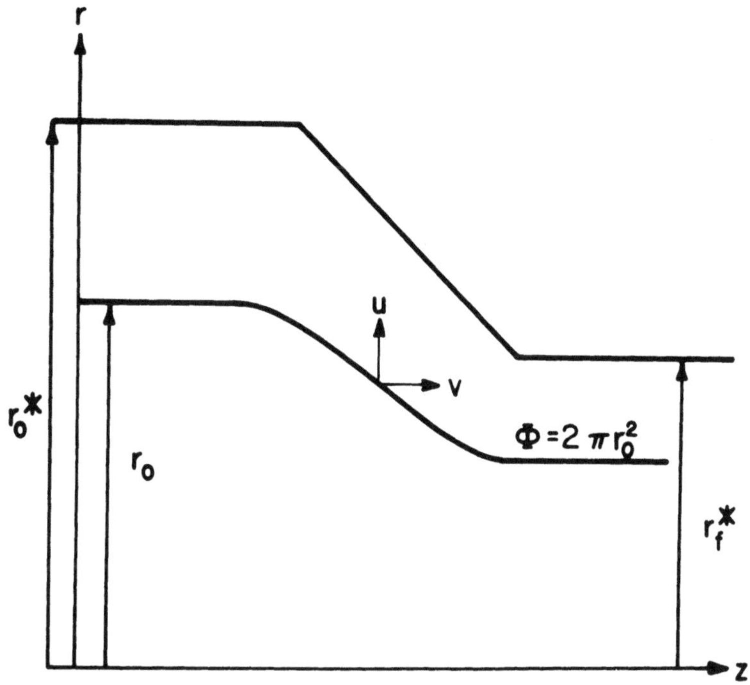

Fig. 1 Spatial coordinate system. The radial and axial coordinates are r and z. The corresponding components of particle velocity are u and v for a material point moving along the flow line. The circumferential coordinate θ is not shown.

tions on its way downstream. Figure 2 indicates schematically the envelope of all circles passing material at one particular mass flow rate; this envelope is a surface of constant Φ. The intersection of a plane containing the axis of symmetry with this envelope is a flow line.

Thus, one important assumption in flow function analysis is that the material is incompressible. The other major assumption is that the flow is time invariant. This implies that the shape of the surface in Fig. 2 does not change with time. Along any flow line the flow function is constant, i.e., $d\Phi = 0$; and if the flow is steady state

$$d\Phi = (\partial\Phi/\partial z)\,dz + (\partial\Phi/\partial r)\,dr = 0 \tag{2}$$

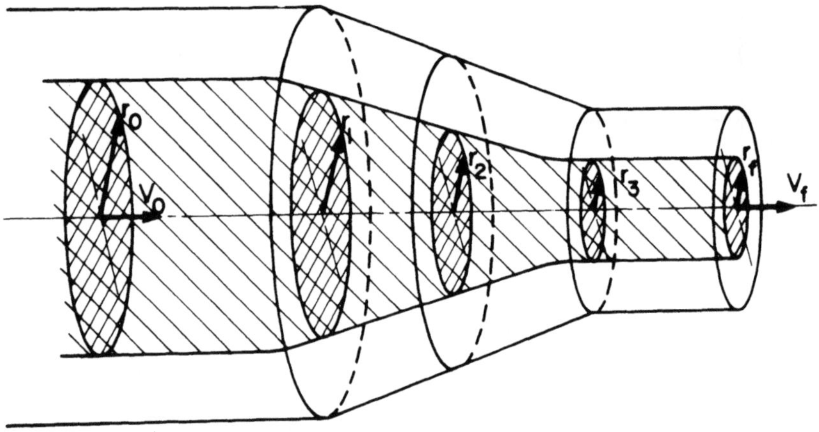

Fig. 2 Diagramatic representation of the flow function. The shaded tube is a surface of constant Φ. The mass rate of flow is constant past any transverse cross section of this surface and, thus, it can be thought of as an imaginary pipe of variable diameter.

Also, along any flow line the ratio of particle velocity components in the radial and axial directions, u, v respectively in Fig. 1, is the slope of the flow line in the r, z plane

$$dr/dz = u/v \tag{3}$$

Combining Eqs. 2 and 3 gives:

$$(\partial\Phi/\partial z)\, v + (\partial\Phi/\partial r)\, u = 0 \tag{4}$$

To appreciate the usefulness of flow function analysis three other equations are required. By differentiating Eq. 1 with respect to r and rearranging we obtain

$$v = (1/2\pi r)\,(\partial\Phi/\partial r) \tag{5}$$

Combining Eqs. 4 and 5 gives

$$u = -(1/2\pi r)\,(\partial\Phi/\partial z) \tag{6}$$

With reference to Fig. 1, assuming that the upstream portion of the billet moves without deformation at the ram speed V_o, the absolute value of the flow function can be obtained for any flow line. Denote by r_o the upstream radius of a flow line; for constant v, Eq. 1 can be integrated to yield

$$\Phi = \pi r_o^2 V_o \tag{7}$$

Equation 7 gives the flow function value along each flow line and thus the function can be easily evaluated throughout the deformation zone by measuring the local positions of flow lines. Eqs. 5 and 6 allow the material particle velocity to be computed from the spatial derivatives of the flow function.

All other kinematical quantities are determined directly from the particle velocity components through the following definitions. For the spatial coordinates r, θ, z the strain rates are

$$\dot{\epsilon}_r = \partial u/\partial r \tag{8}$$

$$\dot{\epsilon}_\theta = u/r \tag{9}$$

$$\dot{\epsilon}_z = \partial v/\partial z \tag{10}$$

$$\dot{\gamma}_{rz} = \partial u/\partial z + \partial v/\partial r \tag{11}$$

Throughout the paper a superposed dot denotes derivative with respect to time. For axially symmetric deformation the remaining two shear strain rates are identically zero.

As a particle moves through the deformation zone along a flow line it rotates. Thus, we introduce orthogonal coordinates a, b and c assumed to be rigidly attached to the material particles. The coordinate axis a is taken to be aligned with an axial grid line (flow line) upstream from the die and to remain tangent to the flow line through the deformation zone. The coordinate axis c is taken to be aligned circumferentially upstream from the die and assumed to remain coincident with the θ - direction because of the axial symmetry. Thus, the rotation of the a, b, c coordinate system, referred to as the material coordinates, is wholly in the r, z plane and can be described by a single parameter. As shown in Fig. 3 this parameter can be taken as the angle between the z and a axes, which we denote by α. The same angle describes the misalignment between the r and b axes.

Applying ordinary transformation theory[8] to Eqs. 8-11 gives the following useful relations

$$\dot{\epsilon}_a = \dot{\epsilon}_z \cos^2\alpha + \dot{\epsilon}_r \sin^2\alpha - \dot{\gamma}_{rz} \sin\alpha \cos\alpha \tag{12}$$

$$\dot{\epsilon}_b = \dot{\epsilon}_z \sin^2\alpha + \dot{\epsilon}_r \cos^2\alpha + \dot{\gamma}_{rz} \sin\alpha \cos\alpha . \tag{13}$$

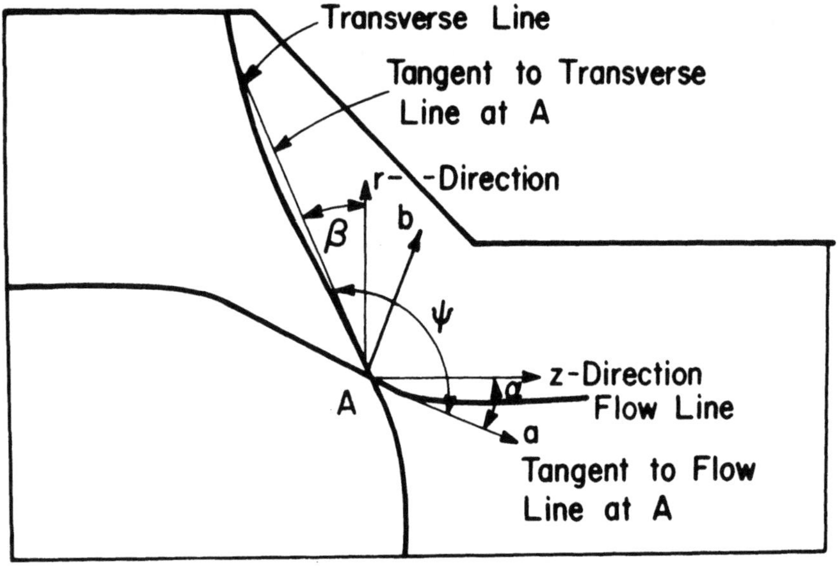

Fig. 3 Material coordinate system. The longitudinal a-axis is tangent to the flow line at the point A under consideration; the b-axis is normal to a and in the r, z plane; not shown is the c-axis which is circumferential. The angle Ψ is between the flow line and the transverse line intersecting at the point. The angle α is the rotation of the a, b axes relative to the z, r axes.

Here $\dot{\epsilon}_a$ can be thought of as the rate of stretching of a flow line and $\dot{\epsilon}_b$ as the rate at which adjacent lines are separating. To determine strains, Eqs. 12 and 13 must be integrated with respect to time, a variable that has not yet been explicitly introduced into the analysis. Indeed it need not be introduced in view of the meaning of particle velocity along a given flow line: v = dz/dt. Thus, the integration with respect to time is replaced by a spatial integration along a flow line

$$\epsilon_a = \int_{z_o}^{z} (\dot{\epsilon}_a/v)\, dz \qquad (14)$$

$$\epsilon_b = \int_{z_o}^{z} (\dot{\epsilon}_b/v)\, dz \qquad (15)$$

Here z_o is some reference station upstream from the deformation zone at which the strain is taken to be zero. Equations 14 and 15 provide the primary deformation measures. They are true finite strains in the direction of the flow line under consideration and in the direction normal to the flow line respectively. In general the b-direction will not coincide with either the radial direction or the direction of transverse grid lines.

Because the transverse lines do not remain perpendicular to the flow lines during deformation, one further parameter of direct interest is the rate of change in the angle of intersection Ψ shown in Fig. 3. As demonstrated previously[7]

$$\dot{\Psi} = -(\dot{\epsilon}_z - \dot{\epsilon}_r)\,[2 \sin\alpha \cos\alpha \sin^2\Psi + \sin\Psi \cos\Psi (\cos^2\alpha - \sin^2\alpha)] + \dot{\gamma}_{rz}\,(2 \sin\alpha \cos\alpha \sin\Psi \cos\Psi - \sin^2\Psi\,(\cos^2\alpha - \sin^2\alpha)] \tag{16}$$

Again, the local values for Ψ along a flow line are found by integration

$$\Psi = \tfrac{1}{2}\pi + \int_{z_o}^{z} (\dot{\Psi}/v)\, dz \tag{17}$$

Here $\frac{1}{2}\pi$ denotes the initial value for Ψ upstream from the deformation zone.

The most important observation to be made is that within the context of flow function analysis the final normal strain components are fixed by the extrusion ratio $R = (r_o^*/r_f^*)^2$ where r_o^* and r_f^* are the radii of billet and extrusion respectively; see Fig. 1. In the regions far from the die we assume that no deformation occurs. Hence, the material undergoes only rigid body motion in these regions, which is characterized upstream by $u = 0$ and $v = V_o$, the ram velocity, and downstream by $u = 0$ and $v = V_f$, the final extruded product velocity.

For constant v, Eq. 1 can be integrated for any flow line and gives

$$\Phi = \pi v r^2 \tag{18}$$

Equating the two values of Φ for the upstream and downstream regions shows that for each surface of constant Φ

$$(r_f/r_o)^2 = V_o/V_f \tag{19}$$

Since Eq. 19 holds for every such surface, it holds for the outer surface of the extrusion. Thus, the final velocity is obtained from Eq. 19 as

$$V_f = V_o R \tag{20}$$

Equation 20 enables 19 to be rewritten as

$$r_f = r_o R^{-\frac{1}{2}} \tag{21}$$

From Eq. 21 and the previous assumption of constant volume deformation, the strain distribution in the extruded product is nearly completely specified.

Consider a tubular element upstream from the die of internal radius r_o, thickness dr_o and length dz_o. According to Eq. 21 its final internal radius and thickness will be $r_o R^{-\frac{1}{2}}$ and $dr_o R^{-\frac{1}{2}}$ respectively. Hence, the final circumferential and radial strains are fixed by R only. Taking true strains as most useful in the present analysis we have for the final circumferential strain

$$\epsilon_\theta^f = \ln(2\pi r_f/2\pi r_o) = -\tfrac{1}{2}\ln R \tag{22}$$

and for the final radial strain

$$\epsilon_r^f = \ln(dr_f/dr_o) = -\tfrac{1}{2}\ln R \tag{23}$$

The condition of constant volume for finite true strains is simply that the sum of three orthogonal normal strains be zero. Hence, the final axial strain is

$$\epsilon_z^f = \ln(dz_f/dz_o) = -\epsilon_r - \epsilon_\theta = \ln R \tag{24}$$

Note that ϵ_z^f, ϵ_r^f, and ϵ_θ^f correspond exactly to the final strain components in the material coordinates ϵ_a^f, ϵ_b^f and ϵ_c^f since the two coordinate systems must coincide downstream from the die.

Because the condition of axial symmetry requires that the shear strain components $\gamma_{\theta z}$ and $\gamma_{\theta r}$ be zero, the only final strain component not fully specified by the extrusion ratio is the shear γ_{rz}^f. Thus, the only features of the final deformation that can be influenced by other aspects of the geometry such as die angle and surface friction are the final shear strain and the final value of "effective strain" $\bar{\epsilon}$, a measure of the cumulative deformation defined by the equations

$$\dot{\bar{\epsilon}} = [\tfrac{2}{3}(\dot{\epsilon}_r^2 + \dot{\epsilon}_z^2 + \dot{\epsilon}_\theta^2 + \tfrac{1}{2}\dot{\gamma}_{rz}^2)]^{\frac{1}{2}} \tag{25}$$

and

$$\bar{\epsilon} = \int_{z_o}^{z} (\dot{\bar{\epsilon}}/v)\, dz \tag{26}$$

Furthermore, by symmetry the shear strain at the axis of the extrusion is zero and therefore the die angle is not expected to have any influence on the final properties of this portion of the extruded product unless strain reversals occur.

The foregoing discussion may appear to suggest that centerbursts never occur in axisymmetric extrusions and that surface finish and choice of lubricants have negligible effects in the extrusion process. Within the context of the flow function analysis these conditions are approximately true. They would be wholly correct except that nothing in the theory prohibits strain reversals, and there is some geometrical effect in the final shear strain.

The computational procedure associated with flow function analysis is detailed in Appendix A. For the purpose of the present discussion it is important to note that the flow function throughout the deformation zone is constructed from measurements of the positions of flow lines only. Experimental positions of transverse lines are not used in the flow function construction and thus an initial check on the accuracy of the flow function is a comparison of the experimental positions with positions computed from the flow function by integration of downstream velocity along each flow line for equal time intervals.

Flow function analysis is thus not a complete and independent theory but merely a theoretical framework within which it is possible to process experimental data. Hence, the concept of comparison with experiment seems contradictory. What is meant is that some quantities that are calculated indirectly from flow function analysis can be compared directly with measurements of the same quantities. The positions of the transverse grid lines represents one such quantity. Other quantities which can be calculated and then checked are the final normal strain values; this represents a twofold test. The theoretical values given by Eqs. 23 and 24 can be compared with experimental values to assess the accuracy of the theory and they can be compared with the accumulated values obtained from Eqs. 14 and 15 to assess the consistency of the numerical procedures. Finally, we can compare calculated and measured values of the final angle of intersection between the flow lines and the transverse lines, the calculated values being given by Eq. 17. If there is good agreement between the computed values and those measured directly for all of these points of comparison, we can conclude that the flow function analysis provides an accurate description of the deformation kinetics, i.e. of the velocity, strain rate and strain fields in the interior regions of the extrusion.

The remainder of this paper will review work carried out by the authors[6,7,9,10] on commercial lead and commercial 2024 aluminum alloy extrusions where the comparisons listed above have been made. It will be shown that there is good agreement between the calculated values and those measured and hence, that the flow function analysis is an accurate description of the deformation kinematics.

EXPERIMENTAL PROCEDURE

Details regarding the lead and 2024 aluminum alloy materials and the extrusion conditions are given in previous papers[6,7,9,10] Briefly, 0.75 in. dia. commercial lead billets were back extruded at room temperature with a ram speed of 3×10^{-4} in. sec^{-1} using 90° conical dies and a reduction ratio of 3:1. The 2024 aluminum billets were 3.5 in. dia. and were forward extruded at $300^{o}C$ to $537^{o}C$ ($572^{o}F$ to $1000^{o}F$) at ram speeds of 0.2 in. sec^{-1} to 13 in. sec^{-1} using a 60° conical die and a reduction ratio of 6:1. In both materials grid lines were applied by the conventional technique of milling slots on one face of the pre-split billet. Following gridding, the billets were rejoined and partially extruded, afterwhich they were removed from the dies, separated and the flow pattern examined and measured. The method used to take the data from the pattern for the flow function analysis is described in the Appendix.

RESULTS

An example of the flow patterns observed for the lead is given in Fig. 4. This is the conventional type of pattern generally reported for extrusion, which consists of a single maximum in the transverse grid line located at the extrusion axis. A similar type of flow pattern was obtained for the 2024 aluminum alloy at a high temperature ($537^{o}C$ [$1000^{o}F$]) and a low ram speed (0.2 in. sec^{-1}). However, for lower temperatures an uncommon, wavy, double maxima pattern occurred; see, for example, Fig. 5. This

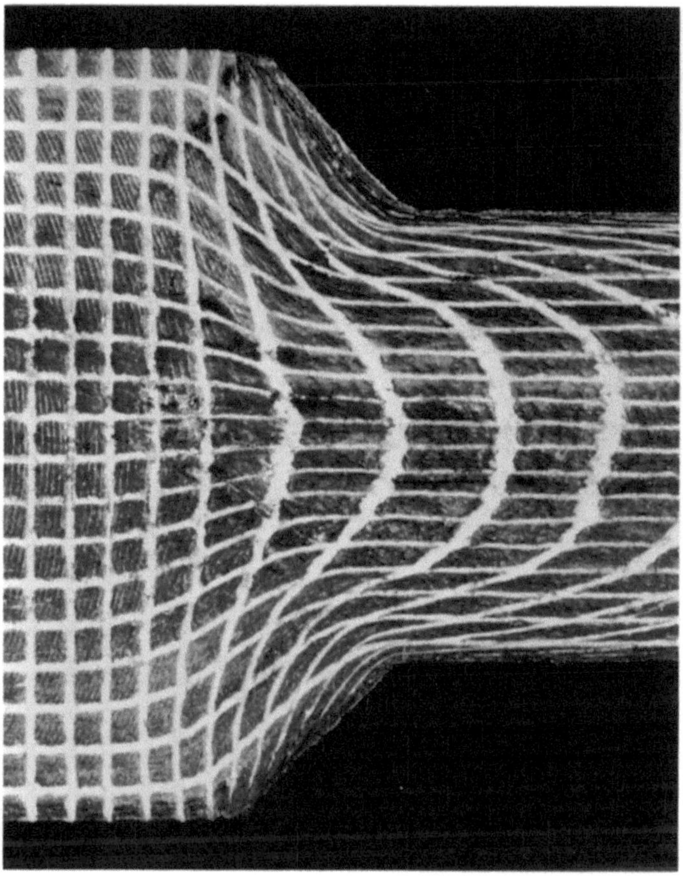

Fig. 4 Experimentally observed flow pattern of lead extruded at room temperature and ram speed of 3×10^{-4} in. sec^{-1} with reduction ratio of 3:1 and conical die angle of 90°.

pattern represents a true "difference in kind" as opposed to a "difference in degree" according to the classification of Pearson and Parkins[11]. The double maxima pattern became less pronounced with increase in temperature or decrease in ram speed; compare, for example, the patterns in Figs. 5 and 6. This differs from the findings of Altan et al.[12] for OFHC copper and 1018 steel where little variations in flow pattern occurred with temperature and ram speed.

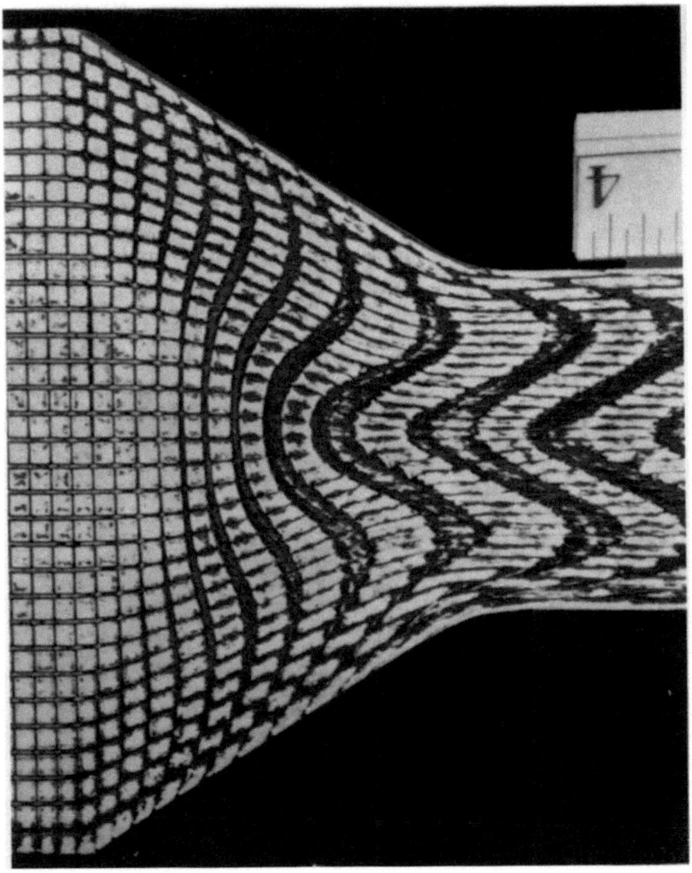

Fig. 5 Experimentally observed flow pattern of 2024 aluminum alloy extruded at 315°C (600°F) and ram speed of 1.1 in. sec^{-1} with reduction ratio of 6:1 and conical die angle of 60°.

It was found[10] that the transition between the single maximum and the wavy double maxima pattern in the 2024 aluminum alloy occurred at a value of approximately 2×10^9 sec^{-1} for the temperature compensated rate parameter Z given by

$$Z = \dot{\bar{\varepsilon}} \exp (H_D/RT) \quad (27)$$

where H_D (= 35 Kcal/mole) is the activation energy for the plastic deformation of aluminum[13] and is approximately the activation energy for self diffusion[14]. $\dot{\bar{\varepsilon}}$ is the mean effective strain rate

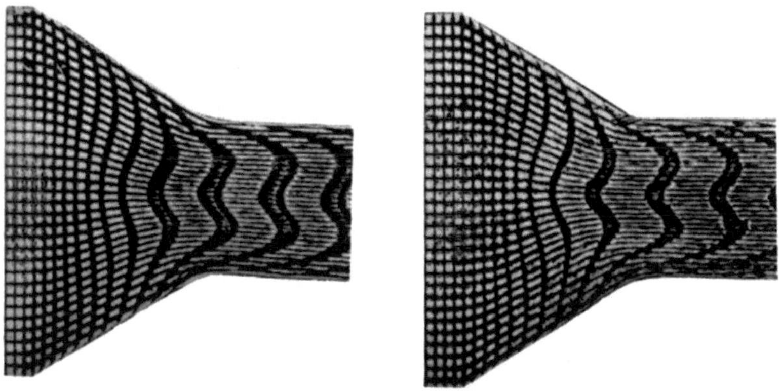

Fig. 6 Experimentally observed flow patterns of 2024 aluminum alloy extruded at 426 $^{\circ}$C (800 $^{\circ}$F) and ram speeds of 11.0 in.sec^{-1} (left) and 1.2 in. sec^{-1} (right) with reduction ratio of 6:1 and conical die angle of 60°.

for the extrusion and is taken to be[15]

$$\dot{\epsilon} = \frac{m\, s\, R^{\frac{3}{2}} \ell n\, R}{D_o (R^{\frac{3}{2}} - 1)} \tag{28}$$

where R is the reduction ratio, D_o the inside diameter of the billet container, s the steady state ram speed, m a geometric factor dependent on the die angle and is 2.82 for a 60° die.

Examples of the flow patterns calculated using the flow function analysis are given in Figs. 7 and 8. It is seen that there exists good agreement between the calculated transverse lines and their experimentally observed counterparts for both types of flow patterns (Figs. 4 and 5).

Calculations of the strains and the angle of intersection Ψ for various stations along the extrusion axis (see Fig. A-1) are presented in Figs. 9 and 10 for the two types of flow pattern. The individual curves in Figs. 9 and 10 represent results for the various stations starting from just before the die entrance (bottom curve) and ending shortly after the die exit (top curve). The radial position is given as the ratio of the distance from the extrusion

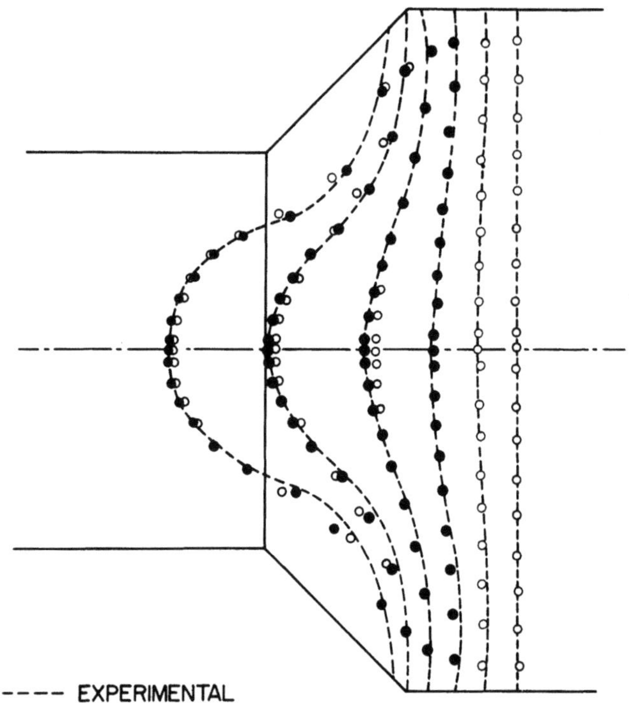

Fig. 7 Comparison of the computed positions of the transverse grid lines from flow function analysis and with experimental positions for lead extruded at room temperature and ram speed of 3 x 10^{-4} in. sec^{-1} with reduction ratio of 3:1 and conical die angle of 90°.

axis to the initial radius of the billet, r^*_o in Fig. 1. Also included in Figs. 9 and 10 are the experimentally measured final strains and angle of intersection. Again, there is good agreement between the calculated and experimental values for both flow patterns.

Comparing the axial velocities for the two types of flow pattern in the 2024 aluminum alloy revealed that the change from the initial to the final velocity at the axis occurs in a shorter distance for the double maxima pattern as compared to the single maximum[9]. The variation in velocity from the extrusion axis to the surface was in accord with that expected from visual observation of the two types of pattern.

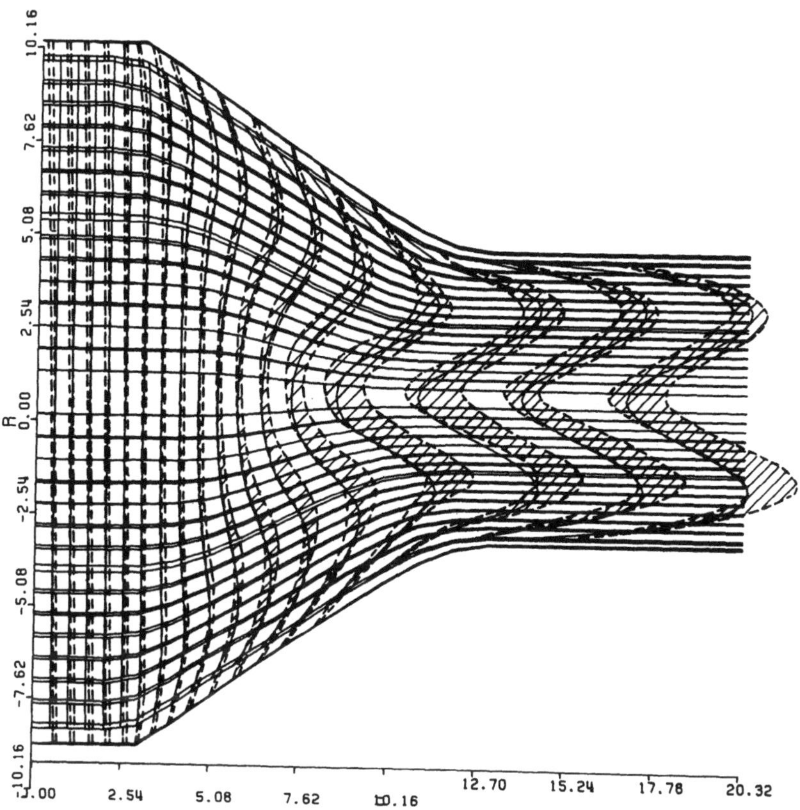

Fig. 8 Comparison of the computed positions (solid lines) of the transverse grid lines from flow function analysis with the experimental positions (dashed lines) for the 2024 aluminum alloy extrusion of Fig. 5.

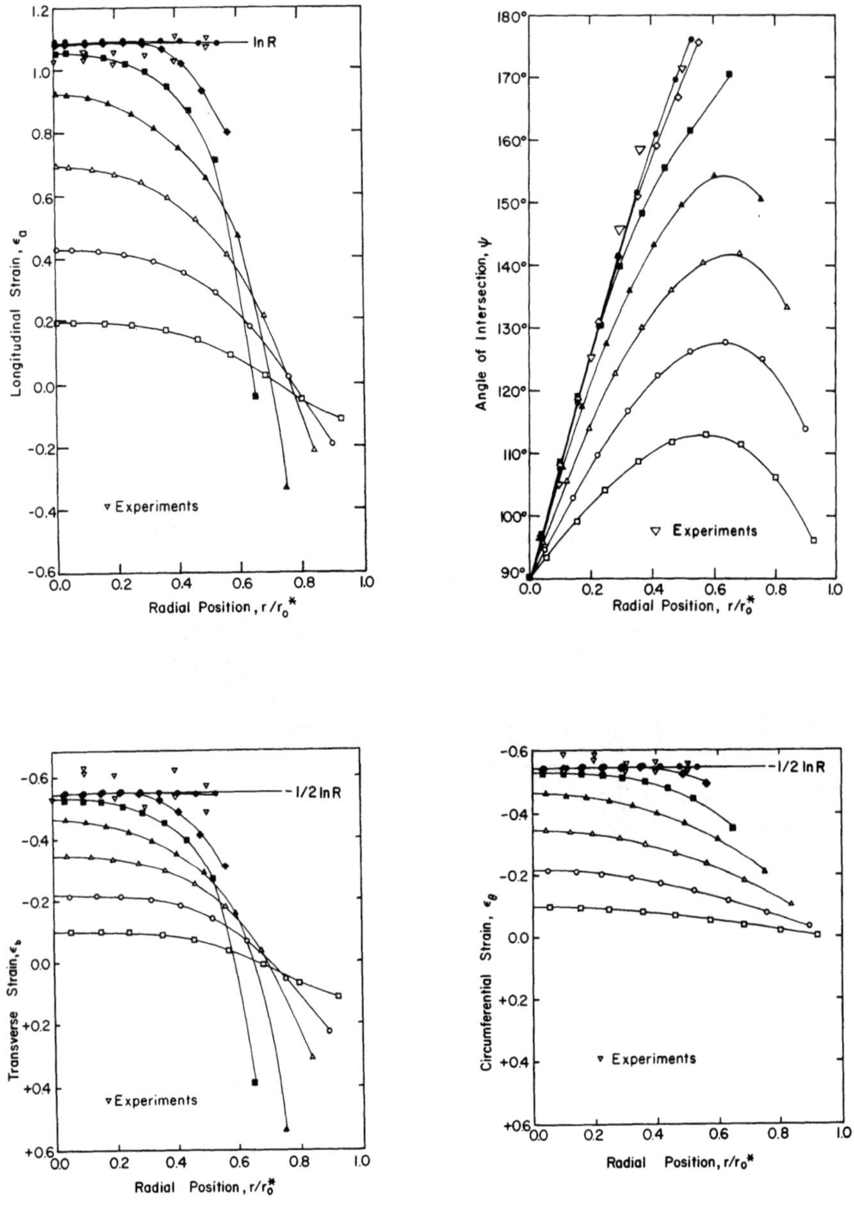

Fig. 9 Computed values at equally spaced stations along the deformation zone and experimental values in the extrusion product of the longitudinal, transverse and circumferential strain and the angle of intersection for the lead extrusion of Fig. 4.

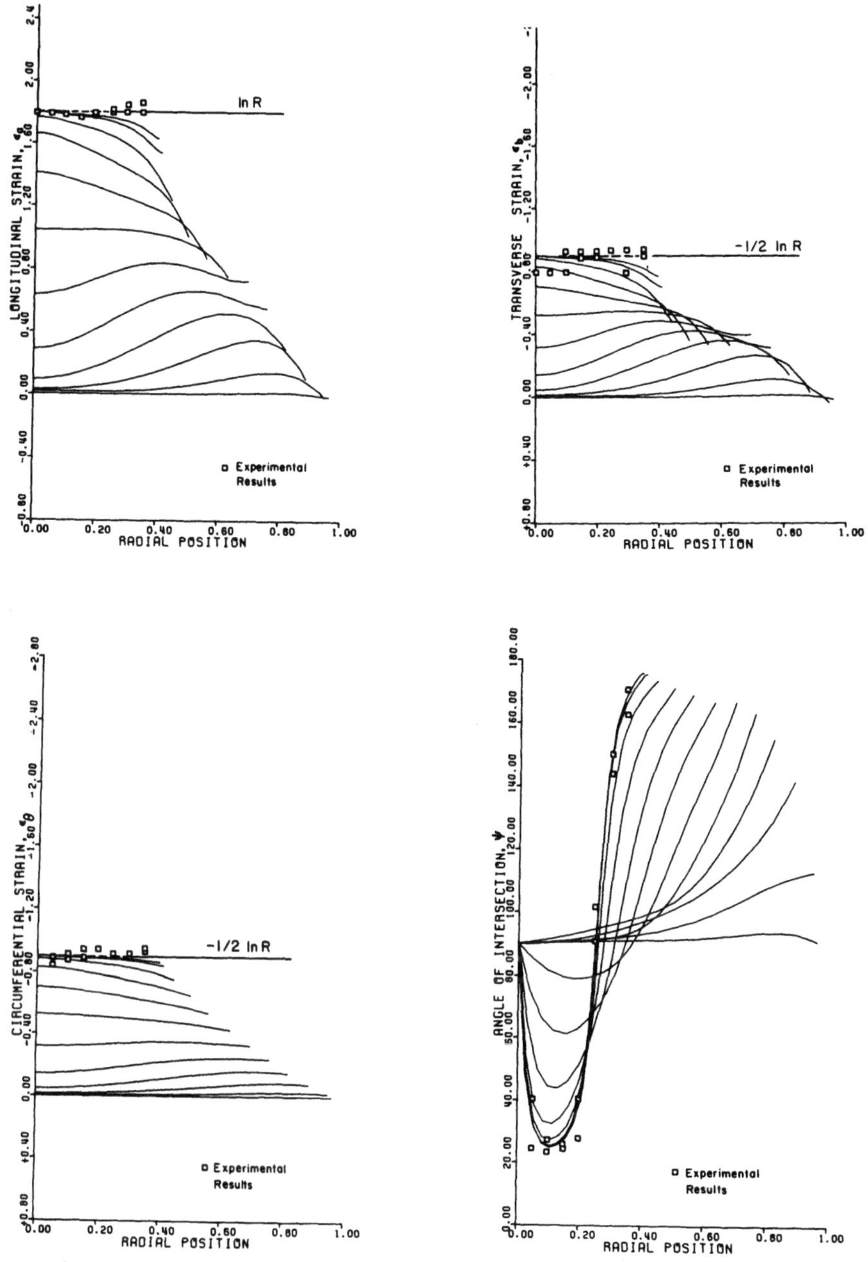

Fig. 10 Computed values at equally spaced stations along the deformation zone and experimental values in the extrusion product of the longitudinal, transverse, and circumferential strain and the angle of intersection for the 2024 aluminum extrusion of Fig. 5.

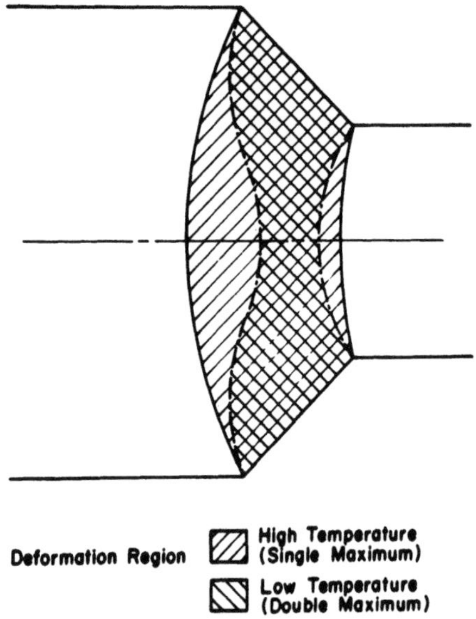

Fig. 11 Schematic of the deformation zone for specimens with different types of flow patterns.

The shear strains for the 2024 aluminum alloy extrusions exhibited a general increase in absolute value as the temperature decreased; i.e. as the degree of the double maxima pattern increased. Moreover, the ratio of the absolute value of the shear strain at the surface to that in the vicinity of the axis increased as the extrusion temperature decreased. There was, however, little difference in the effective strain between the two types of flow patterns in the 2024 aluminum alloy extrusions[9].

The geometry of the deformation zones for the two types of flow patterns was quite different; this is illustrated schematically in Fig. 12. The difference was, however, not constant, rather the zone geometry depended somewhat on temperature and ram speed, in accord with the "kind" and "degree" of the flow pattern.

DISCUSSION AND CONCLUSIONS

In the above it was shown that there exists good agreement between certain quantities (transverse grid lines, final normal strains and final angle of intersection of the horizontal and transverse grid lines) calculated from the flow function analysis and those measured directly from the deformed grids. This agreement occurred for both types of flow pattern and for the various "degrees" of the double maxima pattern. It is therefore concluded that the flow function analysis can provide an accurate description of the deformation kinematics for extremely diverse types of patterns.

Associated with the difference in type of flow pattern there occurred differences in axial velocities and shear strains and in the geometry of the deformation zone. In view of the difference in the deformation zone, any theoretical analysis of the extrusion process which is based on the prior assumption of the shape of the zone cannot be generally applicable to all materials over a wide temperature and effective strain rate range.

The exact reason for the change in flow pattern at a specific value of the temperature compensated rate parameter Z is not clear at this time. Some of the factors which may be important are the shape of the stress-strain curve as a function of Z, localized adiabatic heating and friction.

ACKNOWLEDGMENT

The authors are grateful to Mr. E. C. Beatty of the Kaiser Aluminum Company for providing the 2024 aluminum alloy, to Mr. A. M. Adair of the Air Force Materials Laboratory and Mr. F. Gurney and Dr. A. Male of Westinghouse Electric Company for carrying out the extrusion tests on the aluminum alloy and to Mr. K. Howard, graduate student, University of Kentucky, for assistance in preparing the 2024 aluminum alloy billets and making measurements on them. This work is part of the Themis Project sponsored by the Office of Aerospace Research, United States Air Force under contract F33615-69-C-1027, Mr. A. Adair, technical monitor.

REFERENCES

1. V. DePierre, "Experimental Measurement of Forces During Extrusion and Correlation with Theory", ASME Paper No. 69-WA/Lub-6.

2. A. Shabaik, S. Kobayashi and E. G. Thomsen, Trans. ASME, 89B 503 (1967).

3. A. H. Shabaik and E. G. Thomsen, Trans. ASME, 90B 343 (1968).

4. A. Shabaik and S. Kobayashi, Trans. ASME, 89B 339 (1967).

5. A. S. Sulijoadikusumo and O. W. Dillon, Jr., University of Kentucky, unpublished research (1970).

6. R. E. Medrano and P. P. Gillis, "Visioplasticity Techniques for Determining Velocity and Strain-Rate Fields During Extrusion", submitted to Met. Trans.

7. R. E. Medrano and P. P. Gillis, "Strain Field Analysis in Axisymmetric Extrusion", submitted to Met. Trans.

8. Y. C. Fung, Foundations of Solid Mechanics, Prentice-Hall Inc., Englewood Cliffs, New Jersey (1965) p. 442.

9. C. P. Hinesley, R. E. Medrano, P. P. Gillis and H. Conrad, "Visioplasticity Analysis of the Axisymmetric Extrusion of 2024 Aluminum Alloy", to be submitted to Trans. ASME.

10. C. P. Hinesley and H. Conrad, "Observation of an Unusual Visioplasticity Flow Pattern in Extrusions of 2024 Aluminum", to be submitted to Mat. Sci. and Eng.

11. C. E. Pearson and R. N. Parkins, The Extrusion of Metals, Chapman and Hall, London (1960).

12. T. Altan, S. Kobayashi, V. DePierre and C. M. Pierce, Trans. ASME, 90B 371 (1968).

13. W. A. Wong and J. J. Jonas, TMS AIME, 242 2271 (1968).

14. P. G. Shewmon, Diffusion in Solids, McGraw-Hill, New York (1963) p. 74.

15. B. Avtizur, Trans. ASME, 88B 410 (1966).

APPENDIX A

Procedure for Visioplasticity Data Analysis

After partial extrusion and resplitting of the billet,an enlarged photograph of the deformed grid is made for taking measurements; see Fig. A-1. The centerline of the extrusion is marked on the photo and two lines perpendicular to the centerline are marked shortly before the die entrance and shortly after the die exit respectively. Some convenient integral fraction of the distance between these two lines is selected for a data acquisition mesh.

All of the scribed flow lines are then tranversed, on both sides of the centerline, and the radii at all mesh line intersections are recorded.

These data are computer processed (using the program given at the end of this Appendix) according to the following general scheme. With reference to Fig. A-2 the value of the flow function for each flow line is computed from its upstream radius using Eq. 7. The shape of the line may then be approximated by a sixth-order polynomial $r_j(z) = \sum_{n=0}^{6} b_{jn} z^n$ in which the coefficients b_{jn} are established by least squares fitting of the measured points (z, r) corresponding to the flow line j. These polynomials may be used to obtain interpolated "data" points by subdividing the intervals Δz between axial stations and evaluating all of the $r_j(z)$ for the original and intermediate stations using the polynomials. This latter procedure is approximately equivalent to an initial smoothing of the data and suffers the defect of possibly eliminating some information from the actual data. Thus, we believe it should be avoided whenever possible. It was not used in the calculations for the aluminum alloy extrusions. However, we have found that in some cases standard numerical techniques for differentiation (three-point) and integration (Trapezoidal rule) would not reproduce given functions when sequentially applied. In such cases generation of additional "data" by the foregoing procedure was an adequate remedy. This procedure was applied for the lead extrusions. At present we have insufficient experience to generalize concerning combinations of die angle, number of stations taken and reduction ratios which will require artificial data generation. We presently test each case individually to determine how well successive differentiation and integration will reproduce a given function.

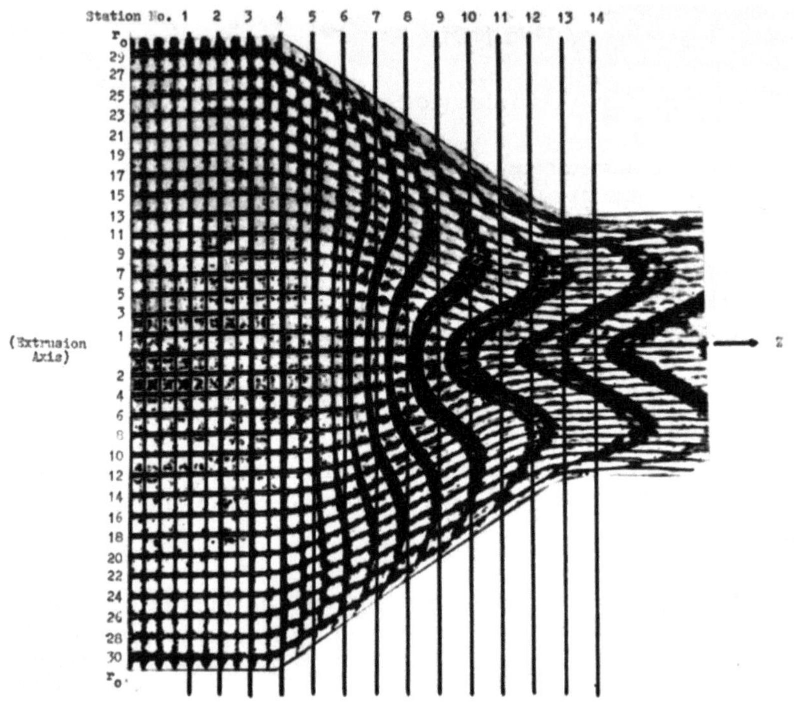

Fig. A-1. Illustration of the procedure used to take data from a grid pattern for the flow function analysis.

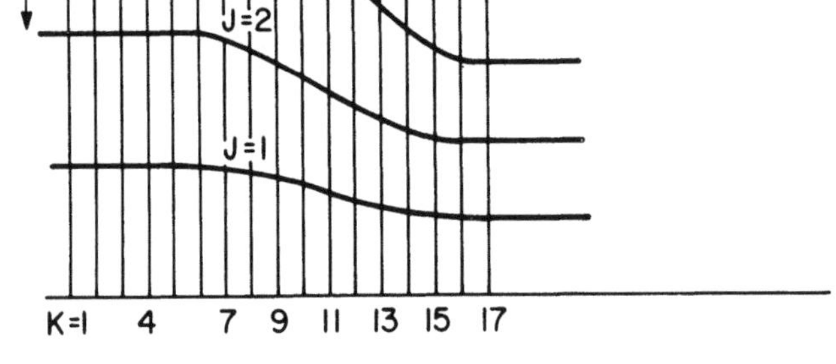

Fig. A-2. Computational mesh formed by the intersection of flow lines having initially equal spacing Δr in the region upstream from the die and radial lines having equal axial spacing Δz.

Next, at each axial station k, i.e., for every z-value in the initial data acquisition mesh, plus any intermediate z-values generated by interpolation, the flow function is approximated by a m-order even polynomial $\Phi_k(r) = \sum_{n=0}^{m/2} a_{kn} r^{2n}$ in which the coefficients a_{kn} are again established by least squares fitting of the measured or interpolated data points (r, Φ).

Without making any modification in the program, it is possible to try different orders of polynomial equal or less than eight and exclude or include the surface boundary as another flow line. The method chosen was that which gave best agreement with the experimental results. In general, the order m of polynomial is greater than four.

The remainder of the listed program is a straightforward computation of velocity, strain rate and strain components according to the equations presented in the Theory section of this paper. Standard STET are employed in the form of library programs for numerical differentiation and integration.

```
C       THIS PROGRAM IS AN APPLICATION OF VISIOPLASTICITY METHODS TO
C       EXTRUSION PROCESSES.  VELOCITY, STRAIN RATE AND STRAIN FIELDS,
C       AND POSITIONS OF TRANSVERSE LINES ARE OBTAINED.  ALSO, THE ANGLE
C       BETWEEN LONGITUDINAL AND TRANSVERSE LINES AND EFFECTIVE STRAIN ARE
C       COMPUTED.
C
C                       ********************************
C
        DIMENSICN RR(21,35),PHI(21,35),PP(7C),Q( 70),ALPHA( 70),BETA(70),W
       2( 70),AL(70),P(70),R(70),PO(70),WN( 70),AN( 70),C(70),PA(70),PB(70
       3),XI(70),U( 70),V( 70),ZN(21,35),RP( 70),S(9),RA(21,35)
     21 FORMAT(' K=',1X,'V=',9X,'U=',9X,'SRZ=',7X,'SRR=',7X,'SRTE=',6X,'GA
       2MMA=',5X,'TANG=',6X,'SRTRANS=',3X,'SRSHEAR=',3X,'ESR=',7X,'SRLONG=
       3')
     22 FORMAT (' J=',I4)
     37 FORMAT(' I=',I2,'ZN=',E13.6,'RA=',E13.6)
     80 FORMAT (' ',I2,11(1X,E10.3))
     95 FORMAT(' K',1X,'LONG=',6X,'TRANS=',5X,'CIRCUMF=',3X,'SHEAR=',5X,
       2'ANGLE=',5X,'ES=')
    100 FORMAT (1X,5I2,2F4.2,2F5.2)
    104 FORMAT (1X,33F4.1)
    105 FORMAT (20F4.1)
C
C                       ********************************
C
C       READING OF INITIAL INPUT GECMETRY OF EXPERIMENTAL FLOW LINES .
C
C       M= NUMBER OF EXPERIMENTAL FLCW LINES.
C       N= NUMBER OF EQUALLY SPACED STATICNS ALONG THE AXIS.
C       LO= NUMBER OF SUBDIVISIONS OF THE D2 INTERVALS.
C       MX= ORDER OF POLYNOMIAL APPROXIMATION.
C       LL= NUMBER OF TRANSVERSE GRID LINES.
C       VO= RAM SPEEC.
C       D2= DISTANCE BETWEEN STATIONS.
C       D3= DISTANCE BETWEEN TRANSVERSE LINES PRIOR TO DEFORMATION.
C       D4= DISTANCE FROM THE NEAREST TRANSVERSE LINE TO THE INITIAL STATION.
C
        READ (5,100) M,N,LO,MX,LL,VO,D2,D3,D4
        WRITE(6,100) M,N,LO,MX,LL,VO,D2,D3,D4
C
C                       ********************************
C
CSTEP 1.  DETERMINATION OF FLOW FUNCTION FIELD USING POLYNOMIAL
C       APPROXIMATIONS AT EQUALLY SPACED STATIONS.
C
C
        M2=M-1
        M3=2*M2
        N1=N-1
        N2=N-2
        MN=MX+1
        MM=MX-1
C
C       THIS LOOP IS PERFORMED TO HAVE INTEGER IN REAL FIELD AND TO GIVE
C       EQUAL WEIGHT TO ALL POINTS IN THE POLYNOMIAL APPROXIMATIONS
C
        DO 5 I=1,70
        XI (I)= I-1
        W(I)=1.0
      5 CONTINUE
C
```

```
C       READ RADIAL POSITIONS OF FLOW LINES PRIOR TO DEFORMATION AND
C       CALCULATE FLOW FUNCTION VALUES.
C       RO= INITIAL RADIUS OF BILLET.
C
        READ (5,105) (PP(J),J=1,M2),RO
        WRITE(6,104) (PP(J),J=1,M2),RO
        DO 4 J=1,M2
        Q(J)=(PP(J)**2)*VO
      4 CONTINUE
        DO 214 K=1,N
C
C       READ RADIAL COORDINATES OF FLOW LINES AT EACH AXIAL STATION.
C
        READ(5,105) (RP(J),J=1,M)
        WRITE(6,104)(RP(J),J=1,M)
        DO 215 J=1,M
        RR(J+1,K)=RP(J)
    215 CONTINUE
    214 CONTINUE
        N3=N1*LO+1
        XLO=LO
        IF (LO.EQ.1) GO TO 107
C
C       FOR WIDE ANGLE DIES ADDITIONAL 'DATA' IS COMPUTER GENERATED BY
C       INTERPOLATION BETWEEN ACTUAL DATA POINTS.
C
        DO 207 J=2,M
        DO 208 K=1,N1
        RP (K)=RR(J,K+1)
        PP(K)=XI(K+1)*D2
    208 CONTINUE
        U(1)=0.0
        U(2)=PP(N1)
        V(1)=RR(J,1)
        V(2)=RR(J,N)
        S(1)=0.0
        S(2)=0.0
C
C       FOLSQ IS USED TO CONSTRUCT SIXTH-ORDER POLYNOMIALS IN Z WHICH
C       APPROXIMATE THE FLOW LINES. EACH INTERVAL D2 IS THEN DIVIDED INTO
C       LO SUBDIVISIONS AND FLOW LINE RADII ARE COMPUTED FOR EACH OF THESE
C       INTERMEDIATE AXIAL STATIONS USING FEVREA AND THE AXIAL POLYNOMIALS
C
        CALL FOLSQ(U,V,2,S,2,PP,RP,N2,W,ALPHA,BETA,AL,P,R,PO,WN,AN,C,2,PA,
       2PB)
        DO 209 K=1,N3
        P (K)=XI(K)*D2/XLO
        CALL FEVREA (BETA,6,P (K),RR(J,K))
    209 CONTINUE
    207 CONTINUE
        D2=D2/XLO
C
C       AT EACH AXIAL STATION - THOSE ESTABLISHED BY THE ORIGINAL DATA AND
C       THOSE ESTABLISHED BY INTERPOLATION - THE VARIATION OF FLOW FUNCTION
C       WITH RADIUS IS FIT USING FOLSQ BY A POLYNOMIAL OF ORDER MX.
C
    107 V(1)=0.0
        U(1)=0.0
        DO 1 K=1,N3
        DO 3 J=1,M2
        RP(J) =RR(J+1,K)
```

```
      RP(M2+J)=-RP(J)
      Q(M2+J)=Q(J)
    3 CONTINUE
      CALL FCLSQ(V,U ,1,RP,Q,M3,W,ALPHA,BETA,AL,P,R,PO,WN,AN,C,MM,PA,PB)
      DO 81 I=1,9
      IF (I-MN) 83,83,84
C     P(I,K)= COEFFICIENT OF ORDER (I-1) AT STATION K.
   83 PHI(I,K)=BETA(I)
      GO TO 81
   84 PHI(I,K)=0.0
   81 CONTINUE
      RR(1,K)=0.0
    1 CONTINUE
C
C                    ********************************
C
CSTEP 2.   DETERMINATION OF THE ROOT OF THE POLYNOMIAL APPROXIMATION
C     MINUS THE FLOW FUNCTION (NEW RADIAL POSITION), TO COMPARE HOW
C     THE PRECEDING STEP, CHANGES THE ORIGINAL DATA.
C
C
      DO 219 J=2,M
      DO 216 K=1,N3
      U(1)=-Q(J-1)
      DO 217 I=2,MN
      U(I) = PHI(I,K)
      PP(I)=0.0
  217 CONTINUE
      S  (1)= RR(J,K)
      V(1)=0.0
      PP(1)=0.0
C
C     AT EACH STATION THE FLOW FUNCTION IS A KNOWN FUNCTION OF RADIUS.
C     MULLP DETERMINES RADII AT EACH STATION, WHICH CORRESPOND TO THE
C     INITIAL FLOW FUNCTION VALUES.
C
      CALL MULLP (U,PP,-MX,S ,V)
      RR(J,K) = U (MN)
  216 CONTINUE
      WRITE (6,104) (RR(J,K),K=1,N3)
  219 CONTINUE
C
C                    ********************************
C
CSTEP 3.   CALCULATION OF THE POSITIONS OF TRANSVERSE LINES.
C
C
C     J= THE NUMBER OF THE FLOW LINE.
C     K= THE NUMBER OF THE AXIAL STATION.
C     ALPHA(K)= RECIPROCAL OF AXIAL VELOCITY AT STATION K.
C     BETA (K)= TIME TAKEN TO REACH STATION K.
C     AL(I)= TIME TAKEN BY LINE I TO GO FROM THE INITIAL TO THE CURRENT
C     POSITION.
C     ZN (J,I)= COMPUTED Z-COORDINATE OF INTERSECTION OF LONGITUDINAL
C     LINE J AND TRANSVERSE LINE I.
C     RA (J,I)= COMPUTED R-COORDINATE OF INTERSECTION OF LONGITUDINAL
C     LINE J AND TRANSVERSE LINE I.
C
C
      DO 30 J=1,M
      WRITE (6,22) J
```

```
      DO 226 K=1,N3
  226 ALPHA(K)=1.0/(PHI(3,K)+2.0*PHI(5,K)*RR(J,K)**2+3.0*PHI(7,K)*RR(J,K
     2)**4+4.0*PHI(9,K)*RR(J,K)**6)
C
C     INTEGRATION OF EQUIDISTANTLY TABULATED FUNCTION BY TRAPEZOIDAL
C     RULE USING QTFE.
C
      CALL QTFE(D2,ALPHA,BETA,N3)
      DO 31 I=2,LL
      AL(I)=(XI(I)*D3+D4)/VO
      DO 32 K=1,N3
      PO(K)=XI(K)*D2
      IF(BETA(K)-AL(I)) 32,33,34
   32 CONTINUE
      IF(BETA(N3).GT.AL(I)) GO TO 31
      ZN(J,I)=VO*RFD*(AL(I)-BETA(N3))+PO(N3)
      RA(J,I)=RR(J,N3)
      GO TO 31
   33 ZN(J,I)=PO(K)
      RA(J,I)=RR(J,K)
      GO TO 31
   34 ZN(J,I)=PO(K)-((BETA(K)-AL(I))*D2)/(BETA(K)-BETA(K-1))
      RA(J,I)=RR(J,K)-((PO(K)-ZN(J,I))*(RR(J,K)-RR(J,K-1)))/D2
   31 CONTINUE
      WRITE (6,37) (I,ZN(J,I),RA(J,I),I=2,LL)
   30 CONTINUE
C
C                    ********************************
C
CSTEP 4.   DETERMINATIONS OF VELOCITIES, STRAIN-RATES, STRAINS, ANGLE
C     BETWEEN TRANSVERSE AND LONGITUDINAL LINES AND EFFECTIVE STRAIN
C     ALONG FLOW LINES. TRAPEZOIDAL RULE IS USED FOR INTEGRATION.
C     A PARENTESIS AT THE END OF EACH STATEMENT DENOTES THE NAME UNDER
C     COMPUTATION.
C
C
C     J= THE NUMBER OF THE FLOW LINE.
C     K= THE NUMBER OF THE STATION.
C     U= R-COMPONENT OF THE VELOCITY.('U')
C     V= Z-COMPONENT OF THE VELOCITY.('V')
C     SRZ= Z-COMPONENT OF STRAIN RATE.('P')
C     SRR= R-COMPONENT OF STRAIN RATE.('R')
C     SRTE= CIRCUMFERENTIAL COMPONENT OF STRAIN RATE. ('WN')
C     GAMMA= SHEAR COMPONENT OF STRAIN RATE.  RELATES TO Z AND R
C     COORDINATES. ('AN')
C     ESR= EFFECTIVE STRAIN RATE.('W')
C     ES= EFFECTIVE STRAIN.('Q')
C     TANG= TANGENT OF ALPHA. ('PP')
C     SRLONG= LONGITUDINAL COMPONENT OF STRAIN RATE.('AL')
C     SRTRANS= TRANSVERSE COMPONENT OF STRAIN RATE.('PO')
C     SRSHEAR= SHEAR COMPONENT OF STRAIN RATE.  RELATES TO ROTATED
C     COORDINATES, A,B,C.('C')
C     LONG= LONGITUDINAL COMPONENT OF STRAIN.('AN')
C     TRANS= TRANSVERSE COMPONENT OF STRAIN. ('R')
C     SHEAR= SHEAR COMPONENT OF STRAIN. ('WN')
C     CIRCUMF= CIRCUMFERENTIAL COMPONENT OF STRAIN.('P')
C     ANGLE= ANGLE BETWEEN TRANSVERSE AND LONGITUDINAL LINES.('RP')
C
C
      DO 2 J=1,M
      WRITE (6,22) J
```

```
      WRITE (6,21)
      DO 59 K=1,N3
      IF (K.GT.1) GO TO 501
      U(1)=0.0
      P(1)=0.0
      R(1)=0.0
      WN(1)=0.0
      AN(1)=0.0
      GO TO 503
  501 IF (K.LT.N3) GO TO 502
      U (N3)=0.0
      P (N3)=0.0
      R (N3)=0.0
      WN(N3)=0.0
      AN(N3)=0.0
      GO TO 503
  502 AN(K)=4.0*PHI(5,K)*RR(J,K)+12.0*PHI(7,K)*RR(J,K)**3+24.0*PHI(9,K)*
     2RR(J,K)**5-((PHI(3,K+1)+PHI(3,K-1)-2.0*PHI(3,K))*RR(J,K)+(PHI(5,
     3K+1)+PHI(5,K-1)-2.0*PHI(5,K)) *RR(J,K)**3+(PHI(7,K+1)+PHI(7,K-1)-
     42.0*PHI(7,K))*RR(J,K)**5+(PHI(9,K+1)+PHI(9,K-1)-2.0*PHI(9,K))*RR
     5(J,K)**7)/(2.0*D2**2)
      U (K)=-((PHI(3,K+1)-PHI(3,K-1))*RR(J,K)+(PHI(5,K+1)-PHI(5,K-1))*RR
     2(J,K)**3+(PHI(7,K+1)-PHI(7,K-1))*RR(J,K)**5+(PHI(9,K+1)-PHI(9,K-1)
     3)*RR(J,K)**7)/(4.0*D2)
      P (K)= (PHI(3,K+1)-PHI(3,K-1)+2.0*(PHI(5,K+1)-PHI(5,K-1))*RR(J,K)
     2**2+3.0*(PHI(7,K+1)-PHI(7,K-1))*RR(J,K)**4+4.0*(PHI(9,K+1)-PHI(9,
     3K-1))*RR(J,K)**6)/(2.0*D2)
      R (K)=-(PHI(3,K+1)-PHI(3,K-1)+3.0*(PHI(5,K+1)-PHI(5,K-1))*RR(J,K)
     2**2+5.0*(PHI(7,K+1)-PHI(7,K-1))*RR(J,K)**4+7.0*(PHI(9,K+1)-PHI(9,
     3K-1))*RR(J,K)**6)/(4.0*D2)
      WN(K)=-(PHI(3,K+1)-PHI(3,K-1)+1.0*(PHI(5,K+1)-PHI(5,K-1))*RR(J,K)
     2**2+1.0*(PHI(7,K+1)-PHI(7,K-1))*RR(J,K)**4+1.0*(PHI(9,K+1)-PHI(9,
     3K-1))*RR(J,K)**6)/(4.0*D2)
  503 V(K)   =     (PHI(3,K)+2.0*PHI(5,K)*RR(J,K)**2+3.0*PHI(7,K)*RR(J,K
     2)**4+4.0*PHI(9,K)*RR(J,K)**6)
      PP(K)=U(K)/V(K)
      PB(K)=( (P(K)-R(K))* 2.0*PP(K)-AN(K)*(1.0-PP(K)**2))/(V(K)*(1.0+
     2PP(K)**2))
      AL(K)=(P(K)+R(K)*PP(K)**2+AN(K)*PP(K))/((1.0+PP(K)**2)*V(K))
      PO(K)=(P(K)*PP(K)**2+R(K)-AN(K)*PP(K))/((1.0+PP(K)**2)*V(K))
      C(K)=((R(K)-P(K))*PP(K)*2.0+AN(K)*(1.0-PP(K)**2))/((1.0+PP(K)**2)*
     2V(K))
      W (K)   = SQRT((2.0/3.0)*(P(K)**2+R(K)**2+WN(K)**2+AN(K)**2/2.0))
      PA(K)=WN(K)/V(K)
      WRITE (6,80) K,V(K),U(K),P(K),R(K),WN(K),AN(K),PP(K),PO(K),C(K),
     2W(K),AL(K)
      BETA(K)=W(K)/V(K)
   59 CONTINUE
      WRITE (6,95)
C
C     INTEGRATION BY TRAPEZOIDAL RULE USING QTFE.
C
      CALL   QTFE(D2,BETA,Q,N3)
C
C     ITERATION LOOP TO OBTAIN ANGLE.
C
      LU=0
  237 CALL QTFE(D2,PB,RP,N3)
      LU=LU+1
      DO 251 K=1,N3
      RP(K)=1.5708+RP(K)
```

```
      ALPHA(K)=COS(RP(K))
      BETA(K)= SIN(RP(K))
      PR(K)=( (P(K)-R(K))*(2.0*PP(K)*BETA(K)**2-BETA(K)*ALPHA(K)*(1.0-
     2PP(K)**2))-AN(K)*(BETA(K)**2*(1.0-PP(K)**2)+2.0*ALPHA(K)*BETA(K)*
     3PP(K)))/(V(K)*(1.0+PP(K)**2))
  251 CONTINUE
      IF (LU.LT.10) GO TO 237
C
C     INTEGRATION BY TRAPEZOIDAL RULE USING QTFE.
C
      CALL QTFE(D2,AL,AN,N3)
      CALL QTFE(D2,PA,P ,N3)
      CALL QTFE(D2,PO,R ,N3)
      CALL QTFE(D2,C ,WN,N3)
      DO 240 K=1,N3
      WRITE(6,80) K,AN(K),R(K),P(K),WN(K),RP(K),Q(K)
  240 CONTINUE
    2 CONTINUE
C
C                    *********************************
C
      RETURN
      END
```

THE EXTRUSION OF RATE SENSITIVE MATERIALS

J. J. Jonas and T. Chandra

Department of Metallurgical Engineering

McGill University, Montreal 112, Canada

ABSTRACT

The principal factors affecting extrusion pressure, viz., temperature, ram speed, die geometry and conditions of lubrication, are discussed briefly. Emphasis is given to the importance of determining the area profile of the die, and through it the strain rate profile. Four different definitions of the mean strain rate are considered, and it is shown that the root mean power strain rate is the appropriate one to use for working load calculations. The rate sensitivity correction factors for strain rate and flow stress are introduced and evaluated as functions of rate sensitivity and extrusion ratio. Experiments involving the indirect extrusion of ice are described in which the rate sensitivity corrections are verified. In an Appendix, an extension of the present analysis to include conditions under which the temperature is also changing is described briefly.

INTRODUCTION

The pressure developed during extrusion is primarily affected by the temperature, ram speed, die geometry and conditions of lubrication. Due to such factors as adiabatic heating, friction, and heat losses to tooling, the temperature does not remain constant during a typical extrusion cycle (1,2). However, an approximate indication of the effect of temperature can be obtained from the relation:

$$P = A_1 \exp(-B\theta) \quad (1)$$

Here θ is the billet or preheating temperature in ^{o}C, B is a material coefficient and A_1 is a constant which depends on material, ram speed, and die geometry. For aluminum and its alloys, B has been found to lie in the range $6{\cdot}4 \times 10^{-3}$ to $8{\cdot}2 \times 10^{-3}$ $^{o}C^{-1}$ (3). It should be noted that the exponential temperature relationship of Eq. (1) is not of the Arrhenius form commonly used for the correlation of high temperature phenomena. The latter approach (2,4) leads to the alternative relation:

$$P = A_2 \exp (C/T) \tag{2}$$

where C is a material constant related to the activation energy and the activation volume for high temperature deformation, and T is the absolute temperature in this case.

The influence of ram speed V is usually determined from log-log plots, which result in power expressions of the form:

$$P = A_3 V^m \tag{3}$$

Here m is an exponent related to the rate sensitivity of the material, and is commonly in the range 0·1 to 0·3 (3). This range is comparable to the rate sensitivity of materials under conditions of creep loading (5,6) and is somewhat lower than the rate sensitivity of superplastic deformation (0·3 to 0·5).

The typical values of m for extrusion suggest that variations in ram speed of the order usually found in practice have in themselves only a minor influence on the pressure for extrusion. Thus the effect of speed will be given no further consideration. By contrast, the effect of local variations in the flow rate within the die zone is of considerable importance and forms the subject of the present paper.

Before discussing this topic in detail, the influence of die geometry will first be considered. The effect of lubrication, although of practical importance, is omitted and the discussion will be focussed largely on the theory and practice of lubricated flow. When the flow is not well-lubricated, the strain rate gradients within the material are accentuated in sharpness and, although theoretical prediction is more difficult, the importance of a rate sensitive approach to the estimation of extrusion pressure is increased.

EFFECT OF DIE GEOMETRY

The most important die design parameter is the extrusion ratio R given by:

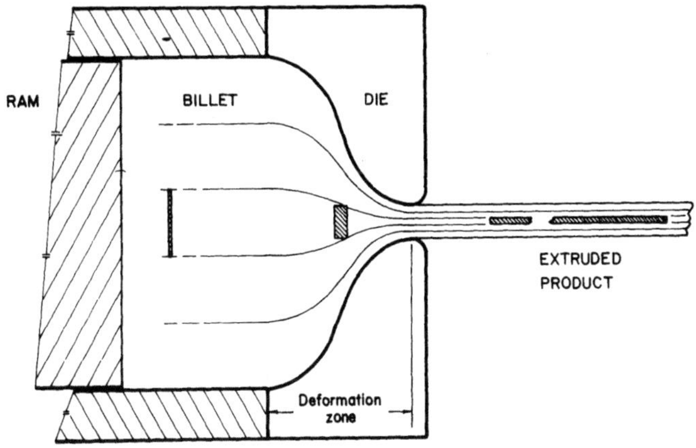

Fig. 1. Longitudinal cross-section of an extrusion container and a die of 99:1 extrusion ratio. The streamlines of flow are depicted schematically for frictionless, homogeneous flow.

$$R = \frac{D^2}{d^2} = \frac{A}{a} \tag{4}$$

where D and d, and A and a are the billet and product diameters and cross-sectional areas, respectively. For the simple case of frictionless, homogeneous flow, the theoretical extrusion pressure P can be estimated from the extrusion ratio and the expression:

$$P = \bar{\sigma} \;\ell n\ R \tag{5}$$

Here $\bar{\sigma}$ is the mean flow stress of the material deformed to a true strain of ℓnR. It is usually calculated from independent tension, torsion or compression tests conducted at the same temperature and mean strain rate.

Eq. (5) is derived by considering the work done on an element of material as it travels along a streamline of flow from the entry to the exit of the die zone (Fig. 1). The work per unit volume done on such an element is:

$$dW = \sigma_i d\varepsilon \tag{6}$$

where σ_i is the instantaneous or local value of the flow stress. σ_i varies along the streamline, and depends on the local values of the strain (work hardening), strain rate (rate sensitivity), and temperature. Thus

$$P = W = \int_0^{\ell nR} \sigma_i \; d\varepsilon \tag{7}$$

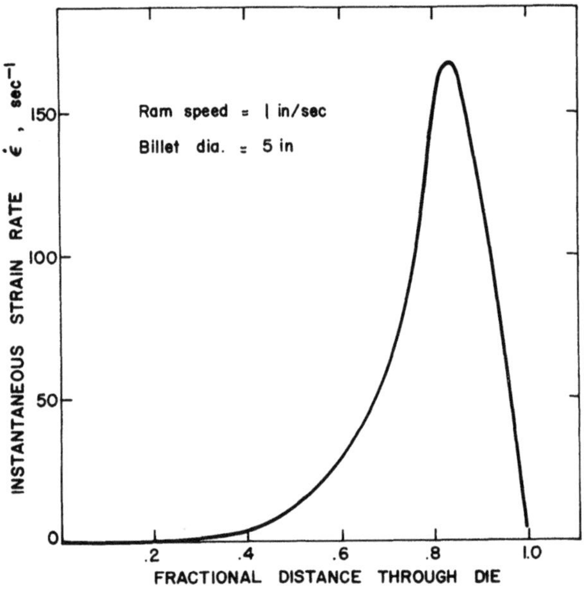

Fig. 2. The local or instantaneous strain rate (4) within the deformation zone as a function of distance from the die entry.

and when the variation of σ_i with local temperature and strain rate is neglected, integration leads to Eq. 5. The local strain rate in extrusion does vary, however, and so does the local value of the flow stress. A more rigorous evaluation of the above integral leads to the true extrusion pressure

$$P_{ext} = \bar{\sigma}_{ext} \ \ell nR \tag{8}$$

where the mean flow stress in extrusion $\bar{\sigma}_{ext}$ is

$$\bar{\sigma}_{ext} = (1/\ell nR) \int_o^{\ell nR} \sigma_i \, d\varepsilon$$

In practice, estimates of extrusion pressure based on Eq. (5) are significantly lower than the industrial or experimental pressures defined by Eq. (8). The lack of agreement is frequently eliminated by introducing into Eq. (5) either a shear factor C' or an efficiency factor η:

$$P' = C' \ \bar{\sigma} \ \ell nR = (1/\eta) \ \bar{\sigma} \ \ell nR \tag{9}$$

C' is typically about 1.5 and η is about 70% (3,7,8) and the factors are considered to correct for the redundant work and other energy losses that arise in industrial practice.

It appears likely, however, that the discrepancy between the predicted and the actual extrusion pressures is not due to the presence of redundant work and the other forms of power loss (9,10). It is due, instead, to the difference between $\bar{\sigma}$, as determined for example in a laboratory compression test, and $\bar{\sigma}_{ext}$, the actual mean flow stress in extrusion. The difference between $\bar{\sigma}$ and $\bar{\sigma}_{ext}$ can in turn be attributed to the problems that arise in calculating the mean strain rate appropriate to the extrusion operation. It appears that conventional estimates of the mean strain rate are far too low and that the pressure predictions are therefore based on tests carried out at inappropriately low strain rates. A more suitable method for the calculation of the mean strain rate will now be described. It requires knowledge of the strain rate profile of the die, and therefore of the area profile, in addition to the extrusion ratio. When a square die is used and a dead zone is formed, the area profile of the dead zone must be used in place of the area profile of the die.

THE STRAIN RATE PROFILE IN EXTRUSION

The instantaneous strain rate of the element of material flowing through the die in Fig. 1 changes continuously along the streamline and is shown in Fig. 2. It can be seen that the strain rate varies by approximately two orders of magnitude. For a die of 4:1 extrusion ratio, a two order of magnitude variation has been observed, using the method of visio-plasticity (7). For dies of 100:1 ratio, an increase of three orders of magnitude has been calculated (10).

The results of such calculations are shown in Fig. 3, in which the local strain rate is plotted as a function of position for a 100:1 conical die of 45° semi-angle. Included in Fig. 3 are curves giving the local strain rate as a function of time and of the accumulated strain. It is clear that the shape of the strain rate profile is very much a function of the flow variable against which it is plotted. Also shown are the three mean strain rates given by:

$$\bar{\dot{\varepsilon}}_{\varepsilon} = \frac{1}{\ell nR} \int_{o}^{\ell nR} \dot{\varepsilon} \, d\varepsilon \qquad (10)$$

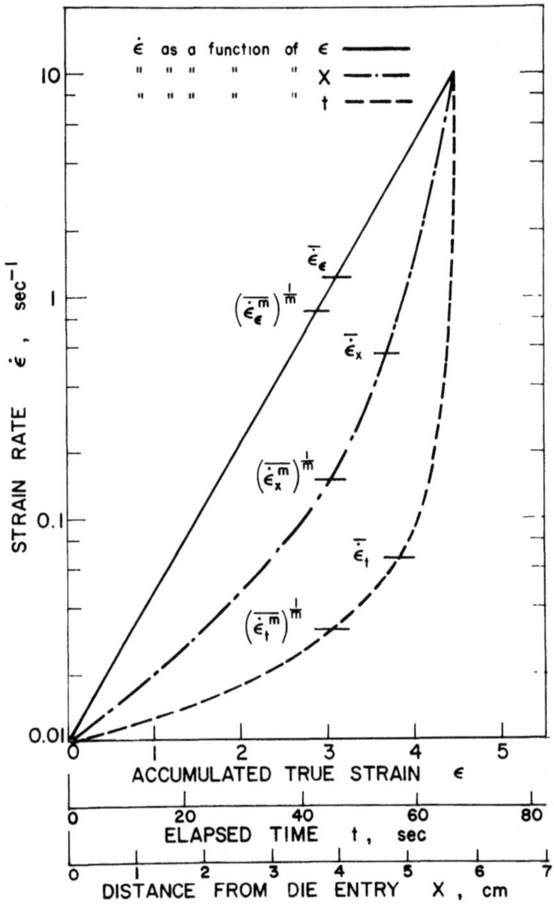

Fig. 3. The local or instantaneous strain rate $\dot{\varepsilon}$ within the deformation zone as a function of accumulated strain (ε), distance from the die entry (x), and elapsed time (t) (10). Also shown are the mean strain rates $\bar{\dot{\varepsilon}}_{\varepsilon}$, $\bar{\dot{\varepsilon}}_{x}$, and $\bar{\dot{\varepsilon}}_{t}$, and the three equivalent RMP strain rates. Note that a log scale is used for the strain rate ordinate and that the local strain rate varies by three orders of magnitude within the die zone. (R = 100, die semi-angle = 45°, ram speed = 1.91 cm/min, container diameter = 12.7cm.)

$$\bar{\dot{\varepsilon}}_{x} = \frac{1}{x_f} \int_{o}^{x_f} \dot{\varepsilon}\, dx \tag{11}$$

$$\bar{\dot{\varepsilon}}_t = \frac{1}{t_f}\int_0^{t_f} \dot{\varepsilon}\, dt \qquad (12)$$

It can be seen that $\bar{\dot{\varepsilon}}_\varepsilon > \bar{\dot{\varepsilon}}_x > \bar{\dot{\varepsilon}}_t$, and it can readily be shown that the above inequality is, with one exception, a general one. The exception is the Constant True Strain Rate (CTSR) die (9), in which the die angle decreases sufficiently with distance through the die to compensate for the increase in local velocity due to the reduction. For such a die, $\dot{\varepsilon}_\varepsilon = \dot{\varepsilon}_x = \dot{\varepsilon}_t$.

THE MEAN STRAIN RATE IN EXTRUSION

The mean strain rate used in extrusion calculations is usually $\bar{\dot{\varepsilon}}_t$ (Eq. 12), where the mean is taken with respect to time. Such a choice is rigorously correct only for CTSR dies, or for materials undergoing small reductions. For example, in cold forming operations such as wire drawing,the "extrusion ratio" is less than 2,so that the error introduced by using $\dot{\varepsilon}_t$ instead of the appropriate mean strain rate is small.

The mean strain rate applicable to extrusion calculations can be derived as follows. It should be recalled that

$$P_{ext} = \int_0^{\ell nR} \sigma_i \; d\,\varepsilon \qquad (7)$$

In order to evaluate the above integral, the function σ_i (ε) must be known. Actually, the flow stress under hot working conditions is given by $\sigma_i = \sigma_i\ (\varepsilon,\ \dot{\varepsilon},\ T)$, so that a full description of σ_i requires knowledge of the variation of both temperature and strain rate with strain. In the present case, the effects of work hardening ($\sigma_i = \sigma_i\ (\varepsilon)$) and of temperature variation ($\sigma_i = \sigma_i(T)$) along the streamline are neglected, and only the rate sensitivity of the flow stress ($\sigma_i = \sigma_i\ (\dot{\varepsilon})$) will be considered.*

The rate sensitivity m varies slightly along the streamline,as it is somewhat dependent on the history of the element, and on the rate of change of strain rate. Hence it cannot be calculated from

* A more complete treatment, in which local variations in temperature are also considered, is outlined in the Appendix.

basic principles and no exact value can be given. However, recent work by Immarigeon (11) suggests that, under transient conditions, m is always less than the steady state rate sensitivity m, and that it is usually within about 10% of the latter. Thus, to a first approximation, m can be given by:

$$m \simeq m_s = \frac{\Delta \log \sigma_s}{\Delta \log \dot{\varepsilon}_s} \tag{13}$$

where σ_s and $\dot{\varepsilon}_s$ are the steady state flow stress and strain rate, respectively, as determined in hot compression for example. We now express σ_i in terms of the instantaneous or local strain rate $\dot{\varepsilon}_i$ as follows:

$$\sigma_i = K \dot{\varepsilon}_i^{\,m} \tag{14}$$

where K is a strength coefficient. (K is the flow stress at a strain rate of 1 $\sec^{-1}$ under steady state conditions.) On substitution in Eq. (7), the following expression is obtained:

$$P_{ext} = K \int_0^{\ell n R} \dot{\varepsilon}_i^{\,m} \, d\varepsilon \tag{15}$$

$$= K \, \overline{\dot{\varepsilon}_\varepsilon^{\,m}} \, \ell n R \tag{16}$$

$$= \bar{\sigma}_{ext} \, \ell n R \tag{17}$$

Here $\bar{\sigma}_{ext} = K \overline{\dot{\varepsilon}_\varepsilon^{\,m}}$ and $\overline{\dot{\varepsilon}_\varepsilon^{\,m}} = (1/\ell n R) \int_0^{\ell n R} \dot{\varepsilon}_i^{\,m} \, d\varepsilon.$

It can be seen from Eq. (16) that the extrusion pressure is actually proportional to the mean power strain rate $\overline{\dot{\varepsilon}_\varepsilon^{\,m}}$, not to a power of $\bar{\dot{\varepsilon}}_t$, $\bar{\dot{\varepsilon}}_\varepsilon$ or $\bar{\dot{\varepsilon}}_x$. Thus when extrusion pressures are estimated from hot compression data, the mean strain rate to be used is $\dot{\varepsilon}_{RMP} = \left\{ \overline{\dot{\varepsilon}_\varepsilon^{\,m}} \right\}^{1/m}$, i.e. the root mean power strain rate.

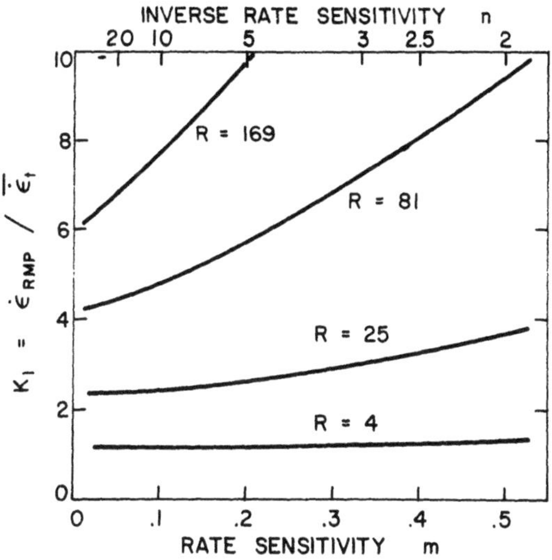

Fig. 4a. The variation in the rate sensitivity correction factor $K_1 = \dot{\varepsilon}_{RMP} / \bar{\dot{\varepsilon}}_t$ with rate sensitivity m.

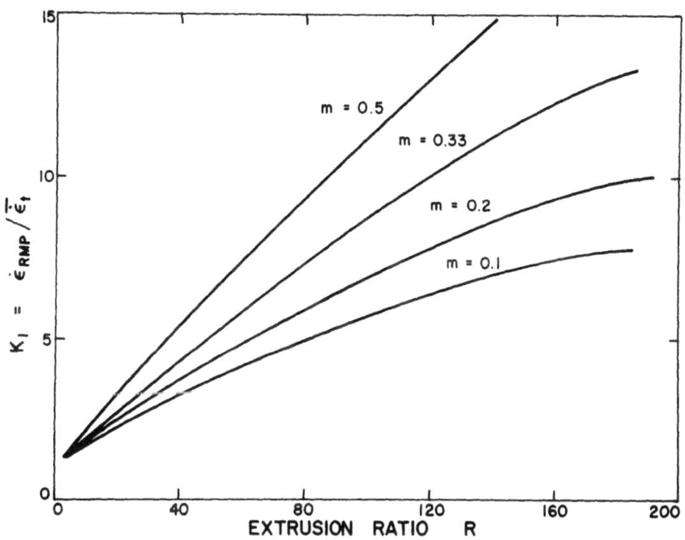

Fig. 4b. The variation in the rate sensitivity correction factor $K_1 = \dot{\varepsilon}_{RMP} / \bar{\dot{\varepsilon}}_t$ with extrusion ratio R.

It should be noted that, as in the case of the mean strain rate, the mean power strain rate can be evaluated with respect to strain ε, displacement x or time t (Fig. 3). The correct base for evaluation is strain, as can be seen from Eq. (15).

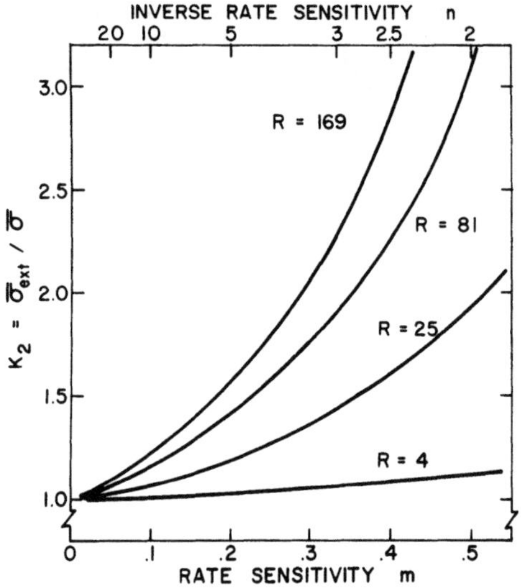

Fig. 5a. The variation in the rate sensitivity correction factor $K_2 = \bar{\sigma}_{ext} / \bar{\sigma}$ with rate sensitivity m.

One problem in the calculation of $\dot{\varepsilon}_{RMP}$ is that, unlike the other mean strain rates, the former is sensitive to the value of m, which varies somewhat from material to material. In Fig. 3, $\dot{\varepsilon}_{RMP}$ is indicated for m = 0.2, the most common value of m. It can be seen that $\dot{\varepsilon}_{RMP}$ is considerably greater than $\dot{\varepsilon}_t$. Thus pressure estimates based on $\dot{\varepsilon}_t$ will always be lower than those based on the more appropriate $\dot{\varepsilon}_{RMP}$.

RATE SENSITIVITY CORRECTION FACTORS

It is now of interest to consider the ratios $K_1 = \dot{\varepsilon}_{RMP}/\bar{\dot{\varepsilon}}_t$ and $K_2 = \bar{\sigma}_{ext}/\bar{\sigma}$, where K_1 and K_2 are the rate sensitivity correction factors for strain rate and flow stress, respectively. K_1 and K_2 represent the ratios by which mean strain rate and extrusion pressure estimates based on conventional analyses must be multiplied in order to allow for the effect of rate sensitivity in the die zone. For conical dies, it can be readily shown that K_1 and K_2 are given by:

$$K_1 = m^{1/m} (2/3 \ \ln R)^{1 + 1/m} [(R^{3/2} - 1)(R^{3m/2} - 1)^{1/m}] \ R^{3/2} \tag{18}$$

$$K_2 = m \ (2/3 \ \ln R)^{1 + m} \ [(R^{3/2} - 1)^m \ (R^{3m/2} - 1)] \ / \ R^{3m/2} \tag{19}$$

It should be noted that relations (18) and (19) are independent of

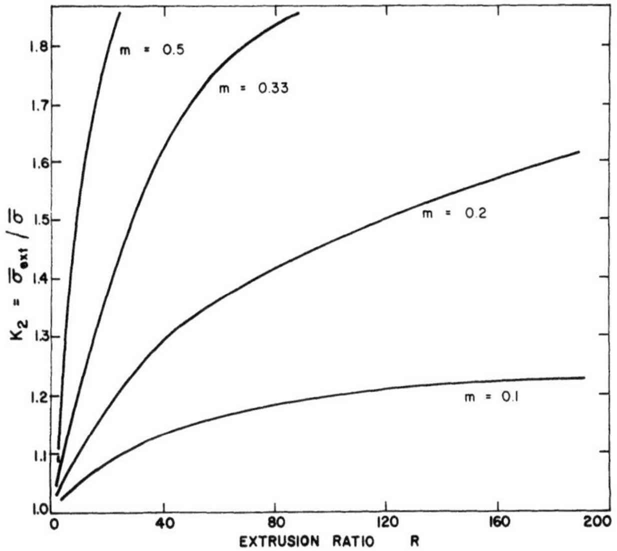

Fig. 5b. The variation in the rate sensitivity correction factor $K_2 = \bar{\sigma}_{ext} / \bar{\sigma}$ with extrusion ratio R.

die diameter and semi-angle and depend only on rate sensitivity m and extrusion ratio R. The dependence of K_1 on m and on R is shown pictorially in Figs. 4a and b. It can be seen that K is appreciable for all values of m and for all extrusion ratios greater than about 25:1. Of greater practical importance is the dependence of K_2 on m and R shown in Figs. 5a and b, as this is the correction factor which applies to working load calculations. In this case, the factor is appreciable for values of $m > 0 \cdot 1$ and for extrusion ratios greater than about 10:1. Thus the correction is important for all hot working operations of large reduction, e.g. extrusion and planetary hot rolling, but is not significant under cold working conditions, or for hot working at small reductions.

EXPERIMENTAL SUPPORT FOR THE RATE SENSITIVITY THEORY

Extrusion experiments carried out to distinguish between alternative flow stress and working load relations are difficult to perform. This is due to the considerable influence of friction, when present, and to the non-uniformities of strain that it introduces. These lead to appreciable contributions to the total work which are difficult to establish with accuracy, and which thus obscure the general trend of the results.

One solution to these difficulties is to extrude by the indirect process and so to eliminate container/billet friction. The

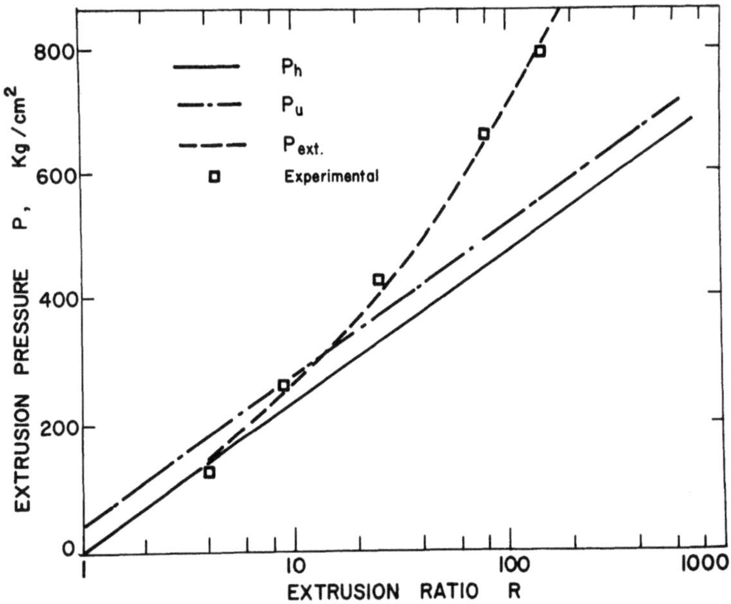

Fig. 6a. Experimental extrusion pressures determined at -10·0°C and $\bar{\dot{\varepsilon}}_t = 1{\cdot}1 \times 10^{-3}sec^{-1}$ compared with the predictions of the homogeneous work theory (P_h), the upper bound theory (P_u) and the present rate sensitive theory (P_{ext}) (10).

remaining source of friction is at the die/billet interface and it can be minimized by the use of suitable lubricants. Particularly low coefficients of friction can be obtained by using ice as the experimental material, in which case water is the lubricant and is formed as a result of friction melting at the interfacial asperities.

Two series of such experiments have recently been completed by the present authors (10), and the results obtained will now be described briefly. In the first series, ram speeds were selected to give the same value of the conventional mean strain rate $\bar{\dot{\varepsilon}}_t$ with each of five dies. The dies were of 45° semi-angle, and had extrusion ratios of 4, 9, 25, 81 and 144. This led to a series of ram speeds in which high extrusion ratios were coupled with low speeds, and vice versa. The constant mean strain rate chosen was $1{\cdot}1\text{x}10^{-3}$ sec^{-1}, which was low enough to avoid the effects of adiabatic heating. The experiments were performed at -10·0°C, so as to avoid pressure melting of the ice. According to the conventional analyses, such a series of experiments should involve a constant mean flow stress $\bar{\sigma} = K\ (\bar{\dot{\varepsilon}}_t)^m$, and a set of extrusion pressures proportional to ℓnR, as given by Eq. (5). However, analysis based on the rate sensitive theory suggests that the extrusion pressures should increase <u>more</u> rapidly than with ℓnR.

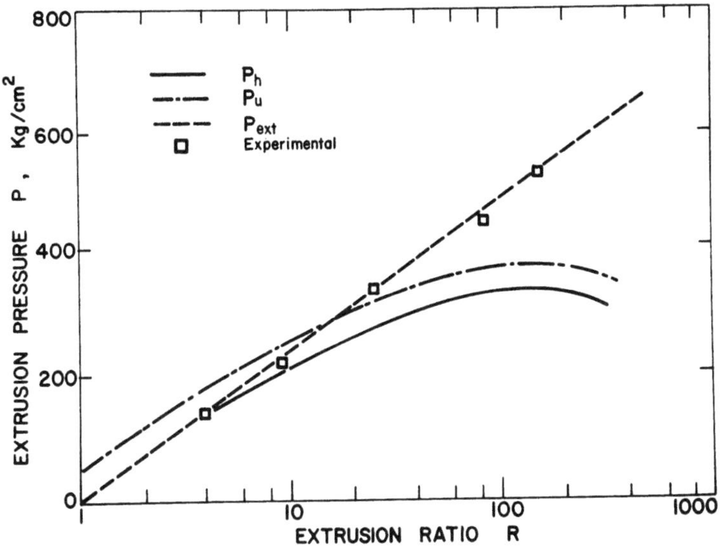

Fig. 6b. Experimental extrusion pressures determined at $-10{\cdot}0^{o}C$ and $\dot{\varepsilon}_{RMP} = 1{\cdot}1 \times 10^{-3}\ sec^{-1}$ compared with P_h, P_u and P_{ext} (10).

In the second series of experiments, the RMP strain rate, $\dot{\varepsilon}_{RMP}$, was set equal to $1{\cdot}1 \times 10^{-3}\ sec^{-1}$. As before, high extrusion ratios were coupled with low ram speeds, but the actual values of the ram speeds required to keep $\dot{\varepsilon}_{RMP}$ constant varied in a different way, and were lower than in the first series.* The second set of ram speeds was calculated from the RMP strain rate equation (9) for m = 0·2, the approximate rate sensitivity of ice in the present experimental range (12). According to the rate sensitive theory, such a series of experiments should involve a constant mean flow stress $\bar{\sigma}_{ext} = K\ \overline{\dot{\varepsilon}_{\varepsilon}^{m}}$ and therefore a set of extrusion pressures proportional to ℓnR, as given by Eq. (8). By contrast, the application of the conventional analysis to the second ram speeds suggests that the extrusion pressures should increase less rapidly than with ℓnR. The two sets of experiments thus provide a critical test of the two theories.

The extrusion pressures developed in the two series of experiments are plotted against extrusion ratio in Figs. 6a and b. Also shown in the figures are the extrusion pressures predicted by the homogeneous work theory (P_h), the upper bound theory (13) for frictionless extrusion (P_u), and the present rate sensitive theory (P_{ext}). It is evident from the diagrams that P_{ext} begins to deviate from P_h at an extrusion ratio of 4, and that the deviations increase markedly with R. However, in the range of extrusion ratios between 4 and

* Both sets of ram speeds are listed in reference 10.

25, P_u is fairly close to P_{ext}. Thus, when experiments are carried out at ratios below 25, it is difficult to distinguish between the upper bound and rate sensitive theories, and experiments at higher R's are required. In the investigation described, the excellent agreement at extrusion ratios greater than 25 between the experimental results and the predictions of the rate sensitive theory can clearly be seen, and thus confirms the necessity for making rate sensitivity corrections.

APPENDIX

Extension to Include Effect of Temperature Variation along the Streamline

The experiments described above were carried out at strain rates low enough to avoid the effects of adiabatic heating. No corrections were therefore made for temperature changes during flow. Under industrial conditions, however, considerable heating occurs, and the present method of calculating extrusion pressure must be modified accordingly. This can be done as follows.

It must first be recalled that, under steady state conditions, the hot working variables of flow stress σ, strain rate $\dot{\varepsilon}$, and temperature T are related by equations of the form:

$$\dot{\varepsilon} = f(\sigma) \exp(-Q/RT) \qquad \text{(A-1)}$$

The stress term $f(\sigma)$ may be of power $(A_1\sigma^n)$, exponential $(A_2\exp(\beta\sigma))$, or power hyperbolic sine form $[A_3\sinh(\alpha\sigma)]^n$, where the A's and n, β and α are constants (14). It must now be assumed that Eq. (A-1) applies at least approximately to conditions when the temperature or strain rate, or perhaps both are changing. Then, for the power law case, and when a Constant True Strain Rate (CTSR) die is used, Eq. (A-1) can be solved for σ to give:

$$\sigma = C \exp(D/T) \qquad \text{(A-2)}$$

where C and D are constants, and are equal to $(\dot{\varepsilon}/A_1)^{1/n}$ and (Q/nR), respectively. The analogous expression for the exponential law is:

$$\sigma = E + F/T \qquad \text{(A-3)}$$

where the constants E and F are equal to $(1/\beta)\ln(\dot{\varepsilon}/A_2)$ and $(Q/\beta R)$, respectively. These expressions can be substituted for σ in Eq. (7), and a suitable integration can be performed as long as the temperature distribution within the die zone can be measured or estimated. Such a calculation leads to a definition of the "mean temperature" which is similar in principle to that of the RMP strain rate described above. When the strain rate and temperature both vary during flow through the die, the complete form of Eqs. (A-2) and (A-3), in which $\dot{\varepsilon}$ is a variable,must be used. If data concerning the rate of temperature increase is lacking, it can of course be calculated for each increment of deformation by assuming that a particular fraction of the full adiabatic temperature increase is produced under the extrusion conditions.

ACKNOWLEDGMENTS

The authors are indebted to the Defence Research Board of Canada for Financial Support under Grant No. 9511-73, to D.J.Delamotte for assistance in some of the extrusion experiments, and to Professor F. Muller for his encouragement and support.

REFERENCES

1. A.R.E. SINGER and J.W. COAKHAM: J. Inst. Metals, 1960-61,89,177.
2. G. GAGNON and J.J. JONAS: Trans.TMS-AIME, 1969, 245, 2581.
3. C.E. PEARSON and R.N. PARKINS: The Extrusion of Metals,Chapman and Hall Ltd., London, 1960.
4. W.A. WONG and J.J. JONAS: Trans. TMS-AIME, 1968, 242, 2271.
5. O. SHERBY and P. BURKE: Prog. Mater. Sci., 1968, 13, 324.
6. A.K. MUKHERJEE, J.E. BIRD and J.E. DORN: Trans. Am. Soc. Metals, 1969, 62, 155.
7. E.G. THOMSEN, C.T. YANG and S. KOBAYASHI: Mechanics of Plastic Deformation in Metal Processing, Macmillan, New York, 1965.
8. W.A. BACKOFEN: J. Metals, 1961, 13, 206.
9. T. CHANDRA and J.J. JONAS: Met. Trans., 1970, 1, 2079.
10. T. CHANDRA and J.J. JONAS: Met. Trans., 1971, 2, in press.
11. J.P.A. IMMARIGEON: M.Eng.Thesis,McGill University,Montreal,1970.
12. T. CHANDRA, F. MULLER and J.J. JONAS: To be published.
13. B. AVITZUR: Metal Forming: Processes and Analysis, McGraw-Hill Book Co., New York, 1968.
14. J.J. JONAS, C.M. SELLARS and W.J. McG. TEGART: Metallurgical Reviews, 1969, 14, 1.

DEFORMATION CRITERIA FOR PREDICTING THE COLD-EXTRUSION PRESSURES OF METALS

L. J. Kashar, United States Steel Corporation, Monroeville, Pa.

R. W. Dunlap, Carnegie Mellon University, Pittsburgh, Pa.

T. E. O'Connell, formerly with United States Steel Corporation, now at the Carnegie Mellon University

ABSTRACT

Equations have been derived previously to predict cold-extrusion pressures from material constants and parameters representative of the extrusion process. To test the validity of these equations over a wide range of conditions, data from backward-extrusion and compression tests obtained on pure lead, aluminum, copper, pure iron, an iron-manganese alloy, and six steels are presented here. The data were obtained over a range of extrusion ratios (1.28 to 2.72) and extrusion speeds (0.005 to 1020 inches/minute). The results confirm the validity of the equations, establishing the fact that the deformation energy, taken as the area under the true-stress—true-strain curve, is the parameter that determines cold-extrusion pressures.

The strain factor parameters used to characterize the extrusion process were significantly different for low speed ($\leq$ 2 in./min) and high speed (960 in./min) extrusion. This reflects a change in the deformation pattern with extrusion speed and is shown to result from heating effects associated with high-strain-rate deformation. Transmission electron microscopy of irons and steels extruded at high speeds show evidence of considerable heating—in the case of pure iron extruded at 986 in./min, the substructure was completely recovered. Predictions of extrusion pressures from theories that neglect this adiabatic heating effect, such as slip-line solutions, may thus be in considerable error.

INTRODUCTION

The performance of a material in a cold-extrusion operation depends largely on its resistance to plastic deformation and the extent to which it can be deformed without fracture. Obviously, these two properties are closely related to the composition and morphology of the material. The factor limiting the cold extrusion of steel (by far the most common material used for cold extrusion) is its high resistance to deformation. In general, alloying elements that exist in solid solution in ferrite increase the resistance of steel to deformation more than elements that form carbides. However, the shape and distribution of the carbides can markedly affect the ease and extent to which steels can be deformed. For example, when large deformations are involved, spheroidized microstructures are preferred and, sometimes, mandatory.

The high resistance of steel to cold deformation results in extremely high extrusion pressures. These pressures place stringent requirements on the tool and die materials. In the backward extrusion or cups, the punches are subjected to the highest pressures developed, at times over 300,000 psi, and, consequently, punch life is very short.

Even if a part can be cold-forged from steel at a pressure below the fracture strength of the tooling, fatigue failures of the tooling are still a major problem. Considering the high production volumes and the high-speed cyclic loading entailed in the cold extrusion of steel, the prevalence of fatigue failures is not surprising. Because of the semilogarithmic relation between the cyclic stress level and the number of cycles for failure of most materials, a small decrease in extrusion pressure can result in a major increase in the fatigue life of the tooling. Thus, even at tool loading below the fracture strength, the extrusion pressure is a major factor in the economics of the cold-forging process. Therefore a method of predicting the extrusion pressure is necessary before an assessment of the economics of the cold extrusion process can be made for a particular application.

By equating the energy involved in uniaxially straining a material in compression to the energy needed to move the extrusion punch during extrusion of the same material, equations for the extrusion pressure, P_E, and punch pressure, P_P, have been derived:[1,2)*]

---

*See References.

$$P_E = \frac{K\,(a + b\,\ell n\,R)^{n+1}}{n+1} \qquad (1)$$

and

$$P_p = \frac{K\,(a + b\,\ell n\,R)^{n+1}}{n+1} \left(\frac{R-1}{R}\right) \qquad (2)$$

where R is the extrusion ratio (the ratio between the original and final cross-sectional areas); a and b are strain factors that reflect redundant and nonhomogeneous work effects;* and K and n are the parameters from Ludwik's stress-strain relation ($\sigma = K\varepsilon^n$).

Thus, to calculate the extrusion pressure needed to cold-forge a part, the following information is required: (1) the true stress—true strain parameters, K and n, of the material being extruded (under the strain rate and temperature conditions existing in the operation); (2) the extrusion ratio; and (3) the strain factors, a and b, which depend primarily on die geometry and friction.

To verify the equations completely, this information was obtained for several different materials at several widely different strain rates. Backward extrusion of cups was chosen as the extrusion process to be investigated. In this mode, lubrication between slug and container was not a critical factor determining the extrusion force (as it would have been in forward extrusion and heading); the extrusion ratio could also be easily varied by changing the punch diameter. Use of the series of punches shown in Figure 1 allowed investigation of this extrusion process at R values from 1.28 to 2.72 for each punch velocity studied. The punch and die designs were the same as those described previously.[1,2]

The punch velocities studied were limited by the availability of equipment with the capability (in terms of force and stroke) to make extrusions. For this reason much of the investigation was performed at speeds (0.005, 0.10 and 2.0 in./min) considerably below those of commercial interest (50 to 1000 in./min); however, the data obtained at these speeds were of great value in testing the hypotheses proposed in developing the extrusion equations. In addition to the slow speed extrusion tests, a considerable amount of data was obtained at a punch velocity of 960 in./min and some information was also obtained at intermediate extrusion speeds (4.0 to 600 in./min).

---

* That is, the actual average strain, $\bar{\varepsilon}$, undergone by the material is given by $a + b\,\ell n\,R$.

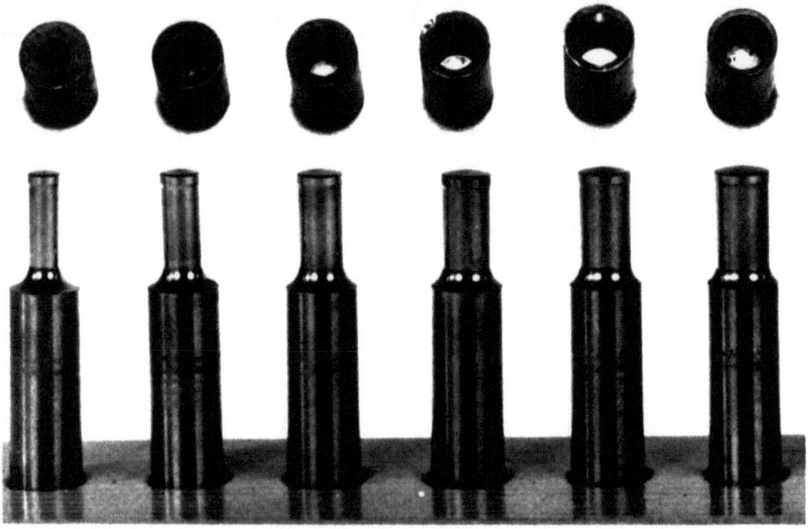

Figure 1. Extruded cups and backward extrusion punches for various extrusion ratios.

Because sufficient data were available at punch velocities of 0.005, 0.10, 2.0 and 960 in./min to warrant a full evaluation of the extrusion equations, the true stress—true strain parameters, K and n, were obtained for each material from compression tests made with the cross-head speed identical to the punch velocity for the comparable extrusion tests.

For a non-work-hardening material, Johnson assumed that the strain factors were independent of the material being extruded, and verified this assumption with a limited set of experiments on lead and aluminum.[3,4] The strain factors, a and b, cannot be derived easily from extrusion or compression test results for a work-hardening material. Because the applicability of the strain factors (and, therefore, of the derived equation) rested on the assumption that the strain factors were independent of strain rate and material, a thorough systematic experimental evaluation of the strain factors was required. By rearranging the terms in Equation 1, the following expression is obtained:

$$\left[\frac{P_E(n+1)}{K}\right]^{\left(\frac{1}{n+1}\right)} = a + b\,\ell n\ R \qquad (3)$$

Then, by using the K and n data from compression tests and the P_E data from extrusion tests (at a series of R values for each material at each cross-head speed), a least-mean-squares linear regression analysis can be made to determine the best values of a and b for each material at each punch velocity. A statistical analysis of the variances of a and b can then be used to test whether these factors are independent of material and punch velocity.

Finally, with the strain factors developed from this procedure and with the K and n values for the materials investigated, the validity of Equations 1 and 2 can be evaluated by comparing predicted extrusion and punch pressures with experimentally measured pressures both for the materials used to obtain the strain factors and for other materials, not employed to develop the strain factors.

MATERIALS AND TEST PROCEDURES

Materials

Because steel is the most difficult to form of those materials commonly cold-forged, an ability to predict the extrusion pressures of steels has great practical value. For this reason the experimental investigations were centered on various steel grades, including 4140 steel (a medium-carbon alloy steel) and several low-carbon steels (both 1008 and 1018) made by various steelmaking practices. To investigate single-phase materials, pure iron, an iron-manganese alloy, and copper were included in the study. An aluminum alloy was also studied and commercial-purity lead was included for historical reasons.

The pure iron, iron-manganese alloy, and six steels used in this investigation, Table I, were obtained from coils of 0.365-inch-diameter wire, supplied by the U. S. Steel Corporation.[5] All the wire had been spheroidize-annealed, coated with zinc phosphate, lubricated with Bonderlube, and cold-drawn with a 5 percent reduction of area to the final size.

Commercial-purity lead (chemical-burning-bar grade) was purchased from the National Lead Company as a coil of 3/8-inch-diameter rod. The chemical composition of the lead is also shown in Table I. Short lengths of the rod were straightened and swaged at room temperature to 0.350 inch diameter.

The aluminum alloy and copper were obtained as 1/2-inch-diameter cold-drawn rods. They were centerless ground to 0.365-inch-diameter and processed similarly to the iron and steels.

Table I

Chemical Composition of Materials Investigated—Weight Percent

Material	C	Mn	P	S	Si	Cr	Mo	Total Al
Pure Iron*	0.007	0.01	0.004	0.004	0.011	0.022	0.005	0.005
Fe-Mn Alloy*	0.007	0.38	0.005	0.004	0.015	0.021	0.005	0.002
1008 VMC*	0.058	0.40	0.018	0.018	0.013	0.002	0.005	0.044
1008 OH**	0.072	0.44	0.006	0.023	0.038	0.023	0.029	0.023
1018 VCD**	0.18	0.74	0.016	0.020	0.21	0.085	0.005	0.037
1018 OH**	0.16	0.83	0.010	0.019	0.23	0.026	0.030	0.035
1018 BOP VCD****	0.20	0.72	0.009	0.013	0.092	0.025	0.011	0.021
4140 OH**	0.37	0.83	0.018	0.022	0.25	0.96	0.21	0.057
	Pb	Fe	Cu	As	Sb	Sn	Ag	Bi
Lead*****	99.92	0.004	0.06	0.01	0.01	0.001	0.002	0.015
	Fe	Mn	Si	Cu	Al	Pb	Mg	
Copper	0.05	0.001	0.001	bal.	0.016	nil	0.002	
Aluminum Alloy	0.21	0.45	0.54	3.87	bal	0.17	0.70	

* Laboratory vacuum-melted and cast irons and steels.
** Commercial open-hearth steels.
*** Laboratory vacuum-carbon-deoxidized steel.
**** Commercial basic-oxygen, vacuum-carbon-deoxidized steel.
***** Commercial-purity lead.

All specimens used for the extrusion and compression tests were prepared in the following manner: steel slugs were cut and surface-ground to a length of 0.500 inch, then lubricated with Molykote (a molybdenum disulfide grease), and coined in an 0.369-inch-diameter die at a pressure of 100 ksi; lead slugs were cut and faced on a lathe to a length of 0.500 inch, lubricated with Molykote, and coined at a pressure of 10 ksi.

Experimental Apparatus

Testing Machines. Both the extrusion tests and the compression tests at a given velocity were made on the same machine. The load-stroke relation for each test was recorded in X-Y fashion either on a high-speed chart connected to a universal testing machine (for the tests at cross-head speeds from 0.005 to 2.0 in./min) or by photographing the screen of a high-gain, dual-beam oscilloscope for the tests at cross-head speeds in excess of 2.0 in./min conducted on either a Krafft Tensile Tester or an MTS unit.

Compression Test Procedure. The compression tests were conducted at the same cross-head speeds as the extrusion tests; because the specimens were 0.500 inch high, the initial strain rates, in inches/inch/minute, were numerically twice the cross-head speed (or punch velocity), in inches/minute. During the test, as the specimen height decreased, the strain rate increased. It is believed that the strain rate in the extrusion tests were about the same as the average strain rate in the corresponding compression tests.

Triplicate tests were usually made for each material at each initial strain rate. The load-strain data obtained from each of these tests were analyzed to give the true stress and true strain at each point chosen from the compression curve, and the elastic components of strain were subtracted. By using Ludwik's relation and transposing the true stress—true strain data to logarithms, the best values of the material parameters were obtained from a least-squares fit of the following equation:

$$\log \sigma = \log K + n \log \varepsilon \quad (4)$$

The K and n values for each material and initial strain rate were then obtained by arithmetically averaging the values obtained from the individual tests.

Table II

Extrusion-Test Results

Extrusion Pressure, ksi

Material	Extrusion Ratio, R							Punch Velocity, in./min
	1.28	1.35	1.49	1.73	2.00	2.34	2.72	
Pure Iron	-	47.2	58.7	76.0	101.2	123.4	135.6	
1008 VMC	-	51.4	66.6	87.0	113.1	130.9	152.0	
1008 OH	-	59.7	70.6	99.6	131.4	161.8	187.5	
1018 VCD	-	72.0	90.2	127.6	138.4	174.8	196.4	0.005
1018 OH	-	72.9	92.0	119.0	146.0	184.0	212.0	
4140 OH	-	86.6	111.5	143.8	170.2	207.6	252.9	
Lead	-	1.6	2.0	2.6	3.2	3.8	4.5	
Pure Iron	-	54.2	65.2	85.3	105.7	125.8	142.6	
Fe-Mn Alloy	-	59.4	74.8	99.2	113.1	156.9	159.0	
1008 VMC	-	53.8	69.7	95.8	124.4	137.9	165.5	
1008 OH	-	63.2	81.9	110.3	130.4	157.6	188.9	0.10
1018 VCD	-	76.7	94.4	128.6	150.1	180.0	214.1	
1018 OH	-	77.6	97.3	129.5	156.2	187.0	215.1	
4140 OH	-	94.4	116.9	152.4	191.7	247.8	271.2	
Lead	-	2.1	2.8	3.6	4.5	5.3	6.0	
1018 OH	-	83.2	101.3	134.5	160.6	192.0	222.0	0.50
Lead	-	2.7	3.4	4.4	5.1	5.9	6.9	
Pure Iron	-	55.6	69.2	87.8	109.9	128.6	154.3	
1008 VMC	-	60.7	78.3	99.9	119.5	148.9	173.9	
1008 OH	60.7	-	83.0	111.3	134.2	161.8	188.9	2.0
1018 VCD	-	83.7	107.1	133.3	161.8	196.4	217.4	
1018 OH	72.8	-	100.0	131.2	158.0	196.0	212.0	
4140 OH	-	98.4	123.5	156.0	191.7	222.8	259.2	
Lead	-	3.0	3.8	5.0	6.0	7.0	7.9	
Pure Iron	-	66.4	84.7	105.2	124.4	145.8	149.6	
Fe-Mn Alloy	58.4	-	82.3	103.2	125.3	137.7	155.2	
1008 OH	65.3	-	94.0	114.3	141.5	161.1	182.4	960
1018 OH	-	84.2	104.3	126.2	149.6	182.3	192.6	
4140 OH	-	103.8	125.5	157.1	185.1	215.4	242.1	
Lead	-	5.6	6.7	8.5	9.6	11.8	12.6	

Table III

Extrusion-Test Results

Punch Pressure, ksi

Material	Extrusion Ratio, R 1.28	1.35	1.49	1.73	2.00	2.34	2.72	Punch Velocity, in./min
Pure Iron	-	182.1	178.4	180.0	202.4	215.5	214.4	
1008 VMC	-	198.4	202.6	206.1	226.3	228.6	240.3	
1008 OH	-	230.3	214.7	236.0	262.8	282.5	296.5	0.005
1018 VCD	-	277.7	274.4	302.5	276.8	305.4	310.5	
1018 OH	-	281.2	279.8	282.0	292.0	321.3	335.3	
4140 OH	-	334.0	339.1	340.7	340.4	361.4	400.0	
Lead	-	6.3	6.1	6.1	6.3	6.6	7.1	
Pure Iron	-	209.2	198.3	202.2	211.3	219.6	225.5	
Fe-Mn Alloy	-	229.0	227.5	232.7	226.3	274.0	251.4	
1008 VMC	-	207.4	211.8	227.1	248.7	240.9	261.7	
1008 OH	-	243.6	248.9	261.5	260.9	275.1	298.7	0.10
1018 VCD	-	295.8	287.2	304.7	300.2	314.3	338.6	
1018 OH	-	299.4	295.7	306.9	312.3	326.6	340.1	
4140 OH	-	364.3	355.4	361.2	383.4	432.7	428.8	
Lead	-	8.1	8.4	8.5	9.0	9.2	9.4	
1018 OH	-	303.0	296.0	303.5	315.0	335.5	354.0	0.50
Lead	-	10.3	10.2	10.3	10.2	10.3	11.1	
Pure Iron	-	214.6	210.0	208.0	219.7	224.5	244.0	
1008 VMC	-	234.4	238.1	236.9	239.0	260.0	275.0	
1008 OH	270.4	-	252.4	263.7	268.4	282.5	298.7	2.0
1018 VCD	-	322.8	325.6	315.8	323.5	342.9	343.8	
1018 OH	326.5	-	304.1	310.9	316.0	342.3	335.3	
4140 OH	-	379.6	377.7	369.6	383.4	384.4	407.8	
Lead	-	11.4	11.4	11.8	11.9	12.3	12.5	
Pure Iron	-	256.1	257.3	249.3	248.8	254.6	236.6	
Fe-Mn Alloy	266.3	-	250.2	244.7	250.6	240.5	245.5	
1008 OH	298.4	-	285.7	270.8	282.9	277.9	288.3	960
1018 OH	-	324.6	317.0	299.2	299.2	318.4	304.6	
4140 OH	-	400.4	382.6	372.3	370.3	372.5	382.0	
Lead	-	21.6	20.5	20.3	19.3	20.6	20.0	

Nonlinearity in this relationship has been noted and discussed by other investigators;[6,7] however, the variations involved are sufficiently small to be insignificant for engineering purposes. However, it will be shown in the discussion that the nonlinearity can be predicted from Equation 1.

For all materials tested, increasing the extrusion ratio from 1.35 to 2.72 increased the extrusion pressure by a factor of about three. The effect of the extrusion ratio on the punch pressure was not as pronounced, and was also more complex than the effect of extrusion ratio on extrusion pressure. The variation of punch pressure over the range of extrusion ratios ranged from 8.1 percent for 4140 OH steel at 960 in./min to 38.1 percent for 1008 OH steel at 0.005 in./min; the average variation was about 15 percent.

The punch-pressure data usually exhibited a minimum in the extrusion-ratio range considered. At the lower punch velocities, 0.005 and 0.10 in./min, the punch-pressure minimum tended to occur at an extrusion ratio of 1.49 or less. As the punch velocity was increased, the position of the minimum tended to move to higher extrusion ratios. At a punch velocity of 2.0 in./min, the minimum was found usually at 1.49 or 1.73. For the tests at 960 in./min, the minimum punch pressure occurred at an extrusion ratio of 2.00 or higher. In the discussion, a calculation that predicts both the position and movement of this minimum is presented.

Effect of Punch Velocity. Figures 4 and 5 illustrate the general relationship between extrusion pressure and punch velocity for pure iron and 4140 steel at different extrusion ratios. For punch velocities up to and including 2.0 in./min, the extrusion pressures either remain constant or increase at a low linear rate with the logarithm of the punch velocity. The pressure increase in this velocity range averaged over all materials and all extrusion ratios, is about 12 percent.

Increasing the punch velocity from 2.0 to 960 in./min introduces effects that depend both upon extrusion ratio and the material being extruded. At low extrusion ratios, the extrusion pressures at 960 in./min are higher than at 2.0 in./min. At high extrusion ratios, the extrusion pressures at 960 in./min are lower than at 2.0 in./min. The extrusion ratio which separate these two phenomena depends upon the material being extruded; materials that are more difficult to extrude exhibit decreasing extrusion pressure at lower extrusion ratios.

To obtain more detailed information on the effects of punch velocity on the cold-forging pressures, a series of extrusions were made at closely spaced velocity intervals in the range 2 to 960

EXPERIMENTAL RESULTS

Extrusion Tests

Reproducibility of the Extrusion-Pressure Results. To obtain some information on the reproducibility of the extrusion results, several tests were replicated from two to six times; because sufficient material was not available, all other extrusion conditions were tested only once. The standard deviation of the replicate extrusion pressures for any one set of conditions (material, punch velocity, and extrusion ratio) ranged from 0.0 to 5.3 percent of the extrusion pressure, with the average variation being 1.5 percent. This indicates that the results of the extrusion tests are highly reproducible, and that the process is highly controlled.

Effect of Extrusion Ratio. The extrusion pressures and punch pressures obtained for the irons and the steels at the various extrusion ratios and punch velocities are given in Tables II and III. Typical relationships between the punch pressure or extrusion pressure and the extrusion ratio are shown in Figures 2 and 3 for pure iron and 4140 OH steel. The relationship between extrusion pressure and the logarithm of the extrusion ratio was approximately linear, although some scatter from linearity in the data was present.

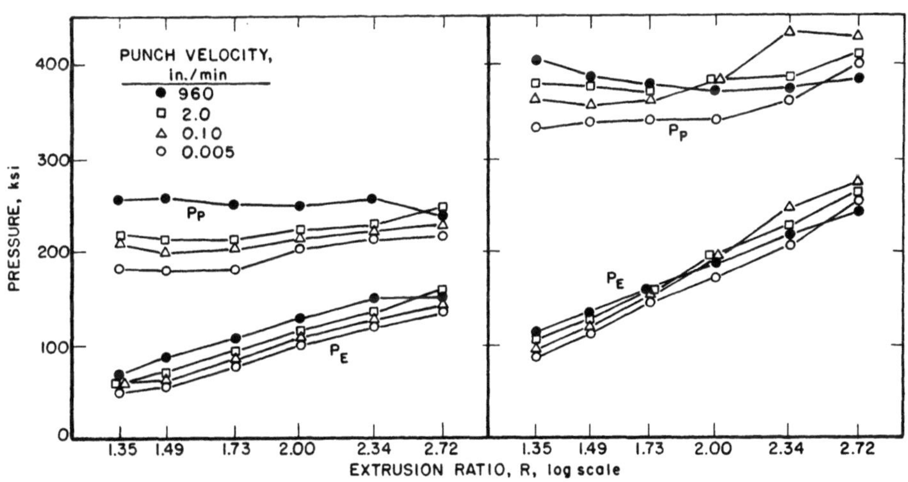

Figure 2. Effect of R on P_P and P_E of pure iron.

Figure 3. Effect of R on P_P and P_E of 4140 OH steel.

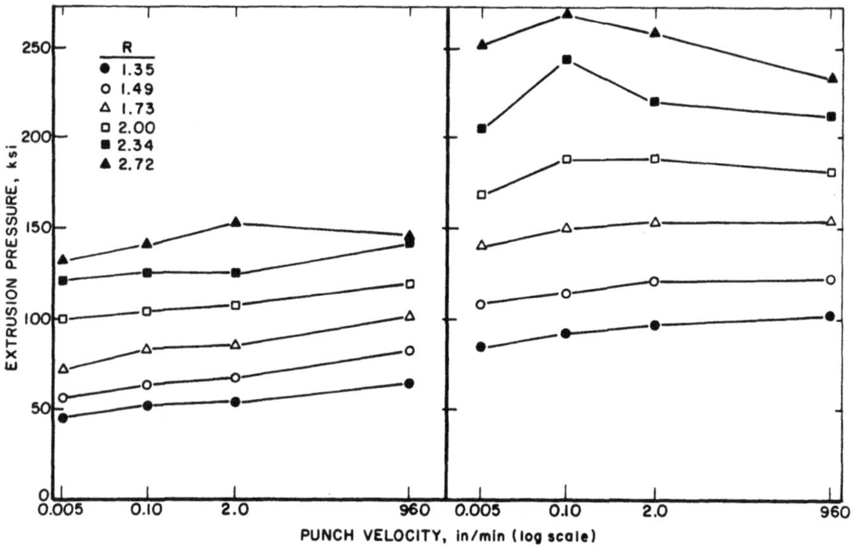

Figure 4. Effect of punch velocity of P_E of pure iron.

Figure 5. Effect of punch velocity of P_E of 4140 OH steel.

in./min with 1018 BOP VCD steel. The results of these tests, illustrated in Figure 6, again demonstrate the complex nature of this effect. At very low or very high punch velocities, the extrusion pressure increases with velocity. However, pressure minimums are found at punch velocities between 10 and 100 in./min at R = 2.00 and at R = 2.72. The magnitude of the decrease in extrusion pressure which occurs between 2 in./min and approximately 50 in./min increases with extrusion ratio.

The complexities in the extrusion pressure-punch velocity relationship are probably associated with the thermodynamic state of the process. At low punch velocities, the deformation rate is slow enough to permit complete dissipation of the heat of deformation, i.e., the process is isothermal. At high punch velocities, the deformation rate does not permit dissipation of the heat of deformation during extrusion, and the process is adiabatic. Thus, a considerable temperature increase occurs during high-speed extrusion. This hypothesis is analyzed more completely in the discussion.

Morphology of the Extrusions. The effects of extrusion ratio and punch velocity on the cold-forging pressures for the irons and steels should be reflected in the structure of these materials. The following section describes the studies of the optical macrostructures and microstructures of the extrusions (in which the extrusion ratio has a major effect, but no punch velocity effect

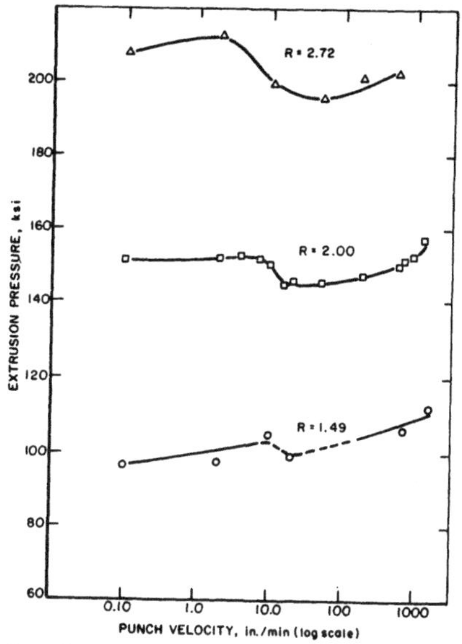

Figure 6. Effect of punch velocity on the extrusion pressure of 1018 BOP VCD steel.

could be found) and the studies of the transmission electron microscopy of thin foils (in which a major effect of punch velocity is observed).

To study the as-extruded structures, the extrusions were mounted in an epoxy-type mounting material to avoid any unnecessary heating of the sample, sectioned longitudinally on a diametral plane, and polished metallographically. The macro-structures of the extrusions examined in this way show essentially no effect of extrusion velocity.

However, the optical microstructure of the pure-iron extrusions, Figure 7, show that the deformation is not homogeneous through the wall of the extruded cups, and that the inhomogeneity is more accentuated at low extrusion ratios. The intensity of the striations near the inside surface of the cups does not seem to be affected by the extrusion ratio; however, the intensity of the deformation near the outside surface of the cups is strongly affected by the extrusion ratio. This effect of extrusion ratio on microstructure is observed for all materials examined in the study.

Optical microscopy does not indicate any effect of punch velocity, as illustrated by the 1018 OH steel extrusions, Figure 8. Because of the heating effect present during high-speed

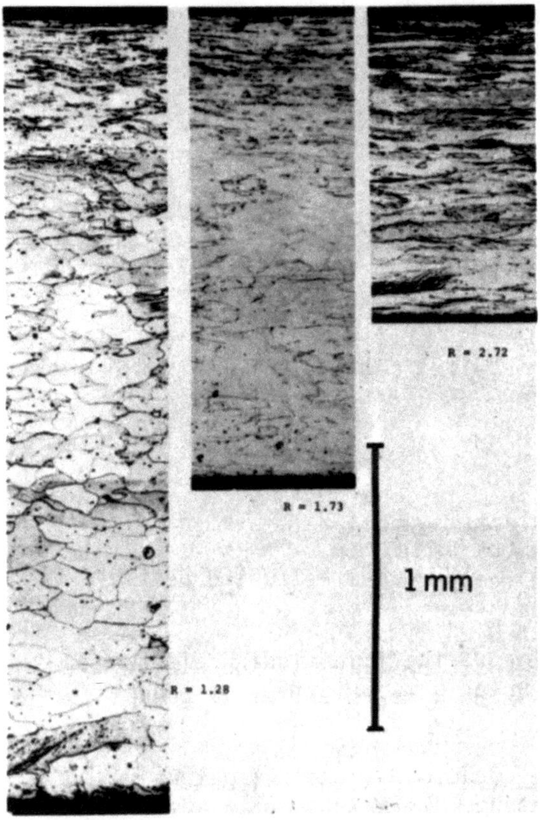

Figure 7. Effect of extrusion ratio on the microstructure of pure iron, showing complete cross-section of extruded wall (inside surface of cup is at the top of each composite). Nital etch.

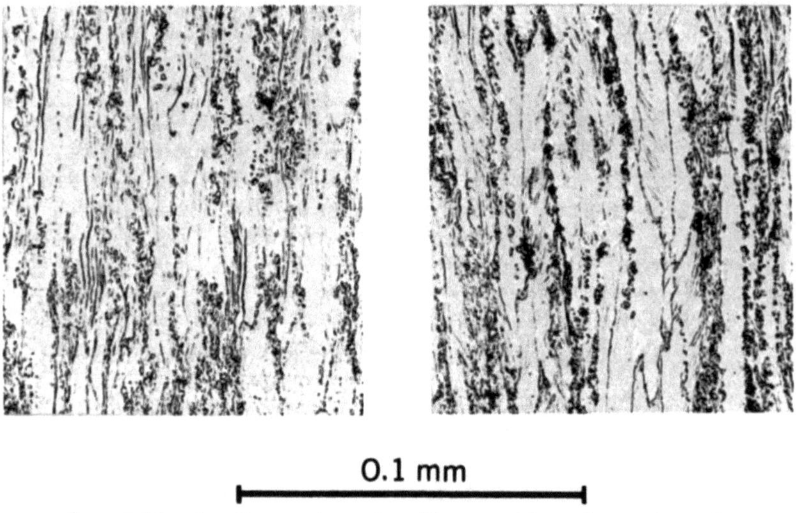

Figure 8. Effect of punch velocity on the microstructure of 1018 OH steel extruded at R = 2.0. Nital etch.

extrusion, a microstructural difference was expected between high- and low-speed extrusions; however, the microstructures at both speeds appear fibrous and "cold worked." A similar result has been observed under light microscopy with commercially pure aluminum that had been extruded both above and below its recrystallization temperature.[8] However, transmission electron microscopy of those aluminum extrusions revealed the presence of subgrain boundaries that became less ragged and thinner as the extrusion temperature increased.[8]

To establish the effect of punch velocity on the morphology of the iron and steel extrusions, electron transmission microscopy was used to study their dislocation substructures. Sections were taken for electron transmission microscopy from unextruded slugs and from the midheight of extruded cups made at R = 2.0 and punch velocities of 0.1 and 960 in./min. In addition, the substructure of some extrusions at other R values were studied.

To examine the material that had been the most severely worked during extrusion (near the inner wall), the outside portions of the extruded wall were removed by hand grinding. The specimens were then mechanically polished to approximately 2 mils thickness and finally were thinned by an electrochemical polishing technique (the Bollman technique) until pinholes developed.

From this examination, it appears that pure iron extruded at 0.1 in./min (R = 2.0) contains a very highly tangled dislocation cell structure (Figure 9). The cells are elongated with an axial ratio of about 2, the major axis approximately 1 micron in length. However, the dislocation structure of pure iron extruded (R = 2.0) at 960 in./min shows that the heating effect in the high-speed extrusions causes the cell walls to collapse into sub-boundaries, Figure 10. Similarly, the dislocation substructure of a commercial extrusion (R = 2.0, V_p = 986 in./min) made from this same material, Figure 11, shows no evidence of a highly tangled dislocation cell structure. The structure contains many subgrains and appears to be fully recovered, an indication that considerable heating occurred during extrusion.

Similar evidence of the effects of adiabatic heating during deformation has also been found in the substructures of the other materials studied in this investigation. However, as expected, the presence of alloying elements and second phase particles in the other materials greatly reduces the rate of recovery.

Compression Tests

Reproducibility of Compression Tests. An analysis of the data from the compression tests shows that the maximum percent

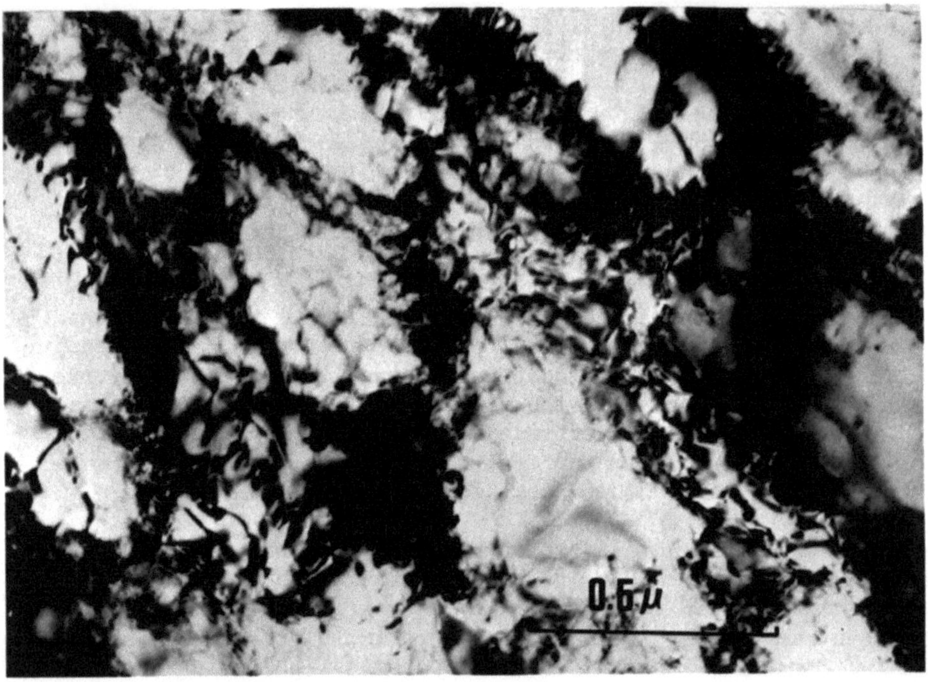

Figure 9. Substructure of pure iron extruded at R = 2.0, V_p = 0.1 in./min.

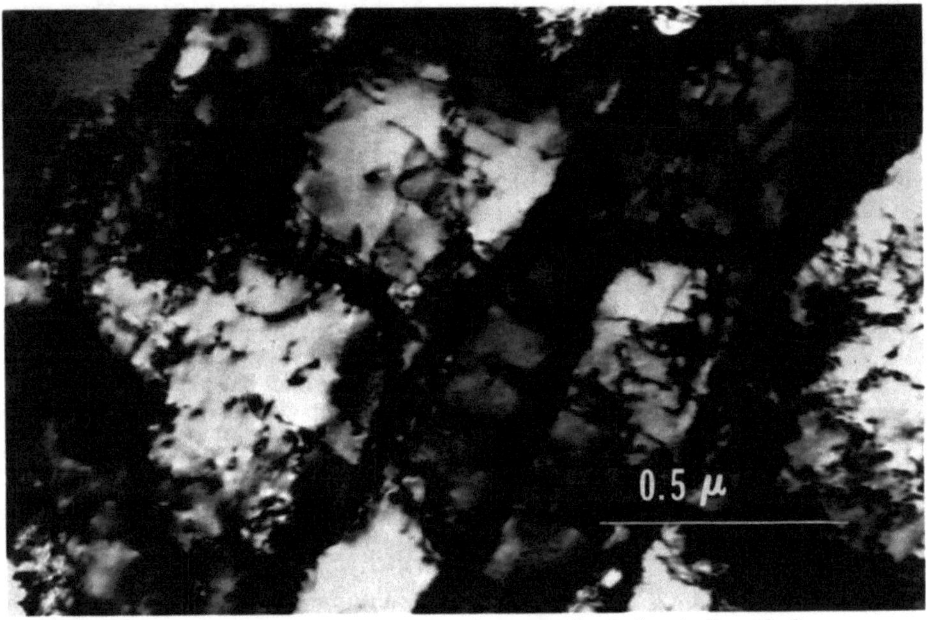

Figure 10. Substructure of pure iron extruded at R = 2.0, V_p = 960 in./min.

Figure 11. Substructure of commercially extruded pure iron; $R = 2.0$, $V_p = 986$ in./min.

standard deviation in strength coefficient, K, among the replicate tests for any one material at a given strain rate is 6.5 percent (for lead at 0.20 min^{-1}); the average standard deviation for all materials and strain rates is about 2.4 percent.

The reproducibility of the work-hardening exponents obtained from the replicate tests must be discussed in terms of actual rather than percentage variations because, for several cases, the values of the exponents are close to zero. The standard deviations do not seem to be affected by the magnitude of the work-hardening exponent. The maximum standard deviation in n is 0.057 (for lead at 1920 min^{-1}), but the average standard deviation is only 0.013, indicating that the reproducibility of the work-hardening exponents is quite good.

Effect of Initial Strain Rate on K and n. The effect of strain rate on the strength coefficients of the irons and steels is not pronounced (Figure 12). Increasing the strain rate (up to $\dot{\varepsilon} \sim 4.0$ min^{-1}) tends to increase the strength coefficient, but a further increase in strain rate to 1920 min^{-1} causes a decrease in K value. The effect of the strain rate on the strength coefficient of lead is extremely pronounced, almost doubling the value of K from $\dot{\varepsilon} = 0.01$ to 4.0 min^{-1} and again almost doubling the value from $\dot{\varepsilon} = 4.0$ to 1920 min^{-1}.

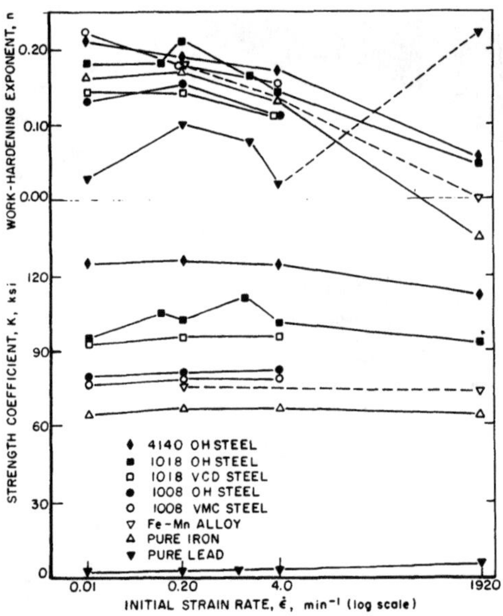

Figure 12. Effect of strain rate on the work-hardening exponent and strength coefficient of lead, irons, and steels.

The strain rate has pronounced effects on the work-hardening exponents, Figure 12. For the irons and steels, n decreases with strain rate, and is close to zero at 1920 min^{-1}. For lead, the effect of $\dot{\varepsilon}$ on the work-hardening exponent is exactly opposite; at low $\dot{\varepsilon}$, n is near zero, but at 1920 min^{-1} lead exhibits a marked work hardening.

VERIFICATION OF THE EXTRUSION EQUATION

Effect of Punch Velocity and Material on Strain Factors

With the use of Equation 3, a linear regression (least-mean-square) analysis was made of each data set for each material at each cross-head speed to obtain the applicable values of a and b, Table IV. Although Johnson assumed that his strain factors were independent of punch velocity and material,[3,4] the a priori assumption must be made that the velocity and material can affect the strain factors, parameters that are determined by the deformation pattern. To determine the effect of punch velocity and material on the strain factors, an analysis of variance of each of the strain factors was made, using standard statistical methods and the "F" test.[9] Data were used from only those materials (pure iron, 1008 OH, 1018 OH, 4140 OH, and lead) for which strain

Table IV

Strain Factors Obtained From Linear Regression Analyses

Punch Velocity (in./min)	Material	Strain Factors "a"	"b"
0.005	Pure Iron	0.29	1.92
	1008 VMC	0.35	1.73
	1008 OH	0.15	2.25
	1018 VCD	0.39	1.79
	1018 OH	0.33	1.94
	4140 OH	0.37	1.70
	Lead	0.27	2.31
0.10	Pure Iron	0.42	1.79
	Fe-Mn Alloy	0.40	1.89
	1008 VMC	0.31	1.88
	1008 OH	0.33	2.00
	1018 VCD	0.36	1.90
	1018 OH	0.43	1.76
	4140 OH	0.31	1.94
	Lead	0.30	1.96
2.0	Pure Iron	0.40	1.92
	1008 VMC	0.34	1.88
	1008 OH	0.32	2.01
	1018 VCD	0.44	1.92
	1018 OH	0.38	1.82
	4140 OH	0.46	1.69
	Lead	0.32	2.10
960	Pure Iron	0.46	2.01
	Fe-Mn Alloy	0.40	1.73
	1008 OH	0.43	1.64
	1018 OH	0.48	1.67
	4140 OH	0.49	1.70
	Lead	0.67	1.48

factors were available at all punch velocities (0.005, 0.10, 2.0, and 960 in./min), Table V.

The analysis shows that punch velocity has a significant effect on a at the 95 percent confidence level, whereas there is no effect of material on this parameter. Similarly, the analysis of variance of b shows that the material has no effect; the punch velocity, although not a significant factor at the 95 percent confidence limits, has a significant effect at the 90 percent confidence limits. Therefore, the two strain factors do not appear to be dependent on the material tested, but are affected by the punch velocity.

This velocity effect is believed to be related to the heating effect at high strain rates, since the main difference in strain factors is between the 960 in./min data and the data at slower velocities. To verify this observation, an analysis of variance was again made, this time using only the data in Table V at the slow punch velocities (0.005, 0.10, and 2.0 in./min). No effect of velocity was found for these data over the slow-velocity range. Therefore, in this range, the strain factors are independent of both material and punch velocity. Considering all the data in Table IV in this velocity range, the average values are $a = 0.35$ and $b = 1.91$.

A further analysis of the high-strain-rate data is revealing. Although no dependence on material is found for the strain factor data as a whole, a separate analysis of the 960 in./min data shows that the strain factors for the irons and steels are not the same as for lead. The strain factors at 960 in./min for lead are $a = 0.67$ and $b = 1.48$, whereas for the irons and steels the average values are $a = 0.45$ and $b = 1.75$. The difference in strain factors obtained for lead and for iron and steel at 960 in./min is probably a result of differences in adiabatic heating and the effects of this heating on mechanical properties.

Correlation of Predicted and Actual Extrusion Pressures

At Low Punch Velocities. Using the average strain factors at the low punch velocities ($a = 0.35$ and $b = 1.91$) and applicable values of K and n, predictions can be made from Equation 1 for the extrusion pressures of the various materials. The predicted extrusion pressures are plotted against the actual extrusion pressures for all the slow-extrusion-speed data, Figure 13. The slope of the correlation is 1.017 with 95 percent confidence limits of ± 0.008; the index of correlation (R x R) was 0.998. An analysis of the correlation shows that the actual extrusion

Table V

Analysis of Variance of Strain Factors

"a"

Material	Punch Velocity, in./min 0.005	0.10	2.0	960
Pure Iron	0.29	0.42	0.40	0.46
1008 OH	0.15	0.33	0.32	0.43
1018 OH	0.33	0.43	0.38	0.48
4140 OH	0.37	0.31	0.46	0.49
Lead	0.27	0.30	0.32	0.67

Source of Variance	Degrees of Freedom	Sum of Squares	Mean Square	"F" Test
Punch Velocity	3	0.129895	0.04329833	7.966
Material	4	0.027557	0.00688925	1.270
Error	12	0.065230	0.00543533	
Total	19	0.222695		

"F" test is significant

with 95% Confidence if $F_{3,12,0.05} \geq 3.49$

if $F_{4,12,0.05} \geq 3.26$

with 90% Confidence if $F_{3,12,0.10} \geq 2.61$

if $F_{4,12,0.10} \geq 2.48$

(Continued)

Table V (Continued)

"b"

Material	Punch Velocity, in./min 0.005	0.10	2.0	960
Pure Iron	1.92	1.79	1.92	2.01
1008 OH	2.25	2.00	2.01	1.64
1018 OH	1.94	1.76	1.82	1.67
4140 OH	1.70	1.94	1.69	1.70
Lead	2.31	1.96	2.10	1.48

Source of Variance	Degrees of Freedom	Sum of Squares	Mean Square	"F" Test
Punch Velocity	3	0.270095	0.090032	2.703
Material	4	0.154170	0.038543	1.157
Error	12	0.399630	0.033303	
Total	19	0.823895		

"F" test is significant

with 95% Confidence if $F_{3,12,0.05} \geq 3.49$

if $F_{4,12,0.05} \geq 3.26$

with 90% Confidence if $F_{3,12,0.10} \geq 2.61$

if $F_{4,12,0.10} \geq 2.48$

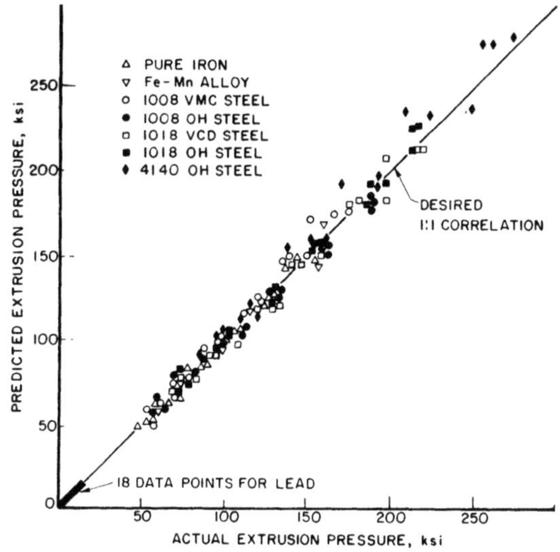

Figure 13. Correlation of predicted and actual extrusion pressures at low punch velocities.

pressure of a slowly extruded material can be predicted by Equation 1 with a standard deviation of 4.4 percent.

At 960 in./min. Using the strain factors developed for irons and steels at 960 in./min, a = 0.45 and b = 1.75, predictions can be made of the extrusion pressures of the irons and steels at 960 in./min, and using the strain factors for lead at 960 in./min, a = 0.67 and b = 1.48, similar predictions can also be made for lead. The correlation of the predicted and actual extrusion pressures for all materials at 960 in./min is shown in Figure 14. Once again the data shows good correlation; the slope of the correlation is 1.007 with 95 percent confidence limits of $\pm$0.012, and the index of correlation (R x R) is 0.998. The actual extrusion pressure of irons and steels extruded at 960 in./min can be predicted by Equation 1 with a standard deviation of 4.7 percent.

Significance of the Correlation. In the two preceding sections, it has been demonstrated that the extrusion pressures can be accurately predicted by Equation 1. It may be argued that this is not surprising because a transposed form of this equation was employed to obtain the constants (a and b) used in the prediction—obviously, the constants obtained will be such that the predicted pressures will be close to the actual values. However, it should be remembered that although all the data were pooled to obtain the

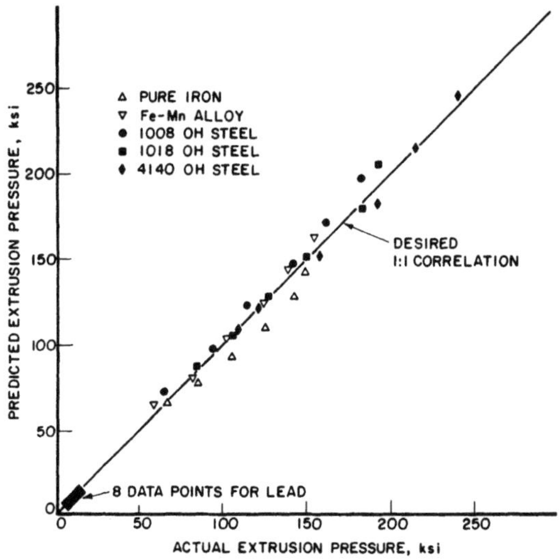

Figure 14. Correlation of predicted and actual extrusion pressures at a punch velocity of 960 in./min.

values of a and b, the values of a and b are not dependent upon material (except for lead at 960 in./min). This means that data for any material could have been excluded from the data pool and the strain factors obtained would have been the same. In addition, the extrusion pressures of materials excluded from the data pool can be accurately predicted by using these strain factors. To demonstrate that this is the case, predictions of extrusion pressures for two materials not previously studied in this investigation are now discussed.

Investigation of an Aluminum Alloy and Copper

Extrusion tests were conducted on the aluminum alloy and copper materials (Table I) at a punch velocity of 0.1 in./min and compression tests were conducted in triplicate at an analogous rate. The results of the tests are given in Table VI.

By using the strain factors previously found for lead, irons, and steels at low velocities ($a = 0.35$ and $b = 1.91$), the extrusion pressures of the aluminum alloy and copper were calculated and compared with the actual pressures (Figure 15). The pressures are predicted quite accurately; the slope of the least-mean-square

Table VI

Extrusion and Compression Test Results on an Aluminum Alloy and Copper

Extrusion Tests (V_p = 0.10 in./min)

Extrusion Ratio	Aluminum Alloy P_E, ksi	Aluminum Alloy P_p, ksi	Copper P_E, ksi	Copper P_p, ksi
1.28	63.6	290.7	42.1	192.4
1.73	116.0	272.6	76.2	179.2
2.00	133.3	266.5	90.2	180.5
2.34	154.8	269.4	106.1	184.8
2.72	170.7	268.2	116.9	183.7

Compression Tests ($\dot{\varepsilon}_i$ = 0.20 in./in./min)

	Hardness, Rockwell B	Strength Coefficient K, ksi	Work-Hardening Exponent, n
Aluminum Alloy	71.0	85.9	0.11
Copper	59.0	55.5	0.08

regression line is 1.050 with 95 percent confidence limits of ±0.042; the R x R of the correlation is 0.997.

This satisfactory prediction of extrusion pressures for copper and an aluminum alloy strongly confirms the significance of the correlations already discussed for lead, irons, and steels. These results show that the extrusion equations are valid for all materials and for extrusion conditions which do not produce fracture.

By using the correct strain factors in Equation 1, a correlation between predicted and actual extrusion pressures was made for all the materials and velocities included in this investigation. The least-mean-square regression line had a slope of 1.014 and 95 percent confidence limits of ±0.007. The correlation indicated that the actual extrusion pressure of all materials investigated can be predicted with a standard deviation of 4.6 percent.

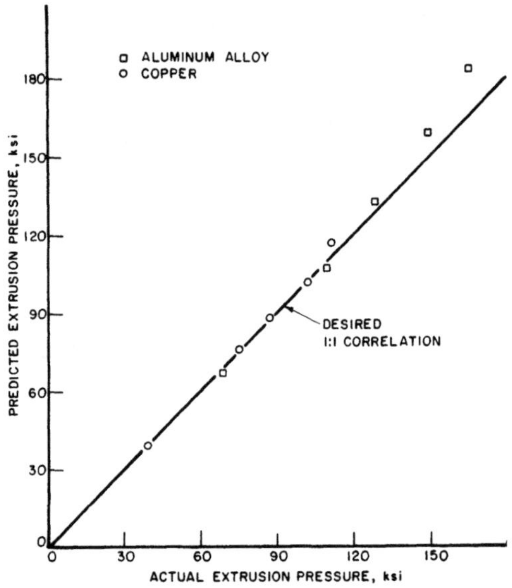

Figure 15. Correlation of predicted and actual extrusion pressures of copper and of aluminum alloy at 0.10 in./min.

The high degree of correlation obtained between the predicted and actual extrusion pressures confirms the validity of Equations 1 and 2 under the conditions investigated and establishes the deformation energy (taken as the area under the true stress—true strain curve) as the criterion that determines the extrusion pressure.

DISCUSSION

Effect of Punch Velocity on the Thermodynamics of the Extrusion Process

Temperature rise during extrusion depends upon the extrusion ratio, the extrusion velocity, and the material being extruded. At a constant extrusion ratio, when the extrusion velocity is too rapid to permit isothermal conditions, the temperature rise increases with increasing punch velocity until the extrusion occurs with sufficient speed so that no energy is dissipated to the surroundings during the extrusion process (that is, the extrusion process is adiabatic). At speeds above that needed for adiabaticity, the extrusion occurs at maximum temperature and an increase in pressure is again observed with increasing punch velocity, reflecting the effect of strain rate on the flow stress.

Singer and Al-Samarrai calculated[10] that, for lead and aluminum, extrusions made at a speed below 0.05 in./min would be completely isothermal and extrusions made above 100 in./min would be completely adiabatic. These velocity bounds were considered to be conservative, that is, the actual transition velocity range should be narrower, encompassed by the calculated limits.

The extrusion-pressure data shown in Figure 6 for 1018 BOP VCD steel support this hypothesis, and it would appear that, for this material, isothermal extrusion occurs at velocities below 2.0 in./min while adiabatic extrusion occurs at velocities above 50 in./min. The velocity bounds obtained on 1018 BOP VCD steel confirm the calculations of Singer and Al-Samarrai and substantiate the hypothesis that adiabatic heating is responsible for the unusual relationship between extrusion pressure and punch velocity.

Effects of Nonhomogeneous Deformation

The extruded microstructures and the strain factors, a and b, obtained in this investigation show that the deformation was inhomogeneous (values of a and b other than zero and one reflect redundant work and nonhomogeneous deformation). Because the deformation during high speed extrusion is both inhomogeneous and adiabatic, the temperature rise in the extruded section cannot be homogeneous. Therefore, the resistance to deformation varies from point to point in the extruded material, depending upon the local temperature and the effects of temperature on mechanical properties. Under such conditions, the strain factors could not be the same as those obtained for isothermal deformation. Hence the difference in strain factors for high and low velocity deformation found in this investigation is believed to be a result of the adiabatic heating effect.

This rationale also explains why the strain factors for high-speed extrusion are not independent of material. At these speeds, the deformation pattern is influenced by the temperature distribution (dependent upon the thermal diffusivity of the material) and the relationship between mechanical properties and temperature. Since the thermal diffusivities and the temperature dependences of mechanical properties are quite similar for the steels considered, the deformation patterns for steels should be similar. Hence, extrusion pressures could be predicted from Equation 1 by using average values of the strain factors derived from the steel data. Use of these values for lead, however, is not warranted when adiabatic heating occurs, because the thermal diffusivity and temperature dependence of mechanical properties are radically different for this material.

The temperature increase in high-strain-rate deformation was not homogeneous for the extrusions considered in this investigation and probably is not homogeneous for many other metal-working processes. As nonuniform temperature distributions cause nonuniform resistance to deformation in the work-piece, slip-line solutions would be invalid because the work-piece is assumed to be homogeneous in slip-line analysis.

In many severe deformation processes the deformation is almost homogeneous, resulting in uniform adiabatic heating; therefore, the slip-line solution will be valid although the elevated temperature properties must be considered in determining the deformation resistance. In light deformation processes, the heating would probably not be sufficient to significantly alter the mechanical properties, and therefore the slip-line solutions are applicable. However, in processes in which the deformation is both severe and nonhomogeneous, such as extrusion, the mechanical properties will not be homogeneous and slip-line solutions will be erroneous.

Corollaries of the Extrusion Equation

Several anomalies have been noted in previous investigations of the extrusion process. Although most investigators agree that the relationship between extrusion pressure and extrusion ratio is approximately semilogarithmic, both positive and negative deviations from this approximate relationship have been found.[7,11] By taking the derivative of Equation 1 with respect to ℓn R, the following equation is obtained:

$$\frac{d\ P_E}{d\ (\ell n\ R)} = Kb\ (a + b\ \ell n\ R)^n \qquad (5)$$

Equation (5) indicates that only for a non-work-hardening material will the relationship be semilogarithmic. Furthermore, the equation indicates that (1) positive deviations will occur if the extruding material work-hardens, or (2) negative deviations will occur if it "work-softens."

By differentiating Equation 2 with respect to R, the minimum in the punch pressure-extrusion ratio curve can also be determined. The extrusion ratio at which the minimum occurs satisfies the relationship:

$$R = \frac{a + b\ \ell n\ R}{b\ (n + 1)} + 1 \qquad (6)$$

Substituting the values in the table below into this equation,

Velocity (in./min)	a	b	n: Pure Fe	1018 OH	4140 OH
0.10	0.35	1.91	0.15	0.21	0.18
960	0.45	1.75	-0.05	0.05	0.05

minimum punch pressures are predicted to occur at the following extrusion ratios:

Velocity (in./min)	R Values for Minimum P_p: Pure Fe	1018 OH	4140 OH
0.10	1.49	1.43	1.46
960	2.08	1.88	1.88

The discussion of the experimental extrusion data has already pointed out that the position of the punch pressure minimum shifts from $R \simeq 1.4$ at low speeds to $R \simeq 2.0$ at high speeds, in general agreement with these predictions.

SUMMARY AND CONCLUSIONS

Compression and backward-extrusion tests were performed on pure lead, an aluminum alloy, copper, pure iron, an iron-manganese alloy, and six steels over a range of extrusion ratios (1.28 to 2.72) and extrusion speeds (0.005 to 1020 inches/minute) to test the validity of equations, developed from an energy balance, for predicting the extrusion pressure, P_E, and punch pressure, P_p:

$$P_E = \frac{K\,(a + b\,\ell n\,R)^{n+1}}{n + 1} \tag{1}$$

$$P_p = \frac{K\,(a + b\,\ell n\,R)^{n+1}}{n + 1} \left(\frac{R}{R - 1}\right) \tag{2}$$

Here, K and n are the strength coefficient and the work-hardening exponent of the material, respectively; a and b are strain factors, primarily dependent on the die design and punch velocity; and R is the extrusion ratio.

The extrusion pressures are found to be related in an approximately linear fashion to the logarithm of the extrusion ratio; the punch pressures generally exhibit a minimum at an intermediate extrusion ratio that appears to depend on the extrusion velocity.

The effect of extrusion speed on the cold-forging pressures is complicated by heating effects at speeds above 2 in./min. As a result of heating, the extrusion pressure exhibits a minimum at velocities of about 50 in./min. Whereas optical microscopy does not indicate any effect of heating, transmission electron microscopy of the extruded materials shows evidence of considerable heating—in the case of pure iron commercially extruded at 986 in./min, the substructure is completely recovered.

A statistical analysis of the strain factors, a and b, obtained for each material and extrusion condition shows that they are independent of both material and extrusion velocity at velocities up to 2 in./min ($a = 0.35$; $b = 1.91$). However, the strain factors at 960 in./min are different from those at the slow speeds, and the strain factors for lead at 960 in./min are different from those of irons and steels at 960 in./min. At 960 in./min, the strain factors for lead are $a = 0.67$ and $b = 1.48$ and for irons and steels are $a = 0.45$ and $b = 1.75$.

By using the correct strain factors, the extrusion pressures can be calculated with the extrusion-pressure equation (Equation 1) and are found to correlate satisfactorily with the actual extrusion pressures. The standard deviation for predicting the extrusion pressure for a material is approximately 4.6 percent.

Thus, the results of this study confirm the validity of the equations under the conditions investigated and establish the deformation energy (taken as the area under the true stress—true strain curve) as the parameter that determines the extrusion pressure. The difference in strain factors for fast and slow extrusions reflects a difference in deformation patterns in the extrusions under these two conditions. This difference results from heating during high-strain-rate deformation. Neglect of this effect, as in the case of slip-line solutions, may lead to considerable error for processes in which the deformation is not homogeneous.

By use of the extrusion equations, several details of the extrusion process are also explained. These include explanations of deviations from linearity of the $P_E - \ln R$ relation and the extrusion ratio at which $P_{p(min)}$ occurs.

ACKNOWLEDGMENTS

The authors acknowledge Dr. Joseph Krafft for his help in obtaining the high-strain-rate test data and Mr. H. J. Tata for his valuable discussions.

REFERENCES

1. L. J. Kashar, "Prediction of Extrusion Pressures in the Cold Forging of Steel," Trans. AIME, 239, 1461-1468 (1967).

2. L. J. Kashar, "Deformation Criteria in the Cold Extrusion of Metals," Doctorate Thesis, Carnegie-Mellon University, Pittsburgh, Pa., 1969.

3. W. Johnson, "Extrusion Through Square Dies of Large Reduction," J. Mech. and Phys. of Solids, 4, 191-198 (1956).

4. W. Johnson, "Experiments in Plane-Strain Extrusion," J. Mech. and Phys. of Solids, 4, 269-282 (1956).

5. H. J. Tata, H. E. Mueller, and T. Kobara, "Effect of Steel Composition and Processing on Punch Pressures in Extruding and Heading Steel Wire," Mechanical Working and Steel Processing VI, AIME, New York, 1969, pp 206-249.

6. H. D. Feldmann, "Cold Forging of Steel," Translated by A. M. Hayward, Hutchinson, London, 1961, pp 64-69.

7. J. McKenzie and A. R. Rodger, "The Extrusion of Metals—Part VIII—Steel Under Cold Impact Conditions," N.E.L. Plasticity Report No. 163, National Engineering Laboratory, 1959, East Kilbride, Glasglow.

8. W. A. Wong, H. J. McQueen, and J. J. Jonas, "Recovery and Recrystallization of Aluminum During Extrusion," J. Inst. Metals, 95, 129 (1967).

9. C. A. Bennett and N. L. Franklin, "Statistical Analysis in Chemistry and the Chemical Industry," John Wiley and Sons, Inc., New York, 1954, p. 377-379.

10. A. R. E. Singer and S. H. K. Al-Samarrai, "Temperature Changes Associated With Speed Variation During Extrusion," J. Inst. Metals, 89, 225 (1960-61).

11. E. Siebel and E. Fangemeier, "Untersuchungen uber den Kraftbedarf beim Pressen und Lochen," Mitt. K.W. Inst. fur Eisenforschung, 13, 29 (1931).

THE EFFECT OF HOMOGENEITY ON THE FORMABILITY OF 7000 SERIES ALUMINUM ALLOYS FOR CARTRIDGE CASES

S. A. Levy and C. Baker

Reynolds Metals Company

Richmond, Virginia

INTRODUCTION

Since World War II various attempts have been made to successfully substitute aluminum for brass in cartridge cases. Only partial success has been achieved; the main problem appears to be burn throughs which infrequently occur in the side wall of the cases. The nature of the failure is such that (i) severe damage occurs to the weapon upon burn-through and (ii) failure analysis is difficult. The general interpretation is that small tears occur during the cupping operation which are obscured by subsequent forming operations.

Considerable effort has been devoted recently to produce homogeneous versions of a number of high strength 7000 series alloys. Much of the theory behind the homogenization practice is due to the excellent work of Flemings and co-workers, (1-3) and Antes et al (4). Three statements summarize the findings of these studies: (i) The ability to decrease gradients of the soluble elements (Zn, Mg, Cu) is related to the dendrite arm spacing (DAS). In fact the time for complete homogenization varies as the square of the DAS. (ii) The DAS depends only on the freezing time, for a given composition. (iii) If an element is present in amounts exceeding the solubility limits, the resulting insoluble particles are essentially unaffected by the homogenization treatment. Thus, to produce homogeneous material it is necessary to freeze the casting rapidly, provide a homogenization time dictated by the resulting DAS, and maintain a low level of insolubles.

The major effect of obtaining highly homogeneous material is greater toughness. This has been demonstrated using percent elongation and percent reduction in area (4). Also, both the Kahn Tear

Test with sheet, and the compact tension specimen with plate (5) indicate substantially higher toughness. The basis for the percent study, therefore, was to attempt to produce an improved aluminum alloys, which by virtue of greater homogeneity, was less susceptible to tearing during fabrication and exhibited a larger critical flaw size for failure in the final age hardened condition.

The characteristics of the alloys were evaluated in both the "O" temper as they affect formability and the T6 and T73 tempers as they affect the service behavior.

EXPERIMENTAL PROCEDURE

The alloys were semicontinuously cast as 4" x 10" x 30" sections. Chemical analysis was conducted spectrochemically and the results are presented in Table I. The ingots were stress relieved overnight at 550°F and then a 23" long section was taken from the center of the ingot for homogenization. The homogenized treatment was performed in dry air (-40 dew point) and consisted of 24 hours at 860°F plus 48 hours at 910°F. The ingots were scalped to 3.25" and hot rolled to 0.300". For rolling the ingots were preheated to 800°F and the general procedure involved 1/4" reductions per pass. The pieces were reheated after 4 passes, or when the temperature dropped to 650°F. The sheets were cold rolled from 0.300" to 0.150".

Table I - Composition of the Experimental Alloys

Sample No.	Other Identification	%Si	%Fe	%Cu	%Mn	%Mg	%Cr	%Ni	%Zn	%Ti	%Zr
S-23879	MP7001	.06	.08	1.87	<.01	2.79	.18	<.01	7.14	.01	<.01
S-23880	MP7075	.05	.06	1.43	.01	2.44	.17	<.01	6.00	.01	<.01
S-23881	MP7178	.05	.06	1.59	.01	2.58	.17	<.01	6.72	.01	<.01
S-23882	HP7075	.02	<.01	1.44	<.01	2.52	.18	<.01	6.06	.01	<.01
S-23883	HP7178	.02	<.01	1.67	<.01	2.79	.17	<.01	6.81	.01	<.01
S-23884	HP7075 + Zr	.02	.01	1.40	<.01	2.45	<.01	<.01	5.99	.01	.12
S-23885	HP7178 + Zr	.02	.01	1.66	<.01	2.65	<.01	<.01	6.78	<.01	.12
S-23886	HP7001	.02	<.01	1.72	<.01	2.73	.18	<.01	7.46	.01	<.01
S-23888	CP7075	.08	.22	1.30	.03	2.32	.16	<.01	6.05	.01	<.01
S-23889	CP7178	.07	.21	1.58	.03	2.53	.15	<.01	6.74	.02	<.01
COMPOSITIONS OF ALLOYS USED TO COMPARE HOT ROLLED TO GAGE VS. COLD ROLLED											
S-26227	CP7075*	.11	.26	1.68	.04	2.43	.20	<.01	5.99	.06	<.01
S-22168	HP7075	.01	.01	1.57	<.01	2.64	.20	<.01	5.58	.03	<.01
S-23795	MP7178	.05	.06	1.39	<.01	2.60	.11	<.01	6.97	.03	<.01

* Plant Produced

Three treatments were performed on different section of the sheet:

A. 775°F for 4 hrs., cool 50°F/hr. to 450°F, hold 2 hrs. at 450°F, and air cool
B. 675°F for 2 hrs., cool 50°F/hr. to 450°F, hold 2 hrs. at 450°F and air cool
C. Solution heat treat 1 hr. at 860°F followed by 1 hr. at 900°F, cold water quench, age 9 hrs. at 450°F

Testing of a modification of practice B indicated better formability and the revised practice will hereafter be referred to as Standard Practice B. It involved:

(i) hold 2 hrs. at 675°F
(ii) cool 75°F/hr. to 475°F
(iii) hold 2 hrs. at 475°F
(iv) air cool

All further annealing for this study will involve only this practice. For "O" temper material the terms hot rolled or cold rolled to gage refer to the rolling practice prior to the final anneal.

Aging Practices

On the basis of aging curves produced for each alloy, two standard practices have been adopted. The material was solution treated for 24 hours at 900°F and cold water quenched. After incubating 5 days at room temperature, the "-T6" material was given 48 hours at 250°F. The "-T73" was given 8 hours at 225°F plus 28 hours at 325°F. A 25°F/hr. heating rate was used to attain all aging temperatures.

Tension Tests on "O" Temper

All of the alloys have been tested in tension in the three "annealed" conditions described above. Tension tests were performed on an MTS machine of 5000 lb. capacity using crosshead speeds of 0.12"/min., 1.2"/min., 12"/min. and 120"/min. These rates correspond to strain rates of approximately 0.06, 0.6, 6, and 60 minutes. Triplicate tests were run at each strain rate and load versus time and displacement versus time recorded using a Brush Oscillographic Recorder. The ram attained full speed before the elastic limit was reached and had a constant speed during testing.

Olsen Cup Tests

In order to rate the formability of the alloys, initially the Olsen cup height was determined. The tests were run on a Tinius Olsen Model 612 Ductomatic using a 7/8" diameter ball and 1-1/4" diameter die and a 2000 lb. hold down pressure. The ball was lubricated with mineral oil before each test and an automatic break detector was used to stop the test at failure.

Limiting Draw Ratio Tests (LDR)

To determine limiting draw ratio, a series of blanks were drawn, of varying diameter, and the drawing force recorded. As the diameter of the blank increased, the amount of friction produced by the hold down pressure increased, as did the drawing force. The limit of the drawing operation occurred when it became easier for the punch to shear through the sheet, than to draw the alloy into the die. The limiting draw ratio is then the limiting blank diameter divided by the punch diameter.

The LDR tests were performed with 0.150" thick sheet, given the Standard B anneal. The blank diameter was varied from 2.18" to 3.15". A hold down force of 500 lbs. was employed and the punch diameter was 1.300" (0.2" radius end). The die contained a 1.81" diameter hole and a 5/16" radius. Both sides of the disk were lubricated with palm oil prior to testing. The initial data were plotted as load versus blank diameter. One line was drawn through the data for the failed cups. A second line was drawn through the data for the successful cups. The limiting diameter was taken as the intersection of the two lines. This is shown schematically in Fig. 1.

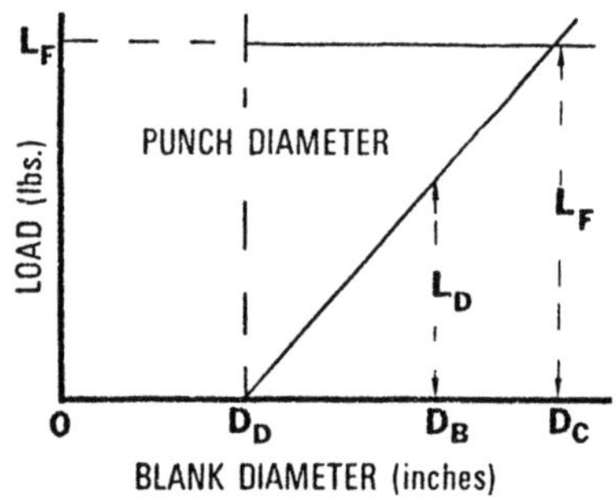

L_F = fracture load
L_D = drawing load
D_D = die diameter
D_C = critical diameter

Figure 1 - Schematic Curve of Drawing Load vs Blank Diameter

"U" Bends for Stress Corrosion Testing

The samples used for "U" bends were removed from 0.063" thick sheet. Half of the material was given the standard "T6" treatment and half the "T73" treatment. Sample blanks were sheared in the long transverse direction.

The cold worked edges, resulting from shearing were removed by pack machining 1/8" from each edge. The final dimensions of the samples were 0.75" x 0.063" x 9". After bending the samples 90° over a mandrel having a radius of 6 times the sample thickness, they were mounted in micarta fixtures. A typical stressed sample is shown in Fig. 2. Half the samples of each aging practice were tested in an industrial atmosphere and half in alternate immersion.

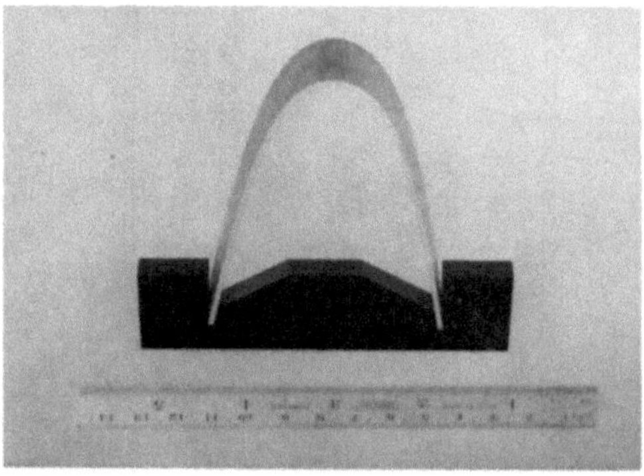

Figure 2 - "U" Bend Stress Corrosion Configuration

Kahn Tear Tests

Kahn Tear Test specimens of the dimensions shown in Fig. 3, were machined from 0.063" and 0.125" sheets. The specimens were pulled in an Instron Testing Machine with an integrator attachment, which records the area of the load elongation curve. The tear strength of the specimens was computed from the tensile component plus the bending moment of the load applied to initiate and propagage the crack to fracture.

$$\text{Tear Strength} = \frac{P}{A} + \frac{Mc}{I} = \frac{P}{bt} + \frac{3P}{bt} = \frac{4P}{bt}$$

P = applied load (lb.)
A = cross section area of the specimen (in.^{2})
M = moment (in. lb.)
C = distance from centroid to extreme fiber (in.)
I = moment of inertia (in.^{4})
b = width at the root of the notch (in.)
t = thickness (in.)

The energy to initiate and the energy to propagate the crack were computed from the areas under the load elongation cruve, as shown in Fig. 4. These values were obtained directly from the integrator which has two channels, one of which can be stopped at maximum load.

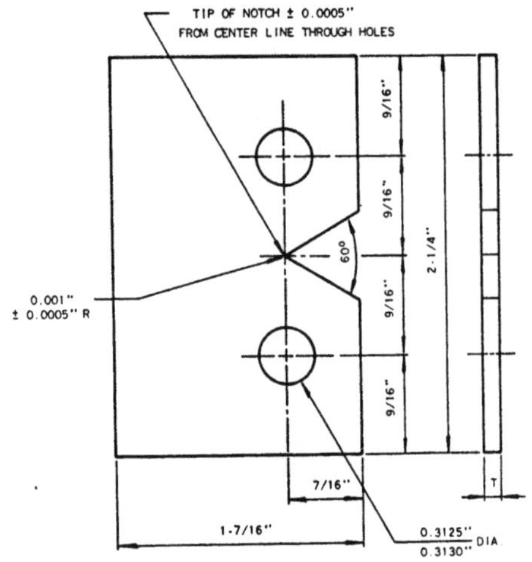

Figure 3 - Kahn Tear Test Specimen

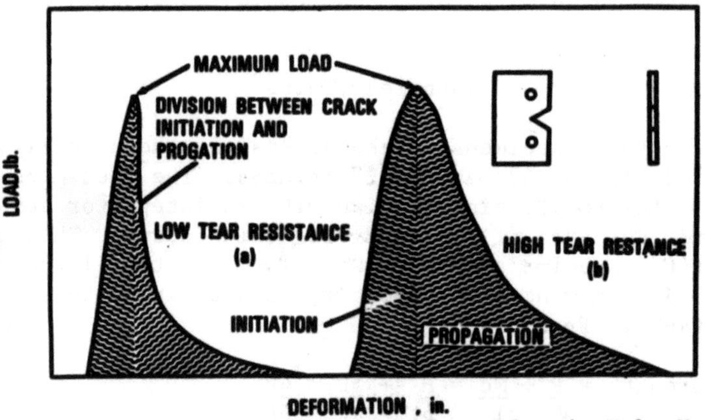

Figure 4 - Representation of Load Deformation in Kahn Tear Test

Tensile Tests - "T6" and "T73"

Tensile tests were conducted on an Instron Testing Machine. Strain rates of 6, 60 and 600 inches per minute were employed with 0.015" sample thickness. As a result of the thinness of the specimens, warping during quenching was encountered. A portion of the data was suspiciously low, and failure of samples outside the gage marks was also noted. The suspect samples will be retested in the future. Samples for tensile tests were also removed from the 0.125" and 0.063" sheet, employed in the toughness tests. These samples were taken from in the longitudinal direction.

RESULTS AND DISCUSSION

Tensile Tests on "O" Temper

The variation of the mechanical properties with strain rate was negligible. The elongation, about which little published data exists, shows no decrease with increasing strain rate. High strain rate cupping and ironing operations should correlate with slower laboratory tests reasonably well. The tensile strength and yield stress show no significant variation over the strain rate range of 0.06 to 60 minutes^{-1} in any of the annealing conditions tested. Typical values of tensile strength, yield strength and elongation for each alloy in the "O" temper is shown in Table II.

Table II - Tensile Data on "O" Temper Alloys

		Ann. @ 775°F			Ann. @ 675°F			SHT & Overaged		
Alloy	#	TS (ksi)	YS (ksi)	El. %	TS (ksi)	YS (ksi)	El. %	TS (ksi)	YS (ksi)	El. %
7075 CP	88	31.6	14.7	19.0	30.6	16.1	18.7	44.2	27.5	12.3
7075 MP	80	31.2	14.6	18.7	33.6	24.9	14.8	45.6	27.5	12.5
7075 HP	82	31.0	14.8	17 3	30.6	17.4	17.5	40.8	27.2	14.8
7178 CP	89	30.8	14.4	19.2	32.1	19.6	18.0	47.8	30.1	12.2
7178 MP	81	31.3	15.1	19.5	31.2	15.9	19.0	48.5	30.7	12.7
7178 HP	83	31.3	14.7	19.7	32.0	18.2	17.3	49.0	31.7	12.5
7001 MP	79	31.0	15.5	19.0	31.4	17.9	18.2	47.4	29.6	12.3
7001 HP	86	31.2	15.4	19.0	30.9	16.2	19.0	48.3	30.5	12.5
7075+Zr	84	37.6	22.2	15.5	30.8	21.1	15.7	48.0	31.6	12.3
7178+Zr	85	34.8	22.8	15.2	31.6	23.4	14.7	51.7	35.3	11.5

A multiple regression analysis of the tensile data was conducted. Two hundred and forty (240) data points and eleven variables were employed. The variables included composition (8 variables), strain rate, thermal treatment, and test direction. No strong correlation was found between any of the variables and the percent elongation. A strong linear correlation was, however, noted between the yield strength and the percent elongation, Fig. 5.

Examination of the data shows that the "C" condition shows considerably higher yield strength and tensile strength values and a lower elongation compared to the "A" and "B" conditions. The strengths of the individual alloys in the "C" condition show the expected order of strengths with the 7001 highest and the 7075 lowest except for the 7178 containing Zr which had the highest yield strength and lowest elongation of all of the alloys. An explanation for this high yield strength could be that this alloy was unrecrystallized and further tests will be made using longer heat treatments to put more zirconium into solution and recrystallize the alloys.

In the "A" and "B" conditions only the alloys containing zirconium (23884 and 23886) show a consistent difference. These alloys have a higher yield strength and tensile strength and a lower elongation compared to the remainder. One anomalous alloy is S23880 in the "B" condition. This alloy shows an inconsistently high yield strength and tensile strength with lowest elongation of any of the alloys in the "A" or "B" condition. The reason for this is not immediately apparent.

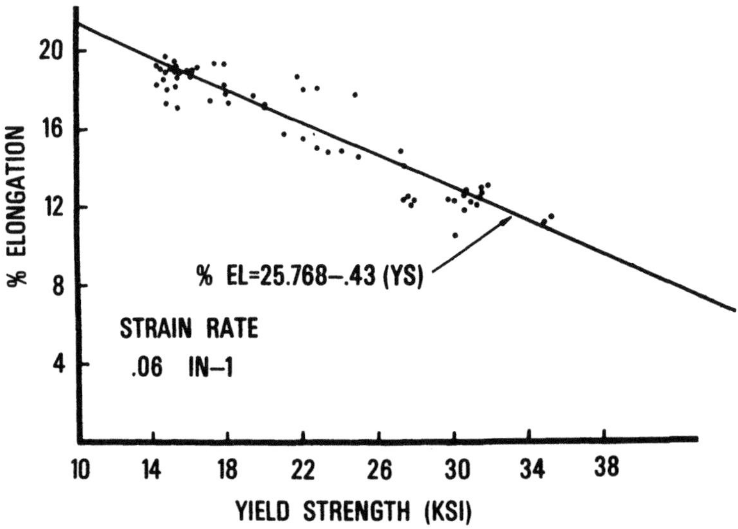

Figure 5 - Correlation between Yield Strength and Elongation for "O" Temper Alloys

Olsen Cup Tests

Table III shows the average cup height and drawing pressure (average of six tests) for each alloy and annealing condition. Fig. 6 shows a plot of the average Olsen cup height for each of the alloys together with the scatter band. As can be seen from the data, the overaged condition has a significantly lower cup height and a higher drawing pressure in all cases. Of the annealed conditions the 675°F anneal shows a greater cup height in all but one of the alloys tested, but the differences in most cases are less than 5%. The greatest differences occur in the zirconium-containing alloys S23884 and S23885. In both of these alloys, the 775°F annealed alloys show considerably lower Olsen cup heights. From the data it would appear that the B condition has better formability. Excluding the zirconium alloy, the differences in cup height between the remaining alloys is less than 5%. Again alloy 23880-B seemed to be an exception. A minor amount of testing was performed in which another section of this alloy was annealed with the B practice, except the cooling rate was increased from 50 to 75°F/hr. The values for the depth and the load were increased from .537 to .570 and 7600 to 8200 respectively. The interpretation, supported by microstructures, was that 50°F/hr. is bordering on a sufficiently slow cooling rate such

Table III - Olsen Cup Data

AVERAGE OLSEN CUP VALUES

Alloy	Olsen Cup Height (in.)	Load
79A	.564	7400
80A	.543	7500
81A	.578	7600
82A	.543	7500
83A	.568	7500
84A	.276	2200
85A	.378	5700
86A	.561	7000
88A	.497	6400
89A	.532	6800
79B	.576	7300
80B	.537	7600
82B	.570	6900
81B	.584	7400
83B	.573	7000
84B	.537	6900
85B	.510	6700
86B	.579	6700
88B	.567	6900
79C	.424	9300
80C	.493	9300
81C	.442	10100
82C	.519	9400
83C	.449	9900
84C	.492	9800
85C	.427	10200
86C	.378	9100
88C	.410	8600
89C	.366	8100

that a continuous network of lathe-like precipitates can occur in the grain boundaries.

The effect of Fe and Si on formability has not emerged in the current test program. It is felt that the cold rolling treatment provided recrystallization to a very fine grain size. The formability was considerably better than was previously experienced with similar alloys hot rolled directly to gage. All of the Cr-bearing alloys, irrespective of the level of Fe and Si, could be redrawn to 0.020" wall cups from the starting 0.150" sheet without intermediate anneals. This result was confirmed by tests on material hot rolled to 0.150 and annealed as shown below.

Limiting Draw Ratio - Cold Rolled Material

As was seen with previous cupping and tensile testing (1), the differences between different alloys and purity levels were very small. The test data are presented in Table IV. It is felt that the values of LDR, for the cold rolled material, are essentially the same within the range of experimental error and that a ranking of the alloys cannot be made on the basis of these data. Again it appears that the final cold rolling treatment given these alloys has improved the formability so substantially as to minimize the effects of purity, observed previously in this laboratory.

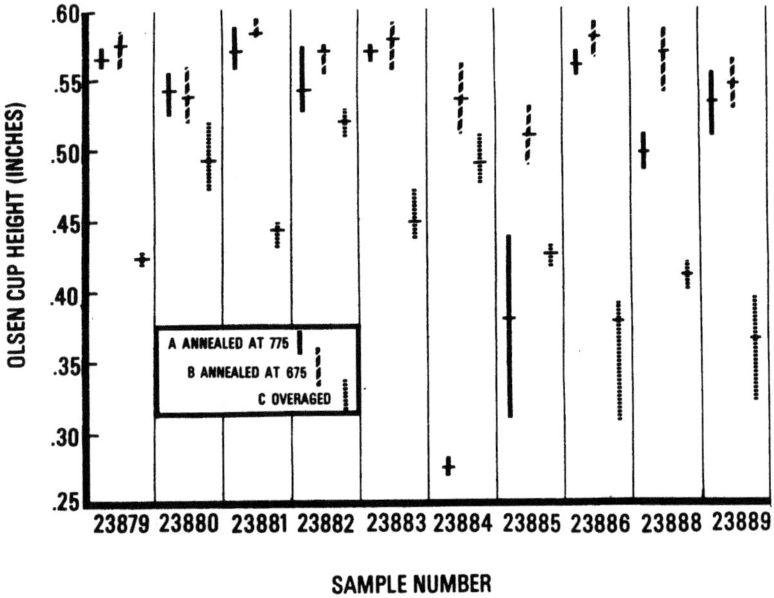

Figure 6 - Olsen Cup Height vs Alloy

Table IV - Summary of LDR Data

S #	Cold Rolled LDR	Hot Rolled LDR
23879	2.176	--
23880	2.173	--
23881	2.188	--
23882	2.182	--
23883	2.191	--
23886	2.182	--
23888	2.222	--
23889	2.176	--
22168	2.133	2.115
23795	2.155	2.118
26227	2.103	1.970

Comparison of Hot Rolled and Cold Rolled Material

In order not to drain the limited amount of material available for this investigation, three other lots of material were tested. The chemical analysis of the three are presented in Table I. Half of each lot was hot rolled directly to 0.150". The remainder was hot rolled to 0.300" and cold rolled to 0.150". All material was

Table V - Olsen Cup Data for Hot Rolled vs Cold Rolled Alloys

		Hot Rolled		Cold Rolled	
	Other Identification	Maximum Load(lbs.)	Depth of Draw(in.)	Maximum Load(lbs.)	Depth of Draw(in.)
26227	CP 7075*	7100	.4625	5963	.5273
23795	MP 7178	6450**	.5333	6037	.5865
22168	HP 7075	7125	.4933	7050	.5833

* Plant Produced

** The individual replicates (4) exhibited the greatest amount of variance for this average value. They ranged from 5800 to 7100 lbs.

then given the standard anneal "B" and then tested for formability. The results are presented in Tables IV and V.

The differences are quite apparent. The cold rolled portions exhibited higher formability and the effect of purity was minimized. The effect of purity was more evident with the hot rolled portion but was somewhat masked by the differences in the levels of the major alloying elements. An explanation of these differences is probably related to the pinning effect which the insolubles exert on grain boundaries. The greater the amount of insolubles the greater the difficulty in obtaining recovery and recrystallization, with small amounts of equivalent cold work. With considerable cold work, the driving force for recrystallization is great enough to reduce the degree of fibering, even with the lowest purity alloys.

A marked difference in earing behavior was also noted. The material hot rolled to gage exhibited relatively large ears, while that cold rolled produced essentially no ears, Fig. 7. This again indicates the presence of a strong rolling texture in the former material and a more isotropic grain structure in the latter material because of recrystallization.

In producing aluminum cartridge cases, the first cupping operation was traditionally the most difficult. The problem was always attribted to the severity of the operation. In light of the present results, the difficulty could also be partially metallurgical. If this first operation was performed with material hot rolled directly to gage, a highly fibered grain structure (leading to inferior formability) would be present. Latter forming operations could appear to be less severe because inter-stage anneals would produce recovery and recrystallization, with a reduction in the degree of fibering.

Figure 7
Earing of Hot Rolled (left) and Cold Rolled (right) Material

Formability of Zr-Containing Alloys After Prolonged High Temperature Treatment

As discussed in the Introduction, the first tests of these alloys indicated poor formability. This was attributed to the strong recrystallization retarding effect of the fine and well dispersed Zr containing particles. A prolonged heat treatment, involving 72 hours at 900°F, cold water quench, the Standard B anneal, roller leveling and the Standard B anneal, was employed. The subsequent testing, Table VI, showed that the formability had been improved. These alloys were still slightly inferior in formability to the Cr-containing alloys.

Table VI - Olsen Cup Results for Zr Containing Alloys

	Initial A Anneal		Initial B Anneal		Initial C Anneal		Retesting*	
S #	L	D	L	D	L	D	L	D
23884	2200	0.276	6900	0.537	9800	0.492	7350	0.541
23885	5700	0.378	6700	0.510	10,200	0.427	7460	0.572
23881(**)	7600	0.578	7400	0.584	10,000	0.442	-	-

** Best performance in initial testing (included for comparison purposes)

L = maximum load (lbs.)

D = maximum depth of draw (inches)

* Retesting conducted after cold rolling 0.300 to 0.150", 72 hrs. at 900°F, cold water quench, followed by standard B anneal, roller leveled followed by standard B anneal.

Tensile Tests on T6 and T73

Aging curves were developed for all alloys with the exception of the two containing Zr. These data were generated with 0.015" thick sheet. The first series of tests yielded a number of quite low values. Other material was processed until reasonable values were obtained.

Similar difficulty was experienced with the 0.015" sheet used to study the effect of strain rate as presented in Table VII. No consistent effect of strain rate was observed. Sections of the specimens having low values were examined using transmission electron microscopy to determine possible metallurgical origin of problem. (As reported earlier, the difficulty was mainly attributed to warping of this thin gage during quenching.) Briefly, it may be stated that these samples exhibited large amounts of coarse $MgZn_2$ particles. It was also noted that the concentration of these particles decreased with lower purity and the dislocation density appeared higher in lower purity alloys.

Table VII - Mechanical Properties of 0.015" Alloys

			"T6"			"T73"		
S No.	Ident.	Strain Rate (IPM)	UTS (KSI)	YS (KSI)	%El.	UTS (KSI)	YS (KSI)	%El.
		6	90.2	83.7	8.0	82.2	72.3	8.5
23879	MP 7001	60	88.5	83.1	8.0	83.4	73.4	8.0
		600	87.3	82.5	8.0	82.8	73.2	8.0
		AVG	88.8	83.1	8.0	82.8	73.0	8.2
		6	85.3	78.6	9.0	75.8	68.5	8.0 *
23880	MP 7075	60	85.8	81.6	9.0	75.8	68.5	9.5
		600	85.3	80.9	9.0	76.3	68.7	8.5 *
		AVG	85.5	80.3	9.0	75.9	68.6	8.7
		6	75.1	71.4	2.5 *	77.5	69.8	8.0
		60	77.8	72.5	9.5	77.5	70.4	8.0
23881	MP 7178	600	78.7	73.5	9.5	78.2	71.1	8.0
		AVG	77.2	72.5	?	77.7	70.4	8.0
		6	86.2	79.8	11.0	80.0	72.2	6.0
23882	HP 7075	60	86.3	80.5	9.5	80.4	72.8	8.0
		600	86.3	81.2	11.5	80.4	72.2	9.0
		AVG	86.3	80.5	10.7	80.0	72.2	8.0
		6	88.7	84.7	9.0	78.9	71.1	9.0
23883	HP 7178	60	89.3	85.3	9.0	80.3	72.4	8.5 *
		600	88.0	84.8	9.0	79.6	71.7	9.0
		AVG	88.7	84.7	9.0	79.6	71.7	8.8
		6	56.6	46.8	6.5 *	70.0	60.8	7.0 *
23886	HP 7001	60	58.4	49.2	6.5 *	71.7	62.8	7.5 *
		600	59.4	48.9	7.0 *	71.4	63.3	7.0 *
		AVG	58.1	48.3	6.8 ?	71.1	62.3	7.2 ?
		6	84.4	77.4	9.5	73.6	65.1	8.0
23888	CP 7075	60	83.7	77.4	8.5	74.8	67.3	7.5
		600	83.1	78.2	8.5	75.5	67.3	8.0
		AVG	83.7	77.6	8.8	74.7	66.8	7.8
		6	66.8	58.7	8.5 *	69.5	59.6	8.0 *
23889	CP 7178	60	67.9	60.6	7.5 *	70.2	60.9	8.0
		600	67.1	60.4	8.0 *	70.8	62.8	8.0
		AVG	67.3	59.9	8.0 ?	70.2	61.1	8.0
		6	72.1	63.4	7.0			
23886	HP 7001	60	70.5	63.8	7.5			
(Retest)		600	73.6	67.6	7.0			
		AVG	72.1	64.9	7.2 ?			

* Failure outside gage marks
? Results suspect (see Discussion section)

An interpretation of these results is that the greater the saturation of the matrix, with respect to Cu, Mg and Zn, the more critical is the quenching operation. In particular, the longer the delay in transfer from the solution treatment furnace to the quenchant, the greater the probability of formation of coarse $MgZn_2$ particles. This situation would reduce strength by decreasing the solute available for the aging reaction. The saturation would be expected to increase with total alloy content (7075 - 7178 - 7001) and purity. The lower the purity the more solute (particularly Cu) removed from the matrix through the formation of insoluble second phase particles. Thus, a successful quench would be most difficult to achieve with the HP 7001, at very thin gages. The driving force for rejection of solute and formation of large $MgZn_2$ particles would be greater for this alloy, particularly in thin gages. Such samples could cool to considerably below the solution treatment temperature, before quenching can be initiated.*

Data for 0.063" and 0.125" samples are presented in Table VIII. There are what appear to be anomalous results when one compares alloys, purities, and thicknesses. For example, the strengths of both 7001 alloys were not as high as expected, and the high purity 7001 was lower in strength than the moderate purity version. Three factors must be borne in mind in interpreting these results.

(i) the driving force for unwanted precipitation, which is related total <u>available</u> solute
(ii) temperature drops which may occur in transit to the quenchant
(iii) the cooling rate possible upon submersion in the quenchant

These same factors probably are responsible for variations in toughness which are discussed in a later section.

"U" Bend Stress Corrosion Tests

Table IX lists the failure observed after 14 days exposure. Only alternate immersion of "-T6" samples has resulted in failures in the data to date. The failure life is observed to be inversely related to the yield strength. All samples which failed in this time period exhibited yield strengths above 74 ksi.

Kahn Tear Tests

Table X shows the data obtained from the Kahn Tear Tests on the eight alloys (excluding the Zr-containing alloys) of 0.063" and

---

*In most of the laboratory work conducted by M. C. Flemings and co-workers, a special furnace-quench arrangement was employed where the samples were dropped into the quench, without opening the furnace.

0.125" thicknesses. These data show some anomalous behavior but trends are evident if the three major variables are considered. The three variables of major importance are purity, yield strength and thickness.

Table VIII - Tensile Results on the Kahn Tear Test Material

TENSILE DATA - 0.063" SHEET

	Purity	S #	"T6" UTS (ksi)	YS (ksi)	% Elong.	"T73" UTS (ksi)	YS (ksi)	% Elong.
7075	CP	23888	81.5	71.8	13.0	73.1	63.3	10.5
	MP	23880	83.6	74.4	13.5	74.7	65.6	10.5
	HP	23882	85.0	74.7	14.0	76.3	66.8	11.0
7178	CP	23889	85.7	75.3	12.0	76.2	66.5	10.0
	MP	23881	88.2	79.4	12.5	79.2	70.4	11.0
	HP	23883	90.4	80.6	13.0	79.2	69.2	10.5
7001	MP	23879	91.5	81.5	13.0	79.2	69.9	11.0
	HP	32886	92.5	80.7	12.0	79.4	67.6	11.0

TENSILE DATA - 0.125" SHEET

	Purity	S #	"T6" UTS (ksi)	YS (ksi)	% Elong.	"T73" UTS (ksi)	YS (ksi)	% Elong.
7075	CP	23888	82.4	72.2	14.0	74.5	65.3	11.5
	MP	23880	80.5	71.3	13.0	76.3	67.2	12.5
	HP	23882	85.0	75.1	15.0	77.0	67.9	13.0
7178	CP	23889	85.5	76.0	12.0	77.0	68.5	11.0
	MP	23881	85.3	76.7	12.0	79.0	68.7	11.5
	HP	23883	87.2	77.4	12.0	81.7	73.3	12.0
7001	MP	23879	91.3	80.9	14.0	80.9	71.1	9.5
	HP	23886	86.1	76.8	11.5	83.3	74.5	11.5

Table IX - "U" Bend Failures After 14 Days

Alternate Immersion

F/N*	Identification	Time to Failure (days)
4/4	MP 7001-T6	1,1,1,1
4/4	MP 7075-T6	12,12,12,13
4/4	MP 7178-T6	3,3,3,3
1/4	HP 7075-T6	14
4/4	HP 7178-T6	1,1,1,3
4/4	HP 7001-T6	1,1,1,3

* Failures/no. Tested

Table IX - Kahn Tear Test Data for T6 and T73 Alloys

(a) 0.125" Thick T73 Temper

	Longitudinal					Long Transverse			
Alloy	UPE	UTE	TS	TS/YS*	YS*	UPE	UTE	TS	TS/YS*
	in. lb./sq. in.		ksi		ksi	in. lb./sq. in.			
HP 7075	455.4	1169	95.5	1.41	67.9	153.5	729.6	86.4	1.27
MP 7075	639.4	1273	87.6	1.30	67.2	-	-	-	-#
CP 7075	341.3	923.8	85.6	1.31	65.3	168.9	606.1	75.2	1.15
HP 7178	154.2	585.4	81.2	1.11	73.3	15.0	429.8	79.9	1.09
MP 7178	289.7	652.7	74.4	1.08	68.7	111.8	401.5	69.3	1.01
CP 7178	247.4	708.6	77.5	1.13	68.5	25.5	359.8	64.9	0.95
HP 7001	109.5	501.4	78.3	1.05	74.5	16.6	418.9	78.9	1.05
MP 7001	277.3	710.3	74.9	1.05	71.1	32.4	399.5	70.0	0.98

(b) 0.060" Thick T73 Temper

	Longitudinal					Long Transverse			
HP 7075	546	1118	94.1	1.41	66.8	485.4	1116	93.1	1.39
MP 7075	689	1229	90.9	1.39	65.6	511	1138	92.2	1.41
CP 7075	454	975	86.7	1.37	63.3	416	916	86.0	1.36
HP 7178	250.3	680.3	87.7	1.27	69.2	243.7	670.1	84.5	1.22
MP 7178	301.2	753.6	84.5	1.20	70.4	177.6	638.1	85.5	1.21
CP 7178	313	716.6	81.2	1.22	66.5	230.6	567.2	76.0	1.14
HP 7001	101.1	420.5	76.0	1.12	67.6	174.0	493.2	75.8	1.12
MP 7001	371.5	795.2	84.0	1.20	69.9	168.6	559	81.7	1.17

(c) 0.125" Thick T6 Temper

	Longitudinal					Long Transverse			
Alloy	UPE	UTE	TS	TS/YS*	YS*	UPE	UTE	TS	TS/YS*
	in. lb./sq. in.		ksi		ksi	in. lb./sq. in.			
HP 7075	92.4	531.8	74.8	1.00	75.1	11.4	310.5	68.5	0.91
MP 7075	216.7	559.7	72.2	1.01	71.3	19.1	380.9	66.4	0.93
CP 7075	228.0	773.5	85.8	1.19	72.2	110.5	603.8	79.8	1.11
HP 7178	37.5	262.7	54.0	0.70	77.4	5.6	160.1	45.7	0.59
MP 7178	15.7	205.7	51.9	0.68	76.7	6.1	168.7	48.5	0.63
CP 7178	49.0	389.9	66.1	0.87	76.0	14.1	247.6	57.2	0.75
HP 7001	19.5	233.4	56.4	0.73	76.8	7.5	204.5	54.1	0.70
MP 7001	41.8	285.3	60.1	0.74	80.9	8.0	225.7	57.1	0.71

(d) 0.060" Thick T6 Temper

	Longitudinal					Long Transverse			
HP 7075	225.0	685.6	90.7	1.21	74.7	192.2	702.5	93.8	1.26
MP 7075	228.0	629.4	83.6	1.24	74.4	94.3	485.1	82.7	1.11
CP 7075	257.3	654.3	84.8	1.18	71.8	210.3	580.1	82.0	1.14
HP 7178	32.4	238.8	66.3	0.82	80.6	13.8	206.8	62.1	0.77
MP 7178	98.1	354.0	69.6	0.88	79.4	18.7	245.5	63.7	0.80
CP 7075	100.4	406.6	74.8	0.99	75.3	32.6	351.7	75.5	1.00
HP 7001	13.2	170.3	56.8	0.70	80.7	6.7	174.9	58.8	0.73
MP 7001	48.8	257.9	61.0	0.75	81.5	9.8	168.7	56.2	0.69

UPE - Unit Propagation Energy
UTE - Unit Total Energy
TS - Tear Strength
TS/YS - Tear Strength to Yield Strength Ratio
YS - Yield Strength

*Due to shortage of 0.125" and 0.060" sheet only the long transverse yield strength was measured and used.

#Due to machine malfunctions this data was not recorded.

Of the measurements obtained from the tear test, the tear strength is probably the more applicable to use as a criterion for selecting an alloy for cartridge cases. The total energy to failure should then be the secondary parameter since it involves the energy to initiate and propagate the crack.

Figures 8 and 9 show plots of tear strength versus yield strength for all of the alloys in the T6 and T73 tempers for the longitudinal and long transverse directions respectively. These show a definite trend towards lower tear strengths with increase in yield strength. The band covering the data is quite wide, however, and the high purity alloys are mostly situated in the top of the band. Figures 10 and 11 show the tear strength plotted in relation to the alloy and purity. In the 7075 alloys, the tear strength increases with increasing purity in all thicknesses, orientations and tempers (except the 0.125" T6 condition). The 7178 alloys in the T73 tempers show a similar behavior but the T6 tempers of 7178 show a decreasing tear strength with increasing purity probably due to the influence of yield strength overshadowing the effect of purity. The 7001 alloys have generally lower tear strengths comparable in general to the lowest values obtained with the 7178 alloys.

The total energy to failure versus purity, Figure 12, shows that the three 7075 alloys with T73 condition have energies which are considerably higher than any of the other alloys or conditions. These high 7075-T73 fracture energies when combined with their high tear strengths as shown in Figure 12 make them the best candidates for use in cartridge case applications. Increasing the strength either by increasing the alloying content (to 7178 or 7001) or by using the T6 temper decreases the fracture toughness of the alloy.

The tests also show up a thickness effect in these alloys. The 0.060" thick 7075 and 7178 in both the T6 and T73 tempers show significantly higher values of tear strength and total energy to fracture than the 0.125" thick alloys.

From these data it appears that the thicker sheets in the more highly alloyed or purer compositions are not responding to quenching and aging as well as the commercial alloys. Transmission electron microscopy is being used to attempt to isolate the cause of these differences but at present no conclusive results have been obtained. An indication of quench sensitivity effects on toughness and importance of the delay in quenching with the thin gage is emerging. The microstructural details are being studied and will be reported in the future.

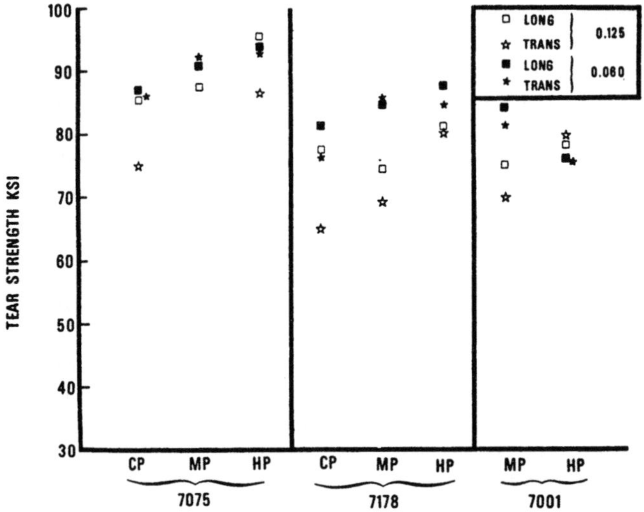

Figure 10 - Tear Strength vs Alloy Composition for T73 Condition

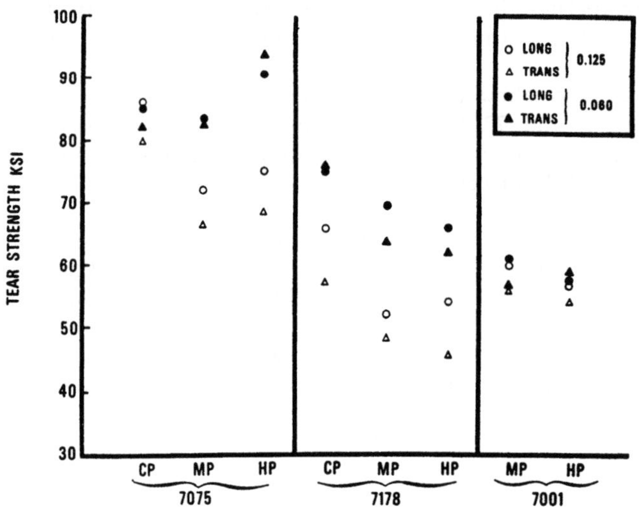

Figure 11 - Tear Strength vs Alloy Composition for T6 Condition

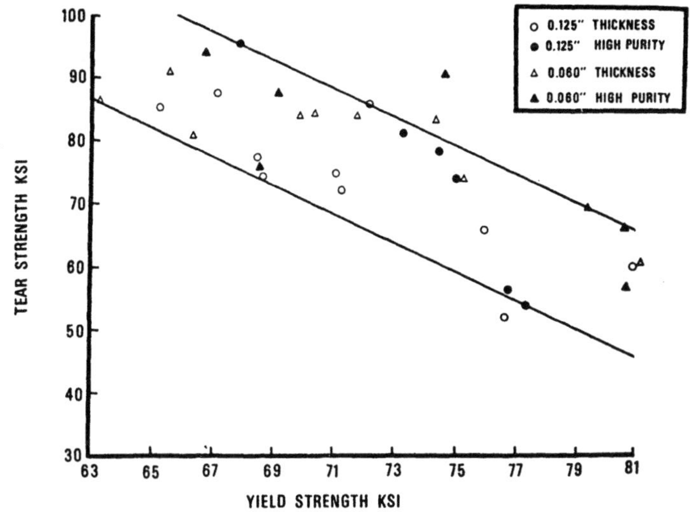

Figure 8 - Tear Strength vs Yield Strength for the Heat Treated Alloys in the Longitudinal Direction

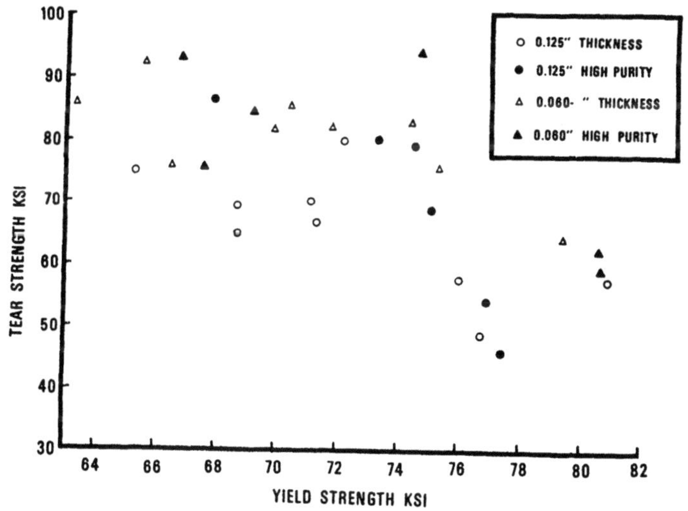

Figure 9 - Tear Strength vs Yield Strength for the Heat Treated Alloys in the Long Transverse Direction

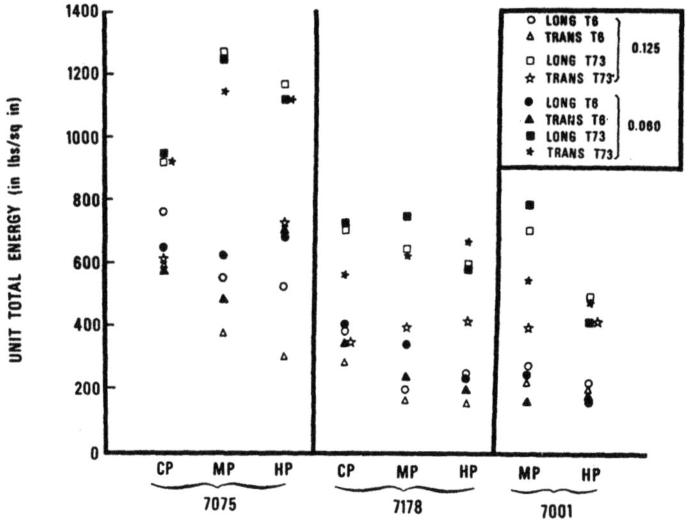

Figure 12
Unit Total Energy vs Alloy Composition for the T73 and T6 Conditions

CONCLUSIONS

Formability of "O" temper material is improved by providing a cold working treatment, rather than rolling directly to gage, prior to the final anneal.

Strain rate has a minimal effect on properties with both annealed and hardened materials.

The formability of the Zr-containing alloys can be substantially improved by forcing recrystallization with a prolonged thermal treatment.

When the cold rolled treatment is employed, purity has a minor effect on formability. The effect is more pronounced with material hot rolled to gage.

Earing is minimized by cold rolling prior to final anneal.

Toughness and stress corrosion resistance are inversely related to yield strength. High purity provides benefits in toughness without strength penalties, mainly with the lower solute alloys.

Thickness effects have been observed with respect to strength and toughness. These effects may be related to quench sensitivity and criticality of the delay in transferring from the solution treatment furnace to the quench. The 0.015" samples are quite sensitive to this delay, for they are able to cool rapidly in air. The 0.063" samples are less sensitive to the delay time, and would experience a higher quench rate than would the 0.125" material. These hypotheses will be evaluated in future studies of structure.

The delay and quench rate effects, discussed above, appear to be dependent upon the total amount of available solute. The greater the saturation of the matrix, the greater the driving force for premature coarse, precipitation. Such precipitation can decrease strength and impair toughness. Homogeneity may increase these effects, for the maximum amount of solute is dissolved in the matrix.

ACKNOWLEDGEMENTS

This investigation was carried out under the sponsorship of Frankford Arsenal Contract No. DAAA 25-70-C-0461 whose permission to publish this paper is acknowledged.

R. E. Moss, K. Roper and many other people at the Metallurgical Research Division of Reynolds Metals Company also contributed to the testing and analysis of the results.

REFERENCES

1. S. A. Levy and C. Baker, "Development of Aluminum Alloys for Cartridge Cases", Contract DAAA 25-70-C-0461, May, 1970.

2. M. C. Flemings, "Application of Solidification of Theory to Large Castings and Ingots", The Solidification of Metals, Iron and Steel Institute publication 110, Session B, 227-287.

3. S. N. Singh and M. C. Flemings, "Solution Kinetics of a Cast and Wrought High Strength Aluminum Alloy", TMS AIME, 245, 1803-1810 (1969).

4. S. N. Singh and M. C. Flemings, "Influence of Ingot Structure and Processing on Mechanical Properties and Fracture of a High Strength Wrought Aluminum Alloy, Op. Cit., 1811-1819.

5. H. Antes, S. Lipson and H. Rosenthal, "Strength and Ductility of 7000 Series Wrought Aluminum Alloys as Affected by Ingot Structure", Ibid, 239, 1634-1643 (1967).

6. S. A. Levy, "Homogeneous High Strength Aluminum Alloys", A paper presented at the Spring Meeting of the Metallurgical Society of AIME, Las Vegas, Nevada, May, 1970.

EFFECT OF PLASTIC ANISOTROPY ON DRAWING CHARACTERISTICS OF ALUMINUM ALLOY SHEET

R. W. Rogers, Jr. and W. A. Anderson

Alcoa Research Laboratories

The production of annealed tempers of commercial aluminum alloy sheet for deep drawing applications has been aimed at providing three characteristics: good drawability, a fine grain size, and minimum earing tendencies. The latter property is one manifestation of anisotropy for which control is attempted by means of fabricating practices designed to balance annealing and rolling crystallographic textures.

While this planar anisotropy has generally been considered undesirable, another form of anisotropy was found to be advantageous in the drawing of commercial low-carbon steel sheet by Lankford et al in 1950 (1). This is normal anisotropy in which the through-thickness or short transverse flow stress is greater than that in the plane of the sheet so that there is a reduced tendency to thin when the sheet is stretched. Since the direct determination of short transverse strength of sheet is not possible, normal anisotropy is measured in a uniaxial tensile test by determining the ratio of width strain to thickness strain which is termed R or the strain ratio. Values of R greater than unity then represent the anisotropic condition whereby longitudinal extension in uniaxial stretching is at the expense of a proportionately greater contraction in width than thickness; thus, in corresponding biaxial stretching thinning is resisted and drawability enhanced.

Figure 1 is from work by R. L. Whitely and illustrates drawability, as expressed by the limiting drawing ratio, in relation to R for a number of commercial sheet metals (2). It will be noted

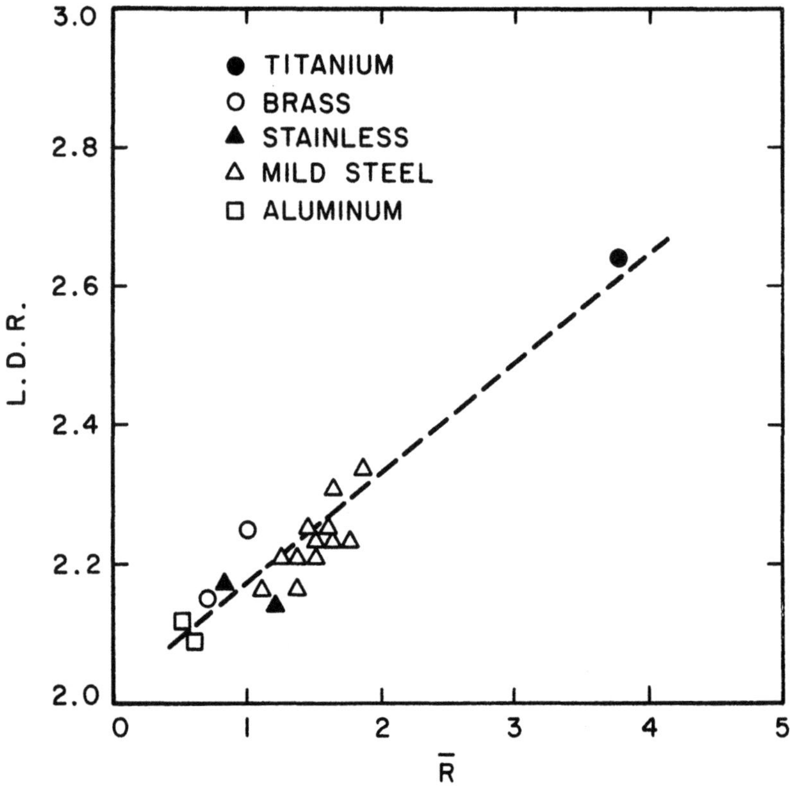

Figure 1 - Limiting Drawing Ratio Vs. $\bar{R}$ for Several Sheet Metals (Whitely).

that aluminum and other FCC metals are at the low end of the range where R is 1 or less while BCC mild steel and HCP titanium have higher R values and greater drawing ratios. The purpose of the present investigation was to explore the relation between the R-value of aluminum and drawability in greater detail and also to study the effects of the strain-hardening coefficient n on the stretching and drawing characteristics of aluminum in much the same type of experimentation that had been done for steel.

The coefficient n appears in the general expression describing the stress-strain characteristics of aluminum during plastic deformation. This equation has been found to hold quite well for many metals:

$$\sigma = K\varepsilon^n$$

In this equation σ is the applied stress, ε is the corresponding strain, while K and n are constants. The significance of n is as a measure of the strain to necking and an indication of the work-hardening characteristics of the metal. The ability to work-harden locally when thinning begins aids in the distribution of strain uniformly over the area being deformed and thus is important in metal working.

PROCEDURE

Two lots of 1100 and one lot of 3003 commercial .040" thick sheet were obtained in each of three as-rolled tempers - H12, H14 and H18. These tempers provided final cold reductions ranging from 11 to 75%.

Part of each lot and temper was annealed for 30 minutes at 650 F (1100) or 800 F (3003) making a total of nine annealed-temper items and nine as-rolled temper items with varying degrees of final cold work or cold work prior to the final anneal. In this way, a variety of properties and crystallographic textures was obtained.

Strain ratio (R) values were obtained at 0°, 45° and 90° to the rolling direction of the sheet by measuring the length and width of tensile specimens before (l_o, W_o) and after (l_f, W_f) straining to 75% of their uniform elongations:

$$R = \frac{\varepsilon_w}{\varepsilon_t} = \frac{\ln e_w}{\ln e_t} = \frac{\ln (W_o/W_f)}{\ln (t_o/t_f)} = \frac{\ln (W_o/W_f)}{\ln (W_f l_f/W_o l_o)}$$

The strain hardening coefficient (n) values were obtained by the procedure developed by Nelson and Winlock for determination of uniform elongation (3).

Planar anisotropy was determined by measuring the direction and degree of earing in 4" diameter flat-bottom and hemispherical-bottom cups.

Drawing characteristics were determined by drawing the 4" diameter flat and hemispherical-bottom cups from blanks of various diameters. Results were reported as the largest successful or "critical blank diameter" (CBD) or as "limiting drawing ratio" (LDR) equal to CBD/punch diameter.

Commercial drawing dies were then employed for further evaluation of drawing characteristics. A melon mold was selected as an application requiring a high degree of stretching and an ingredient cup for purely drawing deformation. These are shown in Figure 2. Evaluation was on the basis of a scoring system from 1 to 10 based on freedom from fracture, necking, buckles and necessity for reducing blankholder pressure or

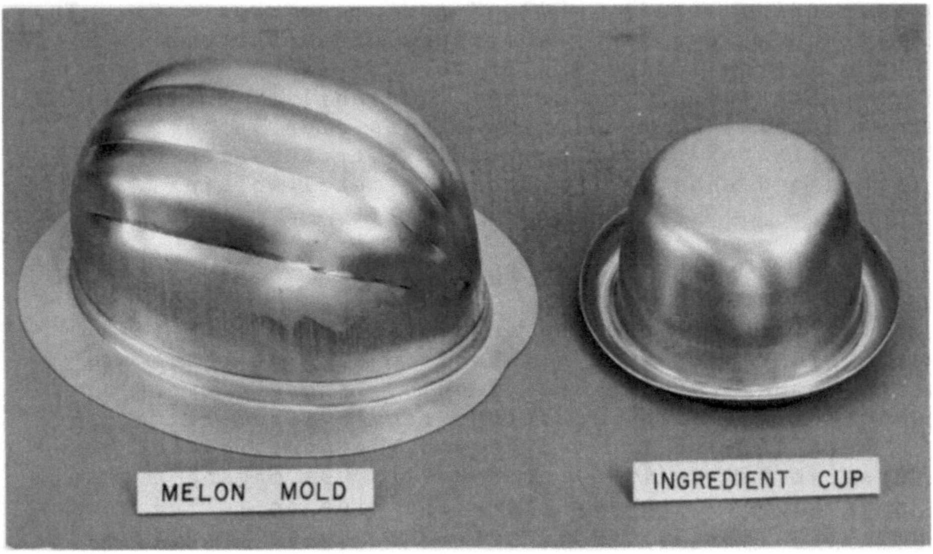

Figure 2 - Shells Drawn on Commercial Dies for Drawability Evaluation

punch depth to accomplish drawing. A score of 10 would be a perfect item.

For further investigation of the relationship between planar and normal anisotropy, additional commercial alloy sheet circles were drawn into 33 mm diameter cups and were tensile tested for determination of R values. These included the following alloys, thicknesses and tempers:

.040" 5052-0
.040" 5052-H32
.040" 6061-0
.040" 6061-T4
.040" 6061-T6
.040" 7075-0

In addition, .064" 5052-H34 and H38 commercial sheet was tested similarly in the as-received condition and after annealing 30 minutes at 700 F to again obtain several varying crystallographic structures.

RESULTS

Figure 3 shows LDR values for flat-bottom cups versus R for 1100 and 3003 in the annealed, intermediate, and hard tempers. For the annealed tempers (filled symbols), the trend line is parallel and close to the line representing Whitely's values. The one point out of line is for 3003 sheet which was subjected to an anneal but did not recrystallize because it was fabricated with a small degree of prior cold work; it, therefore, is actually more comparable to the other as-rolled temper points (open symbols) which do not show any kind of relationship with R. The R and n values are shown as $\bar{R}$ and $\bar{n}$ which are weighted average values at 0°, 45° and 90° to the rolling direction since R and n vary directionally in anisotropic sheet.

Figure 4 shows similar data but for cups with hemispherical bottoms whereby stretching becomes significant. Here a more pronounced dependence of LDR on R is shown for annealed sheet and the influence of n is also illustrated for the as-rolled sheet. Again the out-of-line point is for the 3003 sheet referred to previously which was given an annealing treatment but remained unrecrystallized (n = .13).

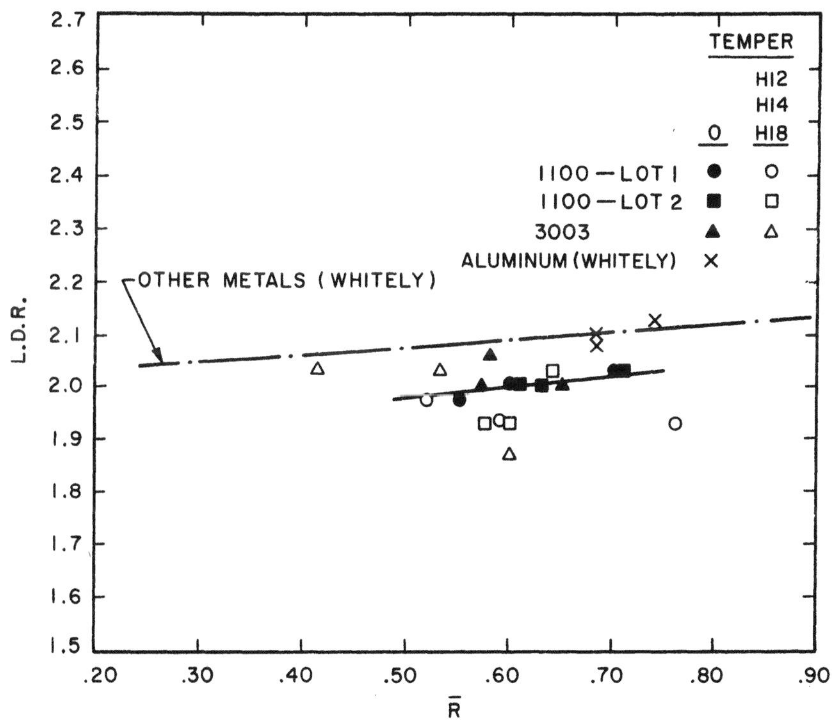

Figure 3 - Limiting Drawing Ratio Vs. $\bar{R}$ for Flat Bottom Cups Drawn from 1100 and 3003.

Figure 5 shows results for 1100 and 3003 sheet in the hydraulic bulge test which is entirely a stretching mode of deformation. As a result, the bulge height versus R data almost duplicate Figure 4 for hemispherical-bottom cup drawing. Another out-of-line point for annealed sheet will be noted which was for annealed 3003 sheet fabricated in a manner developing an extremely coarse grain size which caused early failure in the bulge test.

The role of stretching in flat and hemispherical-bottom cups is further illustrated by Figure 6 which relates bulge height from bulge tests to LDR from cup drawing tests for the two types of cups. It is clear that the stretching mode of deformation makes bulge testing and hemispherical-bottom cup drawing much more discriminating in regard to drawability of sheet than is the drawing of flat-bottom cups.

Figure 7 shows LDR results for the annealed and as-rolled tempers of 1100 and 3003 sheet on an $\bar{R}$ versus $\bar{n}$ plot for both types of cups. For the hemispherical-bottom cups, LDR values tend to increase from the low

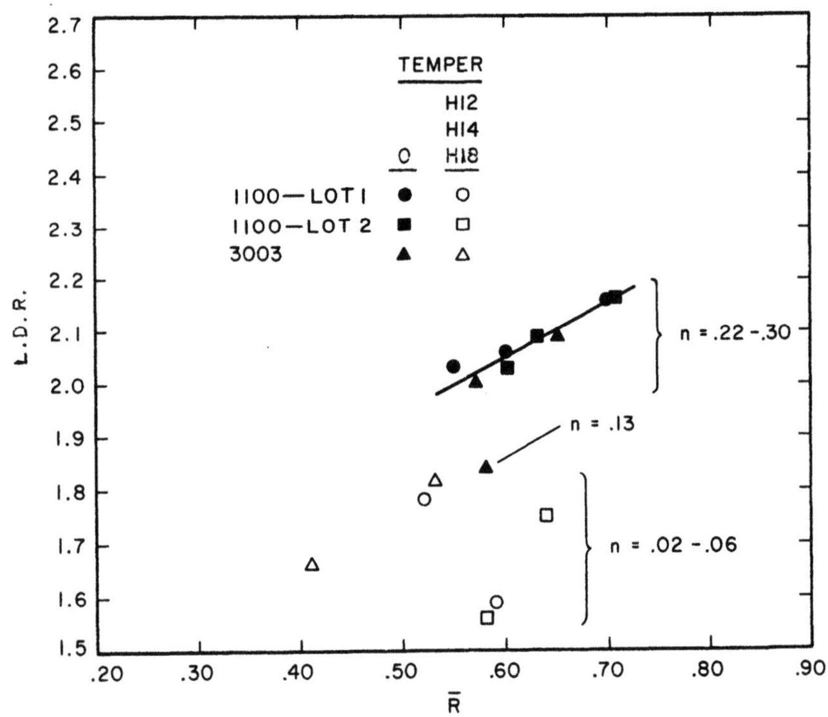

Figure 4 - Limiting Drawing Ratio Vs. $\bar{R}$ for Hemispherical Bottom Cups Drawn from 1100 and 3003.

n, low R corner to the high n, high R area. With the flat-bottom cups, however, the overall relationship is poor; this is largely because there was little difference in LDR values as shown in the previous figure. For annealed material (n > .20), however, the LDR values show a direct variation with R.

These relationships were verified in the commercial die trials. When the drawn products were rated according to such features as freedom from buckles, necking, and fractures, the melon molds with highest ratings of 10 were from material with properties in the high $\bar{n}$ - high $\bar{R}$ region as shown in Figure 8. On the other hand, the ingredient cup ratings displayed the same overall lack of relationship to n and R that the flat-bottom cups of Figure 7 showed.

Figure 9 illustrates a relationship between normal and planar anisotropy noted by Wilson and Butler (4).

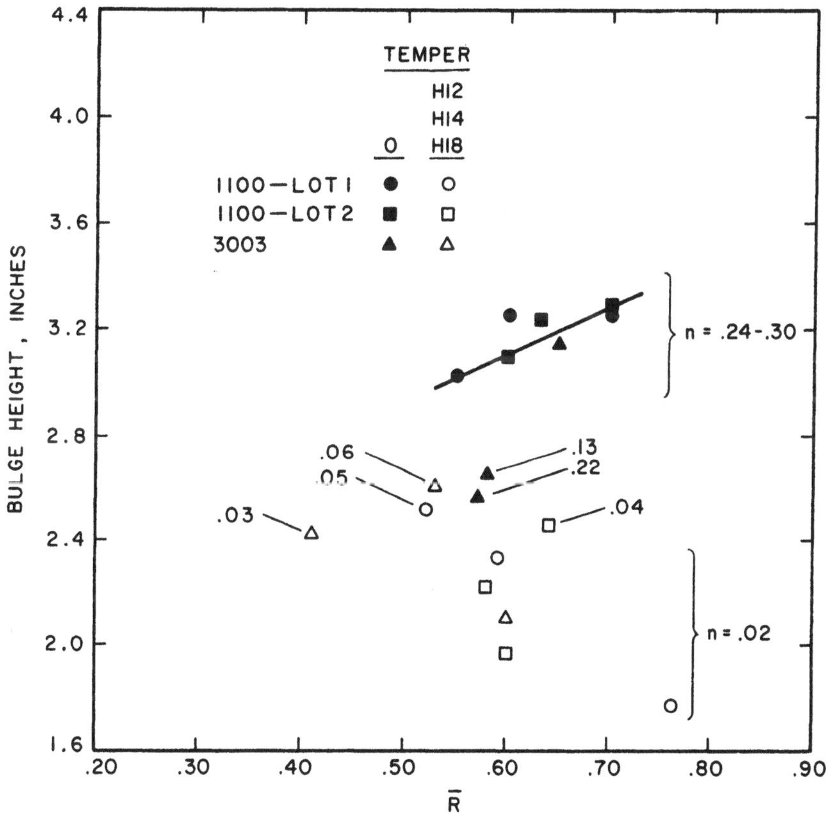

Figure 5 - Limiting Drawing Ratio Vs. $\bar{R}$ for Hemispherical Bottom Cups Drawn from 1100 and 3003.

Here $\Delta R/\bar{R}$ is plotted against % earing for 1100 and 3003 sheet with ΔR being the difference between R in the 0°, 90° directions and R in the 45° directions. When stated in this way, positive values correspond to 0-90° earing and negative values to 45° earing with the regression line passing through the origin or nearly so. The regression line calculated by the least squares method is % ears = 0.438 + 14.9 $\Delta R/\bar{R}$.

In Figure 10 earing results for the other aluminum alloys and tempers evaluated are added to show that the earing versus $\Delta R/\bar{R}$ relationship holds well for a variety of aluminum materials.

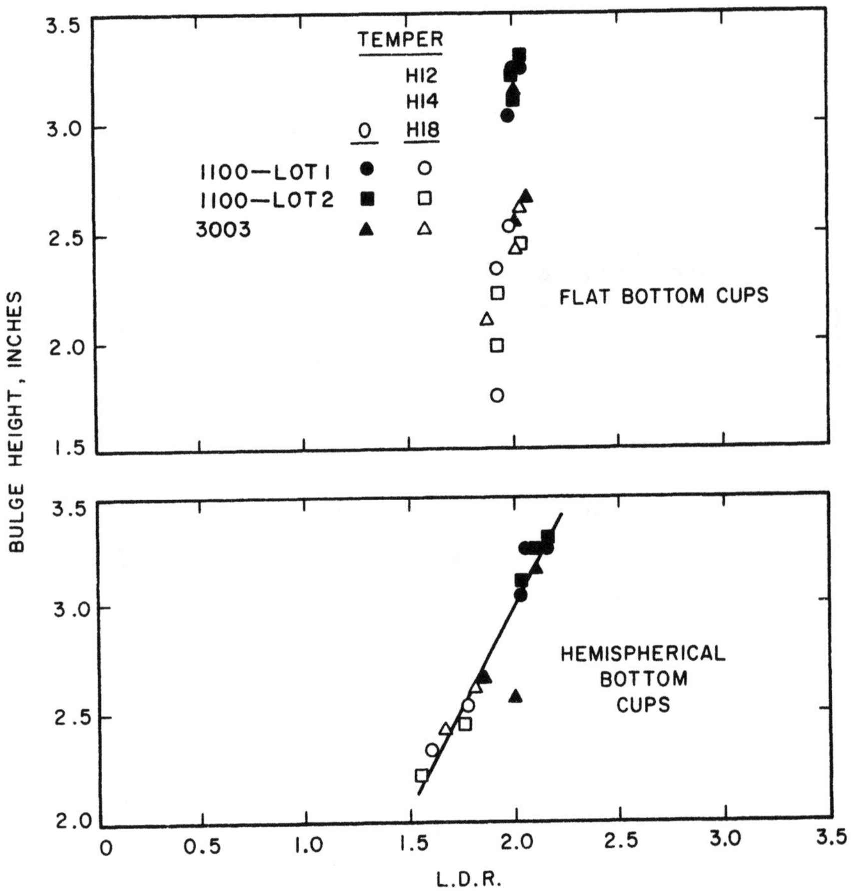

Figure 6 - Bulge Height Vs. Limiting Drawing Ratio for Flat and Hemispherical-Bottom Cups.

DISCUSSION

In work with commercial stampings from low-carbon steel, it has been demonstrated that materials with high R and high n values had substantially better drawability, i.e., less breakage than if these values

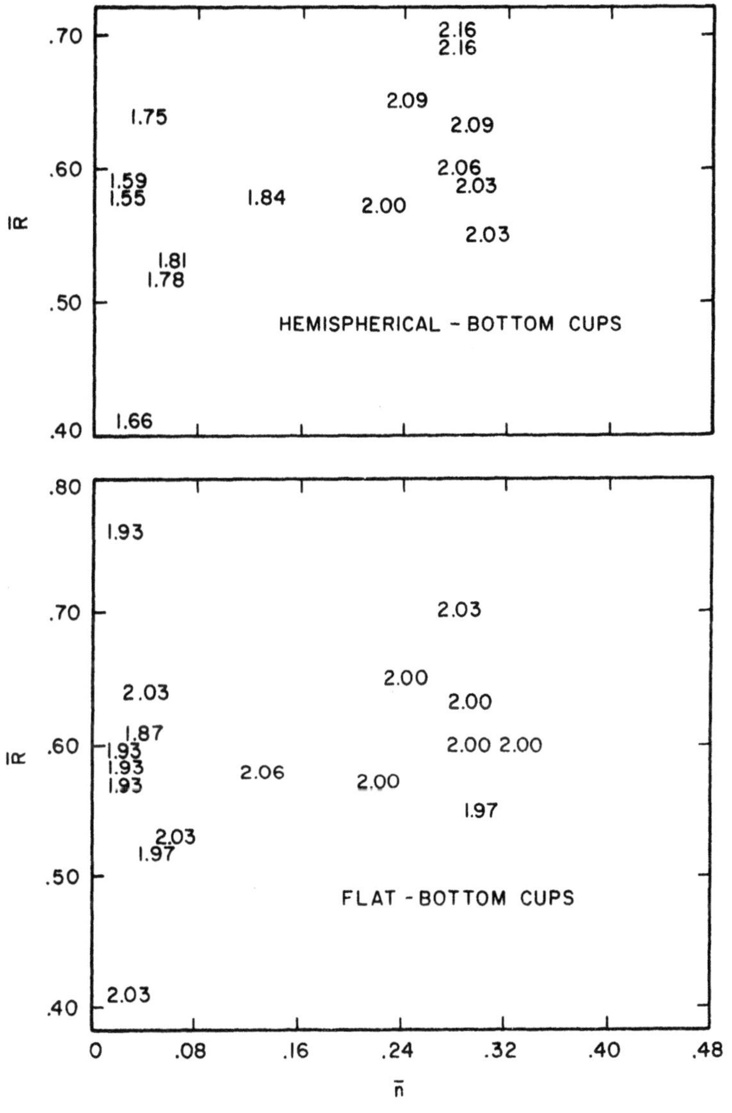

Figure 7 - Limiting Drawing Ratio Values on an $\bar{R}$-$\bar{n}$ Plot for Flat and Hemispherical-Bottom Cups Drawn from 1100 and 3003 Sheet.

were low (1,5). Differentiation according to the type of stamping was also shown: where a high degree of stretching was involved, breakage occurred with low n values and when the deformation was substantially by drawing, breakage was associated with low R values. Best results were achieved when both values were high. The current work with aluminum alloys 1100 and 3003 indicates similar relationships.

For annealed-temper 1100 and 3003 sheet, drawability was shown to be proportional to $\bar{R}$ in flat and hemispherical-bottom cup drawing, melon mold and ingredient cup drawing and hydraulic bulge testing. The relationship was weak where only drawing was involved but was considerably stronger when the mode of deformation was primarily stretching.

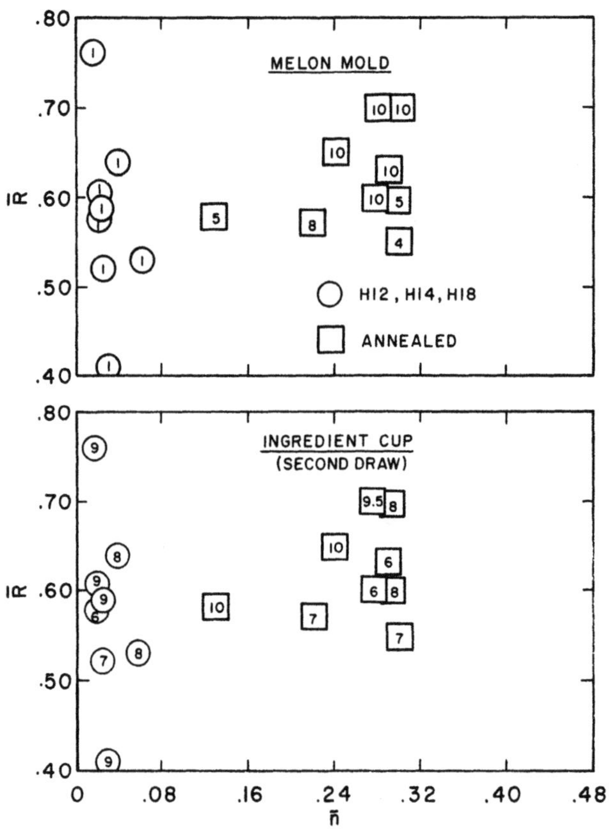

Figure 8 - Drawability Ratings on an $\bar{R}$-$\bar{n}$ Plot for Melon Molds and Ingredient Cups Drawn from 1100 and 3003 Sheet.

The importance of n was also apparent where stretching was involved; deepest drawing of the hemispherical-bottom cups occurred when both n and R were high while the melon mold could not be drawn at all with low-n material (as-rolled tempers).

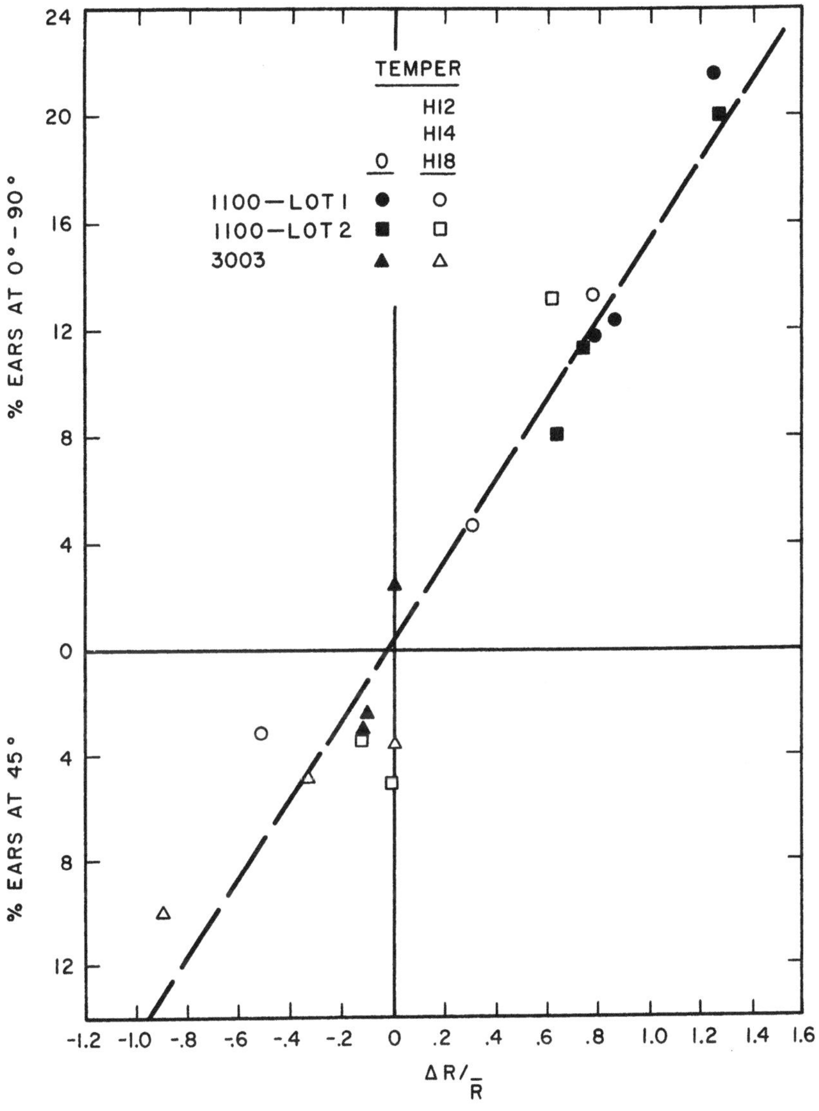

Figure 9 - Relation Between Earing in Cup Drawing and Differential R from Tensile Testing for Flat Bottom Cups.

On the other hand where drawing alone was the mode of deformation as in flat-bottom and ingredient cups, the low-n as-rolled 1100 and 3003 sheet drew about as well as the high-n annealed tempers. Also, considering drawability of the low-n as-rolled tempers only in this type of deformation, there appeared to be little dependence upon R in such material.

The relationship shown in Figures 9 and 10 demonstrates that planar anisotropy represented by earing is the directional manifestation of normal anisotropy. This concept is useful in better under-

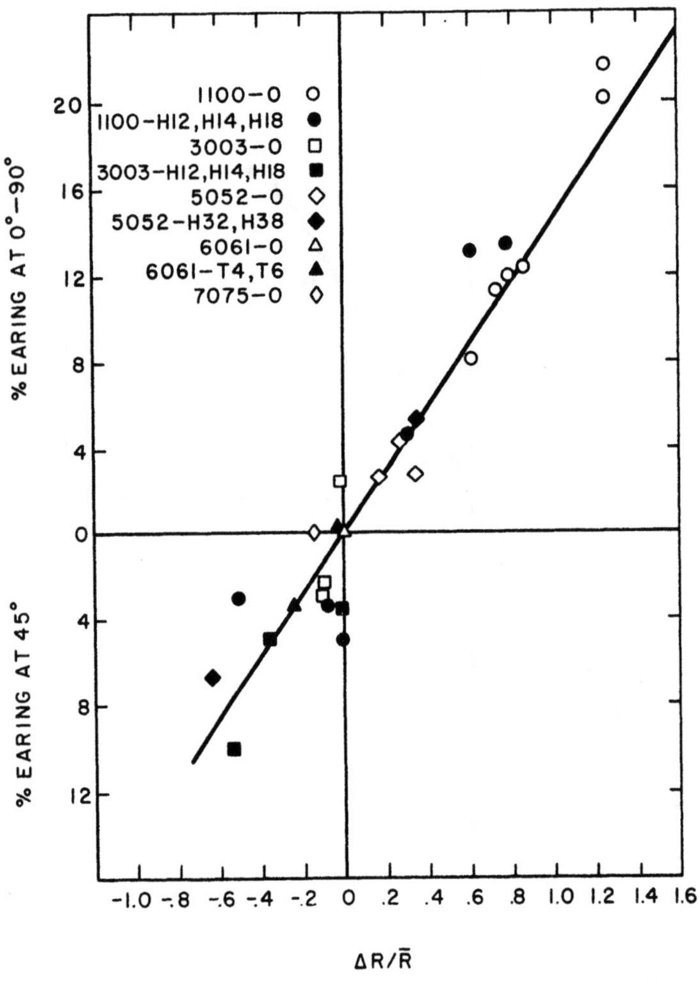

Figure 10 - Per Cent Earing Vs. $\Delta R/\bar{R}$ Ratio for Several Aluminum Alloys and Tempers

standing some of the dimensional changes during drawing and also how fracture due to highly localized stretching may be overcome with sheet of the proper anisotropy.

Where stretching is a significant part of the mode of deformation such as in the recessed head of a beer barrel or in the corners of a rectangular pan, taking advantage of the normal anisotropy may be of substantial help. In the case of a rectangular pan, for example, 45° earing sheet will produce less failures from fractures in the corners since in this high-R direction there is less tendency to thin locally than is the case with 0-90° earing sheet.

The relationship illustrated here also shows that earing is at a minimum when R_0, R_{90}, and R_{45} are equal in magnitude. If these values are small but equal, $\bar{R}$ is small and drawability is poor whereas if they are large but equal, drawability is good; in both cases earing would be absent since ΔR was zero. Thus, minimum earing tendencies are not in themselves indicative of the ability of a material to be deep drawn successfully.

A considerable amount of information about the plastic deformation capabilities of aluminum sheet can be obtained from the uniaxial tensile test. It should be pointed out that n and R by no means provide all the answers about formability. For example, the role of grain size was not considered in this work although in one instance its influence on results was identified. Other influences not considered in this analysis are strength and mechanical textures as well as such operating variables as lubricant, surface finish, die design, etc.

This type of analysis has been found useful in comparing different materials when as many as possible of these other variables are held constant and also in evaluating effects of variations in fabricating procedures and composition upon workability.

CONCLUSIONS

1. Although the spread in R values was small, the drawability of annealed sheet of aluminum alloys 1100 and 3003 varied directly with R. The relationship was stronger when the mode of deformation was primarily stretching rather than drawing.

2. When the mode of deformation was stretching, best drawability was with annealed temper sheet having

high R and high n values. As-rolled temper sheet with low n values drew poorly if at all.

3. When the mode of deformation was drawing alone, low-n sheet (as-rolled tempers) drew satisfactorily and with little dependence on R.

4. Planar anisotropy as evidenced by earing was shown to result from directional differences in normal anisotropy; the degree and direction of earing in drawing can be predicted from R values established by tensile testing.

REFERENCES

1. W. T. Lankford, S. C. Snyder and J. A. Bauscher, Trans ASM, 52 (1950), 1197.

2. R. L. Whitely, Trans ASM, 52 (1960), 154.

3. P. G. Nelson and J. Winlock, ASTM Bulletin (January, 1949), 53.

4. D. V. Wilson and R. D. Butler, J. Inst. of Metals, 90, (1961-62), 473.

5. Report on Cooperative Research Program of USA Committee on International Deep Drawing Research Group, 1964.

ANALYSES OF DEFORMATION AND TEXTURE AS FUNCTIONS OF FABRICATION IN MANDREL-DRAWN TUBING*

D. O. Hobson

Metals and Ceramics Division

Oak Ridge National Laboratory, Oak Ridge, Tenn. 37830

INTRODUCTION

Present-day light-water power reactors require large amounts of high-quality tubing as cladding for the UO_2 pellet fuel presently in use. A typical 1000 MWe boiling water reactor requires almost 500,000 ft of nominally 1/2-in.-diam Zircaloy tubing per core loading. The production of such tubing is potentially a very large-scale operation with the advent of increasing numbers of power reactors. We are presently concerned with the effects that fabrication can have on the mechanical properties of Zircaloy tubing. It has been found that texture, or crystallographic orientation, in this anisotropic material has a large effect on the properties of the material.[1] It has also been shown that texture is profoundly affected by deformation.[2,3] Most commercial tubing manufacturers in this country use combinations of extrusion, tube reducing (rocking), and drawing to produce their tubing. The types of tooling used, the die designs, the reductions per pass, etc., vary among manufacturers and are generally proprietary.

It is the purpose of this paper to present details of the examination of one particular type of fabrication, rod-mandrel drawing, and to analyze the strains inherent in the operation and their effect on tubing texture. This paper is part of a more extensive study covering the effects of other types of tubing fabrication procedures. The material presented is based on Zircaloy tubing, but the discussion of deformation is of general practicality.

---

*Research sponsored by the U.S. Atomic Energy Commission under contract with the Union Carbide Corporation.

BACKGROUND

Zircaloy is the generic name for a family of zirconium-base alloys containing small amounts of tin, iron, and other trace elements. The crystal structure is hexagonal close packed and, characteristic of metals with this structure, Zircaloy exhibits strong anisotropy of mechanical properties. The presence of the strong textures that can develop in tubing causes this macroscopic anisotropic behavior.[1] An example is shown schematically in Fig. 1 taken from the above reference.

The orientations of the grains, and thus the texture of the material, were quite adequately described for this study by mapping the directions in which the basal pole (the "c" axis of the unit cell; the [0001] direction) is oriented with reference to the tubing. A general rule of thumb for zirconium and its alloys is that the basal poles tend to line up nearly parallel to the compressive strain direction. An extension of this rule qualitatively states that tubing fabrication with wall thinning predominating (ironing) will tend to have basal poles near the radial directions of the tubing. Actually, the sheet-rolling texture is obtained with basal pole intensities concentrated 30 to 40 deg on each side of the radial direction toward the transverse direction. Conversely, tubing fabricated with diametral reduction predominating (sinking) will tend to have the basal poles toward the tangential direction. By properly tailoring fabrication schedules, with appropriate ratios of wall strain to diametral strain, texture control can be effected. Conversely, the strain history of the tubing

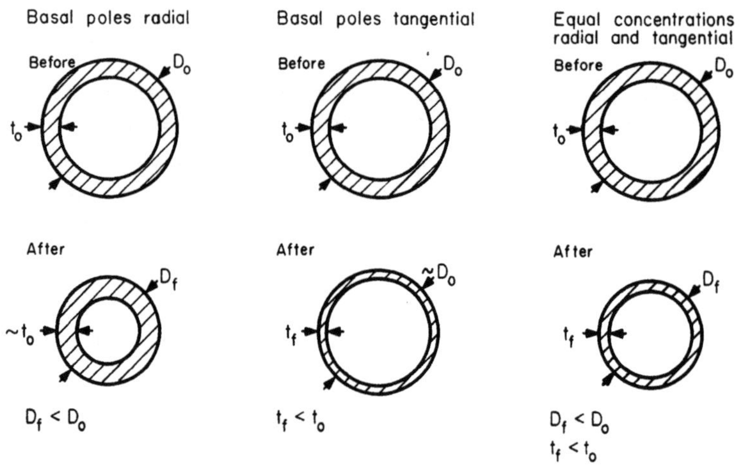

Fig. 1. Postulated effects of uniaxial tensile tests on Zircaloy-2 tubing with the above predominant textures.

during fabrication can be deduced from the textures present in the finished tubing. Papers constituting other portions of the overall study of this subject have been presented orally and are presently published as abstracts.[4,5] Fabrication schedules can be plotted on graphs such as the one shown in Fig. 2. The reductions shown here are given in engineering values, measured in percentage units. Reductions in area are plotted against values of Q ratio (R_W/R_{ID}) with reductions in outside diameter and wall thickness forming a subordinate gridwork within the plot. Given a starting tube size with a ratio of inside diameter to outside diameter (D ratio) equal to 0.8, any point on the graph represents another tube size based on the original one. Figure 3 illustrates this by plotting various tube sizes as a function of position on the graph. In practice, a given graph can be used for any starting tube size with a D ratio near the D ratio on which the graph is based. This is done by minor arithmetic operations on the starting tube size. It is seen that tubing fabricated with Q values greater than unity have greater wall reductions than diameter reductions, and conversely for Q values less than unity. Lines of constant D ratio represent neutral fabrication paths with neither ironing nor sinking predominating. Paths to the right of the neutral lines represent fabrication by ironing and tend to cause rotation of the basal poles toward the radial direction of the tubing. Paths to the left represent sinking and tend to cause migration of basal pole intensity toward the tangential direction.

In practice, the foregoing technique is quite adequate for predicting textures in tubing fabricated by tube reducing or plug drawing techniques and for tubing fabricated by drawing at either high or low Q numbers. The former two techniques provide rigid

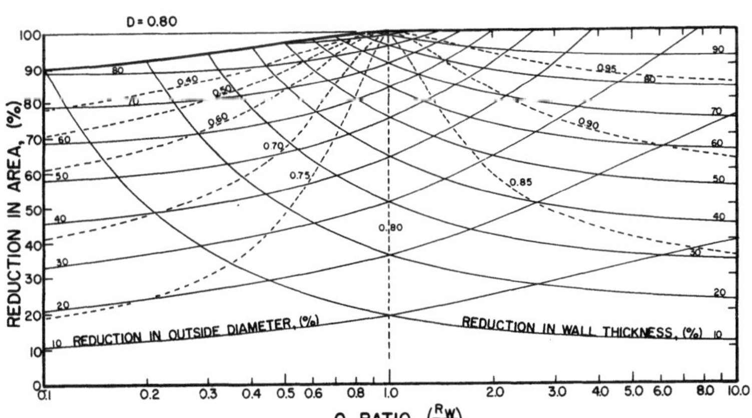

Fig. 2. Fabrication parameter graph based on tubing with an inside diameter to outside diameter ratio of 0.8.

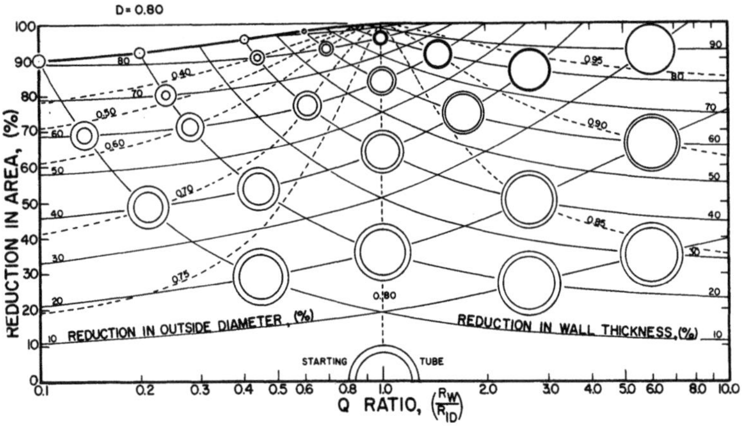

Fig. 3. Repeat of Fig. 2 with the addition of tubing cross sections to illustrate the physical significance of points on the graph.

control of the inside diameter reduction which, in turn, allows the fabrication paths to be plotted as smooth curves on the Q versus R_A graph. The latter technique, as will be discussed in this paper, produces abrupt bends in the tube wall that cause discontinuous changes from ironing to sinking situations at intermediate Q numbers. Heavy sinking or ironing passes tend to overshadow this effect and, thereby, produce the textures expected by theory. This paper will examine the deformations produced by mandrel-drawing tubing at intermediate Q numbers and will relate such deformation to the textures produced.

EXPERIMENTAL PROCEDURE

This paper is part of an extensive study of the effects of fabrication parameters on the texture and strength of Zircaloy tubing. Space does not permit a complete description of the experimental procedures involved in this study. Briefly, the tubing used in this study was mandrel extruded from 4-in.-diam hollow billets into 1.5-in.-OD × 1.0-in.-ID tube shells. These were then tube reduced and drawn to their finished size according to a variety of fabrication schedules incorporating various amounts of ironing and sinking. Textures were determined after each pass by x-ray examination of specimens taken from the outer and inner surfaces and the midwall. The x-ray specimens consisted of rings machined from the tube wall at the appropriate diameter, which were split and thinned to approximately 0.001 in. The specimens were flattened elastically and embedded into a waxed specimen holder for examination by the Schultz reflection method for pole figure determination. Most of the pole figures were corrected for the defocusing that occurs as the specimens are rotated and data are taken farther away from the specimen normal. Such defocusing corrections produce pole figures with results that are quantitative out to the 75-deg limit of data acquisition.

One specific procedure used in the study reported here was the machining of a tube-drawing shell incorporating small, carefully drilled marker holes for use in measuring the relative elongations of layers of the tube wall. Figure 4 illustrates the shell. The starting material was a 1 3/4-in.-diam forged bar of Zircaloy-2. This was machined as shown in the figure. The 30-deg included angle matched the die angle and the diameters of the small ends were machined to go through a 1.250-in. die and over a 0.850-in. mandrel. The marker holes were filled with platinum wire and the entire shell was pickled, given a phosphate coating, and then covered with drawing lubricant. The shell was drawn (37,000 lb drawing force) until the die was situated in the center of the set of marker holes. The drawbench was stopped and the shell was backed out of the die and pulled off the mandrel. The shell was then machined into two halves, exposing the wires, and the wires were stripped from the holes. Measurements made with a toolmakers microscope enabled a scale drawing to be made from which strain values were calculated. Specimens for texture determination were machined from the inner and outer surfaces of the large, undeformed end of the tube and from the inner and outer surfaces of the small, reduced end that had gone through the die.

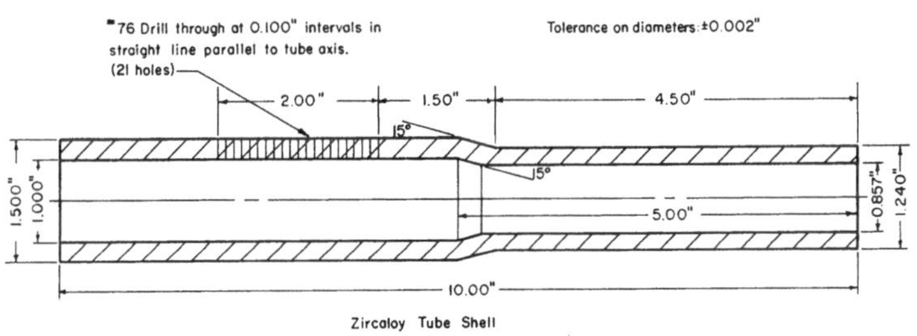

Fig. 4. Longitudinal cross section of as-machined Zircaloy tube shell showing the initial positions of the marker holes.

THEORY

Fabrication of tubing by mandrel drawing or mandrel swaging requires additional analyses of the types of deformation being performed on the tubing. The Q versus R_A graphs discussed earlier recognize, in general, only the initial and final tube sizes for each pass in defining a fabrication path. One could take the effort to calculate and plot incremental steps in a fabrication operation, but this is not necessary as long as the tubing is reduced smoothly from one size to the next. An example would be the tube reduction (rocking) of a tube over a tapered mandrel, where no abrupt changes in wall thickness or tube diameter take place. One would expect the initial and final points to be connected by a smooth curve on the Q versus R_A graph. It has been found that the textures in tube reduced tubing respond quickly and predictably to fabrication path changes. This is not the case for mandrel-drawn tubing. At fabrication steps near Q = 1 or at constant values of D ratio (neutral fabrication), the textures produced in drawn tubing do not coincide with the general theory of fabrication and texture formation. Specifically, texture analysis indicates that the tubing tends to undergo a sinking operation, even at Q numbers approaching Q = 2. These observations are for tubing drawn through dies with 30 deg included angles. Other die angles could conceivably modify the results.

The purpose of this study is to explain the discrepancy between the texture predicted for the fabrication method used and the texture actually produced. The Q ratio is defined as wall thickness reduction divided by inside diameter reduction (R_W/R_{ID}). This ratio, although useful as a measurement of overall deformation pattern, does not give clear information about what is happening within the wall of the tubing as it passes through the die. The latter information can be obtained by the use of true strain values rather than the engineering values previously discussed. We define three strains which produce a change from one tube size to another. These are:

$$\bar{\epsilon}_W = \ln \frac{W_f}{W_i}, \quad (1)$$

$$\bar{\epsilon}_c = \ln \frac{D_f}{D_i}, \quad (2)$$

and

$$\bar{\epsilon}_L = \ln \frac{L_f}{L_i}, \quad (3)$$

where

$\bar{\epsilon}_W$ is true wall strain,

$\bar{\epsilon}_c$ is true circumferential strain,

$\bar{\epsilon}_L$ is true length strain,

W_f is final wall thickness,

W_i is initial wall thickness,

D_f is final diameter,

D_i is initial diameter,

L_f is final length, and

L_i is initial length.

Since, in normal metalworking practice, the volume change is negligible, one can equate

$$\bar{\epsilon}_W + \bar{\epsilon}_c + \bar{\epsilon}_L = 0 . \qquad (4)$$

It is apparent that, for any volume element in the tube wall, a knowledge of the amounts of length strain and circumferential strain leads directly to the corresponding wall strain. This information results, in turn, in a value for the strain ratio $(\bar{\epsilon}_W/\bar{\epsilon}_c)$ which is the microscopic analogue to the Q ratio. One helpful property of true strains is that they are directly additive. If one were to divide a tube length into segments, as will be discussed later, and then draw that tube through a die one could, by direct measurement and calculation, obtain values of $\bar{\epsilon}_c$ and $\bar{\epsilon}_L$ for each segment. From these one could obtain $\bar{\epsilon}_W$ and the strain ratio $(\bar{\epsilon}_W/\bar{\epsilon}_c)$. Assuming subdivision of a tube section into "n" segments, the cumulative strain that an initial volume element undergoes, as it proceeds from one segment position to the next until it reaches segment "n," is:

$$(\bar{\epsilon}_W)_{\text{cumulative}} = (\bar{\epsilon}_W)_{0\to1} + (\bar{\epsilon}_W)_{1\to2} + \ldots\ldots + (\bar{\epsilon}_W)_{n-1\to n} \qquad (5)$$

$$(\bar{\epsilon}_c)_{\text{cumulative}} = (\bar{\epsilon}_c)_{0\to1} + (\bar{\epsilon}_c)_{1\to2} + \ldots\ldots + (\bar{\epsilon}_c)_{n-1\to n} \qquad (6)$$

$$(\bar{\epsilon}_L)_{\text{cumulative}} = (\bar{\epsilon}_L)_{0\to1} + (\bar{\epsilon}_L)_{1\to2} + \ldots\ldots + (\bar{\epsilon}_L)_{n-1\to n} \qquad (7)$$

and

$$\left(\frac{\bar{\epsilon}_W}{\bar{\epsilon}_c}\right)_{\text{cumulative}} = \frac{\sum_{i=1}^{n}\left(\bar{\epsilon}_W\right)_i}{\sum_{i=1}^{n}\left(\bar{\epsilon}_c\right)_i} \qquad (8)$$

As stated in the Background section of this paper, the basal poles in Zircaloy tend to line up with the compressive strain direction. Therefore, any increment of reduction of a tube as it passes through a die results in a strain ratio. That strain ratio determines the type of texture that starts to form during that reduction and the amount of reduction through which that ratio acts determines how far the texture change will evolve. Since the cumulative strain and the cumulative strain ratio depend upon prior strains, so must the texture depend upon prior history. A succession of segments of reduction which each involve a strain ratio greater than unity will evolve a texture characteristic of ironing (basal poles radial). On the other hand, another succession of reduction segments in which the strain ratios in the first half are less than unity and in the second half are greater will cause the basal poles to first migrate toward the tangential direction and then reverse and tend toward the radial direction. Their final position would depend upon relative amounts of reduction and the magnitudes of the strain ratios involved.

RESULTS AND DISCUSSION

As stated earlier, there existed an apparent breakdown between theory and practice when attempts were made to predict texture formation in mandrel-drawn tubing with Q numbers near unity. This discrepancy has been resolved and the results will be presented; first, in an analysis of a specially designed drawing experiment, and, second, in an analysis of a tubing fabrication sequence that started with an as-extruded tube shell and ended with finished, drawn tubing.

The machined tube specimen shown in Fig. 4 was drawn as described previously, and the resulting longitudinal cross section is shown in Fig. 5. A scale drawing is shown in Fig. 6 together with the average true area strain. It is immediately apparent that the tube began to respond to the approach of the die some distance away from first contact, specifically at the dotted positions

Fig. 5. Photograph of a longitudinal cross section of the tube shell after drawing the marker holes to the middle of the die.

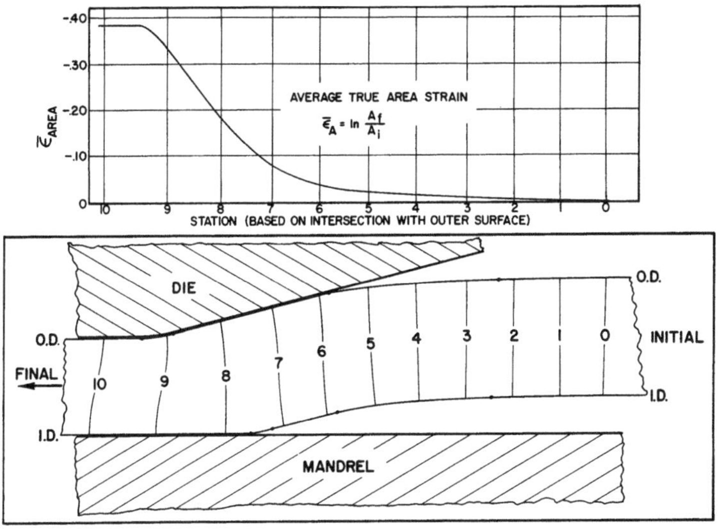

Fig. 6. Scale drawing of the as-reduced section with the center lines of the marker holes shown as curved lines and numbered as stations. Average true area strain is plotted versus axial position.

between stations 2 and 3. Both the inner and outer surfaces began to reduce at approximately the same time. The outer surface contacted the die at the dotted position just before station 6. From that point both surfaces remained parallel until the inner surface began to respond to the approaching mandrel. Macroscopically, the tube was undergoing sinking up to this point. At the inner surface point just past station 7 wall thinning started to occur and continued until the outer surface cleared the die land between stations 9 and 10. It should also be noted that there was a tendency for the drilled holes to remain perpendicular to the tubing surfaces. This involved a metal flow opposite, in the case of stations 6 through 9, to the direction the metal would flow if friction were causing shearing of the metal. No evidence was found for shearing or smearing of the metal as it traversed the die.

A third point to be noted is that axial elongation of the tube occurred in two major steps. Minor elongation took place during the initial sinking of the tube, before mandrel contact. The first major elongation occurred in the inner half of the tube wall, following mandrel contact. This is shown most clearly between stations 8 and 9, where the foot of station 9 has moved farther away from station 8 than the top has. The second major elongation occurred in the outer half of the wall as the die continued to reduce the outer diameter while the inner diameter remained essentially constant over the mandrel. This is noticeable between the tops of stations 9 and 10 whose spacing now approximates their foot spacing.

As shown in Eq. (4), true wall strain values can be obtained from true circumferential and elongation strains. These latter strains are available by either direct measurement or by calculation from the specimen or the scale drawing. The results of such operations are shown in Fig. 7 as cumulative strains calculated from Eqs. (5) through (7). Positive values denote cumulative elongation or thickening and negative values represent cumulative contraction or thinning. Length strains were always positive, and circumferential strains were always negative, as would be expected. Wall strain, however, varied from positive to negative, both with position through the wall and with axial position. Specifically, wall expansion took place at stations 4, 5, 6, and 7 on the inner surface and stations 5, 6, and 7 on the outer surface. Incremental strains at any station are the differences between the cumulative strains at that station and the corresponding cumulative strains at the preceding station, again as related by Eqs. (5) through (7).

The circles in each segment represent the centroids of those areas and the numbers are the segment volumes. Theoretically the volumes should be equal. Since the drilled holes were not parallel originally, due to drill leadoff, the volumes showed some variation, but not more than 5% from the 1600 mm^3 average volume.

Figure 8 is a plot of cumulative strain ratio versus station for the inner and outer surfaces. It is obvious that the tube underwent heavy sinking during most of its axial travel through the die. The inner surface, during its approach and contact with the mandrel, received increasingly high strain ratios ($\bar{\epsilon}_W/\bar{\epsilon}_C$) whereas the outer surface was heavily ironed only in the final segment, between stations 9 and 10. It was expected that the texture of the outer surface would have basal pole concentrations closer to the tangential directions and that the inner surface would have the basal poles concentrated closer to the radial direction. The textures found in this tube, both before and after passage through the die, agreed with theory and expectation. Figure 9 illustrates the initial and final textures for both surfaces. The initial textures are spotty and ill-defined, characteristic of forged and annealed Zircaloy. The final textures, even after the small amount of reduction in area the tube underwent, are relatively sharp. They show, beyond question, the effects of the different strain ratios. The texture of the inner, more heavily ironed, surface has a peak basal pole intensity concentrated approximately 15 to 30 deg from the radial direction toward the tangential. The outer surface

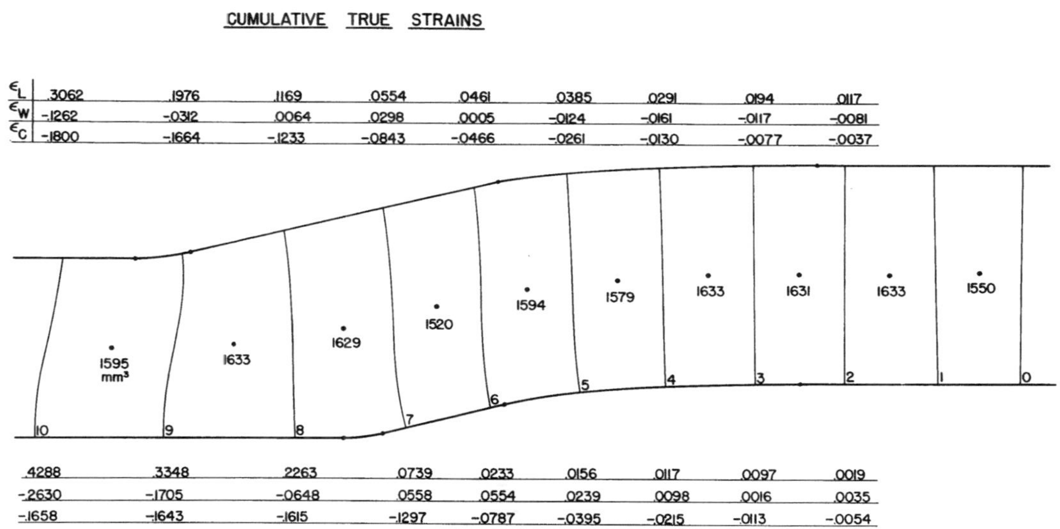

Fig. 7. Scale drawing of the section with cumulative inner and outer surface strains listed.

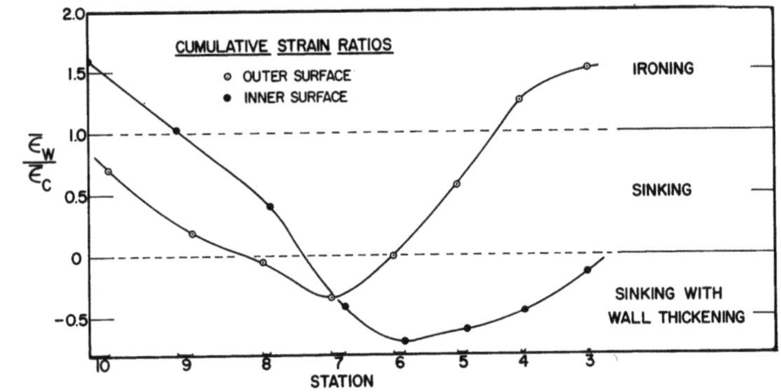

Fig. 8. Cumulative surface strain ratios plotted against axial position. The three types of fabrication are shown.

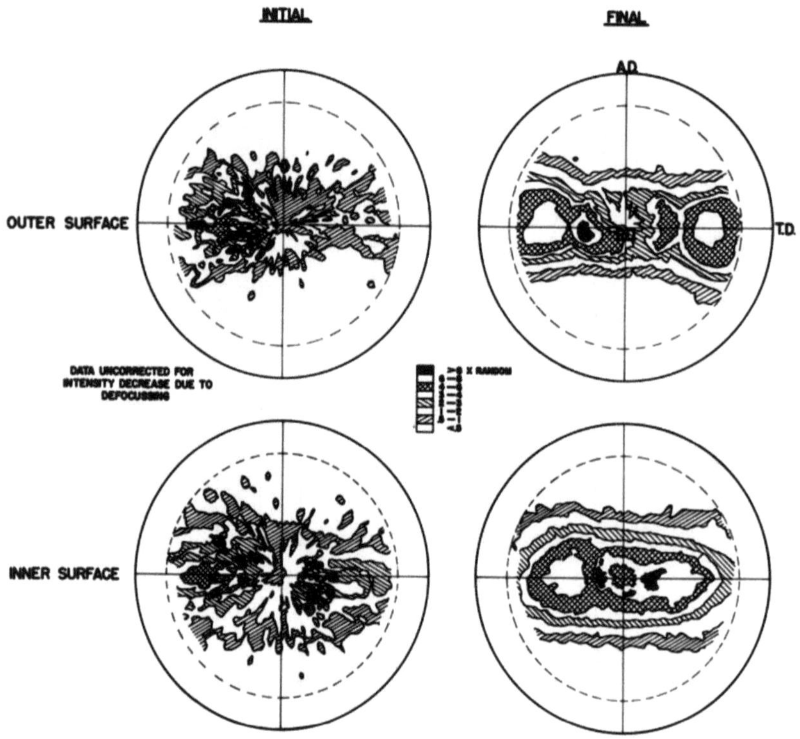

Fig. 9. Pole figures showing the textures present in the inner and outer surfaces of the tube shell before and after drawing.

texture shows the effects of sinking; its basal pole concentration is 30 to 50 deg from the radial direction toward the tangential.

General fabrication-texture theory, which utilizes the Q versus R_A graphs, and mandrel-drawing theory, as discussed in this paper, were combined to analyze a complete fabrication schedule, which included extrusion, tube reduction and drawing.

The extruded tube shell was found to have basal poles concentrated very strongly in the tangential direction throughout the wall, a very common extrusion texture in Zircaloy. The first of the two tube reduction passes was performed at an average wall-to-diameter strain ratio of 2.07 and the poles were found to peak approximately 15 deg from the tangential direction. Some migration toward the radial direction had occurred. It is felt that most of the texture change resulting from this pass represented consolidation of the somewhat diffuse extrusion texture. The second tube reduction pass, at a strain ratio of 7.62, caused massive migration of basal poles into peaks approximately 30 deg on each side of the radial direction. This is the stable ironing texture for Zircaloy and is analogous to the sheet-rolling texture.

The tube resulting from the second tube reduction pass was used as starting material for a series of drawing passes performed under supposedly ironing conditions. Figure 10 illustrates the texture data obtained from this series. The data are plotted as intensity (times random) versus the distance, in degrees, from the radial direction (RD) to the tangential direction (TD) of the tubing along the equator of the pole figure. The data represent averages of both sides of the conventional pole figure. These intensity data were corrected for the dropoff toward the tangential direction that is characteristic of the Shultz method and are completely quantitative.

From previous discussion of the drilled tube, it would be expected that the inner and middle layers of the tubes given the first three drawing passes would approach the stable ironing texture (±30 deg from the radial direction). The outer surface, which has been demonstrated to undergo a large amount of sinking during fabrication at overall strain ratios between unity and two, would be expected to show basal pole migration away from the stable position toward the tangential direction. This was found to be true and is shown in the first three sections of Fig. 10. The final drawing pass was a moderate sinking pass and is shown in the fourth section of the figure. The total movement of the outer surface intensity peaks during the four drawing passes was from a ±25 deg position to a ±48 deg position. A movement from ±25 to ±38 deg occurred during the three ironing passes. The path of the outer surface texture peak is shown as a dotted line in the fourth section.

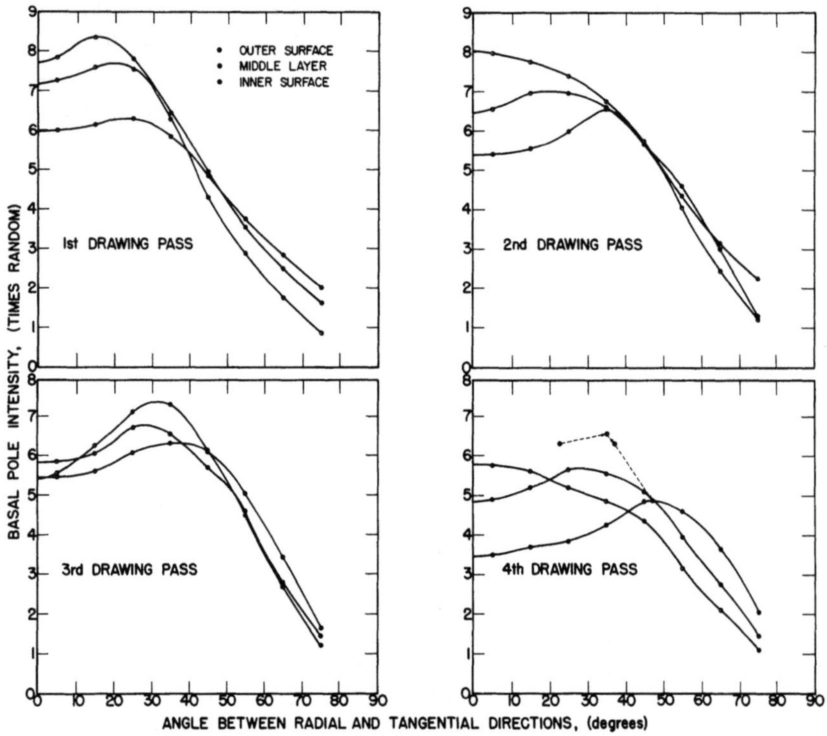

Fig. 10. Basal pole intensity plotted as a function of position, in degrees, from the radial to the tangential direction. Textures for three positions in the tube walls are shown. Texture data were taken following each of four drawing passes.

CONCLUSIONS

It is concluded from this study that textures in mandrel-drawn Zircaloy tubing can be predicted from analyses of the deformation the tubing undergoes. It has been shown that drawing schedules that seem to be moderately ironing in nature can produce large amounts of sinking deformation in the inner layers of the tube wall.

This study was prompted by a need to develop methods for controlling texture in Zircaloy tubing. However, although the study was confined to Zircaloy, the deformation analyses that have resulted are of general use to designers of fabrication schedules that involve mandrel drawing. The large strain reversal that occurs as the tube contacts the mandrel could be responsible for inner surface cracking in materials of moderate to low ductility. A series of rod-mandrel drawing passes, even if each pass were an overall ironing operation, would cause repeated reversals from wall thickening, $(\overline{\epsilon}_W/\overline{\epsilon}_c) < 0$, to wall thinning, $(\overline{\epsilon}_W/\overline{\epsilon}_c) > 0$, in the incremental layer near the inner surface. Such reversals could be eliminated by the use of either fixed or floating plug mandrels. By suitably tapering the exit end of the plug, the inside diameter reduction could be controlled and the strain reversal eliminated. In such a case the fabrication step could be plotted on a Q versus R_A graph such as the one shown previously in Fig. 2.

The fact that the first part of the reduction in a rod-mandrel drawing pass is by sinking is not original to this study. What we have attempted to show is the magnitude of that sinking reduction in terms of true strains and strain ratios and correlate those with the textures that form. The texture formation characteristics of Zircaloy are a useful tool in such an examination since they provide a strong indication, by their migration, of the strains the tubing is undergoing.

REFERENCES

1. D. O. Hobson and P. L. Rittenhouse, "Nonhomogeneous Deformation in Zircaloy Tubing," Trans. Met. Soc. AIME 245(4), 797–800 (April 1969).

2. D. O. Hobson, "Texture Changes Produced During Zircaloy-4 Tubing Fabrication: From Forged Billet to Finished Tubing," pp. 37–49 in Applications-Related Phenomena for Zirconium and Its Alloys, Spec. Tech. Publ. 458, American Society for Testing and Materials, Philadelphia, 1969.

3. E. Tenckhoff and P. L. Rittenhouse, "Texture Development and Texture Gradients in Zircaloy Tubing," Ibid., pp. 50–67.

4. D. O. Hobson and E. Tenckhoff, "Effects of Fabrication Parameters on Textures Produced in Zircaloy Cladding," (summary) Trans. Am. Nucl. Soc. 12(1), 134 (June 1969).

5. D. O. Hobson, "Texture Control in Zircaloy Tubing," (summary) Trans. Am. Nucl. Soc. 12(2), 571–572 (November 1969).

EXPERIMENTAL EVALUATION OF THEORETICALLY IDEAL DRAWING DIES

M. L. Devenpeck

Edgar C. Bain Laboratory for Fundamental Research

United States Steel Corporation, Monroeville, Pa. 15146

ABSTRACT

On the basis of plasticity theory, Richmond et al. [1,2,3] have derived strip and wire drawing die profiles which require minimum work and result in minimum tensile stresses and strains during a given reduction. The fact that these 'ideal' dies do give a minimum tensile stress and strain suggests that they also may give least structural damage. To test this possibility, stainless steel strip and wire were drawn through dies of various contours and the structural damage was evaluated in terms of density decrements. Drawing work efficiencies and tensile properties of the drawn material also were evaluated. The drawings consisted of 50% single reductions of annealed 410 stainless steel 0.375 in. (0.952 cm) strip using nine dies, and 65% two pass reductions of annealed 305 and cold worked 416 stainless steel 0.250 in. (0.635 cm) diameter wires using five dies.

With few exceptions, the ideal dies produced less density decrement and had greater drawing work efficiencies than the other dies. These improvements were large where the die profiles differed most from the ideal profiles, but they were small for linear approximations of the ideal profiles. A convex circular arc profile (similar to rolling) gave the largest density decrement and least work efficiency for the 410 strip. For wire, the greatest decrement occurred at large angles whereas the least efficiency occurred at small angles, the optimum angle for least decrement being smaller than that for the least work. The drawn wire showed a weak correlation of increasing tensile elongation with decreasing density decrement. This small effect may indicate the existence of greater effects in fatigue properties (not tested). The zero entrance angle condition of ideal dies was found to be significant, and it is suggested that the exit should also be smooth and have a zero angle.

INTRODUCTION

This study was undertaken to evaluate the performance of strip and wire drawing dies which are contoured with theoretically determined ideal profiles as given by Richmond et al. [1,2,3] These ideal profiles are obtained from rigorous and complete slip-line solutions of nonhardening plasticity theory for a frictionless draw. They are 'ideal' in that they theoretically require minimum work to make a draw and impart minimum and uniform strain to the drawn piece.

Previous strip drawing experiments [4] demonstrated that an ideal 'sigmoidal' profiled die [1] did indeed require the least work and imparted the least strain among a number of dies of equal axial length. Of more significance however was the finding that the fatigue life was greatest for the material drawn through the ideal die. The reason for this was not specifically determined but it seems reasonable that any structural damage generated during the draw would be detrimental to fatigue properties.* Indeed, Rogers [5] has found that for strip there can be an appreciable effect of structural damage upon the fatigue properties of drawn and rolled materials, the magnitude depending strongly upon the material. Rogers et al. [5,6,7] and Coffin and Rogers [8] have established that in the drawing of strip, structural damage of the strip increases with the 'hydrostatic' tensile stress that exists within the deforming region. Thus, since theory indicates that the ideal dies develop minimum 'hydrostatic' tensile stresses for a frictionless draw, the hypothesis arises that structural damage should be minimal when such dies are used. The main purpose of the present work was to establish whether such is the case, i.e., do ideally profiled dies minimize structural damage in the drawing of strip and wire? To test this hypothesis, 410 stainless steel strip, and 305 and 416 wire were drawn through dies with various contours and density decrements were determined to assess the structural damage. To further evaluate the performance of the dies, drawing work efficiences and the tensile properties of the drawn material also were determined.

---

*) Subsequent density measurements did not detect any density decrements beyond the $\pm$ .003% precision of the measuring technique. Consequently, no correlation could be made between fatigue properties and density decrements.

EXPERIMENTAL DESIGN

Strip. Nine tool steel dies which separately effected a 50% reduction on a 0.375 in. (0.952 cm) thick strip in a single pass had the following profiles:

Ideal:	Sigmoidal [1]
Ideal 'Sigmoidal' Approximations -	
Linear:	$2\theta = 19.29°$
Concave Circular Arc:	R = 1.671 in.(4.244 cm)
Convex Circular Arc:	R = 1.671 in.(4.244 cm)
Ideal:	Minimum Length [3,9]*
Ideal 'Minimum Length' Approximations -	
Linear:	$2\theta = 29.13°$
Concave Circular Arc:	R = 0.740 in.(1.880 cm)
Convex Circular Arc:	R = 0.740 in.(1.880 cm)
Linear, Exit Angle of 'Minimum Length':	$2\theta = 39.75°$

where 2θ is the total included angle of the linear dies and R is the radius of curvature of the circular arc dies. As shown in Fig. 1, the 'sigmoidal' die was curved throughout and had zero entrance and exit angles whereas the 'minimum length' die began with a zero entrance angle curved section and ended with a 19.88° linear section. Consequently, the 'minimum length' die was about one-third shorter than the 'sigmoidal' die. The approximation profiles passed through the entrance and exit points of the ideal profiles and thus had the same axial length excluding any land. The concave and convex circular arc approximations were respectively tangent to the strip at the entrance and exit. The 'sigmoidal' die and the three approximations thereof had no exit lands, but all other dies had a 0.015 in. (0.38 mm) exit land. There were no entrance lands.

---

*) Ref. 3 describes the development of the 'minimum length' die for axisymmetric (wire) drawing. For 2-dimensional (strip) drawing, Morrison [9] has determined the 'minimum length' die profiles by transforming the axisymmetric field equations into the 2-dimensional case.

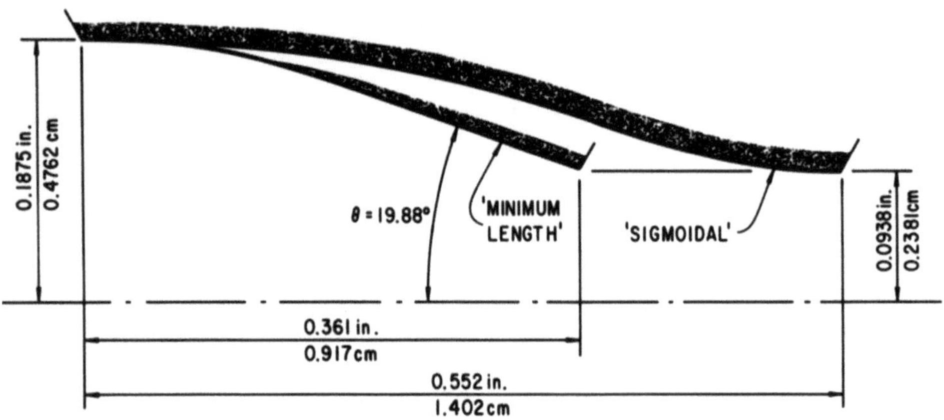

Fig. 1 - Ideal 'Sigmoidal' and 'Minimum Length' Strip Drawing Die Profiles.

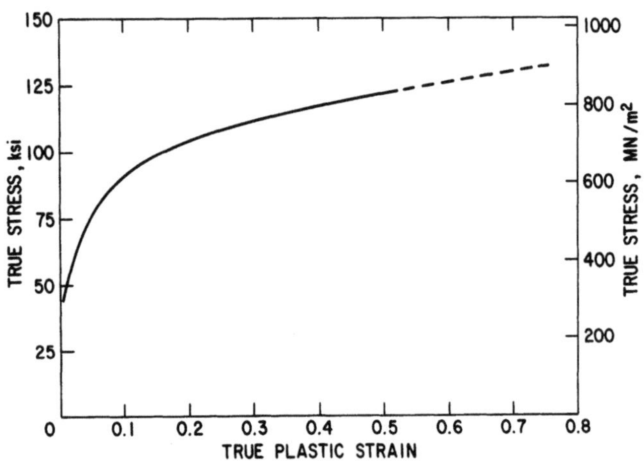

Fig. 2 - True Stress-True Plastic Strain Curves for Original 410 Stainless Steel Strip in Compression.

The drawing material originated from a 0.40 in. (1.0 cm) thick stable ferritic 410 (0.14 C, 12.0 Cr) stainless steel plate which was received in a hot rolled, annealed, and pickled condition. The stress-strain curve for this plate material, Fig. 2, was determined from the compression of a 0.175 in. (0.429 cm) diameter by 0.350 in. (0.858 cm) length (from the thickness) sample which was lubricated the same as the strip drawing (described below). Strips, 16 in. (0.4 m) long, were cut parallel to the rolling direction and heat treated at 1350°F (732°C) for 30 min. and air cooled. The strips were then longitudinally surface ground with a 220 wheel to a 0.375 in. (0.952 cm) thickness by a 1.000 in. (2.54 cm) width. The strips were indented, assembled with the dies and drawn as described in Ref. 4 with two exceptions. First, lubrication was made by spraying the strips with a resin bonding PTFE (polytetrafluoroethylene) resin coating and wetting with SAE 30 motor oil. Second, only single strips were pulled and were free to expand within the width of the die during drawing. Approximately 12 in. (30 cm) of drawn material was secured at a rate of 1 in./min. (0.4 mm/sec) for each die.

Wire. Two stainless steel 0.25 in. (0.64 cm) diameter wires with stable structures, austenitic 305 (0.033 C, 17.6 Cr) and ferritic 416 (0.091 C, 13.9 Cr, 0.81 Mn, 0.38 S) were drawn to a total of 65% reduction in area in a two-pass, 42% reduction each, sequence. The die contours used consisted of 7, 14, and 35° (total angle) blended cones and the ideal 'minimum length' profile. [3] As shown in Fig. 3, the ideal 'minimum length' die for wire drawing is curved throughout its entire length and has a zero entrance angle but a nonzero exit angle. The 14 and 7° dies are conical approximations to the 'minimum length' and 'sigmoidal' dies respectively. 'Sigmoidal' dies were not constructed as their additional length gives significantly greater surface area and thus increases the effects of friction.* The conical dies had lands of about

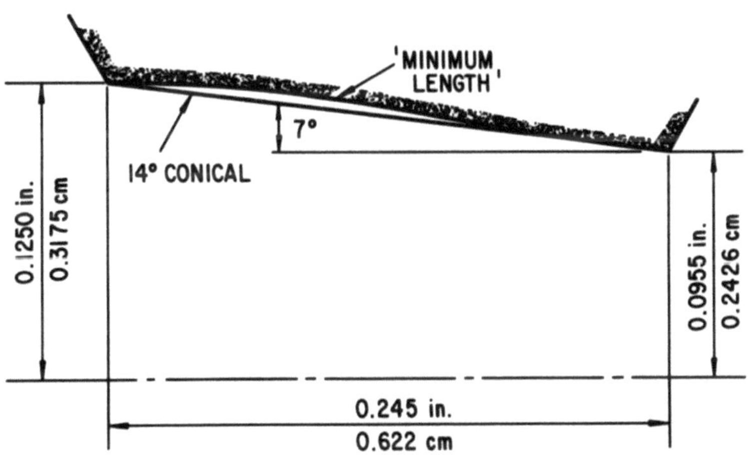

Fig. 3 - Ideal 'Minimum Length' and 14° Conical Wire Drawing Die Profiles.

*) To reduce the total frictional force by reducing the length of the die was the motivation for developing the 'minimum length' die. Also, frictional considerations are much more important for wire than for strip since the surface area to volume ratio is much greater for wire.

0.04 in. (1 mm) whereas the ideal dies had virtually no lands. All wire dies were constructed of tungsten carbides, the ideal die contours being made by electrical discharge machining. The greatest deviation of the machined ideal profiles from the calculated was at the entrance. The actual entrance profiles deviated from the theoretical by about 4° which led to maximum radial deviations of .003 in. (80 μm).

The 305 wire was obtained in an annealed and oxidized coil. The lubricant completely covered the oxide during drawing and thus the oxide was not removed. The 416 wire was received in 12 ft. (3.7 m) straight lengths, cold drawn and stretched. It contained approximately 30% prestrain (26% reduction in area). The stress-strain curves for the as received wires as obtained in tension at a 0.02/min. nominal strain rate are shown in Fig. 4.

The drawing operation consisted of pointing the wire, cleaning with methanol, coating with a water solution of sodium soap, allowing to dry, and drawing at a rate of 20 in./min. (0.85 cm/sec) in a horizontal screw driven universal testing machine. The wires were relubricated between passes. The drawing load was recorded continuously with time as approximately 6 ft. (2 m) lengths of material were drawn through each die. Single lengths of material were drawn for each condition except for the 416 wire where multiple lengths were drawn to evaluate the effects of the variation that existed among the different stock wire lengths, effects that were not fully appreciated at the onset of testing.

Densities. The densities of strip and of wire were determined by the method of "hydrostatic weighing," with monobromobenzene as the heavy liquid. The apparatus is described in Ref. 10. The

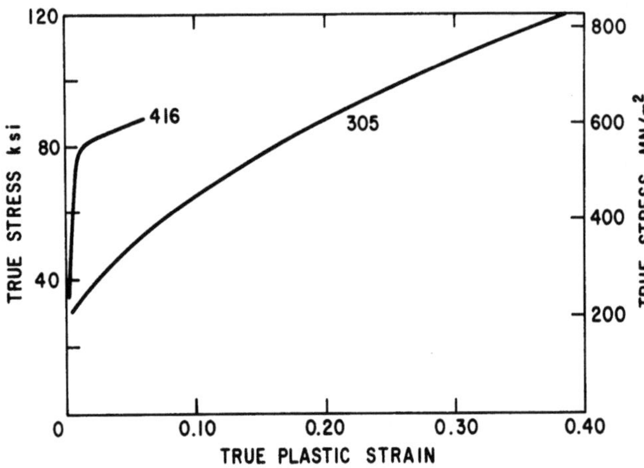

Fig. 4 - True Stress-True Plastic Strain Curves for as Received 305 and 416 Stainless Steel Wires in Tension.

density values were corrected to vacuum, to 25°C and for any deviation of the density of the heavy liquid from its nominal value. For strip, the determinations were made on samples taken from the center of the width which contained the full thickness of the drawn strip. These samples, 0.19 in. (4.8 mm) thick by 0.38 in. (9.5 mm) wide by 3.5 in. (9 cm) long, weighed 35 g. To reduce the possibility of air entrapment, they were machined with a suspension hole through a thin fin at one end and lightly polished. The density of the original material was determined from a single specimen containing the full thickness of the undrawn strip. For wire, the specimens, which were 3.5 in. (9 cm) long, weighed 7 g for drawn wire and 20 g for original wire. At the end of each specimen, which was polished to remove surface imperfections, a short length was machined to a 0.05 in. (1 mm) diameter, polished, and bent to make a suspension hook. For each draw, a density decrement was calculated and related to the decrement obtained with the ideal 'sigmoidal' die for strip or with the ideal 'minimum length' die for wire.

Drawing Work Efficiencies. To evaluate the work performance of the dies, drawing work efficiencies were determined by comparing the estimated theoretical drawing work with the actual drawing work, per unit volume (specific work). The actual specific drawing work is equal to the drawing force times the drawing velocity divided by the volume flow rate, which reduces to simply the draw stress. The theoretical specific drawing work is equal to that which is required to make the same shape change in simple tension or compression and is equal to the area under the stress-strain curve of the original material from zero strain to the strain of the shape change. The equivalent strain* was used for the strip reductions to account for the effect of spreading. The stress-strain curve for the strip was obtained from the axisymmetric compression of the original material (Fig. 2) and for wire was obtained by extrapolating the tensile results of the original wire (Fig. 4) to that of the drawn wire, the drawn wire tensile strain being displaced by the shape change strain.

Tensile Properties. Tensile properties of the drawn material were determined from specimens tested in tension at a nominal strain rate of 0.02/min. For strip, single longitudinal tensile specimens were taken from the center of the width and contained the full thickness of each drawn strip. The gage sections were 0.250 in. (0.635 cm) wide by 1.00 in. (2.54 cm) long. An extensometer of magnification 500X was attached to the gage lengths to

---

*) Equivalent strain: $\bar{\varepsilon} = \sqrt{\frac{2}{3}} \sqrt{\varepsilon_1^2 + \varepsilon_2^2 + \varepsilon_3^2}$

evaluate the 0.2% offset yield strength and elongations were read from 1.00 in. (2.54 cm) stamped gage marks. For wire, tensile properties for each die geometry were determined from five samples of wire as drawn. To obtain a 0.2% offset yield strength, a 1.00 in. (2.54 cm) gage length 250X extensometer was attached to the specimens at the center of a 3 in. (7.6 cm) length exposed between the faces of the grips. Elongations were read from 2.00 in. (5.08 cm) stamped gage marks. In addition to the 0.2% offset yield and elongation, tensile strength and reduction in area of fracture were determined.

RESULTS

The density results are given in Table I for all cases and the density decrements are given in Fig. 5 for the 305 and 416 wire. The drawing conditions and work efficiencies are given in Table II for the 410 strip and in Table III for the 305 and 416 wire, the work efficiencies for the 305 and 416 wire being illustrated in Fig. 6. The tensile properties of the drawn material are given in Table IV.

Density Decrements. In Table I, the relative density decrements were calculated with respect to the following 'ideal' cases: the 'sigmoidal' for the 410 strip; the 'minimum length' for the 305 wire; and the average of the 'minimum length' values for the 416 wire. For the 410 strip, single density specimens were taken from single lengths drawn through each die. For the 305 wire, multiple density specimens were taken from single lengths drawn through each die. For the 416 wire, multiple density specimens were taken from multiple lengths drawn through each die. The densities given in Table I are for each specimen, and are averages where more than one determination was made. Twenty-three redeterminations were made, of which, only three were beyond the stated precision.* The average density decrements for each 305 and 416 wire are shown in Fig. 5 with brackets indicating the combined effects of the precision of the measurement and the longitudinal variations in the wire. In some 416 wires, actual variations along the wire were greater than the precision of the measurements. Density decrements were not calculated where the base density was unknown except for the 35°, 416 wire case, where it was necessary to do so as no known base wire existed. This calculated value is subject to unknown error, the variation for known bases being indicated with brackets in Fig. 5.

*) For density, the stated precision corresponds to 90% confidence limits.

Table I

Mass Densities (g/cm^3) of Drawn 410, 305, and 416 Stainless Steel

Material	Die	Original	Final	Average Decrement	Relative Avg.Decr.
		± .0001	± .0002	± .003	± .03
410 (Strip)	Ideal Sigmoidal	7.7262	7.7198*	0.083%	1.00
	Concave R = 1.67 in	↓	7.7200*	0.080%	0.96
	Linear 2θ = 19°	↓	7.7197	0.084%	1.01
	Convex R = 1.67 in	↓	7.7188*	0.096%	1.16
	Ideal Min. Lngth	↓	7.7193*	0.089%	1.07
	Concave R = 0.74 in	↓	7.7186	0.098%	1.18
	Linear 2θ = 29°	↓	7.7195	0.087%	1.05
	Convex R = 0.74 in	↓	7.7182*	0.104%	1.25
	Linear 2θ = 40°	↓	7.7187	0.097%	1.17
		± .0003	± .0007	± .008	± .08
305 (Wire)	Ideal	7.9254	7.9170* 7.9171	0.105%	1.00
	7°	↓	7.9150 7.9156*	0.127%	1.21
	14°	↓	7.9139 7.9140* 7.9143*	0.143%	1.37
	25°	↓	7.9157 7.9163*	0.119%	1.13
	35°	↓	7.9139	0.145%	1.38

Table I (continued)

Mass Densities (g/cm^3) of Drawn 410, 305, and 416 Stainless Steel

Material	Die	Original	Final	Average Decrement	Relative Avg.Decr.
		± .0003	± .0007	± .008	± .015
416 (Wire)	Ideal	7.6547*	7.6138	0.534%	1.05
		7.6551	7.6141		
			7.6141		
	Ideal	7.6527*	7.6147	0.483%	0.95
			7.6167		
	Ideal	7.6515*	7.6127	0.505%	1.00
			7.6129		
			7.6130		
	7°	7.6547*	7.6130	0.536%	1.06
		7.6550*	7.6148		
	7°	unknown	7.6141		
	14°	unknown	7.6136		
			7.6160		
	14°	7.6527	7.6105	0.540%	1.06
			7.6109*		
			7.6115		
			7.6125		
	14°	7.6543*	7.6135*	0.525%	1.03
			7.6137*		
			7.6151*		
	25°	unknown	7.5986		
	25°	7.6549	7.5959	0.771%	1.52
	35°	unknown	7.5752	(1.02%)**	(2.0)**

*) More than one determination made.

**) Calculated using average base, unknown error.

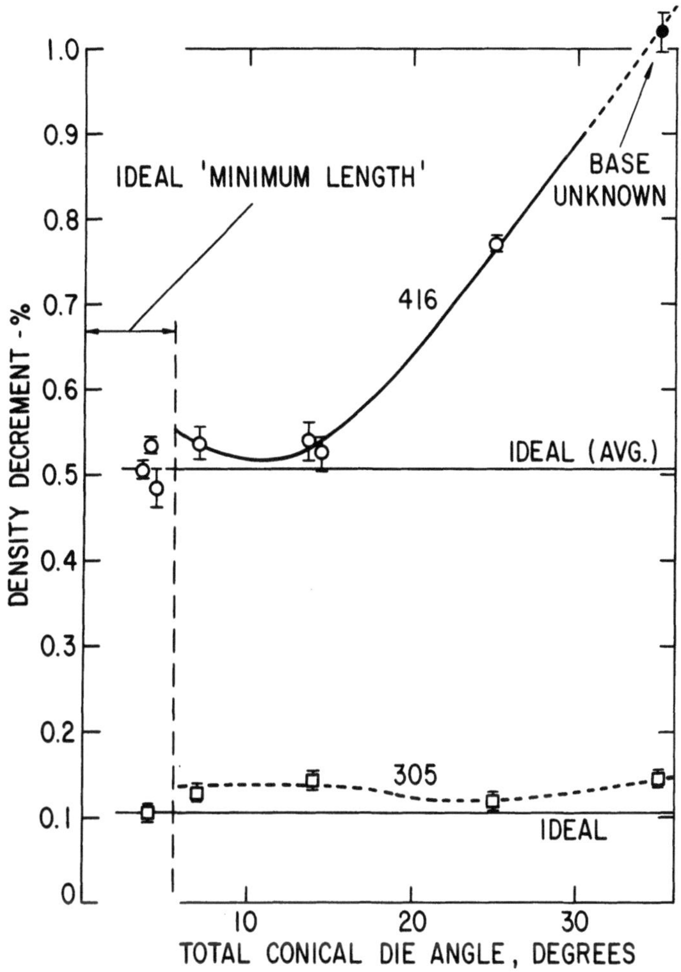

Fig. 5 - Density Decrement vs. Conical Die Angle for 305 and 416 Drawn Wirc.

The linear and conical dies all produced greater density decrements than did the selected ideal dies. The differences between the ideal and the best linear and conical cases observed, however, were small in comparison to the variation of decrement with angle. In general, the decrement increased with angle, the 416 data indicating a possible minimum decrement between 7 and 14° (Fig. 5). For the 305 wire, the variation in density with angle in some cases was no greater than the ± .0007 g/cm^3 precision of the measurements and thus no definite trends could be detected. These data do however, show the greatest difference in relative density decrement between the ideal and the best conical cases. The 416 wire experienced the largest decrements (0.5 to 1.0%) and was the most sensitive to die angle. Cursory metallographic examination revealed the presence of fractured and separated stringer particles, presumed to be MnS.

For strip, the density decrements increased, in general, with increasing deviation from the respective ideal die contour, as listed in Table I. The ideal 'minimum length' die and the concave circular arc approximation thereof had notably greater decrements than the 'sigmoidal'. The concave circular arc approximation of the 'sigmoidal' profile die gave the least decrement, whereas both convex circular arc profiles (similar to rolling) gave the greatest decrements.

Drawing Work Efficiencies. The drawing work efficiencies are given in Tables II and III for the strip and wire respectively. The error range, ± 1, refers to the precision of the efficiencies and not to the accuracy. In all cases, the drawing work efficiencies were the greatest for the ideal dies (based on averages for the 416 wire). The drawing work efficiencies for the 410 strip as given in Table II* indicate only a minor loss of efficiency for the approximation profiles except for the convex circular arc cases.

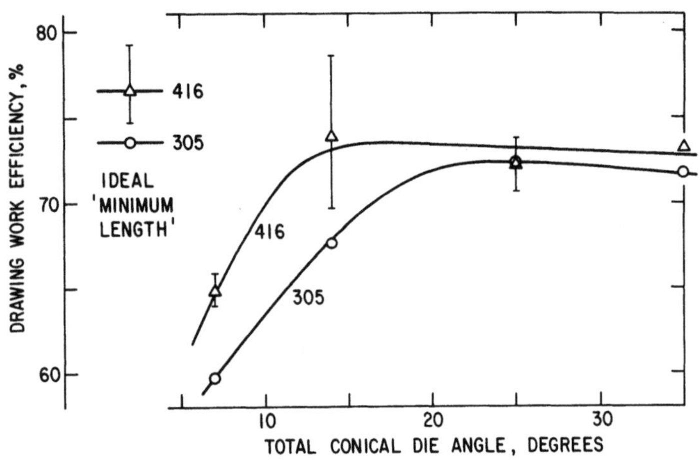

Fig. 6 - Drawing Work Efficiency vs. Conical Die Angle for 305 and 416 Wire Drawing.

---

*) The author is well aware that the 100% plus efficiency calculations recorded do not indicate perfect forming operations. Rather, they indicate 1) the inability to compare the work of two different deformation modes: drawing with spreading and axisymmetric compression, 2) the lack of isotropy, and 3) the inability to eliminate friction in compression testing (note that the flow stress at 75% strain in Fig. 2 is 132 ksi compared to a tensile strength in the drawn material of 122 ksi).

In Fig. 6, the efficiencies for the 305 and 416 wire are plotted against die angle, where the variations in the 416 wire are indicated with brackets about average values. This figure illustrates a sharp increase and then a slight decrease of drawing work efficiency with die angle. The variations of efficiency with the 416 wire samples for a given die, in general, correlate with the variations in density decrement, i.e., where the efficiency is the greatest, the decrement is the least.

Table II
410 Stainless Steel Strip Drawing

Drawing Speed: 1 in/min; Lubricant: Resin Bonded PTFE + Oil;
Size: 0.375 x 1.000 in.

Die*	Draw Load klbs. ± .1	Draw Stress ksi ± .6	Red.in Thickness % ± .1	Incr.in Width % ± .1	Equiv. Strain % ± .2	Efficiency % ± 1
Ideal Sigmoidal	16.1	77.5	47.4	4.7	71.8	102
Concave R = 1.67 in	16.4	79.2	47.7	5.4	72.1	100
Linear 2θ = 19°	16.4	78.8	47.5	5.3	71.8	100
Convex R = 1.67 in	17.7	87.5	47.6	2.9	73.2	92
Ideal Min.Lngth	16.2	81.4	48.3	2.7	75.2	103
Concave R = 0.74 in	16.7	84.7	48.6	2.3	75.9	100
Linear 2θ = 29°	16.9	83.9	48.0	3.4	73.9	98
Convex R = 0.74 in	19.0	97.4	48.5	0.9	76.3	87
Linear 2θ = 40°	17.2	87.1	48.5	1.7	75.7	97

*) See text for further description

1 klbs = 4.45 kN

1 ksi = 6.89 MN/m^2

Table III
305 and 416 Stainless Steel Wire Drawing

Drawing Speed: 20 in/min; Lubricant: Dried Sodium Soap Coating

Mat'rl	Die	Drawing Load 1st Pass	Drawing Load 2nd Pass	Wire Diameter Original	Wire Diameter 1st Pass	Wire Diameter 2nd Pass	Draw Stress 1st Pass	Draw Stress 2nd Pass	Efficiency
		klbs ± .015	klbs ± .015	in. ± .0002	in. ± .0002	in. ± .0002	ksi ± .7	ksi ± 1	% ± 1
305	Ideal	2.14	2.03	0.2540	0.1920	0.1472	73.9	119	73.6
	7°	2.70	2.42	0.2540	0.1890	0.1470	96.2	143	59.7
	14°	2.32	2.20	0.2540	0.1914	0.1472	80.6	129	67.7
	25°	2.16	2.06	0.2540	0.1912	0.1474	75.2	121	72.3
	35°	2.23	2.11	0.2540	0.1917	0.1450	77.3	128	71.8
416	Ideal	1.95	1.41	0.2495	0.1915	0.1469	67.7	83.2	74.4
	Ideal	1.88	1.30	0.2498	0.1916	0.1470	65.2	76.6	79.3
	Ideal	1.98	1.35	0.2494	0.1914	0.1469	68.8	80.4	75.1
	7°	2.17	1.67	0.2495	0.1889	0.1468	77.4	98.7	63.8
	7°	2.16	1.59	0.2495	0.1889	0.1465	77.1	94.3	65.6
	14°	1.97	1.43	0.2499	0.1910	0.1473	68.8	83.9	73.4
	14°	1.89	1.31	0.2497	0.1912	0.1472	65.8	77.0	78.3
	14°	2.12	1.48	0.2494	0.1911	0.1473	73.9	86.8	69.5
	25°	2.02	1.48	0.2495	0.1905	0.1469	70.9	87.3	70.6
	25°	1.99	1.40	0.2496	0.1905	0.1465	69.8	83.1	73.6
	35°	2.09	1.40	0.2496	0.1899	0.1444	73.8	85.5	73.1

1 klbs = 4.45 kN; 1 in. = 2.54 cm; 1 ksi = 6.89 MN/m^2

Table IV

Tensile Properties of Drawn 410, 305, and 416 Stainless Steel

Mat'rl	Die	Pass	0.2% Yield ksi ± 1.0	Tensile Stngth ksi ± 0.5	% Elong. in 1 in.(strip) 2 in.(wire) ± 1	Reduction in Area % ± 1
410 (Strip)	Ideal Sigmoidal	1st	120	121.4	13	a
	Concave R = 1.67 in	1st	117	120.4	15	
	Linear 2θ = 19°	1st	118	120.5	13	
	Convex R = 1.67 in	1st	115	122.9	a	
	Ideal Min.Lngth	1st	120	122.3	14	
	Concave R = 0.74 in	1st	112	123.2	14	
	Linear 2θ = 29°	1st	113	120.7	14	
	Convex R = 0.74 in	1st	116	122.8	a	
	Linear 2θ = 40°	1st	113	120.9	13	
305 (Wire)	Ideal	1st	129	140	13.8	65
	7°	1st	138	143	11.0	64
	14°	1st	132	140	13.1	65
	25°	1st	131	142	13.0	63
	35°	1st	128	141	13.0	64
	Ideal	2nd	167	177	7.3	60
	7°	2nd	180	182[b]	b	56[b]
	14°	2nd	170	176	4.0	57
	25°	2nd	168	177	7.7	57
	35°	2nd	173	179	6.3	57
416 (Wire)	Ideal	1st	103	108	8.1	52
	7°	1st	104	107	7.6	51
	14°	1st	101	106	8.0	52
	25°	1st	103	107	7.6	51
	35°	1st	101	106	8.0	48
	Ideal	2nd	116	120	6.3	44
	7°	2nd	121	122[b]	b	41[b]
	14°	2nd	114	119	6.0	43
	25°	2nd	115	119	6.0	41
	35°	2nd	116	121	5.5	38

a - Not read
b - All specimens broke in grips

1 ksi = 6.89 MN/m^2

Tensile Properties. There was only a small effect of die contour on the tensile properties of the drawn material as given in Table IV. For the 410 strip, the 0.2% offset yield strength was greatest for the ideal cases whereas there was little difference in the tensile strengths and elongations. Reductions in area were not read as the specimens split at the center plane during fracture. The wires, in general, indicated a weak correlation of decreasing elongation with increasing mass density decrement and decreasing tensile strength with increasing drawing work efficiency. For the 416 case, the tensile results for the first pass are for the first wire and for the second pass for the second wire listed in Tables I and III with the exception of the 35° die. The tensile specimens of both wires drawn through the 7° die, two pass sequence, broke in the grips, and thus no reading could be made of the elongation. Also, the tensile strength determined was necessarily less than the actual and the reduction in area was subject to unknown error.

DISCUSSION

The improvement in reducing the density decrement by the use of dies with ideal profiles was small but prevailed for all conditions tested. The differences in decrement had only a small effect upon the longitudinal tensile properties of the drawn material. Rogers [5] has reported an effect of decrement on the transverse tensile properties and fatigue properties of drawn strip, but neither of these were measured in this study. Any extrapolation of the effects found by Rogers would be speculative, particularly in regard to fatigue, as the effects varied greatly with material. It has been reported by Melville [11] that fatigue properties of wire are appreciably affected by the conical angle of the drawing dies. For 85% total reduction (27% per pass) of a 0.266 in. (0.676 cm) 1085 steel MH patented rod, the wire drawn through a 10° die series had a 50% greater fatigue ratio in reverse bending than the wire drawn through either a 5 or 15° geometry. This peak in fatigue ratio may correspond to a minimum in decrement as suggested by the 416 data, Fig. 5. (The ideal 'minimum length' conical approximation for this reduction sequence is 9°.) Furthermore, the variations in elongations found here may lead to greater variations in fatigue life. Thus, significant effects upon fatigue properties of wire are possible and most likely increase with increasing total reduction.

The improvement in drawing work efficiency with the use of dies with ideal profiles was small but, as for the density decrement, prevailed for all conditions tested.* The variation of efficiency with angle, indicating an "optimum" angle is well known, resulting from the compromise between frictional work, predominate at small angles, and redundant work,‡ predominate at large angles, (for instance see Yang [12], Green and Hill [13], and Caddell and Atkins [14]). The "optimum" angle for minimum work however, was larger than that suggested for minimum density decrement for the 416 wire case (compare Figs. 5 and 6). This corresponds with the results that the drawing work efficiency decreased the most for small angles and the density decrement increased the most for large angles.

As previously mentioned and shown in Table IV, the various die geometries for strip and wire had only a small effect on the longitudinal tensile properties of the materials drawn. The following arguments are presented to aid in the interpretation of this result: 1) Increasing redundant work increases the strain imparted to the drawn material. For a workhardening material this increases the strength of the drawn piece. 2) Increasing structural damage decreases the strength of the drawn material. 3) Both redundant work and structural damage in a drawn piece increase with the angle of the die used for a frictionless draw. 4) Increasing nonuniformity of strain increases residual stress within the drawn material. 5) Increasing friction increases residual stress within the drawn material. 6) Increasing residual stress decreases the 0.2% offset yield whereas it has little effect upon the tensile strength. Thus, the ideal dies may not produce the greatest tensile strength in a material unless the structural damage becomes significant.† They may, however, produce greater 0.2% offset yield strengths where workhardening and friction are low. Furthermore, the ideal dies should produce greater elongations under any conditions. In general, the results of this study follow these expectations.

---

*) The fact that the conical dies had lands whereas the ideal dies had virtually none, detracts from this conclusion. Yang [12] has studied the effects of lands, and calculations based upon Eq. 25 of Ref. 12 indicate that for no lands, there would be a decrease in work of 1 and 2% for the 416 and 305, 14° cases respectively.

‡) Work imparted to the drawn piece which produces strains that do not contribute to the shape change.

†) For very high and maximum attainable reductions, this could very well be the case. In fact, the ideal dies should give the maximum attainable reduction.

Some other characteristics of the results obtained in this program which warrant discussion are as follows. Characteristics of the die profile which appear to give improved results can be seen from the strip tests. The concave circular arc approximation of the 'sigmoidal' die was only 2% less efficient and produced 4% less density decrement than the 'sigmoidal'. Further, the convex circular arc contours required the greatest work and produced the greatest density decrement. The linear dies fell in between. Thus, the zero entrance angle condition is a significant factor in lowering the work and density decrement.* This is consistent with the fact that all existing ideal die solutions require zero entrance angles. On the other hand, the concave circular arc approximation to the 'minimum length' die yielded a relatively large density decrement. This may have resulted from the extreme exit angle (29°) of this die. Thus, limitation of the exit angle appears advisable. The 'minimum length' strip die itself had a rather large exit angle (20°), and this may have been a source of redundant work and increased 'hydrostatic' tension. The angle at the exit can be reduced to zero for an ideal die by choosing a streamline for the die contour nearer to the center line in the theoretical flow field (see Ref. 3). The resulting die profile would then be sigmoidal, i.e., S-shaped, with zero entrance and exit angles, as in the original 'sigmoidal' die profile [1], however, these new profiles would still be shorter than the original 'sigmoidal' profile. In general, the longer the die, the more gradual will be the curvature of the die at the exit. Selection of the best of these profiles could be determined only by trial and error and undoubtedly will depend upon the specific conditions of material and the drawing operation.

The differences in density decrement and efficiency among the multiple 416 wires for a given die are believed to be due to the inability to reproduce the lubricant coating as well as possible variations in the initial prestrain and composition of the various wire lengths. These variations unfortunately introduced unknown errors that precluded any precise comparison of the results.

CONCLUSIONS

Based upon the results of this study in which 410 stainless steel strip and 305 and 416 stainless steel wires were drawn through dies with ideal [1,3,9] and other profiles, the following conclusions were drawn:

---

*) The fact that the actual ideal wire drawing dies did not meet the zero entrance angle condition may have detracted from the possible improvement obtainable in those cases.

1) Dies with theoretically ideal profiles minimize structural damage during drawing over a wide range of testing conditions. The difference between the ideal and the other cases, in general, increases with increasing departure from the ideal profile; however, the difference between the ideal and the best of the other contours is small (Fig. 5 and Table I).

2) Dies with theoretically ideal profiles have greater drawing work efficiencies than other contours over a wide range of testing conditions (Fig. 6 and Tables II and III).

3) Differences in die profile and in the density decrements resulting therefrom, have a small effect upon the longitudinal tensile properties of the drawn material (Table IV), the ideal dies giving some improvement in elongation.

4) For conical dies, the 'optimum' angle for minimizing structural damage is smaller than that for minimizing work (compare Figs. 5 and 6). Furthermore, structural damage is greatest at large angles and work is greatest at small angles.

5) A significant characteristic of die profiles that minimizes structural damage and work is a zero entrance angle. Reducing the angle of the exit also appears desirable, and thus, consideration should be given to ideal dies which have zero exit angles and which are slightly longer than the 'minimum length' die (Ref. 3).

ACKNOWLEDGMENTS

The author gratefully acknowledges the cooperation of the members of the U.S. Steel Corporation Fundamental and Applied Research Laboratory staff who aided in this study. In particular, he acknowledges H. A. Wriedt and H. A. Hughes for consultations concerning and timely procurement of density determinations, J. H. Davis for careful assistance throughout the entire program, N. Makrides (ARL) for consultations and for providing wire and wire dies including the 'ideal' dies, L. T. Hurat (ARL) for providing wire dies, G. Langford for providing the horizontal testing (drawing) machine, and O. Richmond for consultations.

REFERENCES

1. O. Richmond and M. L. Devenpeck: Proc. 4th U.S. Natl. Congr. Appl. Mechs., 1962, pp. 1053-1057.

2. O. Richmond: Mechanics of the Solid State, pp. 154-167, U. Toronto Press, 1968.

3. O. Richmond and H. L. Morrison: J. Mechs. Phys. Solids, 1967, vol. 15, pp. 195-203.

4. M. L. Devenpeck and O. Richmond: J. Eng. Ind. Trans. ASME B, 1965, vol. 87, pp. 425-428.

5. H. C. Rogers: "Structural Damage in Metal-Forming Processes," Final Report, Contract No. N00019-68-C-0147, Naval Air Systems Command, March 1969.

6. H. C. Rogers and L. F. Coffin, Jr.: "Investigation of the Nature of Structural Damage in Metal-Forming Processes," Final Report, Contract NOw-66-0546-d, Bureau of Naval Weapons, June 1967.

7. H. C. Rogers, R. C. Leach, and L. F. Coffin, Jr.: "Investigation of the Nature of Structural Damage in Metal-Forming Processes," Final Report, Contract NOw-65-0097-f, Bureau of Naval Weapons, Nov. 1965.

8. L. F. Coffin and H. C. Rogers: Trans. Quar. ASM, 1967, vol. 60, pp. 672-687.

9. H. L. Morrison: "Private Communication."

10. F. Garofalo and H. A. Wriedt: Acta Met., 1962, vol. 10, pp. 1007-1012.

11. T. Melville: "Private Communication."

12. C. T. Yang: J. Eng. Ind. Trans. ASME B, 1961, vol. 83, pp. 523-530.

13. A. P. Green and R. Hill: J. Mechs. Phys. Solids, 1952, vol. 1, pp. 31-36.

14. R. M. Caddell and A. G. Atkins: J. Eng. Ind. Trans. ASME B, 1969, vol. 91, pp. 664-672.

THE APPLICATION OF THE AVITZUR UPPER BOUND EQUATION TO TUNGSTEN WIRE DRAWING AND ITS USE IN DIE–LINE DESIGN

James A. Mullendore

Sylvania Electric Products Inc., Chemical & Metallurgical Division

Towanda, Pennsylvania 18848

INTRODUCTION

The die line used in the drawing of tungsten wire has evolved, as is the case for most materials, through a trial and error process. No known attempts have been made to pattern the die line to the drawing characteristics of tungsten or to its mechanical properties. It was the purpose of this study to measure the force required for the drawing of tungsten wire and from those measurements and from the tensile properties, determine what an ideal die line should be.

In carrying out this analysis, the Avitzur equation (1,2) was selected as a working model since it did not involve the use of empirical data in its derivation as is the case for other equations that have been developed for wire drawing. A brief description of this equation and the terms used in this paper are given in the Appendix. For a detailed derivation the reader is referred to the references cited.

EXPERIMENTAL PROCEDURE

The equipment used for the draw force measurements is shown in Figure 1. The force is measured by means of the pivoted die block which exerts a compressive load on the load cell. The signal from the load cell is recorded on a strip chart. The pressure pad was used to provide enough back tension to hold the wire in place under the preheat burners. During each run the amount of back tension was determined by releasing the pressure pad and measuring the instantaneous increase in the load. The apparatus was calibrated by suspending a fifty-pound weight from a wire which went over a pulley and was attached to the die block at the die position. The wires used for the measurements were lamp-doped tungsten (LDT) and tungsten-3.25% rhenium (WR). A graphite-in-water suspension was used as the lubricant.

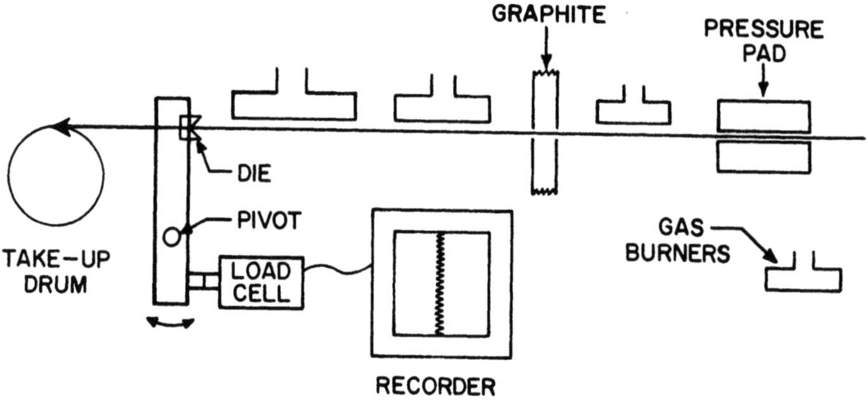

Fig. 1. Schematic of Draw Force Measuring Apparatus

The draw force measurements were made on a variety of die lines between .065 in. and .039 in. diam. The reduction in area per pass varied from 5% to 50%. For the LDT wire, temperatures between 920 and 950°C at the die entrance were used and for the WR wire, temperatures of 880 to 920°C. During any given drawing pass the temperature was maintained within ±10°C. The drawing temperatures were measured with a micro-optical pyrometer and are reported as brightness temperatures.

A draw speed of 12 fpm was used on all passes except the pass to .039 in. where a speed of 16 fpm was used. No difference in draw force could be found between these two speeds. The dies had tungsten carbide nibs in a steel casing. Die semi-cone angles were 6, 8 and 10°. The die angle and bearing length of all dies were measured from a plastic impression. In all cases the measured die angle was within ±.5° of the nominal value. The bearing lengths varied from .5 to .8 times the diameter. However, in every case, the bearing length was not parallel to the die axis but formed a cone with the apex toward the entrance side of the die and having a cone angle of 1 to 2°.

After each drawing pass the wire was sampled for tensile testing. The wire was heated by using a split-tube furnace with a flowing nitrogen atmosphere. The grips were located outside of the furnace so there was no well-defined gauge length. However, for the calculations, a 4-inch gauge length was used since this was the length of the hot zone. The testing was done at a temperature of 900°C and at a crosshead speed of .2 in./min. The tensile tests were run by loading the sample up to the maximum load in the load-elongation curve and then releasing the load. The wire was cleaned and the diameter measured at various points using a knife-edge dial micrometer which could be read to .0001 in. The initial and neck diameters were obtained from these measurements. The stress at strains of .005 and .01 (assuming uniform elongation) was determined from the load-elongation chart. Using these two points and the stress and strain obtained from the maximum load and measured neck diameter, a stress at a plastic strain of .02 was determined. This has been used as the flow stress, σ_F, in subsequent calculations.

RESULTS

In Figure 2 the measured values of the flow stress have been plotted against the true strain as defined by

$$\phi = 2 \ln D_o/D$$

where D_o is an arbitrary reference diameter, in this case, .065 in. For the die angles used, $f(\alpha)$ (see Appendix) takes on values from 1.00023 to 1.00064 and therefore use of the effective strain $\overline{\phi}(1)$ does not alter the plot significantly. If $\overline{\phi}(2)$ is used, then the differences in the effective strain at a given diameter due to drawing through a different number of dies or to using different die angles are not reflected in the strain hardening. It was therefore assumed that the shear on the surfaces of velocity discontinuity is not important.

For the case of the LDT wire two coils (labeled A and B) with significantly different tensile strengths were used. In addition, it was found that, for the LDT wire, the strain hardening rate for wire drawn through three or more dies between .065 and .039 in. was less than for wire drawn through two dies. This behavior was not observed in the WR wire.

In making draw force measurements, it was found that short range variations of 1% due to erratic lubrication were always present. In addition, long range variations of 3 to 5% sometimes were present and were believed due to variability in the flow stress of the wire. It was found that during the course of any drawing pass that minimum values of the draw force were always reproduced. Since these values represented optimum lubrication, they were the values that are reported. Tables 1 - 6 give the calculated values of the draw stress (σ) along with other parameters to be defined below.

DRAW FORCE ANALYSIS

For this analysis the Avitzur equation is written as

$$\frac{\sigma}{\sigma_E} = [f(\alpha) + \frac{m}{\sqrt{3}} \cot\alpha] \; 2\ln D_1/D_2 + \frac{2}{\sqrt{3}} [g(\alpha) + \frac{2mL}{D_2}] \qquad (1)$$

or

$$\sigma/\sigma_E = 2A \ln D_1/D_2 + B = B - A \ln (1 - R.A.) \qquad (2)$$

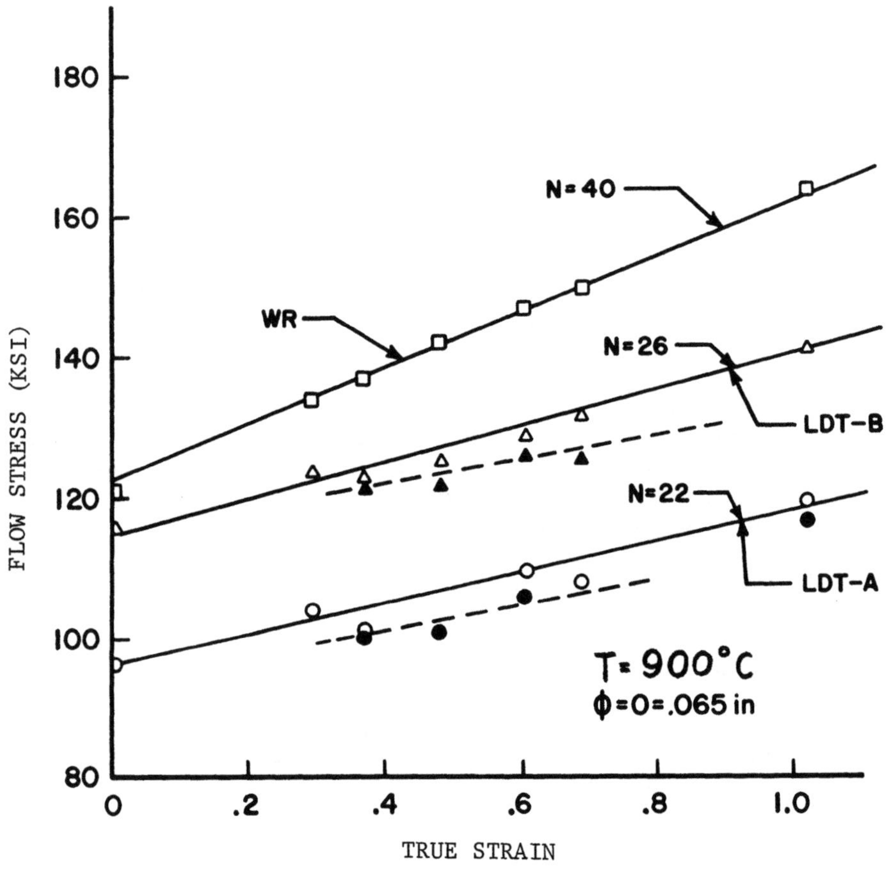

Fig. 2. Flow Stress of Wires

TABLE 1

LDT Results; Alpha = 6°; Coil A

D_1 (in.)	D_2	RA	σ (ksi)	ϕ	ϕ_T	σ/σ_E
.0654	.0473	47.7	141.1	.6480	.6480	1.37
.0473	.0392	31.3	92.0	.3757	1.0237	.81
.0654	.0513	38.4	96.8	.4857	.4857	.96
.0513	.0390	42.2	125.6	.5483	1.0330	1.13
.0654	.0548	29.8	79.7	.3537	.3537	.79
.0548	.0388	49.9	142.1	.6905	1.0442	1.28
.0654	.0566	25.1	71.5	.2890	.2890	.72
.0566	.0479	28.4	79.4	.3338	.6228	.76
.0479	.0466	5.3	22.3	.0550	.6778	.21
.0466	.0391	29.6	87.4	.3510	1.0288	.80
.0566	.0513	17.8	54.2	.1966	.4856	.52
.0513	.0465	17.8	50.1	.1965	.6821	.47
.0465	.0391	29.3	87.4	.3467	1.0288	.81
.0566	.0545	7.3	27.9	.0756	.3646	.27
.0545	.0478	23.1	59.6	.2623	.6269	.57
.0478	.0392	32.7	97.8	.3967	1.0236	.90

TABLE 2

LDT Results; Alpha = 8°; Coil B

D_1 (in.)	D_2	RA	σ (ksi)	ϕ	ϕ_T	σ/σ_E
.0653	.0458	50.8	197.3	.7094	.7094	1.62
.0458	.0392	26.7	103.6	.3112	1.0206	.77
.0653	.0481	45.7	152.4	.7126	.7126	1.26
.0481	.0392	33.6	120.1	.4092	1.1218	.91
.0653	.0511	38.8	130.7	.4904	.4904	1.09
.0511	.0391	41.5	137.4	.5353	1.0257	1.04
.0653	.0547	29.8	93.6	.3542	.3542	.79
.0547	.0390	49.2	167.4	.6766	1.0308	1.29
.0653	.0560	26.4	76.3	.3073	.3073	.65
.0560	.0483	25.6	96.1	.2958	.6031	.79
.0483	.0457	10.5	42.7	.1107	.7138	.34
.0560	.0510	17.1	59.7	.1871	.4944	.49
.0510	.0461	18.3	64.7	.2020	.6964	.52
.0560	.0547	4.6	25.5	.0470	.3543	.21
.0547	.0483	22.0	76.4	.2489	.6032	.62
.0483	.0391	34.5	112.4	.4226	1.0258	.87
.0483	.0391	34.5	110.0	.4226	1.0258	.85

TABLE 3

LDT Results; Alpha = 10°

D_1 (in.)	D_2	RA	σ (ksi)	ϕ	ϕ_T	σ/σ_E
.0653*	.0470	48.2	169.3	.6577	.6577	1.35
.0470*	.0383	33.6	112.8	.4094	1.0671	.84
.0653*	.0487	44.6	151.9	.5866	.5866	1.26
.0487*	.0385	37.5	137.4	.4700	1.0566	1.03
.0653*	.0548	29.6	97.5	.3506	.3506	.82
.0548*	.0382	51.4	181.5	.7217	1.0723	1.38
.0654	.0546	30.3	82.4	.3610	.3610	.82
.0546	.0382	51.1	145.7	.7144	1.0754	1.29
.0654	.0562	26.2	72.6	.3032	.3032	.73
.0562	.0485	25.5	70.4	.2947	.5979	.67
.0485	.0463	8.9	41.6	.0928	.6907	.38
.0562	.0546	5.6	34.2	.0578	.3670	.33
.0546	.0486	20.8	62.0	.2328	.5938	.58

*Coil A, remainder are Coil B.

TABLE 4

WR Results; Alpha = 6°

D_1 (in.)	D_2	RA	σ (ksi)	ϕ	ϕ_T	σ/σ_E
.0645	.0488	42.7	155.0	.5579	.5579	1.17
.0488	.0388	36.8	152.2	.4586	1.0165	.99
.0635	.0488	40.9	152.4	.5266	.5266	1.16
.0488	.0389	36.5	159.9	.4534	.9800	1.06
.0645	.0507	38.2	131.3	.4814	.4814	1.01
.0507	.0388	41.4	177.6	.5350	1.0164	1.16
.0645	.0543	29.1	101.5	.3443	.3443	.80
.0543	.0461	27.9	113.8	.3274	.6717	.82
.0461	.0388	29.2	126.9	.3447	1.0164	.84
.0645	.0563	23.8	84.4	.2719	.2719	.67
.0563	.0543	7.0	41.0	.0723	.3442	.31
.0543	.0488	19.2	72.2	.2136	.5578	.52
.0488	.0390	36.1	150.7	.4483	1.0061	.99

TABLE 5

WR Results; Alpha = 8°

D_1 (in.)	D_2	RA	σ (ksi)	ϕ	ϕ_T	σ/σ_E
.0640	.0474	45.1	158.6	.6005	.6005	1.19
.0474	.0386	33.7	136.7	.4107	1.0112	.87
.0635	.0474	44.3	153.0	.5848	.5848	1.15
.0474	.0386	33.7	128.2	.4107	.9955	.82
.0645	.0510	37.5	127.3	.4697	.4697	.97
.0510	.0388	42.1	160.7	.5468	1.0165	1.04
.0640	.0510	36.5	129.7	.4541	.4541	.99
.0510	.0388	42.1	148.0	.5468	1.0009	.96
.0510	.0388	42.1	152.2	.5468	1.0009	.99
.0510	.0387	42.4	165.8	.5519	1.006	1.08
.0640	.0540	28.8	91.7	.3398	.3398	.71
.0540	.0455	29.0	110.7	.3425	.6823	.78
.0455	.0390	26.5	117.2	.3083	.9906	.76
.0640	.0559	23.7	81.5	.2706	.2706	.64
.0559	.0540	6.7	39.3	.0691	.3397	.29
.0540	.0467	25.2	87.6	.2904	.6301	.62
.0467	.0388	31.0	131.1	.3706	1.0007	.85

TABLE 6

WR Results; Alpha = 10°

D_1 (in.)	D_2	RA	σ (ksi)	ϕ	ϕ_T	σ/σ_E
.0635	.0476	43.8	143.3	.5764	.5769	1.07
.0476	.0385	34.6	141.7	.4243	1.0007	.90
.0645	.0476	45.5	151.7	.6077	.6077	1.13
.0476	.0385	34.6	137.4	.4243	1.032	.87
.0635	.0542	27.1	91.0	.3167	.3167	.71
.0542	.0455	29.5	113.8	.3499	.6666	.80
.0455	.0388	27.3	118.5	.3186	.9852	.77
.0635	.0554	23.9	80.9	.2729	.2729	.64
.0554	.0542	5.7	41.2	.0581	.3310	.30
.0542	.0476	22.9	87.1	.2597	.5907	.60
.0476	.0381	35.9	144.7	.4452	1.0359	.92

Here D_1 and D_2 are the initial and final diameters and R.A. is the reduction in area. A plot of σ/σ_E vs (ln(1-RA) should give a straight line the slope of which will give the shear factor, m. This is done to avoid the errors associated with the uncertainty in the effective bearing length. In obtaining the values of σ_E from Figure 2 it was assumed that no temperature drop occurred across the die. This error will be corrected for later. Figure 3 shows the plots for each die angle.

The values for A and B were obtained by regression analyses and are given in Table 7. No significant difference in A and B could be found between the LDT wire and the WR wire. The values of m calculated from the slopes are also given in Table 7.

TABLE 7
Results of Regression Analysis
$\sigma/\sigma_E = \text{B-A ln (1-RA)}$

α	-A	B	m	L/D_2(Calc.)	L/D_2(Meas.)
6	1.888	.140	.16	.15	.53
8	1.658	.197	.16	.24	.59
10	1.480	.258	.15	.36	.73

The agreement among the three values is excellent. With those values of m, the ratios of effective bearing length to diameter can be calculated from the intercepts and these are also given in the table along with the average values measured on the plastic impressions. As seen, the calculated values are all considerably less than the measured values but are reasonable in view of the fact that the bearing length is not parallel to the die axis. The fact that the calculated and measured values each increase in the same order with respect to the die angle adds credence to the results.

The next step in the analysis is to correct for the fact that there is a temperature drop in the wire of about 200°C between the entrance and the exit sides of the die. To do this we will assume a linear drop in the temperature and an effective stress given by

$$\sigma_{E2} = 1/2\ (\sigma_{11} + \sigma_{22}) \tag{3}$$

These terms are defined in Figure 4. The temperature dependence of the flow stress is written as

$$\sigma_{22} = \sigma_{21}\ (1+\Theta)$$

where Θ is the temperature coefficient of the flow stress.

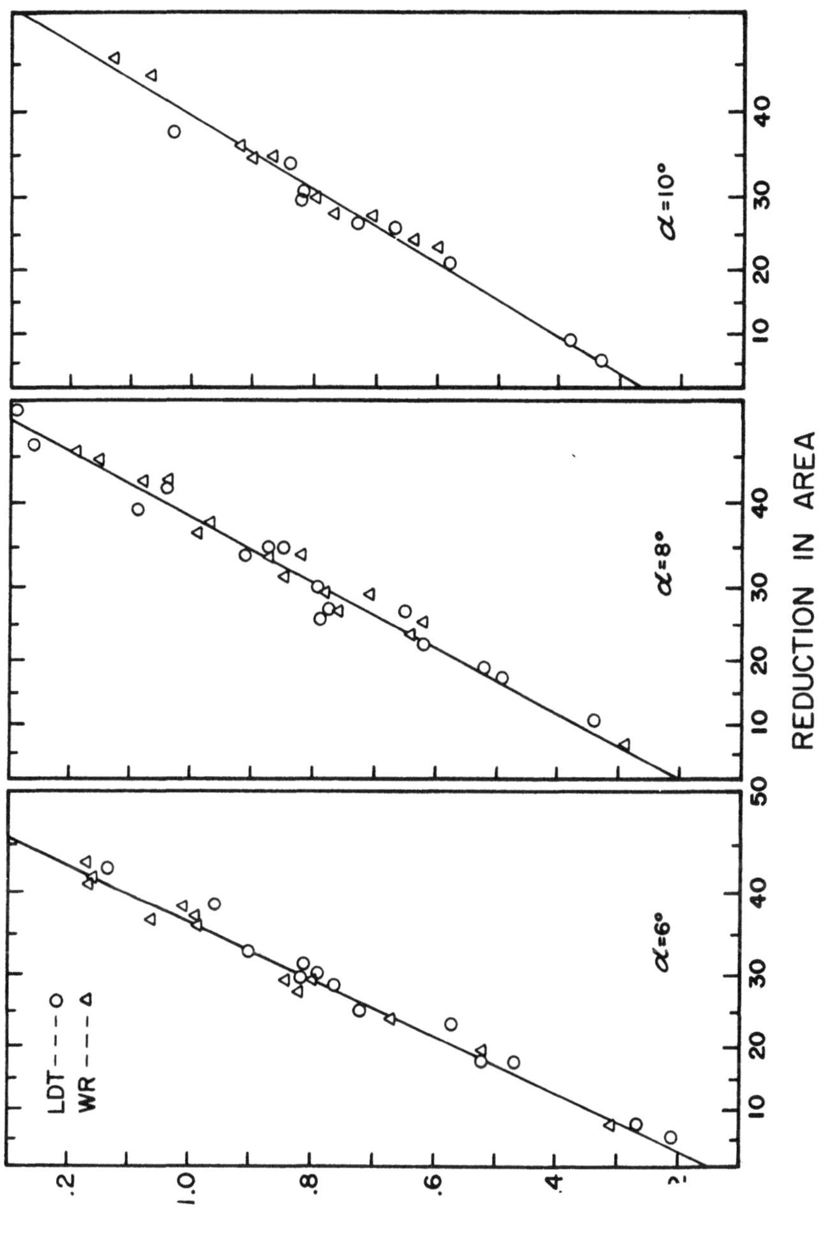

Fig. 3. Effect of Reduction on The Reduced Draw Stress

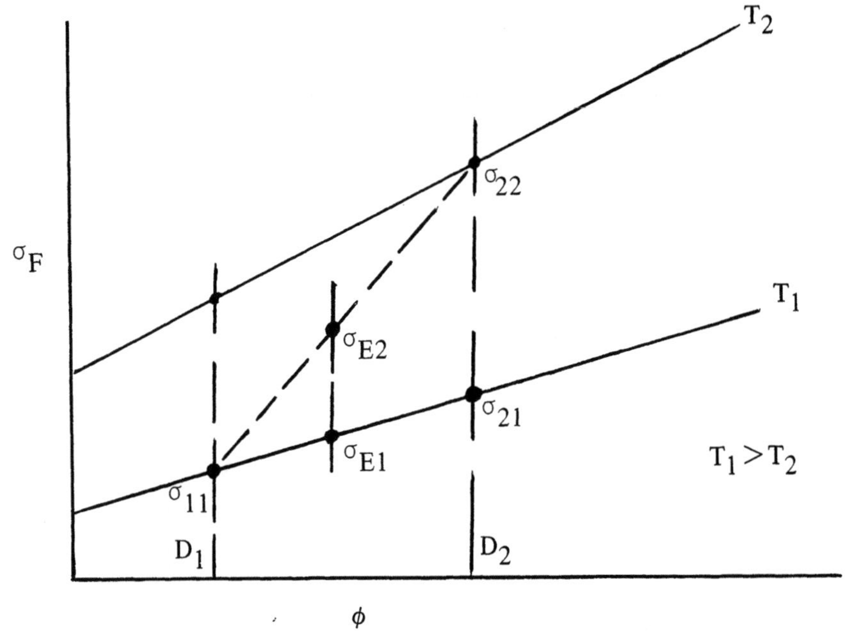

FIGURE 4

Equation 3 is then

$$\sigma_{E2} = 1/2\ [\sigma_{11} + \sigma_{21}\ (1+\Theta)]$$

Since

$$\sigma_{11} = \sigma_{E1} - N \ln D_1/D_2$$

$$\sigma_{21} = \sigma_{E1} + N \ln D_1/D_2$$

We then have

$$\sigma_{E2} = 1/2\ [\sigma_{E1}\ (2+\Theta) + N\ \Theta \ln D_1/D_2] \qquad (4)$$

For the wires used, Θ is given by

$$\Theta = .064 \ \frac{900\text{-}T_2}{100}$$

For a 200°C temperature drop and a value of N of 26, Equation 4 becomes

$$\sigma_{E2} = 1.06\ \sigma_{E1} + 1.66 \ln D_1/D_2$$

When this correction is applied to the plots in Figure 3, the results are as shown in Table 8.

TABLE 8
Corrected Values for Draw Force Constants

α	A	B	m
6°	1.785	.132	.14
8°	1.565	.186	.14
10°	1.395	.243	.12

Applying the correction has lowered the shear factor by .02.

DIE-LINE CALCULATION

The approach to the die-line design is simply one of avoiding excessive draw stresses. We will state that the draw stress, σ, be some fraction, q, of the flow stress of the wire on the exit side of the die. Thus

$$\frac{\sigma}{\sigma_{E2}} = \frac{q\sigma_{22}}{\sigma_{E2}} \qquad (5)$$

Equation 3 can be written as

$$\sigma_{E2} = \sigma_{21} \ (1+\frac{\Theta}{2}) - N \ln D_1/D_2 \qquad (6)$$

and Equation 5 then becomes

$$\frac{\sigma}{\sigma_{E2}} = \frac{q\sigma_{21}(1+\Theta)}{\sigma_{21} \ (1+\Theta/2) - N \ln D_1/D_2} \qquad (7)$$

Equating (7) and (2) gives

$$\ln D_1/D_2 = \frac{1}{2A} \ [\frac{q \ (1+\Theta)}{(1+\Theta/2)-N/\sigma_{21} \ln D_1/D_2} - B] \qquad (8)$$

This equation can then be used for calculating the reduction ratios for a die line. In the case of tungsten wire the equation can be simplified by examining the denominator inside the brackets. Values for the factor N/σ_{21} are from .2 to .5. D_1/D_2 will have values in the range 1.1 to 1.3. Thus the term

$$(1+\frac{\Theta}{2}) - \frac{N}{\sigma_{21}} \ln D_1/D_2 \cong 1$$

and the die-line equation is

$$\ln D_1/D_2 = \frac{q(1+\Theta)-B}{2A}$$

We then have a very simple equation that can be used to calculate a die line. The values for A and B can be obtained from Table 8 or calculated from the Avitzur Equation for other die angles. Note that the shear factors reported represent optimum lubrication. For tungsten wire drawing, an increase in m of 20% is required to allow for normal variations in the lubrication. A value for q is the most difficult to select and depends on the degree of diametral uniformity that is required. Values ranging from .85 down to .60 have been used. Θ can be evaluated experimentally although caution must be used since the temperature drop is strongly dependent on the heat transfer conditions that exit.

Our work has shown that use of this analysis does in fact give good die lines. If we assume that our standard die line must be close to correct since it is based on years of experience, then the analysis is good since it shows that, for the most part, our old die line is in fact the correct one. For certain passes where the calculated die line differed from that being used, experience has shown that these have been troublesome over the years.

Probably of more value than the actual die-line calculation is the fact that use of the Avitzur equation permits a more quantitative evaluation of the wire-drawing process than has been possible before. This has proved to be very valuable in our attempts to make improvements in our process.

REFERENCES

1. B. Avitzur; Metalforming: Processes and Analysis; McGraw-Hill, New York, 1970.

2. B. Avitzur; J. Eng. Ind., Vol. 89, Series B, No. 3, Aug. 1967, p.556-562.

APPENDIX

The Avitzur equation utilizes the upper bound solution for energy expended to cause flow through a conical converging die. The assumed velocity field is shown below.

An element of volume travels parallel to the wire axis with a velocity υ_o in Zone I. At the spherical boundary, B_1, whose origin is at the apex of the cone of the die, the velocity changes discontinuously. In Zone II the element is directed toward the apex of the cone and the velocity is $\upsilon_o \cos \Theta$. At the spherical boundary B_2, the velocity again changes discontinuously. In Zone III the element is again moving parallel to the wire axis but now with a velocity υ_f. At the surfaces B_1 and B_2, the changes in velocity result in shear over these surfaces. The derivation of the equation involves minimizing the power consumed in causing

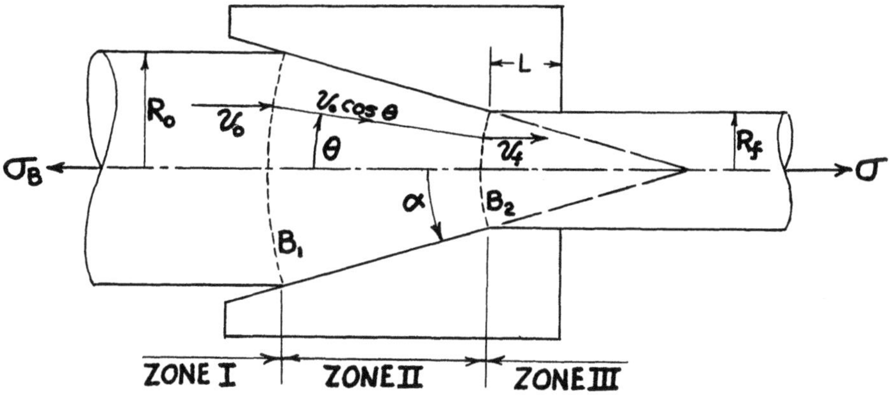

FIG. 1A VELOCITY FIELD FOR FLOW THROUGH DIE

the reduction in cross-section, in the shear at the surfaces of velocity discontinuity and in overcoming friction at the interfaces. The original derivation (1) assumed the material obeyed the Von Mises criteria and thus did not strain harden. This equation was

$$\frac{\sigma}{\sigma_1} = \frac{\sigma_B}{\sigma_1} + 2f(\alpha)\ \ln R_o/R_f + \frac{2}{\sqrt{3}}\ \{g(\alpha) + m\ [\cot\alpha\ \ln R_o/R_f + \frac{L}{R_f}\ ]\} \qquad (1A)$$

Where σ_1 is the flow stress of the wire and $f(\alpha)$ and $g(\alpha)$ are functions of α only. m is the shear factor defined by

$$\tau = m\ \frac{\sigma_1}{\sqrt{3}}$$

Where τ is the shear stress at the interface. In a later paper (2) the upper bound solution was derived for the case of a strain hardening material which obeyed the relation

$$\sigma_F = \sigma_1\ (1 + \beta\bar{\phi}) = \sigma_1 + N\bar{\phi}$$

Where the second equation defines the strain hardening coefficient used in Figure 2. The effective strain $\bar{\phi}$, is given either by

$$\bar{\phi}(1) = 2f(\alpha)\ln R_o/R_f$$

or

$$\bar{\phi}(2) = 2f(\alpha) \ln R_o/R_f + \frac{2}{\sqrt{3}} g(\alpha)$$

$\bar{\phi}(1)$ applies where shear on the surfaces of B_1 and B_2 are not important and $\bar{\phi}(2)$ applies where they are important. It was shown that in all cases except those involving a combination of large values of β, α, and R_o/R_f that Equation 1A still held except that an effective stress, σ_E, used instead of σ_1 where

$$\sigma_E = \sigma_1 (1+1/2\ \beta\phi) = \frac{\sigma_1 + \sigma_2}{2} \tag{2A}$$

COMPUTER SIMULATION TO PREDICT LOAD, STRESS, AND METAL FLOW IN AN AXISYMMETRIC CLOSED-DIE FORGING

Taylan Altan

Metalworking Division, Columbus Laboratories,

Battelle Memorial Institute, 505 King Ave., Columbus, Ohio

ABSTRACT

The design of a closed-die forging process requires the estimation of maximum forging load and the necessary forging energy. To determine the forging energy, the forging load at various stroke positions must be estimated.

In the past, empirical methods have been used with varying degrees of success. The present study attempts to predict the forging load and stresses through relatively basic analytical methods. Using the example of an axisymmetric forging, consisting of a flange and a shaft, the slab or Sachs method has been applied to develop a computer-simulation technique. The forging process is analyzed in small steps of deformation. The stress distribution, the load, and the magnitude of filling of the die and the flange have been estimated at each deformation step. The theoretical predictions have been compared with experimental results in forging the part from both lead and aluminum to various stroke positions with a hydraulic press.

The computer program simulating the axisymmetric forging process, applied to an example in the present study, can be extended to other shapes and be used for various billet sizes, part dimensions, temperature, ram speeds, and friction.conditions.

INTRODUCTION

Closed-die forging is an extremely complex forming process from the point of view of deformation mechanics. The nonsteady-state and nonuniform metal flow, the interface friction, and the heat transfer between the deforming material and the tooling are difficult to analyze. However, by making some simplifying, but acceptable assumptions, it is possible to predict the stresses, the forging load, and the metal flow within useful approximations.

To be most useful, the analysis of the forging process must include the estimation of maximum stress distributions on the dies, the maximum load required by the equipment, and the total energy necessary to complete the deformation. The forging energy is given by the surface area under the load-displacement curve of the forging process of interest. This curve is determined by estimating the forging load at various positions of the deformation stroke. To establish and illustrate the method of analysis, the axisymmetric part seen in Figure 1 is selected.

METAL FLOW AND DEFORMATION STAGES

Three main stages of deformation must be distinguished during the deformation of the forging shown in Figure 1. The metal flow and the load variation is illustrated for these stages in Figure 2.

(1) Upsetting. In the beginning of forging the axisymmetric slug is compressed between the upper and lower dies and the material flows (a) outward to form the flange, (b) inward to extrude into the shaft, Figure 3.

(2) Filling. When the lower cavity is essentially filled (except maybe at the corners), the flash starts to form; at this stroke position the shaft is not necessarily entirely filled. Then flow or metal toward the flash is restricted and the metal is forced to extrude into the shaft as seen in Figure 4.

(3) End of Forging. At this final stage, the lower and the upper dies are completely filled. However, the flat die surfaces are not yet in contact. As seen in Figure 5, the metal extrudes into the flash and the load increases until the dies contact each other.

A reasonable theoretical model should simulate all three stages of the forging as described above. No complete analysis of the present forging operation appears to exist at this time.

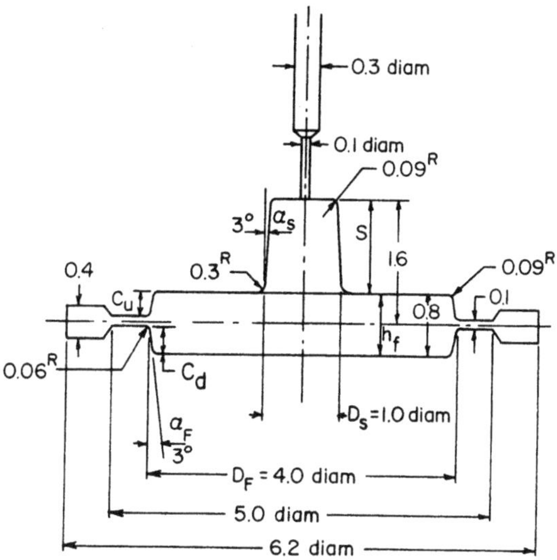

FIGURE 1. SKETCH OF THE AXISYMMETRIC DIE USED IN CLOSED-DIE FORGING STUDIES

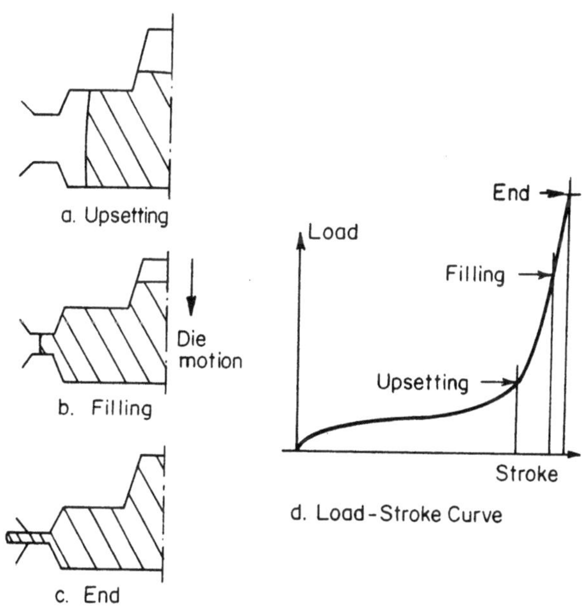

FIGURE 2. ILLUSTRATION OF METAL FLOW AND LOAD-STROKE CURVE IN FORGING IN DIES SEEN IN FIGURE 1

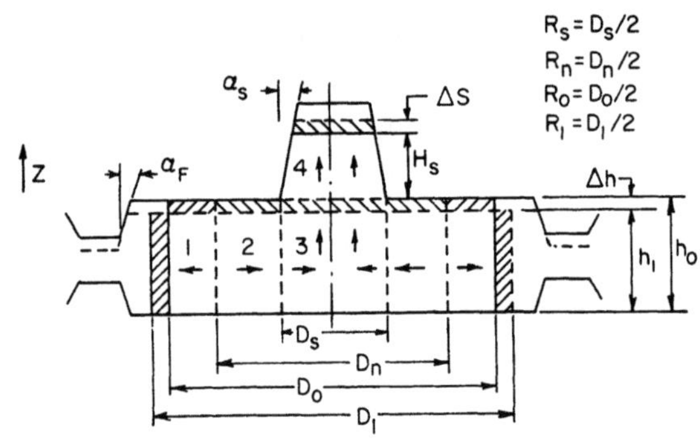

FIGURE 3. UPSETTING STAGE DURING FORGING IN THE DIES SEEN IN FIGURE 1

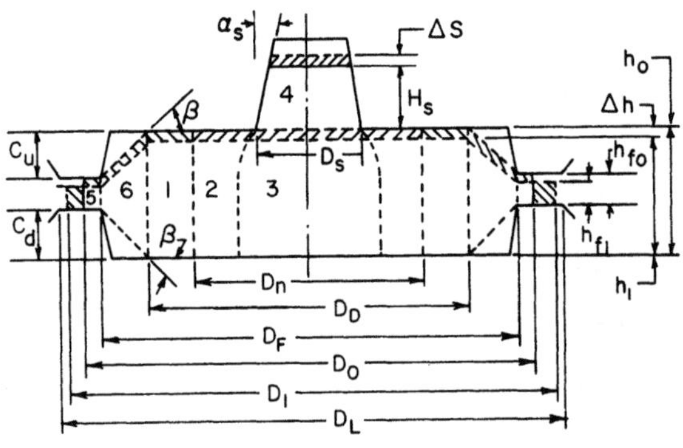

FIGURE 4. DEFORMATION ZONES AND METAL FLOW DURING THE FILLING STAGE

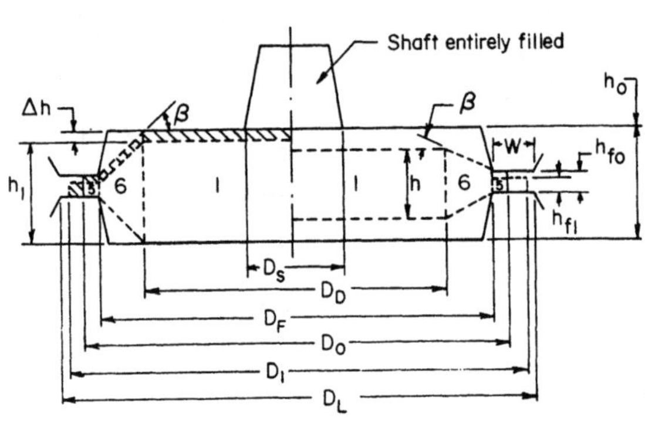

a. Metal Flow by Sliding b. Metal Flow by Shearing

FIGURE 5. DEFORMATION ZONES AND METAL FLOW DURING THE END STAGE

Several workers, however, analyzed the different stages of the process by using the slab (or Sachs') method of analysis.[1-6]

Estimation of Stresses and Loads by the Slab Method

The slab (or Sachs') method assumes that the stresses on a plane perpendicular to the flow direction are in principal directions and that the deformation is homogeneous throughout the deformation zone studied. A slab of infinitesimal thickness is selected and a force balance is made on this slab. The resulting differential equation of static equilibrium is solved with the existing boundary conditions.

The following usual assumptions are made:

(a) the material is isotropic and incompressible
(b) the elastic deformations are neglected
(c) the inertia forces are small and neglected
(d) the plane surfaces in the material remain plane
(e) the dies do not deform elastically
(f) the flow stress $\overline{\sigma}_i$ is constant at the interior of the deformation zone "i" studied, however, it does not have necessarily the same value in another deformation zone
(g) the friction shear stress is expressed by $\tau_f = f_i\, \overline{\sigma}_i$, where f_i = friction factor at the tool-material interface of zone "i" of the forging. $(o \leq f_i \leq 0.577)$, σ_i = flow stress in the zone "i" of the forging.

The variations of the flow stress, $\overline{\sigma}$, due to strain, $\overline{\epsilon}$, strain rate, $\dot{\overline{\epsilon}}$, and temperature, θ, can be approximately considered by estimating $\overline{\sigma}_i$ for each separate zone of deformation. The expression $\tau_f = f_i\, \overline{\sigma}_i$ is approximate and greatly facilitates the computations.[1,4]

Analysis of Stresses and Loads for "Unit Deformation Zones"

In Figures 3, 4, and 5, it is seen that the entire forging can be divided into various "unit deformation zones". Thus, the stresses and loads can be calculated for each zone by considering that the stress distribution must be continuous, i.e., the value of the axial forging stress must be the same at the interface of two adjacent zones. In Figure 3, for example, the stress calculations can be conducted by starting from the free boundaries: the Zone 4 in the shaft, and the Zone 1 in the flange, outside of the neutral surface. The neutral surface is then determined from the condition that stresses, calculated by starting from both sides must be equal at the neutral surface. In order to facilitate the calculations the "unit deformation zones", which occur in the forging of Figure 1, are described and analyzed below.

Converging or Diverging Flow in Longitudinal Direction*. This type of converging flow occurs in the shaft, Zone 4 of Figure 3. Using the symbols given in Figure 3, the axial stress distribution in the shaft is given by:

$$\sigma_{z4} = K_4 \ln \frac{R_s - z \tan \alpha_s}{R_s - H_s \tan \alpha_s} \tag{1}$$

where

$$K_4 = \frac{2\,\bar{\sigma}_4 \left[f_4 \,(1 + \tan^2 \alpha_s) + \tan \alpha_s \right]}{\tan \alpha_s} \tag{2}$$

$z = 0$, at the entrance to the shaft.

The load, P_s, at the upper surface of the flange, necessary to extrude the shaft is,

$$P_s = \pi R_s^2 \, K_4 \ln \left(\frac{R_s}{R_s - H_s \tan \alpha_s} \right) \quad . \tag{3}$$

The equations (1), (2), (3) are valid also for longitudinal diverging flow by replacing $(+\alpha_s)$ with $(-\alpha_s)$.

Parallel Flow in Longitudinal Direction. This type of flow occurs in Zone 3 of Figure 3, where the metal flows upward, in axial direction, by shearing along a cylindrical surface. The axial stress, σ_{z3}, increases towards the lower die according to:

$$\sigma_{z3} = \sigma_{zB} + \frac{4\bar{\sigma}_3 \, z}{\sqrt{3}\, D_s} \quad , \tag{3}$$

where

σ_{zB} = axial stress at the upper surface of the deformation zone

$\bar{\sigma}_3$ = flow stress inside of the deformation zone

D_s = diameter of the deformation zone

z = 0, at the upper surface of the deformation zone.

---

* Longitudinal direction is the direction of die motion, while lateral direction is perpendicular to the die motion.

Parallel Flow in Lateral Direction. Inward Flow. This type of flow occurs at the flange, inside of the neutral surface, as seen in Zone 2 of Figure 3. The axial stress, σ_{z2}, is given by

$$\sigma_{z2} = \frac{2\,\bar{\sigma}_2\, f_2}{h_o}\,(r - R_s) + \sigma_{z3\ \mathrm{ave}} \qquad (4)$$

where

$\bar{\sigma}_z$ = flow stress inside the Zone 2

f_2 = friction factor at the interface of the dies and Zone 2

$\sigma_{z3\ \mathrm{ave}}$ = the averaged magnitude of axial stress at the boundary of Zones 2 and 3.

The axial load, P_2, over Zone 2, Figure 3, is obtained by integrating:(3)

$$P_2 = 2\pi \int_{R_s}^{R_n} r\,\sigma_{z2}\,dr \qquad (5)$$

where

R_n = radius of the neutral surface.

Outward Flow. This type of flow exists at the flange, outside of the neutral surface, as seen in Zone 1 of Figure 3. The axial stress, σ_{z1}, is given by:

$$\sigma_{z1} = \frac{2\,\bar{\sigma}_1\, f_1}{h_o}\,(R_o - r) + \bar{\sigma}_1 \qquad (6)$$

where

$\bar{\sigma}_1$ = flow stress inside Zone 1

f_1 = friction factor at the interface of dies and Zone 1.

The axial load, P_1, is obtained from integration:(3)

$$P_1 = 2\pi \int_{R_n}^{R_o} r\,\sigma_{z1}\,dr \quad . \qquad (7)$$

Converging Flow in Lateral Direction. Outward converging flow in lateral direction is seen in Zone 6 of Figure 4. The axial stress, σ_{z6}, is:

$$\sigma_{z6} = -\frac{K_2}{K_1}\ln\left(\frac{K_3 + K_1\, r}{h_{fo}}\right) + \sigma_{z5}\,(r = R_F) \quad , \quad . \qquad (8)$$

where

σ_{z5} $(r = R_F)$ = axial stress at $r = R_F$ calculated from Zone 5

$K_1 = - 2 \tan \beta$

$K_2 = - K_1 \bar{\sigma}_6 + 0.577 \bar{\sigma}_6 \, 2 \, (1 + \tan^2 \beta)$

$K_3 = h_o + 2 \cdot \tan \beta \cdot R_D$

$R_D = R_F - C_u / \tan \beta$

$\bar{\sigma}_6$ = flow stress in Zone 6.

All other symbols are illustrated in Figure 4.

The load P_6, is obtained by integration:

$$P_6 = 2\pi \int_{R_F}^{R_D} r \, \sigma_{z6} \, dr \quad . \tag{9}$$

UPSETTING STAGE

The Geometry of the Forging During Deformation

The variables to be studied during various stages of forging are shown in Figure 3 and the corresponding symbols are described in the text or in the Figures 3, 4, and 5.

The neutral surface, defined by the diameter D_n in Figure 3, defines the idealized separating surfaces between Zones 1 and 2. As the upper and the lower dies come together, the material outside of the neutral surface flows outward and increases the diameter of the forging, while the material inside the neutral surface flows toward the center and fills the shaft. The diameter of the neutral cylinder, D_n (i.e., the location of the neutral surface) depends upon several process variables and varies continuously during forging. The analysis must, therefore, take place in infinitesimal, or for practical purposes in very small steps of the stroke, Δh, Figure 3. The variations of the neutral diameter, D_n, are neglected for a small step of the stroke, Δh. Average dimensions are used over the surfaces which barrel (surfaces defined by the diameters D_o and D_1, and the upper surface of the metal flowing into the shaft, Figure 3). Considering the metal flow during a step, Δh, we can determine the instantaneous workpiece geometry.

Inside the neutral surface, the volume of metal that flows toward the shaft, ΔV_4, is equal to the volume displaced inward in Zones 2 and 3 as the die advances, ΔV_{23}. Thus, using the symbols given in Figure 3, we have:

$$\Delta V_4 = \pi \Delta S \; (R_s - H_s \tan \alpha_s)^2 \tag{10}$$

and

$$\Delta V_{23} = \pi R_n^{\,2} \; \Delta h \quad . \tag{11}$$

Since the deforming metal is incompressible, the volumes calculated by Equations (10) and (11) must be equal. Thus,

$$\Delta V_{23} = \Delta V_4, \text{ or } \Delta S = \frac{R_n^{\,2} \cdot \Delta h}{(R_s - \tan \alpha_s \cdot H_s)^2} \quad . \tag{12}$$

Similarly, volume displaced outside of the neutral surface:

$$\Delta V_1 = \pi \; (R_o^{\,2} - R_n^{\,2}) \; \Delta h \quad . \tag{13}$$

Volume of additional upset in the flange cavity:

$$\Delta V_f = \pi \; (R_1^{\,2} - R_o^{\,2}) \; h_1 \quad . \tag{14}$$

Incompressibility again gives:

$$\Delta V_f = \Delta V_1 \text{ or,}$$

$$R_1 = \left[(R_o^{\,2} - R_n^{\,2}) \frac{\Delta h}{h_1} + R_o^{\,2} \right]^{1/2} \quad . \tag{15}$$

With Equations (12) and (15), the geometry of the workpiece at a certain step of the stroke can be determined provided the location of the neutral surface (i.e., D_n) is known.

To determine D_n, it is necessary to estimate the stresses and loads acting upon different zones of the forging. The axial stresses in the flange area will have their maximum value at the neutral surface. Thus, the axial stress increases from the free end of the upset sample toward the neutral surface. As seen in Figure 6, the axial stress decreases again from the neutral surface towards the axis of the forging. Consequently, the stresses calculated at both sides of the neutral surface must be equal at the neutral surface.

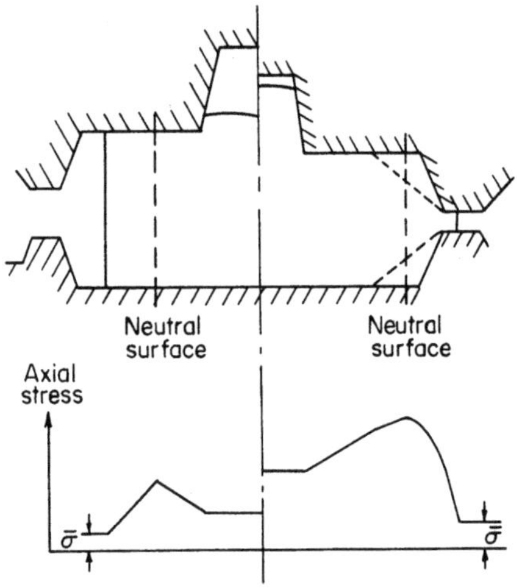

FIGURE 6. REPRESENTATION OF AXIAL STRESS DISTRIBUTIONS IN UPSETTING AND FILLING STAGES DURING FORGING

Loads and Stresses at Unit Deformation Zones

The calculation of stresses and loads in Zones 1, 2, and 4 of Figure 3 was discussed earlier. At the center of the flange, where the metal is being extruded into the shaft, the metal flow can be approximately illustrated by three possible flow models as shown in Figure 7. The actual flow model is the one which minimizes the amount of deformation energy. Therefore, at a given forging stage, the actual flow can be approximated with a flow model that results in minimum magnitude of axial stress, σ_{zA}, acting at the surface of the lower die cavity, Figure 7.

In the flow models of Figure 7a and 7c, the stress, σ_{zA}, is calculated by adding the axial stresses calculated in the two adjacent deformation zones (converging or diverging and parallel longitudinal flow zones). For a given flange height to shaft diameter ratio, h_o/D_s, the flow model which gives the minimum axial stress at the bottom die, σ_{zA}, has been determined and the results are given in Figure 8. The details of the derivations are given elsewhere.(3) Thus, the model valid for metal flow is established by determining the diameter of the shear surface, D_c, from Figure 8.

Once the flow model and the axial stress distribution along the height of the flange is determined, Figure 3, an average value

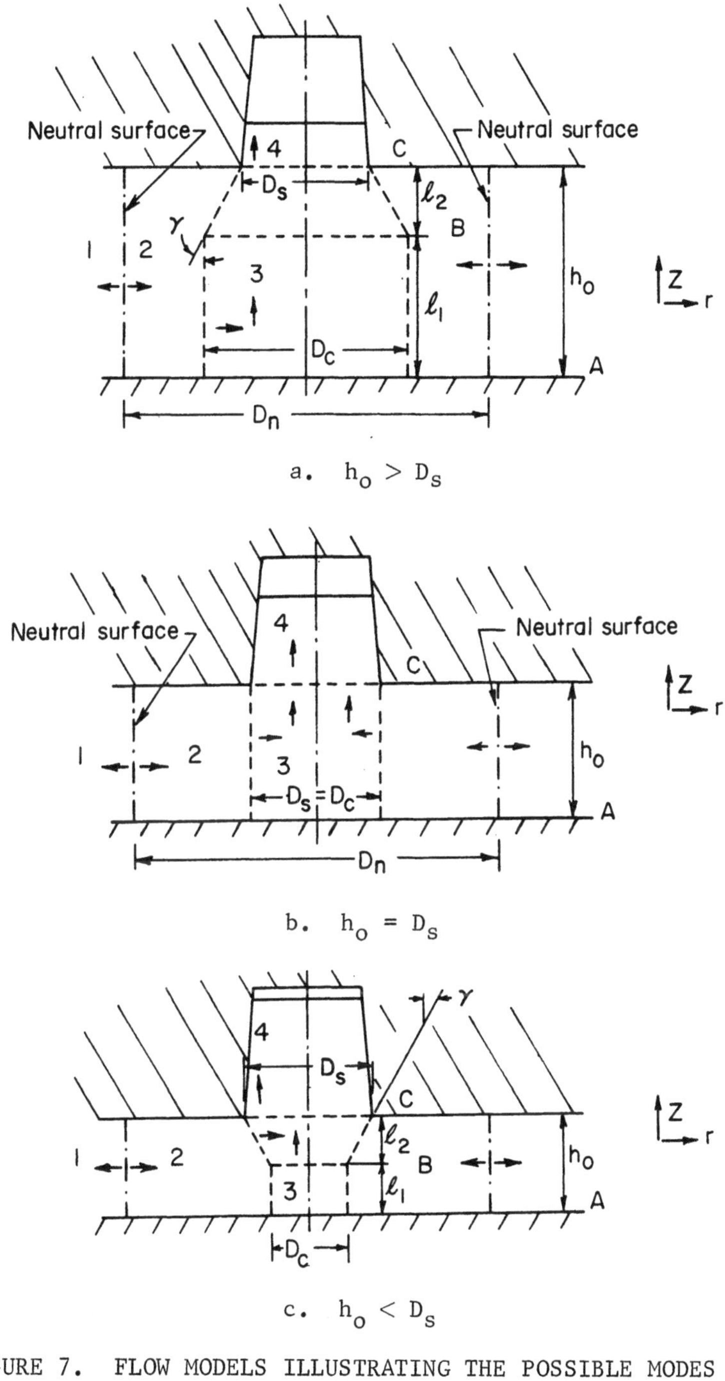

FIGURE 7. FLOW MODELS ILLUSTRATING THE POSSIBLE MODES OF METAL FLOW INTO THE SHAFT

of the axial stress, $\sigma_{z3\ ave}$, is calculated. This stress, $\sigma_{z3\ ave}$, is used as boundary condition for calculating the stresses in the Zone 2 of Figure 3.

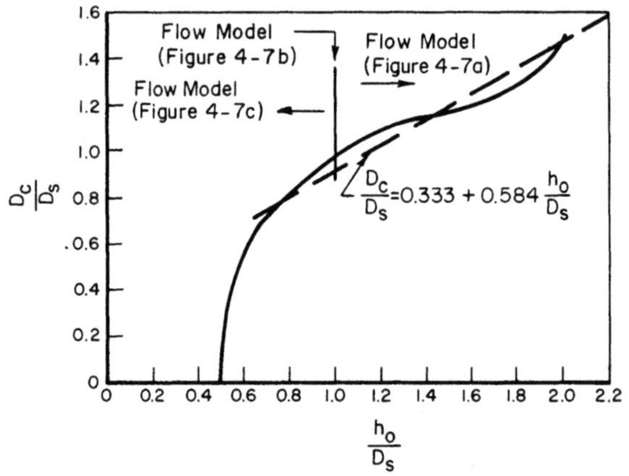

FIGURE 8. DETERMINATION OF THE CYLINDRICAL SHEAR-SURFACE DIAMETER D_c OF THE FIGURE 7 FOR GIVEN INSTANTANEOUS HEIGHT h_o AND FOR KNOWN SHAFT DIAMETER

Determination of the Location of the Neutral Surface

The neutral surface is the boundary between the Zones 1 and 2 of Figure 3. At the neutral surface, for $r = R_n$, both axial stresses, σ_{z1}, and σ_{z2}, must be equal. From this condition:

$$R_n = \frac{1}{\bar{\sigma}_2 f_2 + \bar{\sigma}_1 f_1} \left[\bar{\sigma}_1 f_1 R_o + \bar{\sigma}_2 f_2 R_s + \frac{h_o}{2} (\bar{\sigma}_1 - \bar{\sigma}_{z3\ ave}) \right] . \quad (16)$$

Total Forging Load for Upsetting Stage

The total forging load at the upsetting stage is the sum of the loads necessary to extrude the shaft, P_s, to upset Zone 1, P_1, and to upset Zone 2, P_2, Figure 3. Thus,

$$P_t = P_s + P_2 + P_1 \quad . \quad (17)$$

To determine P_1, P_2, and P_s at each stage of deformation, it is necessary to know the position of the neutral surface R , the forging diameter D_o, the height of the shaft H_s, and the thickness

h_o. All these dimensions vary with the forging stroke. It is therefore necessary to start with the computations from the start of the deformation and calculate step by step. At each increment the entire load and volume relationships will be evaluated and the results will be used for the next step and so on. This procedure is programmed and thus the forging is simulated on a digital computer.

FILLING STAGE

Figure 4 illustrates the approximate geometry of the forging during the filling stage.

The Geometry of the Forging

For a small step of the stroke, Δh, the additional fill in the shaft is given by Equation (12),

$$\Delta S = \frac{R_n^2 \, \Delta h}{(R_s - \tan \alpha \, H_s)^2} \quad .$$

The symbols used are given in Figure 4. As seen in Figure 4, the volume displaced outside of the neutral zone ΔV_o is:

$$\begin{aligned} \Delta V_o &= \Delta V_5 + \Delta V_6 + \Delta V_1 \\ &= \pi \Delta h \left(R_o^2 - R_n^2\right) \quad . \end{aligned} \tag{18}$$

The volume of additional extruded flash is:

$$\Delta V_f = \pi \left(R_1^2 - R_o^2\right) h_{f1} = \pi \left(R_1^2 - R_o^2\right)\left(h_1 - C_u - C_d\right) \quad . \tag{19}$$

Because of volume constancy, the metal displaced outside of the neutral zone must be equal to the additional extruded flash, i.e.:

$\Delta V_f = \Delta V_o$, or using Equations (18) and (19)

$$R_1 = \left[\left(R_o^2 - R_n^2\right) \frac{\Delta h}{(h_1 - C_u - C_d)} + R_o^2\right]^{1/2} \quad . \tag{20}$$

Equations (12) and (20) describe the geometry of the forging during the filling stage.

Estimation of Stresses and Loads. Toward the periphery of the flange, Figure 4, the metal flows into flash by shearing along a tapered surface. The position of this tapered surface, seen in

Figure 4, is obtained from angle β. The angle β must be such that the axial stress, σ_n, at the neutral surface must be minimum. Thus, the expression describing, σ_n, is derived and minimized with respect to tan β. This minimization gives:

$$\tan \beta = \left[1 - \frac{(R-1)\sqrt{3}\, f_1}{R \ln R}\right]^{1/2} \quad , \tag{21}$$

where

$R = h_o/h_{fo}$

f_1 = friction factor at the flange.

The location of the tapered surface is then given by:

$$R_D = R_F - C_u/\tan \beta \quad . \tag{22}$$

The location of the neutral surface, R_n, is again determined from the condition that for $r = R_n$; $\sigma_{z1} = \sigma_{z2}$. Thus,

$$R_n = \frac{1}{\bar{\sigma}_2 f_2 + \bar{\sigma}_1 f_1}\left[\bar{\sigma}_1 f_1 R_D + \bar{\sigma}_2 f_2 R_S + \frac{h_o}{2}\left(\sigma_{z6}\,(r = R_D) - \sigma_{z3\ ave}\right)\right] \quad . \tag{23}$$

The stresses and the loads are now calculated in the same way as it was done for the upsetting stage.

Under certain conditions of geometry and friction, the flow model seen in Figure 4 might not exist. In this case, deformation Zone 1 is eliminated (i.e., $R_n = R_D$) and the neutral surface is the boundary between Zones 6 and 2 of Figure 4. However, the load calculations are the same as for the general metal-flow model. Angle β is equal to 45 degrees since this angle would give the minimum stress distribution in Zone 6.

Total Forging Load in Filling Stage

The total forging load in the filling stage is obtained by adding the forging loads acting upon each deformation zone.

$$P_T = P_1 + P_2 + P_{34} + P_5 + P_6 \quad . \tag{24}$$

The calculations of individual loads P_1, P_2, P_{34}, P_5, and P_6 are obtained by integrating the axial stress distributions.

END OF FORGING

The end stage of the forging process is illustrated in Figure 5. At this stage the shaft is entirely filled and the neutral surface has moved to the center of the forging. The excess metal present in the die cavity is being extruded into the flash. The load and energy necessary during this end stage of forging help only to reach the final dimensions of the forging in the axial direction. In an ideally designed forging process, this stage should be eliminated, or minimized.

The Geometry of the Forging

As seen in Figure 5, during a small step of the forging stroke, Δh, the total volume of displaced metal, ΔV_T, is extruded into the flash. Thus,

$$\Delta V_T = \pi R_o^2 \Delta h \quad . \tag{25}$$

The volume of additional extruded flash, ΔV_f, is:

$$\Delta V_f = \pi \left(R_1^2 - R_o^2\right) h_{f1} \quad . \tag{26}$$

With volume constancy, we have $\Delta V_T = \Delta V_f$ or,

$$R_1 = \left[\frac{\Delta h \ R_o^2}{h_{f1}} + R_o^2\right]^{1/2} \quad . \tag{27}$$

Equation (27) determines the external radius of the flash, R_1, and the geometry of the forging at a given stroke position.

Stresses and Load in the Die Cavity

All the metal volume displaced from the die cavity is extruded into the flash. The neutral surface is at the center of the forging. This type of metal flow is extensively analyzed by Zünkler[6] and his results are summarized elsewhere.[3] In accordance with the die dimensions and with the friction conditions, there are two possible types of metal flow.

(1) <u>Complete Shearing in the Cavity</u>. In this case, the height of the flange, h_o, is such that the metal flows easier, i.e., by consuming less energy, by forming lateral shear surfaces. This case is illustrated at the right side of Figure 5. In this case the geometry of the flow model is given by:

$$\frac{h}{h_{fo}} = 0.8 \left(\frac{RF}{h_{fo}}\right)^{0.92} \frac{h_o}{h_{fo}} \text{ , and} \tag{28}$$

the shear angle, β, is given by:

$$\tan \beta = \left[1 - \frac{R - 1}{R \ln R}\right]^{1/2} , \tag{29}$$

where

$R = h/h_{fo}$

h = height of the shear deformation zone, Figure 5

h_{fo} = instantaneous flash thickness, Figure 5.

(2) Shearing Only Along the Tapered Surfaces. In this case, the height of the flange, h_o, is such that the condition expressed by Equation (28) is not satisfied. The flow model is illustrated at the left side of Figure 5. In Zone 1 the metal flows by sliding at the die-material interface.

Total Forging Load

The total forging load is obtained by adding the loads determined for all deformation zones. The end stage of forging is also studied in small steps in order to obtain the variation of the forging load during the stroke. The load determined for the finished forging, i.e., when the flange and the flash thicknesses have reached their design value, will be the maximum forging load. The equipment capacity must be selected on the basis of this maximum load.

COMPARISON OF THEORETICAL AND EXPERIMENTAL RESULTS

The equations for calculating the stress distribution, the total forging load, and the forging geometry at each position of the stroke have been programmed for Battelle's Control Data 6400 computer. From flow-stress data of lead and 6061 aluminum alloy, and assuming reasonable friction factors, the computer program has been used to calculate the forging load and the geometry of the forging during the forging stroke. The calculations have been performed in small steps of the stroke. For the calculations, the length of a step was 0.1 inch during the upsetting stage (no flash extruded yet) and 0.01 inch during filling and the stages after flash started to form. The theoretical predictions and experimental results are compared below for a lead and a 6061 aluminum forging forged in the die shown in Figure 1.

Comparison of Computer Predictions With Experimental Results in Forging Lead

Lead forgings obtained at various ram-stroke positions are illustrated in Figure 9. The lead samples of 3-inch diameter x 1.69-inch height were forged at room temperature under a 700-ton hydraulic press at 80-in./min nominal speed.

To determine the flow stress of the lead used in present experiments, well lubricated upset tests were carried out under the same hydraulic press and at the same nominal speed setting of 80 in./min. In the first series of experiments, lead samples of 3-inch diameter x 1.69-inch height were upset to approximately 0.6-inch final height.

In the second series of upset tests, the 0.6-inch-high samples upset in the first series of tests were machined to perfect cylindrical form and further upset to about 0.1-inch final height. These tests were conducted to simulate the deformation of lead in the die cavity and in the flash by using approximately the same strain and strain-rate conditions. The flow stresses obtained from these tests did not vary with strain. The flow stress for lead in the die cavity was $\overline{\sigma}_c$ = 7,000 psi while for the flash the flow stress was $\overline{\sigma}_f$ = 7,700 psi. These values were used in the computations.

The comparison of the experimental load-displacement curve with the load-displacement curve obtained through computer simulation is seen in Figure 10. A constant and uniform friction factor $f = 0.1$ was used in these calculations since the lead samples and the dies were well lubricated with machine oil prior to each experiment. In Figure 10, it is seen that the agreement between theoretical and the experimental results is good except in initial stages of deformation.

The actual and theoretical dimensions of the lead forging at two different stroke positions, 0.5 inch and 0.125 inch before complete die closure, is illustrated in Figure 11. It is seen that the predictions, especially at the initial upsetting stage, do not agree well with experiments. At 0.5 inch before die closure, the actual fill into the shaft has already started while the theory predicts that no metal flow into the shaft has yet occurred. This discrepancy might be due to the difficulty of predicting friction at the die-material interface. In theory, a constant and uniform friction factor $f = 0.1$ was assumed. In reality, however, the friction factor is expected to vary not only during deformation but also along the die-material interface.

FIGURE 9. LEAD FORGING OBTAINED AT VARIOUS RAM STROKE POSITIONS AT R.T., BILLET 3 x 1.69 IN., RAM SPEED 80 IN./MIN, LUBRICATED. From left to right: initial stock; 0.5 inch, 0.25 inch, 0.125 inch, 0.063 inch, 0.0 inch before die closure.

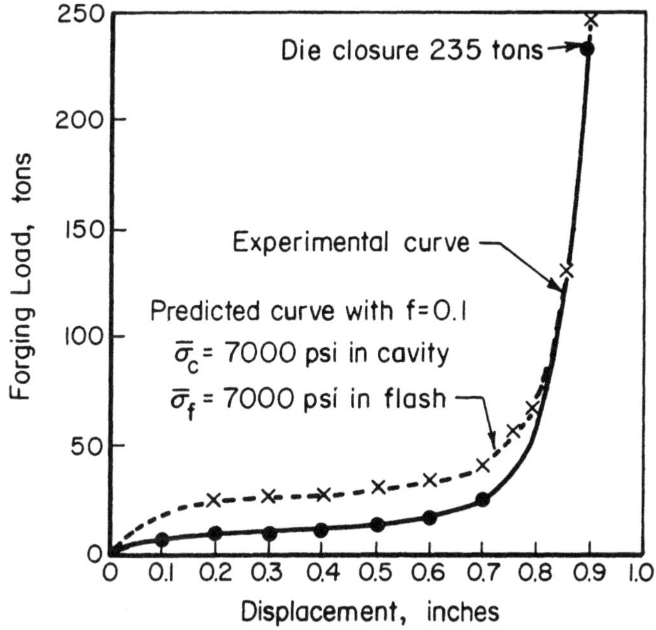

FIGURE 10. COMPARISON OF EXPERIMENTAL LOAD-DISPLACEMENT CURVE WITH PREDICTIONS FROM COMPUTER SIMULATION IN FORGING LEAD WITH LUBRICATION (SAMPLE 3-IN. DIAM X 1.69 IN. HIGH)

Comparison of Computer Predictions With Experimental Results in Forging 6061 Aluminum Alloy

The flow stress of 6061 aluminum alloy was determined under Battelle's 700-ton hydraulic press at 80-in./min ram speed. Thus, isothermal, lubricated upset tests were conducted at 800 F ($\bar{\sigma}$ = 7,000 psi). Since the forging experiments and the upset tests were conducted at approximately the same range of strain rates under the same press, for the purposes of the present study, the strain-rate effects were neglected and the flow-stress values given above were used for 6061 aluminum.

The hydraulic press used in present studies was relatively slow for the size of part to be hot forged in a closed die. Consequently, the effects of die chilling upon the friction at the die-material interface and upon the flow stress of the forged material could not be neglected. The experimental load-displacement curve obtained in forging 6061 aluminum in the dies seen in Figure 1, is shown in Figure 12. It is seen that the load increases in an extremely steep way toward the end of forging. This increase in load is larger than that expected in a practical closed-die forging.

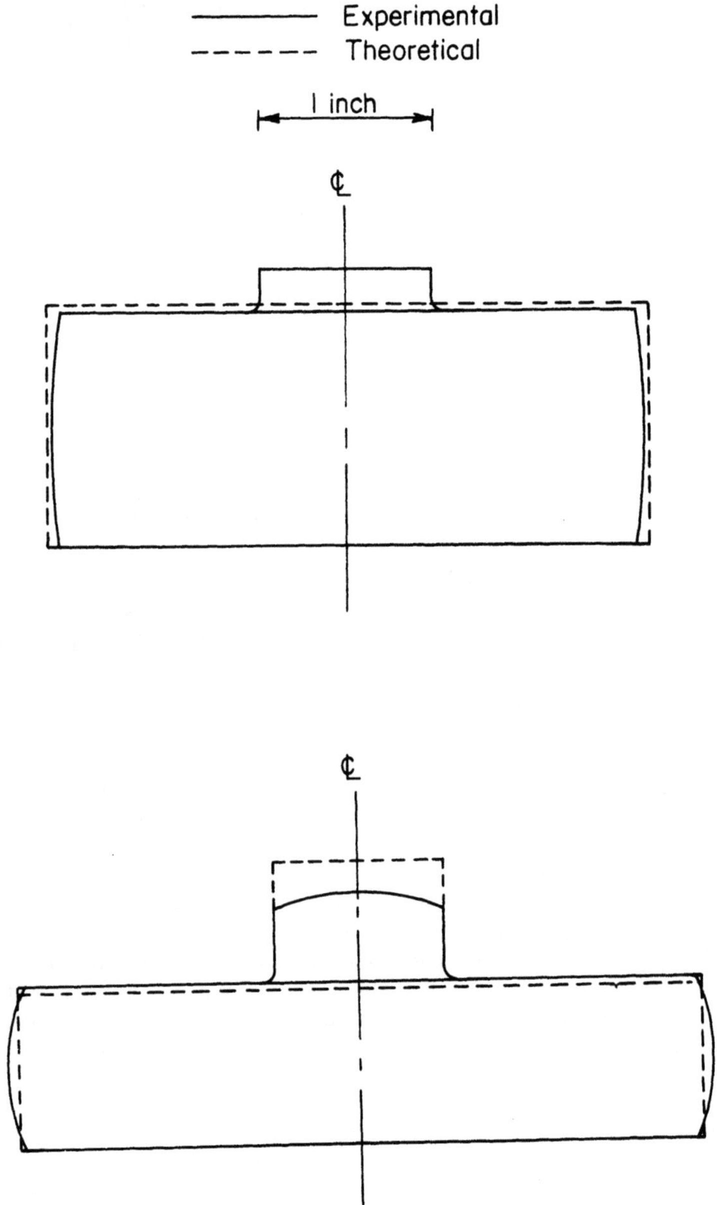

FIGURE 11. COMPARISON OF EXPERIMENTAL AND THEORETICAL DIMENSIONS OF LEAD FORGING AT TWO DIFFERENT STAGES OF DEFORMATION (ASSUMED f = 0.1, $\overline{\sigma}_c$ = 7,000 PSI IN CAVITY, $\overline{\sigma}_f$ = 7,700 PSI IN FLASH)

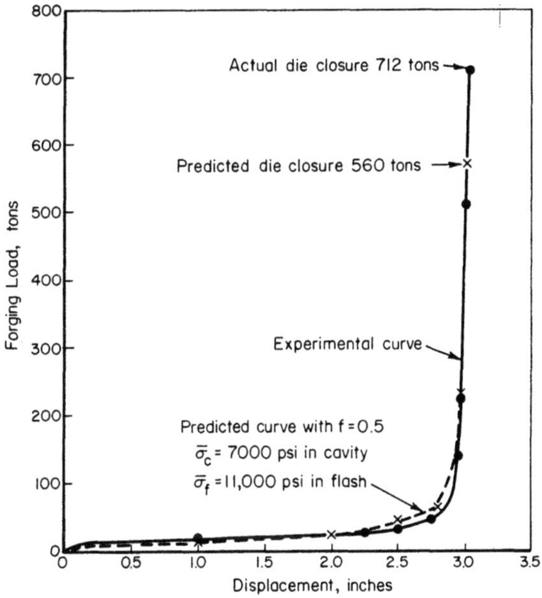

FIGURE 12. COMPARISON OF EXPERIMENTAL LOAD-DISPLACEMENT CURVE WITH PREDICTIONS FROM COMPUTER SIMULATION IN FORGING 6061 ALUMINUM (SAMPLE 2 IN. DIE X 3.8 IN. HIGH, SAMPLE TEMPERATURE 800 F, DIE TEMPERATURE 350 F)

To select reasonable values of flow stress in the cavity, $\bar{\sigma}_c$, and in the flash, $\bar{\sigma}_f$, the die chilling is considered and the temperature in the flash is calculated as seen below. The temperature in the die cavity was assumed to remain unchanged. This assumption is reasonable since the metal in the die cavity has a relatively large volume and volume-to-surface ratio. Therefore, the cooling in the cavity will not be as pronounced as in the flash.

The temperature gradients are neglected and the flash is considered as being a thin plate, with an average uniform temperature, cooled symmetrically from both sides[7]. Thus, the average flash temperature during cooling is given by:

$$\theta = \theta_1 + (\theta_0 - \theta_1) \exp\left(- \frac{\alpha \tau 2w}{bt}\right) , \quad (30)$$

where

θ_0 = initial temperature = 800 F = 430 C

θ_1 = die temperature = 350 F = 180 C

α = heat-transfer coefficient between dies and flash
= 15,000 kcal/m^2 h C, based on Klafs' results in forging 314 stainless steel(8)

c = heat capacity of aluminum = 0.211 kcal/kg C

ρ = specific gravity of aluminum
= 2.71 g/cm^3

w = actual flash width = 0.5 in. = 12.7 mm

b = flash width in the die = 0.5 in. = 12.7 mm

t = average height of flash = 0.162 in. = 4.1 mm (t is not the final flash thickness which is 0.1 in.)

τ = average cooling time of flash = 0.24 sec (τ is obtained from oscillograph recordings, it is the time from start of flash formation until die closure).

The evaluation of Equation (30) gives:

$$\theta = 180 + (430 - 180) \exp(-0.43)$$
$$= 342 \text{ C} = 650 \text{ F}.$$

Based on the calculations made above and for approximate predictions conducted here, we can consider the average flash temperature to be about 650 F (342 C). Thus, the flow stress in the flash, $\bar{\sigma}_f$, is about 11,000 psi.

Using the values $\bar{\sigma}_c$ = 7,000 and $\bar{\sigma}_f$ = 11,000, the load-displacement curve has been calculated through computer simulation. A high value for friction factor, f = 0.5, is assumed since in these experiments die chilling was considerable. The theoretical and experimental load-displacement curves are compared in Figure 12. It is seen that, although the agreement is good at most stroke positions, the predicted maximum forging load (560 tons) is 27 percent lower than the experimental value. This result suggests that the flash temperature was probably lower than estimated by approximate calculations. The theoretically and experimentally determined dimensions of the forged part at two stroke positions, 0.5 inch and 0.125 inch before closure, are illustrated in Figure 13. In these cases, the agreement between theory and experiment is considered good.

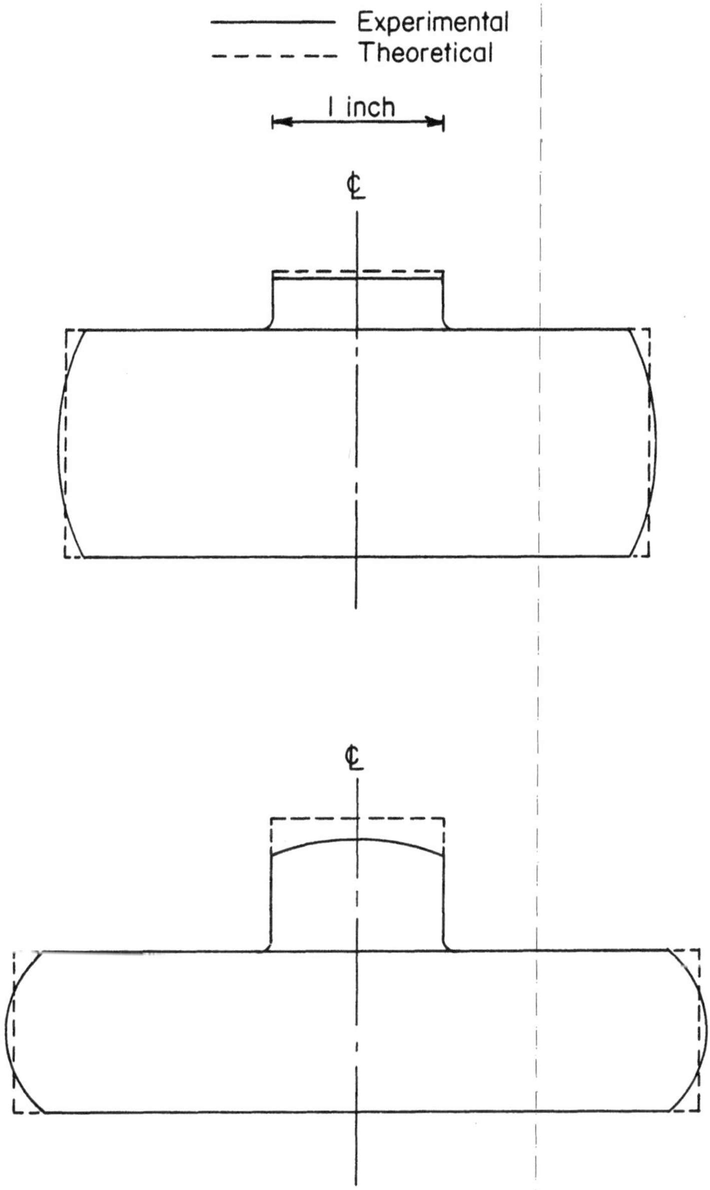

FIGURE 13. COMPARISON OF EXPERIMENTAL AND THEORETICAL DIMENSIONS OF A 6061 ALUMINUM FORGING AT TWO DIFFERENT SLOPES OF DEFORMATION (ASSUMED $f = 0.5$, $\overline{\sigma}_c = 7{,}000$ PSI IN CAVITY, $\overline{\sigma}_f = 11{,}000$ PSI IN FLASH)

CONCLUSIONS

The slab method of analysis has been used for determining the stresses, the load, and the part dimensions at various positions of the stroke for the forging illustrated in Figure 1. For this purpose three stages of the forging have been considered:

(1) Upsetting, where metal flows laterally in the flange and longitudinally into the shaft.

(2) Filling, where metal flows laterally into the flash and longitudinally into the shaft.

(3) End of forging, where the shaft and the flange are filled and metal flows only laterally into the flash.

The equations were derived for all the zones of deformation in the forging. In order to conduct the analyses, a theoretical flow model was determined, whenever necessary, at each small step of deformation. The procedure was computerized and thus, the forging process was simulated.

Using estimated values of the friction factor and experimentally determined flow-stress values, the theoretically predicted loads and part dimensions have been compared with experimental results. From these comparisons, the following conclusions are drawn:

- The computer-simulation technique can be used in predicting load-displacement curves and energies in forging axisymmetric parts.
- In order to conduct the calculations, the flow stress of the forged material must be determined for the ranges of strain and strain-rate that occur in actual forging.
- The value of the friction factor must be estimated. The friction factor can be estimated within acceptable approximations, from data given in literature, or on the basis of experience in predicting forging loads. A better method of determining the friction factor would be to conduct a ring test under the same equipment, that will be used in forging, and by using rings having approximately the average thickness of the forging.
- The temperature variations in different locations in the forging, especially in flash, must be considered. Die-chilling effects are extremely significant when forging

under slow equipment such as hydraulic presses. Therefore, contact times under pressure must be estimated, or measured, for a given press, and must be used for estimating variations of flow stress due to die chilling.

- The use of a digital computer is essential for conducting detailed calculations as described in this study. In the future, it would be useful to develop computer subprograms for various deformation zones. Thus, it would be possible to assemble these subprograms, in a building block manner, to develop a new large computer program for a given forging.

REFERENCES

(1) Altan, T., et al, "Forging Loads and Stresses in Closed-Die Forging — Part One", Third Interim Topical Report to AMMRC on Contract No. DAAG46-68-C-0111, Battelle Memorial Institute.

(2) Altan, T., et al, "Forging Loads and Stresses in Closed-Die Forging — Part Two", Fourth Interim Topical Report to AMMRC on Contract No. DAAG46-68-C-0111, Battelle Memorial Institute, April 30, 1969.

(3) Altan, T., et al, "The Use of Analytical Methods in Predicting Loads and Stresses in Closed-Die Forging", Chapter 3 of the final report of the same project as (1) and (2) above.

(4) Tarnovskiy, I. Ya., "Filling of Annular Dies" (in Russian) Sverdlovsk, Uralskiy politikhnicheskiy Institut Trudy 48, Trans. Moskva, 1953.

(5) Burgdorf, M., "On the Calculation of Axial Stress Distribution and Forming Load in Pin Forging" (in German) Industrie-Anzeiger, 89, 1967, p. 182 and p. 1558.

(6) Zünkler, B., "Determination of Stresses and Loads in Plane Strain Closed-Die Forging" (in German) Industrie-Anzeiger, 84, 1969, p. 67.

(7) Sonkin, E. A., "Calculation of Flash Temperature in Closed-Die Forging" (in Russian) Kuznecno-Stampovocnoe Proizvodstvo, 1961, No. 3, p. 8.

(8) Klafs, U., "Ein Beitrag zur Bestimmung der Temperaturverteilung in Werkzeug und Werkstuck beim Warmumformen" (A Contribution to Determination of Temperature Distributions in Tool and Workpiece in Warmforging), Doctoral Dissertation, Technical University, Hannover, 1969.

THE VALIDITY OF SIMULATING TESTS IN EVALUATING LUBRICANTS FOR DEFORMATION PROCESSES

John A. Schey

Department of Materials Engineering

University of Illinois at Chicago Circle

One of the most significant stumbling blocks in establishing correlation between theory and application in deformation processes has been the uncertainty attached to the magnitude of interface friction. Even the simplest theory must account for the effects of friction on forces, power requirements and material flow, and the magnitude of this friction must be known-together with the correct value of the flow stress-if the validity of a theory is to be checked. While there is still a great scarcity of data on flow stresses at relevant strain rates and temperatures, a beginning has, nevertheless, been made by determining flow stresses in plastometers. No comparable development occurred with regards to friction. There is no universal, basic method of determining friction under conditions applicable to deformation processing; instead, a numerical value representing friction (in the form of a coefficient of friction, or an interface shear strength) is usually derived from experimental data through the utilization of a theory, often at the same time when proof of the validity of the very same theory is sought. Under these circumstances, there is a danger of friction becoming an adjustable, variable "constant" chosen at convenience. It would be highly desireable, therefore, that friction should be determined in simple tests that are readily evaluated, rely on a minimum of theoretical or simplifying assumptions, yet simulate actual deformation processing conditions sufficiently to make the results relevant, and to allow lubricant evaluation with a measure of confidence. This paper aims at clarifying to some degree the suitability of some existing test techniques for these purposes.

SIMULATING TESTS

The number of simulating tests that have been developed throughout the last forty years are too numerous to mention here. Many have been applied to a variety of situations with great enthusiasm only to be dropped when their limitations became all too evident.

While no unanimity can be expected on this controversial subject, we will proceed with the assumption that a critical review published elsewhere [1] offers a reasonable appraisal of various test methods. On this basis, all tests that involve purely elastic contact conditions will be dismissed for the simulation of bulk deformation, primarily because they do not generate new surfaces typical of metal deformation processes (with the exception of sheet metalworking, not considered here).

Among the test methods that involve some limited bulk plastic deformation of the softer (workpiece) member, the twist compression test has found application particularly for adhesion studies and for lubricant investigations involving severe, typically boundary contact conditions. Some earlier results obtained with this test will be introduced here for comparison; the present work, however, utilized techniques characterized by bulk plastic flow of the workpiece material:

(a) The ring compression test is essentially a small scale upsetting operation, is rather sensitive to squeeze-film formation, but has the advantage over the axial compression of cylinders that the relative magnitude of friction may be evaluated purely from the geometry of the deformed specimen and a knowledge of the flow stress is not required even for a quantitative evaluation [2].

(b) Plane strain compression has been shown to provide a useful simulation of hydrodynamic effects [3] and the flow stress of the workpiece material can be eliminated as an unknown by performing tests with two selected geometries [4].

(c) Wire drawing at slow speed is one of the most convenient small scale deformation processes and was used here as a simple means of checking the relevance of the simulating tests, simply by establishing the order of merit of lubricants as judged from draw stress and surface quality.

EXPERIMENTAL MATERIALS

Three workpiece materials, all in the annealed condition, were chosen to represent a variety of interface conditions:

3003-aluminum alloy shows a high adhesion typical of aluminum and should be sensitive to the presence of boundary lubricants. Its almost linear strain hardening characteristics make it a desireable material for experimental work.

7075-aluminum alloy is known to have a lesser adhesion to steel than 3003 alloy and also generates much higher interface pressures. The oxide formed on annealing the sheet specimens was stripped in an acid etch, and a natural oxide was allowed to form by storing for 2 days.

Unleaded 70/30 (cartridge) brass generates high interface pressures, it is less responsive to boundary lubrication and may, possibly, reveal the effects of EP additives on the tool and/or workpiece material. The oxide was removed in an acid bright dip.

Lubricants were selected partly to give a broad spectrum of lubricating mechanisms, and partly to reveal the effect of small additions to a mineral oil. Thus, it would transpire whether the test methods are capable of discriminating between lubricants that are known to give clearly distinguishable performance under practicable conditions. Characteristics of the lubricants are given in Table I; they were always applied to both the specimens and the die surfaces with a brush.

Table I
Composition and Properties of Lubricants

Description	Composition	Viscosity cs at 38°C	Viscosity cs at 100°C
M.O.	Highly naphthenic refined mineral oil	78.7	8.2
M.O. + O.A.	1% oleic acid in the above M.O.	70.5	---
M.O. + C.P.	2% C.P. in M.O.	81.0	---
M.O. + Gr.	1% graphite of 6μ particle size in M.O.	---	---
O.A.	Oleic acid (iodine No. 88)	21.6	4.7
C.P.	Chlorinated paraffin of 50% Cl content	---	47.0
Gr.	Spray graphite in volatile carrier	---	---

Die material and surface finish are known to have a marked influence on lubricant performance. Unfortunately, it was not possible to standardize these conditions throughout all experiments, but at least all dies were made of tool steel (Table II).

Table II
Die Materials

Process	Die Material AISI-SAE	R_c	Surface Finish μ in. RMS
Ring Compression	4340	50-55	4-6
Plane Strain Compression	4340	50-55	6-8
Wire Drawing	M4	60-65	2-3

EXPERIMENTAL TECHNIQUES

Originally it was intended that all experiments be performed with tools warmed to a temperature above the boiling point of water, in order to eliminate the possible disturbing effect of adsorbed films. However, because of difficulties encountered in maintaining a constant temperature, experiments had to be conducted with dies at ambient temperature (22°C) except for some wire drawing tests, in which the dies were kept at 120°-140°C.

In the ring compression tests, rings of 1.25 in. O.D., 0.625 in. I.D., and 0.417 in. height, turned to give end faces of typically 25 to 30 μ in. RMS finish, were upset to 50% reduction on a 100 ton press at a press speed of typically 2 ipm. For a ranking of lubricants, the change in internal diameter was measured; for a quantitative evaluation of the coefficient of friction, the calibration by Male and Cockroft [2] was used, which in turn is based on an analytical evaluation of upsetting forces by Schroeder and Webster [5].

Plane strain specimens of 0.175 in. and 0.158 in. thickness and 2 in. width were indented on the same press with anvils of 0.500 and 0.250 in. width (length), thus giving after 50% reduction an L/h ratio of 7 and 3, respectively, permitting a determination of the coefficient of friction conveniently from recorded compression forces, utilizing the approximate solution given by Alexander [4]. Since this solution is based on the same assumptions as the one used in ring compression, some of the bias attributable to the particular form of theory employed is eliminated, although there is, of course, still no proof of the absolute validity of the derived coefficient of friction values. After approximately 30% reduction, shear cracks developed in the 7075 alloy specimens and caused a drop in the recorded forces, therefore, data were extracted for 25% reduction and the calculations were made for L/h ratios of 4.5 and 2. Each result represents the average of two impressions, with a spread seldom exceeding one percent.

Wires of 0.187 in. diameter were drawn at a speed of 20 fpm on a 2000 lb. capacity hydraulic drawbench with dies of 6° half angle. Since regular polishing of the working die surfaces resulted in slightly increasing diameters, two draws were always made, one at typically 25 and the other at typically 35% reduction, and the force for a nominal 30% reduction was obtained by linear interpolation. When the drawing force was not steady, the minimum and maximum recorded during drawing a 5 ft. length were taken. The draw stress given is referred to the exiting cross-section.

The die surfaces were redressed between each experiment by lapping or polishing with 3μ diamond paste, followed by degreasing with mineral spirits. Specimens were degreased with mineral spirits except when otherwise shown. The surface of the dies and specimens was inspected after deformation and the presence and severity of die pick-up and scoring on the specimen were visually assessed and classified as N(none), L(light), M(medium) and H(heavy). Heavy pick-up and scoring were severe enough to be objectionable in practice from the beginning of a run; while no information relevant to longer production runs was generated, it is reasonable to assume that even medium pick-up and scoring would be intolerable.

EXPERIMENTAL RESULTS

Since each testing technique showed peculiarities of its own, results will be discussed here by the test method rather than by the workpiece material.

Ring Compression

Repeatability of experiments conducted on workpiece materials degreased with mineral spirits was very good, provided that the surface preparation of the anvil was kept constant. Thus, the results shown in Table III for dies cleaned and prepared in the standard manner (lapping with 3μ diamond compound, followed by degreasing with mineral spirits) show in general trends one would anticipate from prior knowledge.

Table III
Results of Ring Compression Experiments

Lubricant	Workpiece Materials 3003 Al			7075 Al			70/30 Brass		
	%	μ	pick-up	%	μ	pick-up	%	μ	pick-up
None	+33.0	0.2	L-L	+23.1	0.13	L-L	+23.0	0.13	N-N
	+47.9	0.35	L-L						
	+52.0	0.4	L-L				+40.0	0.25	N-N
M.O.	+1.76	0.06	L-L	-1.9	0.052	L-H	-18.0	0.028	N-N
	+1.44	0.06	L-L						
	-5.0	0.046	L-L	-1.0	0.053	L-H			
M.O.+O.A.	-8.17	0.04	L-L	+4.0	0.065	L-H	-20.0	0.025	N-N
				-5.4	0.046	L-H			
M.O.+C.P.	-3.85	0.048	L-L	+1.6	0.06	L-H	-22.0	0.023	N-N
				-5.8	0.045	L-L			
M.O.+Gr.	-4.65	0.046	L-M	-4.65	0.047	H-H	-22.0	0.023	N-N
O.A.	-11.7	0.032	L-L	+1.5	0.058	H-H	-29.0	0.019	N-N
				+4.0	0.064	H-H			
C.P.	-14.4	0.030	L-L	-4.5	0.047	M-M	-10.0	0.037	N-L
				-6.0	0.044	L-M			
Gr.	-4.0	0.048	L-L	-3.5	0.048	L-L	-20.0	0.025	N-N
Water	+0.8	0.055	L-L	-1.9	0.052	N-L	-4.0	0.048	L-L

Note: I.D. change is shown in percent; underlined data refer to dies degreased with mineral spirit, followed by an alkaline cleaner, water rinse and ethyl alcohol rinse.

With dry anvils, friction is highest with the 3003 aluminum alloy known for its high adhesion to steel, and is lower with the inherently less adhesion-prone 7075 aluminum alloy and the brass. With 3003 aluminum rings, the addition of oleic acid to the mineral oil markedly reduced friction, and even chlorinated paraffin and graphite were quite effective. In the neat, undiluted form, the higher viscosity of chlorinated paraffin is noticeable, but the rather low friction obtained with the solid graphite film must be considered unusually good. This led to further experiments with graphite dispersed in water which, quite unexpectedly, gave the lowest friction of the entire series, and directed suspicion on water: to our surprise, it gave extraordinarily low friction. It was surmized that some residual films left by the preparation method may have been responsible for the results; this prompted experiments in which the anvils were prepared in the regular manner but, after washing with mineral spirits, they were also immersed in an aqueous alkaline cleaner, followed by a water and an ethyl alcohol rinse. Data thus obtained are underlined in Table III; the higher friction value registered without a lubricant seemed to indicate that the surface was actually cleaner, yet it also affected, in some instances quite markedly, the performance of various lubricants. Nevertheless, the very low friction registered by water was almost entirely retained. That the low friction obtained with water could not be attributed to residual films of the ethyl alcohol either was proven by tests run with a pool of alcohol; the internal diameter of the brass ring increased by 4% (μ=0.065). When the alcohol rinse was omitted, the dry specimen again gave the high friction previously found, and when the aqueous alkaline cleaning solution was used as the lubricant, the internal diameter increased by 16% (μ=0.1). The apparently very low friction with water was finally explained when Dr. Male suggested in a discussion of these results that water may actually have filled up the hole of the specimen. Following this suggestion, the experiments were repeated, with the excess water carefully removed. This time friction was more reasonable (Table III) but still quite low. Hydrodynamic pockets observed on the surface indicated that even the low viscosity water was entrapped at the interface, despite the low press speed.

The results obtained with 7075 aluminum alloy fall less into the anticipated pattern. On the regularly prepared anvils, the low friction registered with mineral oil would appear anomalous, although anvils treated with the alkaline cleaner and alcohol rinse show the effect of additives as anticipated. Clearly, changes in surface preparation could readily invalidate the results and, from the practical point of view, anvils prepared by degreasing in mineral spirits should give the more relevant information. It is conceivable that viscosity overshadowed the effect of the boundary additive.

Compared to the two aluminum alloys, friction was always very low in the lubricated compression of brass and, viewed in isolation, it is difficult to pin-point trends in ring compression tests other than the beneficial effect of viscosity.

Inspection of the upset ring surfaces revealed that whenever a liquid lubricant including water was present, the original edge of the ring specimen was clearly discernible, and when pick-up and surface damage was noted, it was attributable to pick-up on the die face originating at these edges. The effect was particularly noticeable on the outer circumference. Viscous liquids left impressions (hydrodynamic pockets) in the surfaces, which were most marked and numerous on the softest 3003 alloy, and were few and isolated on the hardest 70/30 brass. These pockets, however, were limited to the original end face, which expanded substantially when friction was low. Outside this original end face the surfaces were bright, indicative of boundary contact. These bright rings could not be eliminated by any change in lubricant application technique. Experiments (not shown in Table III) proved that the bright ring would persist and the original ring edges would cause lubricant breakdown and pick-up, irrespective of whether lubricant was applied only to the anvil, to the anvil and the specimen end face, or to all surfaces of the specimen. This phenomenon, investigated in some detail by Pearsall and Backofen [6], thus appears to be an inevitable consequence of upsetting and indicates that deformation proceeds, at least partially, through folding over of the side faces, even in the presence of effective lubrication.

Plane Strain Compression

In plane strain compression the only freely deforming surface is the edge of the specimen and while the magnitude of lateral strains and the shape of the spread-zone give a clue regarding the efficiency of lubrication [7], a quantitative evaluation of the test is possible only if compression forces are measured. Ideally, this would require a workpiece material with a very flat strain hardening curve, however, sufficiently strain-hardened material quite often would not take the further cold worked imparted by the friction experiment itself. Therefore, all material was annealed, but it was then compressed to the same percentage reduction, thus largely eliminating the effect of strain hardening.

When the coefficient of friction is calculated following the suggestion of Alexander [4], the ratio of two stresses is taken and a friction value can be derived (μ in Table IV) as long as stress at the higher L/h ratio is greater than at the lower ratio. Most of the results appear quite reasonable: friction is very high on

Table IV
Results of Plane Strain Compression Experiments

Lubricant	3003 Al p_7 kpsi	3003 Al p_3 kpsi	3003 Al μ	3003 Al μ*	3003 Al pick-up	7075 Al $p_{4.5}$ kpsi	7075 Al p_2 kpsi	7075 Al μ	7075 Al μ*	7075 Al pick-up	70/30 Brass p_7 kpsi	70/30 Brass p_3 kpsi	70/30 Brass μ	70/30 Brass μ*	70/30 Brass pick-up
None	87.0	44.4	>0.5	>0.5 0.23	L	110.2	68.3	0.25	>0.5 0.23	M	---	119.7	---	--- 0.165	L
M.O.	34.5	32.5	0.032	0.03 0.03	L	56.5	57.4	---	0.01 0.06	L	107.3	97.0	0.053	0.035 0.03	L
M.O.+O.A.	32.9	32.2	0.013	0.015 0.025	L	57.8	55.9	0.03	0.015 0.04	M	103.7	93.9	0.05	0.02 0.007	L
M.O.+C.P.	33.8	32.4	0.025	0.025 0.03	L	57.8	56.3	0.025	0.015 0.045	L	104.7	95.6	0.045	0.025 0.02	L
M.O.+Gr.	33.7	32.0	0.03	0.025 0.02	L	61.6	55.1	0.09	0.05 0.025	L	104.4	96.2	0.045	0.025 0.025	L
O.A.	34.4	32.5	0.03	0.03 0.03	L	58.3	54.9	0.05	0.02 0.025	L	97.7	93.9	0.025	0.01 0.007	M
C.P.	32.3	32.2	0.007	0.015 0.025	L	59.8	56.7	0.05	0.035 0.055	L	113.1	97.3	0.07	0.05 0.03	L
Gr.	41.5	31.7	0.12	0.08 0.015	L	70.8	60.4	0.12	0.11 0.12	L	107.3	98.0	0.048	0.035 0.035	L
Water	86.0	40.4	>0.5	>0.5 0.18	M	---	---	---	---	---	---	106.0	---	--- 0.09	L
Plane-strain yield stress assumed:	31.0	31.0				56.0	54.0				96.0	93.0			

the unlubricated 3003 aluminum, somewhat lower on 7075 aluminum, and quite low for the brass (even though forces became too high to complete compression at L/h=7). Friction with the liquid lubricants seems to be of the right order of magnitude, although some values are rather too low to be readily believed (e.g. mineral oil on 7075 aluminum). Friction with the solid graphite was quite high and was coupled with localized damage of the workpiece surface where the graphite coat broke up as a result of surface extension. Plain water is as ineffective as one would expect for this process.

An inspection of the data shows that the results are extremely sensitive to the history of the surface. If we take the not unreasonable assumption that the best lubricants used in these tests could be regarded to give the plain strain yield stress of the workpiece material with a good approximation, we could choose yield stress values (shown in the last row of Table IV) for the various L/h ratios. Using Alexander's [4] solution, it is possible then to calculate a coefficient of friction, shown with an asterisk in Table IV for the two L/h ratios (friction for the higher L/h ratio is always shown in the top line, and the value for the lower ratio in the bottom line). In most instances, the agreement with the coefficient of friction calculated from the ratio of the two stresses is quite satisfactory. The reason for some of the anamolies noted is now seen to be due to a situation in which higher friction was generated at the lower L/h ratio than at the higher one; this could be due to a genuine change in the lubrication mechanism, and inspection of the deformed surfaces supports this assumption. The impressions obtained with viscous lubricants at high L/h values (anvil width L=0.5 in.) show a much higher density and depth of hydrostatic (hydrodynamic) pockets, than the narrower impressions produced at lower L/h values(L=0.25 in.). This must be taken as an indication of more efficient, predominantly hydrodynamic lubrication, originating in a better entrapment of lubricant when the strip is indented with a relatively wide anvil. Indeed, force-displacement curves rise then much more gradually than when a narrow anvil is used. This difficulty could not be overcome by using the same anvil on strips with thicknesses chosen so as to give the desired L/h ratios(say, 7 and 3) after equal percentage reduction, since the significant factor- for the entrapment mechanism- is not L but L/h. While this phenomenon does impair the accuracy of calculated friction values, it is actually an aid in identifying lubricants that have useful hydrodynamic properties.

It may well be that some of the true friction effects were overshadowed by random variations in the hardness of the workpieces. Even if the workpiece hardness were constant to ± 2%, results could be affected. In practice, such close tolerances on workpiece hardness could be hardly expected even though the specimens were all rolled and annealed with the greatest care and uniformity. This must be regarded as a fundamental weakness of the plane strain compression test, especially when lubricants giving friction values close to zero are investigated.

Wire Drawing

Wire drawing at low speeds (20 fpm) is really more typical of industrial bar drawing practice, however, lack of time did not allow to repeat experiments at higher speeds representative of industrial wire drawing practices. Thus, the results are undoubtedly biased towards lubricants that are effective in boundary situations, and hydrodynamic agents and additives that become active at higher surface temperatures are given unfair treatment. Nevertheless, the results are useful in that surface sliding velocities approach fairly well those existing in the ring compression and the plane strain compression test.

Table V
Results of Wire Drawing Experiments

Lubricant	3003 Al		7075 Al		70/30 Brass	
	Stress kpsi	pick-up	Stress kpsi	pick-up	Stress kpsi	pick-up
None	13.0 →	H	23.2 →	H	41.6-51.0	H
M.O.	12.1-12.7	N	18.0-18.4	L	30.5-31.7	N
M.O.+O.A.	12.2-12.5	N	18.3-19.3	N	34.5-37.0	L
M.O.+C.P.	12.6-13.8	N	18.0-19.3	N	32.0-33.8	N
M.O.+ Gr.	12.0-12.3	N	17.2-18.2	L	34.2-34.6	N
O.A.	12.0-13.1	N	16.6-19.8	N	30.5-34.5	L
C.P.	14.1-14.4	N	19.8-20.2	N	38.8-39.5	L
Gr.	12.3-14.0	N	18.8-24.0	M	28.0-44.0	M
Water	13.5-23.0	M	30.0 →	H	37.0-45.0	L
Water + Gr.	15.2-19.0	M	----	-	39.0-42.0	M

It is gratifying to see (Table V) that some correlation does indeed exist. As would be expected from the high friction found without a lubricant, none of the aluminum alloys drew, and brass generated high enough forces and heavy enough pick-up to lead one to believe that continued drawing would have resulted in failure too. The forces measured with the two aluminum alloys follow similar but by no means fully anticipated trends. Oleic acid did not prove to be the effective additive one would have expected it to be. The performance of the neat oleic acid could possibly be explained by its lower viscosity, but then the poor results given by the neat chlorinated paraffin are difficult to explain. The high stresses and the surface damage observed with water and water-graphite mixture was expected on the aluminum alloys.

The obvious conclusion that mineral oil is only degraded by the addition of a fatty acid or a chlorinated paraffin in the drawing of brass is in line with results found in earlier unpublished work by the present author, but contradicts some generally held beliefs and also some industrial practices (although the foundation of these practices is doubtful). The high forces developed with the chlorinated paraffin are particularly noteworthy in view of the very high viscosity of this lubricant.

Some of the spread of results with the dry graphite film must be attributed to uneven coverage; even though the wires appeared to be well coated prior to drawing, film thickness must have been variable as evidenced by visible thinning and breakdown in localized areas.

DISCUSSION

In attempting to evaluate the relative value of the two simulating tests investigated here, the purpose of the simulation must be kept in mind.

Both ring compression and plane strain indentation are suitable for deriving coefficient of friction values. Ring compression is very attractive in its simplicity, and in the ease of converting measured deformation into an equivalent coefficient of friction. In plane strain indentation, the effort invested in the test and in its evaluation is substantially larger, and minor changes in the yield stress of the material could easily invalidate the absolute magnitude of friction calculated. It is, therefore, mandatory that specimens from the same batch of material should be used for comparative purposes. Even though the same basic assumptions are involved [4,5] in the calculations utilized here for evaluating the two test methods, the friction coefficient derived from ring upsetting is usually larger than the one derived from the plane strain test in the compression of the aluminum alloys (Fig. 1 and 2).

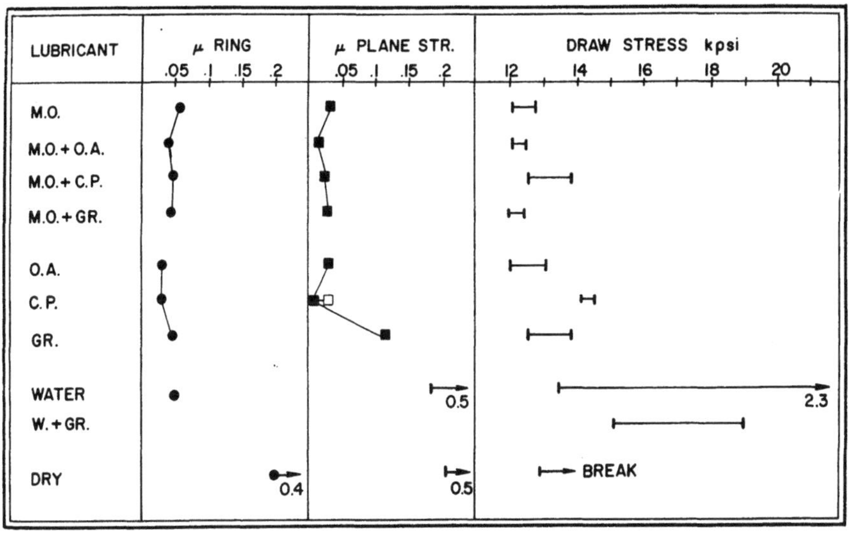

FIG. 1 -- Comparison of lubricant performance in ring compression, plane strain indentation and wire drawing of 3003 aluminum alloy.

For the brass specimens (Fig. 3), the converse is true. The reason is perhaps to be found in the history of the specimen during testing: the development and breakdown of the squeeze film has a marked effect on the ring test results and the test measures a cumulative effect [8]. In plane strain compression, with the ratio of interface pressures measured at two different L/h ratios used for the determination of friction, the differences attributable to greater lubricant entrapment in indentation with large initial L/h ratios are sufficient to lead to sometimes quite irrealistically low, even negative coefficient of friction values. A friction coefficient derived from the absolute magnitude of the interface pressures (values marked with an asterisk in Table IV) are then more realistic, however, their absolute value will then be even more sensitive to the magnitude of the yield stress. Judged objectively, both tests yield acceptable relative friction values, but none of them is capable of generating coefficient of friction data that would be universally applicable, mainly because of the marked effect of the lubricating mechanism. Of course, this is also a feature of actual deformation processes and no universal friction coefficient should really ever be used.

Ring upsetting appears to be rather forgiving as far as lubricant performance is concerned. Some of this behavior is no doubt due to the rather limited sliding at the tool/workpiece interface

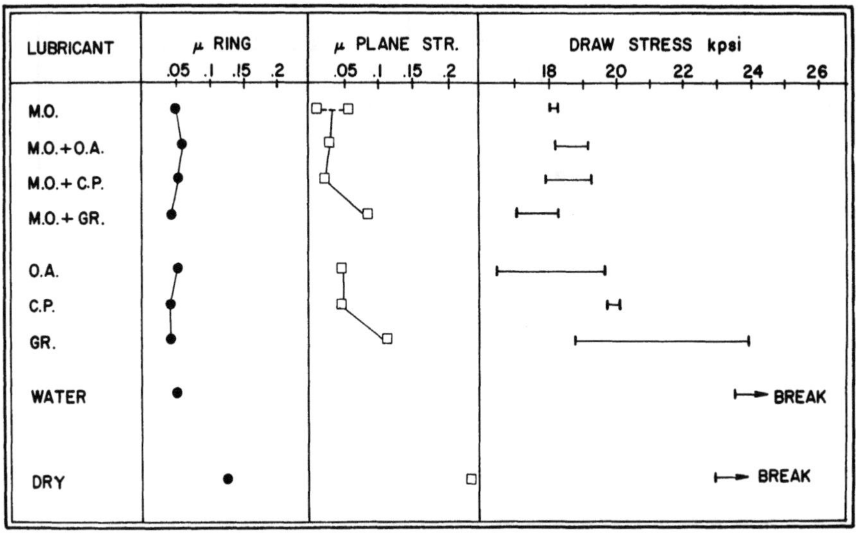

FIG. 2 -- Comparison of lubricant performance in ring compression, plane strain indentation and wire drawing of 7075 aluminum alloy.

and, combined with squeeze-film formation, this could well explain the lack of discrimination found in some earlier work [9] and, to some extent, also the absence of very high friction with a lubricant as poor as water. For deformation processes in which sliding along the tool/workpiece interface is significant, plane strain compression offers much better discrimination. The ability of a lubricant to extend with the deforming surface is more apparent; thus, it is clearly visible (Figs. 1 and 2) that a dry graphite film applied to the workpiece surface can actually rupture unless process geometry is favorable. Because the material is effectively extruded from the zone between the indentation anvils, protection of new surfaces against adhesion and possible surface damage is much more critically tested too. This is probably the reason why plane strain compression properly predicts a poor lubricating performance for water under these conditions.

Both tests appear to be able to register relatively minor changes in lubricant composition, although the change in internal diameter in the ring compression test is a much more sensitive indicator than friction values calculated from either of the simulating tests. The meaning of such observed effects is, however, by no means clear. If one chose the effect of additives in a mineral oil in the forming of 7075 aluminum alloy, one could argue that ring

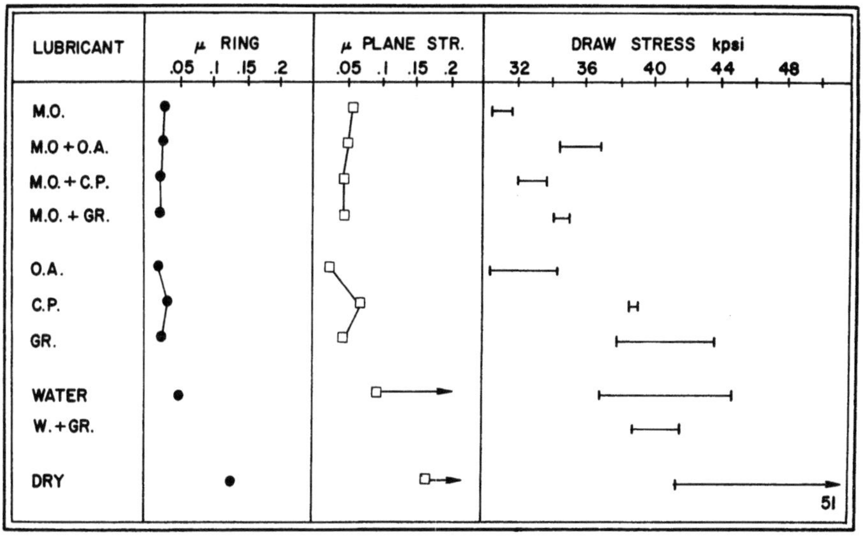

FIG. 3 -- Comparison of lubricant performance in ring compression, plane strain indentation and wire drawing of 70/30 brass.

compression will accurately predict the performance in low-speed wire drawing. If this favorable correlation would then be taken further, disastrous results would follow if pure water were used as a drawing lubricant. Similar examples could be selected from Figs. 1, 2 and 3 for other workpiece materials or lubricants and for both simulating tests. It is now obvious that a mineral oil of high viscosity was not the best choice as the base lubricant for tests conducted at room temperature, because it could well be that much of the boundary and E. P. action was masked by the predominantly hydrodynamic lubrication prevailing by virtue of the high base-oil viscosity. This explanation does not, of course, suffice to account for the occasionally very poor performance offered by the neat chlorinated paraffin. From an inspection of the deformed surfaces it would appear that the chlorinated paraffin was the most viscous even at the pressures prevailing in the deformation zone, and the high drawing force (and coefficient of friction in plane strain indentation) on the brass (Fig. 3) is difficult to account for.

Neither of the tests used here are particularly suitable for assessing tendencies towards pick-up and surface damage to the workpiece. A closer inspection of the rather heavy damage noted on compressed 7075 ring specimens (Table III) revealed that this was actually caused by an exhaustion of ductility in this material.

Because of excessive shear, the original edges of the rings opened up circumferentially, and these cracks then combined with tool pick-up to give the appearance of heavy damage. Plane strain indentation is somewhat more sensitive because material streaming past the sharp edges of the indenter will show signs of scoring whenever tool pick-up is present. The effect of repeated or continued sliding, however, is not readily revealed in either test, unless a number of specimens are deformed on the same tool surface and thus the cumulative effect of pick-up formation is allowed to manifest itself. In this respect, the twist compression test would appear to be a much more favorable choice because continued rotation rapidly reveals any tendency to form die pick-up as a consequence of lubricant breakdown [9].

It was unfortunate that in the present experiments dies had to be kept at room temperature, because it is known from earlier work that the effect of additives both in reducing friction and in preventing tool pick-up would become much more significant if the dies were heated to a slightly elevated temperature, typical of steady-state production conditions. At these temperatures the viscosity of the base oil would drop sufficiently to allow additive effects predominate.

SUMMARY AND CONCLUSIONS

In a series of experiments aimed at elucidating the relative validity of ring compression and plane strain indentation tests, 3003 and 7075 aluminum alloy and 70/30 brass workpieces were deformed at slow press speeds with substances representative of hydrodynamic lubricants (a high viscosity mineral oil), boundary agents (oleic acid), E. P. additives (chlorinated paraffin), solid lubricants (graphite) and also with compounded mineral oils, with water, and without intentionally applied lubrication. The same program was repeated with low speed (20 fpm) wire drawing as a means of providing some comparison with a widely practiced metalworking process.

Ring compression was again found very appealing in its simplicity and sensitivity; however, it indicated unusually low friction for substances (such as water) that otherwise would be known as ineffective lubricants. It appears that the mode of deformation creates relatively mild interface conditions, particularly, an absence of significant sliding, and this could lead to a favorable judgement on lubricants that are otherwise useless for most purposes. It would appear, therefore, that this test must be used with circumspection even for lubricant screening purposes, even though it can be a valuable and simple tool of investigation if a careful check on the validity of conclusions is made, preferably through occasional tests in the process to be simulated.

Plane strain compression demanded much more care and effort in conducting the experiments, and artificially low friction values were sometimes derived if the lubricant had substantial hydrodynamic properties. Because of sliding on the tool faces, the ability of the lubricant to expand with the workpiece surface and to protect newly developed surfaces is much more critically tested, and for the same reason tendencies to form tool pick-up are also more readily observed. The test does, however, require workpieces of very uniform and reproducible yield strength and accidental errors could easily distort results.

True correlation could not be established between wire drawing and either of the two simulating tests. Plane strain compression would appear to be more reliable as a preliminary screening test because it at least predicts breakdown behavior that would be catastrophic in processes involving substantial interface sliding such as wire drawing. Occasionally, general trends in the effectiveness of lubricants both in wire drawing and in the simulating tests could be discerned, however, there were a sufficient number of exceptions to make generalizations impossible.

In summary, this investigation confirmed that no generally applicable simulating test exists as yet. It is also obvious that too little is known about lubricating mechanisms in various deformation modes, and it is for this reason that transfer of results from one deformation process to another is so difficult. The need for developing a more generally applicable test, with proven relavance to a number of practical deformation processes , still remains.

ACKNOWLEDGEMENTS

This work was performed in the laboratories of the Department of Materials Engineering, University of Illinois at Chicago Circle, Chicago, Illinois. The author is indebted to Messrs. P. A. Abramowitz, T. F. Restivo and D. K. Fung for the care and interest shown in conducting the experiments.

REFERENCES

[1] J. A. Schey, Metal Deformation Processes: Friction and Lubrication, Marcel Dekker Inc., New York, 1970, pp. 267-295.

[2] A. T. Male and M. G. Cockroft, J. Inst. Metals, 93, 1964-65, pp. 38-46.

[3] J. A. Schey, P. W. Wallace and K. M. Kulkarni, to be published.

[4] J. M. Alexander, J. Mech. Phys. Solids, 3, 1955, pp. 233-245.

[5] W. Schroeder and D. A. Webster, Trans. ASME, Ser. E(J. Appl. Mech.), 16, 1949, pp. 289-294.

[6] G. W. Pearsall and W. A. Backofen, Trans. ASME, Ser. B(J. Eng.Ind.),85, 1963, p. 329.

[7] J. A. Schey, in Friction and Lubrication in Metal Processing, ASME, New York, 1966, pp. 20-38.

[8] A. T. Male, J. Inst. Metals, 94, 1966, pp. 121-126.

[9] J. A. Schey and J. A. Newnham, Lubric. Eng., 26, 1970, pp. 129-137.

A NEW METHOD FOR THE DETERMINATION OF MATERIAL FLOW STRESS VALUES UNDER METALWORKING CONDITIONS

George Saul*, Alan T. Male† and Vincent DePierre*

Air Force Materials Laboratory* and Westinghouse Electric Corporation†

I. INTRODUCTION

For the adequate mathematical understanding of metalworking processes and the theoretical prediction of deformation loads, there is a precise need for accurate knowledge of the basic flow stress behavior of the workpiece material under particular conditions of temperature and strain rate. This information can be obtained by means of tension, torsion, or compression testing, but because of major drawbacks with the first two methods, compression testing is most widely used.

In normal upsetting of cylindrical specimens the deformation stress is composed of the stress necessary to overcome friction at the tool/specimen interfaces and the additional stress caused by non-homogeneous deformation in addition to the basic flow stress of the specimen material. Polakowski (1) has developed a technique of compression testing which effectively eliminates these two additional stresses, the deformation stress then being the flow stress of the material. This technique involves the use of a cylindrical specimen with a ratio of height to diameter of approximately 1.5 to 2.0 and requires that the specimen be periodically remachined back to this ratio to eliminate any tendency for barrelling during deformation. Such a technique, although very accurate, is very tedious to perform and can only be used effectively at low strain rates and ambient temperature.

For obtaining data on a material's resistance to deformation in compression at elevated temperatures and high strain rates, the past philosophy [2, 3] has been to minimize frictional and non-homogeneous deformation effects by judicious choice of the initial specimen geometry and lubrication, and then to assume that these effects are non-existent. Such a philosophy can lead to the generation of data which may be in error by as much as 30%. [4]

It is proposed that a much more accurate method of obtaining this data would be to use flat ring shaped specimens of the general type used for friction studies [5, 6] rather than a simple cylindrical specimen. The change in shape of the ring upon deformation would give a measure of the interface friction, and measurement of the deformation load and contact area would give a value for the total average applied pressure. With a suitable mathematical theory for the compression of a ring, it should be possible to use this data to compute the flow stress of the material at that strain under the particular deformation conditions. The process could be repeated using a number of ring specimens compressed to different strains to obtain a flow stress-strain curve.

The object of the present work was to investigate this technique as a possible means of determining flow stress data and to compare the results so obtained with other data generated using the Polakowski technique.

II. THEORY

Avitzur [7] has analyzed the problem of the axial compression of flat ring-shaped specimens between flat dies using an upper bound technique and making the following assumptions:-

(1) The ring material obeys Mises' stress-strain rate laws, implying no strain hardening effect, no elastic deformation and no volumetric change.

(2) At the die/ring interface, for a given die and material under constant surface and temperature conditions, there is a constant friction factor, m, such that the interfacial shear stress, τ, is given by $\tau = m\sigma_o/\sqrt{3}$ where σ_o is the flow stress of the ring material.

(3) The ring is sufficiently thin for the effect of the surface frictional retardation stresses to be transmitted uniformly throughout the thickness of the ring, i.e. no barrelling.

Avitzur's theoretical solutions are summarized as follows:-

When $R_n \leq R_i$, and

$$m\frac{R_o}{T} \leq \frac{1}{2\left(1-\frac{R_i}{R_o}\right)} \ln\left[\frac{3\left(\frac{R_i}{R_o}\right)^2}{1+\sqrt{1+3\left(\frac{R_o}{R_i}\right)^4}}\right]$$

Then

$$\frac{P}{\sigma_o} = \frac{1}{1-\left(\frac{R_i}{R_o}\right)^2}\left[\sqrt{1+\frac{1}{3}\left(\frac{R_n}{R_o}\right)^4} - \sqrt{\left(\frac{R_i}{R_o}\right)^4+\frac{1}{3}\left(\frac{R_n}{R_o}\right)^4} + \frac{2}{3\sqrt{3}}\, m\frac{R_o}{T}\left\{1+\left(\frac{R_i}{R_o}\right)^3\right\}\right] \quad (1)$$

$$\text{where}\left(\frac{R_n}{R_o}\right)^2 = \frac{\sqrt{3}}{2}\,\frac{1-\left(\frac{R_i}{R_o}\right)^4 X^2}{\left\{X(X-1)\left[1-\left(\frac{R_i}{R_o}\right)^4\right]X\right\}^{1/2}} \quad (2)$$

$$\text{and } X = \left\{\frac{R_o}{R_i}\exp\left[-m\frac{R_o}{T}\left(1-\frac{R_i}{R_o}\right)\right]\right\}^2 \quad (3)$$

When $R_i \leq R_n \leq R_o$, and

$$m\frac{R_o}{T} \geq \frac{1}{2\left(1-\frac{R_i}{R_o}\right)} \ln\left[\frac{3\left(\frac{R_o}{R_i}\right)^2}{1+\sqrt{1+3\left(\frac{R_o}{R_i}\right)^4}}\right]$$

or when $m\frac{R_o}{T} > \frac{1}{2}$

Then

$$\frac{P}{\sigma_o} = \frac{1}{1-\left(\frac{R_i}{R_o}\right)^2}\left[\sqrt{1+\frac{1}{3}\left(\frac{R_n}{R_o}\right)^4} - \sqrt{\left(\frac{R_i}{R_o}\right)^4+\frac{1}{3}\left(\frac{R_n}{R_o}\right)^4} + \frac{2}{3\sqrt{3}}\, m\, \frac{R_o}{T}\left\{1+\left(\frac{R_i}{R_o}\right)^3 - 2\left(\frac{R_n}{R_o}\right)^3\right\}\right] \tag{4}$$

where R_n/R_o is found by successive approximations from

$$2\, m\, \frac{R_o}{T}\left(1+\frac{R_i}{R_o} - 2\,\frac{R_n}{R_o}\right) + \ln\left\{\left(\frac{R_i}{R_o}\right)^2\left[\frac{\left(\frac{R_n}{R_o}\right)^2 + \sqrt{3+\left(\frac{R_n}{R_o}\right)^4}}{\left(\frac{R_n}{R_o}\right)^2 + \sqrt{3\left(\frac{R_i}{R_o}\right)^4 + \left(\frac{R_n}{R_o}\right)^4}}\right]\right\} = 0 \tag{5}$$

Nomenclature

σ_o = flow stress of ring material

P = average forging pressure on ring

τ = average shear stress at die/ring interfaces

m = interface friction factor

R_i = internal radius of ring

R_o = external radius of ring

R_n = radius of metal flow divide (no-slip radius) in ring.

It is thus possible to calculate values of P/σ_o at the instant deformation stops in terms only of the ring geometry and the interfacial shear factor, m. In these equations, neither the basic yield stress of the material, σ_o, nor the interfacial shear stress, τ, appear in terms of independent absolute values, only as the ratio, m. The basic assumption in the analysis is that this ratio remains constant for the material and deformation conditions. If the analysis is carried out for a small increment of deformation, σ_o and τ can be assumed to be approximately constant for this increment and the solution is valid. Thus, if the shear factor, m, is constant for the whole operation, it would appear justifiable to continue the mathematical analysis in a series of small deformation increments using the final ring geometry from one increment as the initial geometry for the subsequent increment and so on. As long as the ratio of the interfacial shear stress, τ, and the material flow stress, σ_o, remained constant it would not be of consequence if the ring material strain hardened during deformation provided that the increase in work hardening in any one single deformation increment could be neglected. The progressive increase in interfacial shear stress accompanying strain hardening would also be of no consequence provided that it could be assumed to be constant over the entire die/ring interface during any one deformation increment. Thus it is possible that the analysis could be justifiably applied to real materials even though it was initially assumed that the material would behave according to the Mises' stress-strain rate laws, provided that the assumption of a constant interfacial shear factor, m, is correct.

DePierre and Male [8], using the above reasoning, have solved the Avitzur equations for several different starting ring geometries over the complete range of m values from zero to unity. These computer solutions can also yield values for the ratio of P/σ_o for the various values of m and amounts of deformation. Using ring specimens which are sufficiently thin to validate Avitzur's assumption, the computer solutions presented in the graph, Figure 1, can be used in the following manner:-

a) A ring specimen is deformed under the required conditions and the deformation load noted.

b) The present change of the hole diameter is indexed with the percent deformation of the ring on the graph (Fig 1). The corresponding ratio of pressure to flow stress P/σ_o is read from the horizontal axis.

c) Measurement of the area of the ring surface formerly in contact with the die, together with knowledge of the deformation load allows the calculation of P, and hence the value of the material flow stress, σ_o, for that amount of deformation.

d) Repeating this process with other ring specimens over a range of deformation would allow the generation of a complete flow stress-strain curve for the material under those particular deformation conditions.

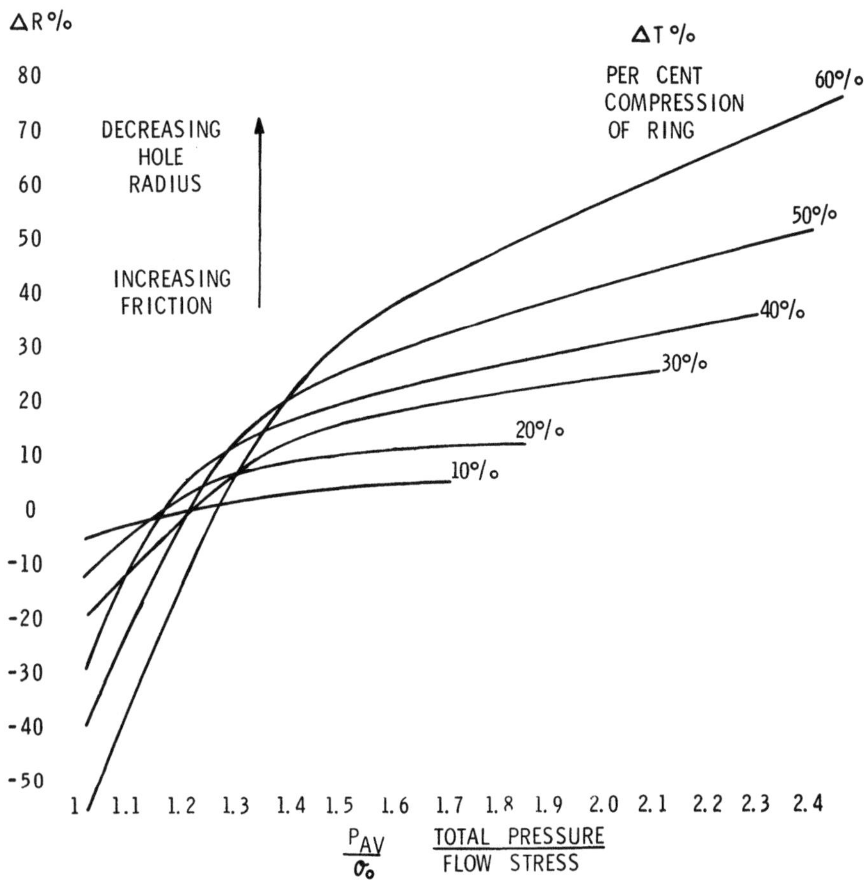

Figure 1. Change of Inside Diameter Versus Ratio of Total Pressure to Material Flow Strength for Ring with O. D. to I. D. Ratio of 2

The maximum ring thickness (compared with the inner and outer diameters) which can be used and still meet Avitzur's assumption of "thin specimen" conditions will vary to a degree dependent upon the actual friction conditions. DePierre and Male [8] showed that, under conditions of maximum friction, the largest specimen height that could be used was obtained with rings of geometry 6:3:0.5 (Outer dia.: Inner dia.: Height). Under conditions of low friction, it is greatly probable that somewhat thicker specimens could be used and still meet the assumption.

III. EXPERIMENTAL WORK AND RESULTS

A. Materials

The materials used in the bulk of this investigation were commercially pure aluminum, O. F. H. C. copper and 1018 steel. The first two materials were used in the partially cold worked condition whilst the 1018 steel was given a normalizing treatment before use.

Cylindrical and ring-shaped specimens of various dimensions were machined from the same bar of each of the three materials. The actual specimen dimensions used were as follows:-

Specimen shape	Specimen dimensions	Ratio of OD:ID: Height
Ring	0.750 in. O.D. x 0.375 in. I.D. x 0.250 in. height	6:3:2
	0.750 in. O.D. x 0.375 in. I.D. x 0.125 in. height	6:3:1
	0.750 in. O.D. x 0.375 in. I.D. x 0.062 in. height	6:3:0.5
Cylinder	0.50" dia. x 1.00" height	

B. Correlation Studies

These tests were designed to initially determine the flow stress behavior of the three materials by using the well proven, but tedious, Polakowski technique [1] and then to use this data for comparison with data obtained from the compression of rings under similar conditions and computed in the manner outlined previously. Deformation was carried out between polished dies on a 50,000 pound capacity Tinius Olsen testing machine operating at a constant crosshead speed of 0.1 ins/min. A roll-type recorder fitted to the testing machine allowed continuous monitoring of load and specimen deformation when coupled to two standard displacement gauge units mounted diametrically opposite between the two dies.

The flow stress behavior of the three materials was first obtained using the Polakowski technique of the compression of tall cylinders (height/diameter $\simeq 2$) and periodically remachining to remove any tendency for barrelling in order to prevent the generation of a triaxial state of stress. Cylinders, 0.50'' dia. x 1.00'' height, of each material were lubricated with dry graphite film (Miracle dgf aerosol spray), as were the dies, and given approximately 20% reduction in height. The specimens were then carefully machined back to a height/diameter ratio = 2, relubricated and compressed a further 20% reduction. This process was repeated until the specimens had been given a cumulative total deformation of 60% reduction in height. The load vs reduction in height data for each step was then recalculated into stress vs reduction in height values by taking into account the progressive increase in area during deformation. These determinations were carried out in triplicate for each material and the results for all three materials are given in Figure 2, where the best single curve has been drawn through each triple set of points and is assumed to be the basic flow stress behavior of the particular materials.

Determination of flow stress values by means of the ring-test technique was carried out using specimens of 0.75'' outside diameter but varying thicknesses and using three different frictional conditions; no lubricant, lubricated with dry graphite film (dgf), and lubricated with teflon sheet. Specimens were compressed over a range of deformation and the deformation loads noted. Using a toolmakers microscope, the outer and inner diameters of the flat faces were carefully measured in several different directions and an average taken. This allowed the contact area, and hence the average deformation pressure to be calculated. Measurement of the change in shape of the rings upon deformation and using the computer solutions of Figure 1 allowed values of the interfacial friction factor to be determined, together with the corresponding ratio of the average deformation pressure, P, to the material flow stress, σ_0. Hence the materials flow stress was determined over a range of deformation.

The values obtained in this way are given in Figure 2, together with the results of the Polakowski-type determinations. These show that random scatter was observed when using the 6:3:1 and 6:3:0.5 ratio rings, with the majority of points falling within the scatter level obtained with the Polakowski technique. This scatter is thought to be principally due to the present

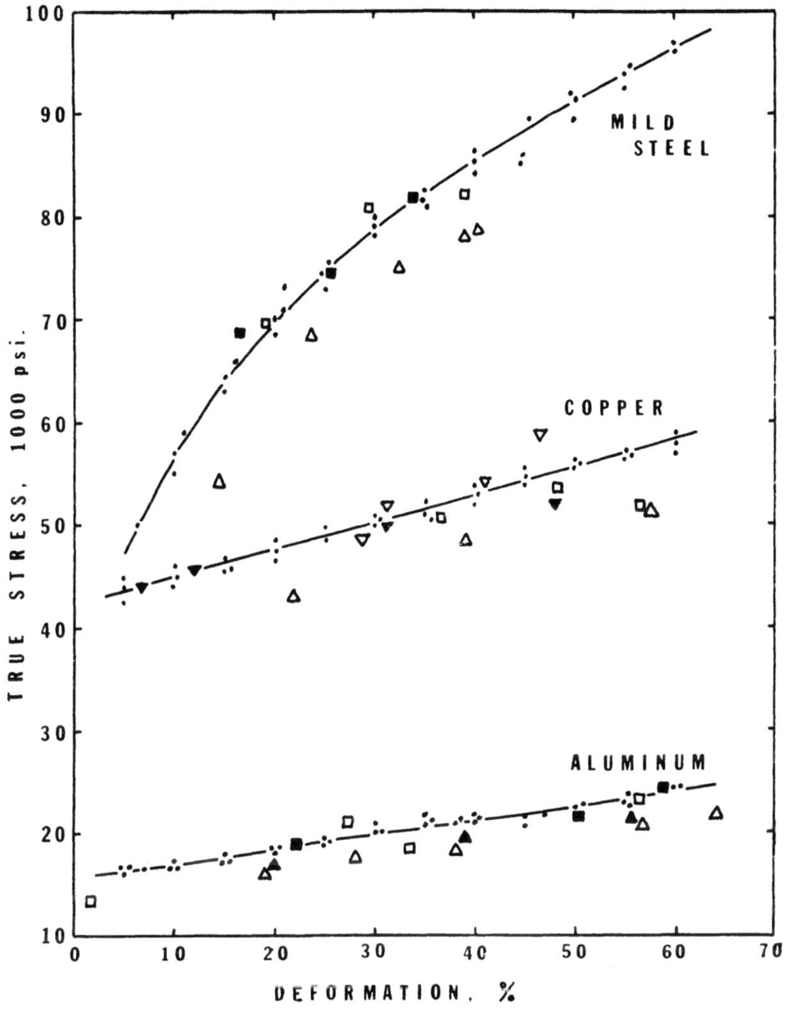

Figure 2. Flow Stress Measurements as a Function of Deformation for Aluminum, Copper and Steel; Obtained Using the Ring Compression Technique and the Polakowski Technique.

▲ 6:3:2; no lubricant

△ 6:3:2; graphite lubricant

■ 6:3:1; no lubricant

□ 6:3:1; teflon lubricant

▼ 6:3:0.5; no lubricant

▽ 6:3:0.5; teflon lubricant

● Polakowski Technique

technique of estimating the final specimen contact area from averages of several measurements of the internal and external diameters. The flow stress values obtained when using rings of initial geometry 6:3:2 were consistently lower than the Polakowski curves. A selection of ring test results is given in Table I to show the general effect of the friction level upon the average pressures necessary to compress rings to various reductions in height, and how this effect is removed during the calculation of the flow stress values.

TABLE I

(a) Deformation of rings of initial geometry 6:3:0.5

MATERIAL	LUBRICATION	DEFORMATION (%)	LOAD (pounds)	P (ksi)	m	σ_o (ksi)
Copper	Dry	6.4	20,000	57.7	0.19	43.7
	Teflon	28.6	25,000	55.0	0.04	48.2
		30.8	30,000	65.6	0.09	52.0
		40.7	40,000	72.4	0.09	54.4
		46.5	50,000	82.0	0.09	59.0

(b) Deformation of rings of initial geometry 6:3:1

MATERIAL	LUBRICATION	DEFORMATION (%)	LOAD (pounds)	P (ksi)	m	σ_o (ksi)
Copper	Dry	12.0	20,000	55.9	0.20	45.8
		30.1	30,000	65.6	0.23	49.7
		48.0	50,000	80.3	0.27	51.1
	Teflon	36.0	30,000	59.7	0.09	50.6
		48.0	40,000	64.4	0.08	53.6
		56.5	50,000	64.3	0.08	51.4
Steel	Dry	16.3	30,000	81.6	0.14	69.2
		25.4	40,000	92.5	0.17	74.6
		34.4	50,000	104.5	0.16	82.3
	Teflon	18.9	30,000	74.5	0.03	69.9
		29.4	40,000	85.5	0.03	80.7
		39.3	50,000	97.4	0.09	81.7
Aluminum	Dry	22.3	10,000	24.6	0.29	18.9
		50.4	20,000	31.2	0.18	20.2
		58.4	35,000	43.0	0.24	24.7
	Teflon	27.0	10,000	23.6	0.06	21.2
		33.0	10,000	22.0	0.07	17.7
		55.5	20,000	24.2	0.06	23.0

C. Application Study

In order to make some necessary calculations concerning the production of Zircaloy 4 tubing by cold deformation processes, a knowledge of the flow stress behavior of this material was required. Ring test specimens of 0.750 in. O.D. x 0.375 in. I.D. x 0.125 in. height were machined from Zircaloy 4 strip in the required metallurgical condition (Courtesy of Westinghouse Electric Corporation, Research and Development Department). Compression of these rings was carried out at ambient temperature between polished, flat dies at an average strain rate of 10^{-3} sec.$^{-1}$ and using Johnson's Wax as a lubricant. Sequential compression of a number of specimens in the manner outlined previously, allowed the generation of the data shown in Figure 3.

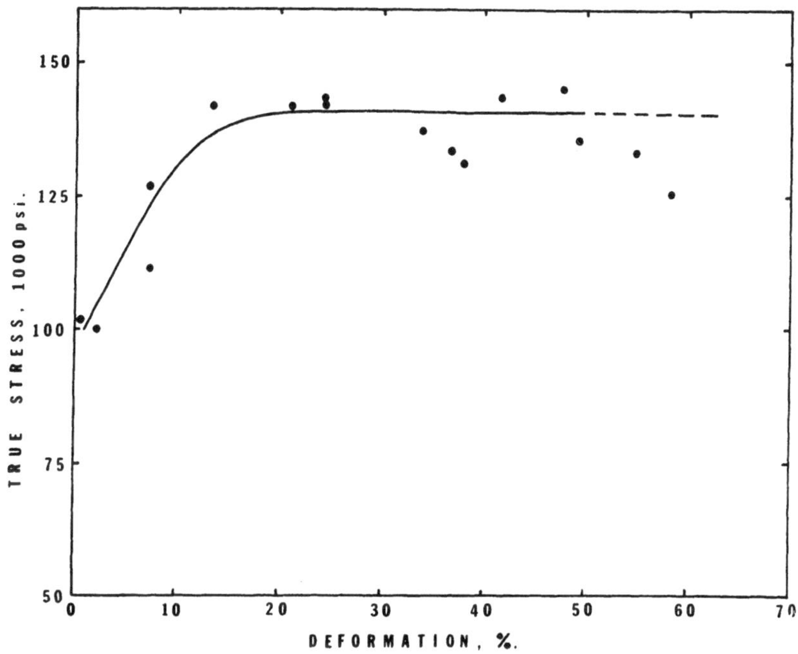

Figure 3. Flow Stress Measurements as a Function of Deformation for Zircaloy 4 Alloy; Obtained Using the Ring Compression Technique.

Random scatter was observed to approximately the same degree as in the work on aluminum, copper and steel. At deformations above 50% reduction in height, the material was prone to cracking and the low values computed for the flow stress at the high reductions was attributed to this cause.

IV. DISCUSSION

Use of the Polakowski technique for the determination of flow stress-strain data for metallic materials is tedious to perform but does yield accurate data. However, the very nature of the technique - requiring stepwise deformation and intermediate machining operations - effectively eliminates its use for deformation conditions other than compression at ambient temperature and low strain rates. Thus it is impossible for this technique to be used under typical metal working conditions. This investigation has shown that use of the proposed technique of ring compression yields results which are comparable with those obtained by the Polakowski technique, provided that the theoretical assumptions are adhered to in the practical situation.

A major assumption in the theoretical analysis of the deformation of a flat ring is that the interfacial friction stresses are transmitted uniformly throughout the ring thickness. This is a simplification of actual conditions and would be expected to give increasing error with increasing friction stress and increasing ring thickness. Earlier investigations [8] have shown that for this assumption to be valid for all friction stresses up to and including full sticking, the initial ring geometry should be approximately 6:3:0.5 (Outer dia. : Inner dia. : Thickness). For deformation under low friction conditions, a somewhat greater thickness ratio should still meet the theoretical requirement.

The results presented in Figure 2 show that flow stress values obtained from the compression of rings of initial geometries 6:3:0.5 and 6:3:1 are in general agreement with data obtained on the same material using the Polakowski technique. Thus, under conditions of low friction, the 6:3:1 geometry ring still justifies the theoretical assumptions. However, when using rings of initial geometry 6:3:2, the flow stress values obtained were consistently lower than those obtained with the Polakowski technique. This suggests that, with this ring geometry, the theoretical analysis is over-estimating the interfacial friction effects, an observation which is in agreement with the results of work performed earlier.[8]

This investigation, carried out at ambient temperature and a low strain rate, has shown that the use of ring compression specimens of the correct geometry, and the appropriate theory, is capable of giving realistic information about a materials flow properties. It thus appears that this technique would be useful for the determination of basic flow properties under typical metalworking conditions of elevated temperatures and high strain rates. Further experimentation is in hand to positively verify this assertion.

V. CONCLUSION

This investigation has shown that realistic data on the basic flow properties of a material deformed under conditions typical of many metalworking operations can be generated by use of compression specimens in the form of flat rings. This technique has a "built-in" measure of interfacial friction which can then be taken into account using the theoretical treatment due to Avitzur in order to obtain a truly "friction-free" value of flow stress. For such determinations, the theoretical assumptions made with regard to ring specimen thickness must be fulfilled. Provided the determinations are carried out under low friction conditions, specimens of initial geometric ratio 6:3:1 (O. D. : I. D. : Thickness) are adequate. Specimens of smaller thickness ratio may also be used.

REFERENCES

1. N. H. Polakowski, "The Compression Test in Relation to Cold Rolling," J. Iron Steel Inst., 163, 250, 1949.

2. J. F. Alder and V. A. Phillips, "The Effect of Strain Rate and Temperature on the Resistance of Aluminum, Copper, and Steel to Compression," J. Inst. Metals, 83, 80, 1954-55.

3. R. R. Arnold and R. J. Parker, "Resistance to Deformation of Aluminum and Some Aluminum Alloys - Its Dependence on Temperature and Rate of Deformation," J. Inst. Metals, 88, 255, 1959-60.

4. J. A. Schey, "The More Common Fabrication Processes," Chapter 34 in Vol. 1, Part 3, of Techniques of Metals Research, Interscience Publishers, New York 1968.

5. A. T. Male and M. G. Cockcroft, "A Method for the Determination of the Coefficient of Friction of Metals Under Conditions of Bulk Plastic Deformation," J. Inst. Metals, 93, 38, 1964-65.

6. A. T. Male, "The Effect of Temperature on the Frictional Behavior of Various Metals During Mechanical Working," J. Inst. Metals, 93, 489, 1964-65.

7. B. Avitzur, Metal Forming: Processes and Analysis, McGraw-Hill, Inc., New York, 1968.

8. V. DePierre and A. T. Male, "Mathematical Calibration of the Ring Test for Friction Studies in Flat Forging Operations," U. S. A. F. Tech. Rep. No. AFML-TR-69-28, October 1969.

FORCE REQUIREMENTS AND FRICTION IN WARM WORKING OPERATIONS

John T. Berry and Malcolm H. Pope

Department of Mechanical Engineering

University of Vermont

1. INTRODUCTION

A clear understanding of the force requirements in metal-working operations is still not entirely available to the manufacturing engineer, in spite of the fact that we are now more than forty years beyond some of the original publications of Siebel(1) and Sachs(2).

It is not necessary to emphasize why such information is of strategic importance, or why progress in this area has been pitifully slow. Anyone who has had to specify press capacity to suit a particular metal-forming task is aware of the gargantuan differences in capitalization that are at stake.

Similarly anyone who has had to design tooling for a similar operation will also be aware of both cost and performance oriented questions which were answered by gross over-design.

Our progress in the theoretical area has been slow because, for all but simple shapes, our mathematics have been complex and also because the materials and friction dependant parameters fed into our calculations have often been inadequate.

Advocacy for using the model-material technique has not gone unheard as the papers of Heuer(3), Brill(4), Altan and coworkers(5) attest. Likewise the so-called slip-line field(6) and the visio-plasticity method(7) have also been pressed into use with some very practical aspects of metal-working.

However, all of these techniques have demanded some particularized, perhaps somewhat esoteric skills to make them work.

Thus, we have seen in industry, a frequent reliance on either rules of thumb or upon some of the analysis developed in the nineteen twenties and thirties.

Very recently, however, we have all become aware of the power of computer-aided numerical methods, either in integrating the previously unintegratable, or perhaps in the basic formulation stages of a problem where we have adapted a finite difference, or possibly the finite element approach.

It is interesting to draw a parallel here with the problem of solidification kinetics in the foundry. Whereas ten to fifteen years ago we were limited to looking at the very simple shapes(8)(9), we can now extend our knowledge to the solidification patterns of much more complex two and three dimensional castings(10)(11).

I believe we are almost at the same stage in metalworking, where we are about to see the full impact of computer aided numerical methodsin predicting force requirements and tooling configuration. (See for example the paper by Dr. Altan in this session which deals with the computer simulation of axisymmetric forging).

However, like the solidification problem alluded to above, we will undoubtedly be limited by the validity of the materials properties and knowledge of boundary conditions fed into our calculations.

The present paper looks at two specific investigations in metal-forming, performed under very different conditions, which illustrate the principal arguments here:

A. An investigation involving established classical analysis where materials and friction parameters were not immediately available.

B. An investigation involving a new analysis and numerical integration, and where materials and friction parameters were to hand.

The first investigation is concerned with the back extrusion of low alloy steels in the warm working range. The second concerns the drawing of cups in titanium and nickel base materials, also in the warm working range.

The warm working range will be defined as one where the workpiece will not be subject to anything more than a minimal oxidation or contamination during the preheating and forming stages of manufacture.

The range is of practical interest, clearly because of attractively lower force requirements than those at room temperature. The minimum scale incurred in induction or rapid resistance type heating make it preferable to true hot working.

However, in terms of our forecasting press capacity and tooling requirements, we are still sadly deficient in data on both friction coefficients and materials parameters.

The two cases detailed in the present paper indicate the degrees of success one can expect working in this area and particularly how they depend upon the availability of the above data.

II. REVIEW OF PROBLEMS INVESTIGATED

1. Warm backward extrusion

Here the basic problem facing the authors was to predict press capacity and provide tooling design information for warm backward extruding two low alloy steels. Mild steel was also included for base line experimental data purposes. Experiments were also extended over a fairly wide range of temperature (RT to 2200 F) to provide further reference data.

Lubricants at the lower temperatures were either the usual phosphate-soap or chlorinated hydrocarbon systems. Above 800 F, a graphite grease was used. Dies were preheated to 400 F, where forming temperatures above that level were used. Punches and dies were constructed of a high speed steel, the designs involved following the standard commercial practice for cold forming. The punch head possessed the usual land and supporting radius, whilst the die possessed a minimal taper. The experiments were conducted on an instrumented mechanical type press. At least two samples of each steel were extruded at each temperature. The majority of the experiments were conducted with an area reduction ratio of 50%, a few experiments involved a slightly higher or lower reduction ratio (62% and 40%).

Table I gives details of the workpiece materials.

TABLE I

Workpiece Materials - Backward Extrusion

Nominal Composition %	Condition	Billet size used, in. Diameter	Height
1.0 C - 1.2 Cr	Spheroidize annealed	1.0	0.75
0.15 C - 1.8 Ni - 025 Mo	Spheroidize annealed	1.0	0.75

Figure 1 displays the results of the various measurements for the 1.0 C - 1.2 Cr steel, while Fig. 2 presents the results obtained with the 0.15 C - 1.8 Ni - 0.25 Mo steel. Figure 3 compares the above sets of results with some obtained with mild steel. Figure 4 presents the affects of area reduction for the 1.0 C - 1.2 Cr steel.

2. Warm Cup Drawing

In this instance our problem was concerned with determining press requirements to execute a cup forming operation on AISI 304 stainless steel, Inconel X-750 (a nickel base alloy) and Ti-6Al-4V, all in sheet form (Table II).

TABLE II
Workpiece Materials - Cup Drawing

Designation and Nominal Composition, %	Condition	Thickness range, in.
AISI 304 0.04C - 18 Cr - 10 Ni	Annealed	0.32 - 0.0505
Inconel X-750 15Cr-7Fe-2.5Ti-0.8Al-0.9Cb Bal. Ni	Annealed	0.052 - 0.066
Ti-6Al-4V 6Al - 4 V - Bal. Ti	Annealed	0.0325 - 0.068

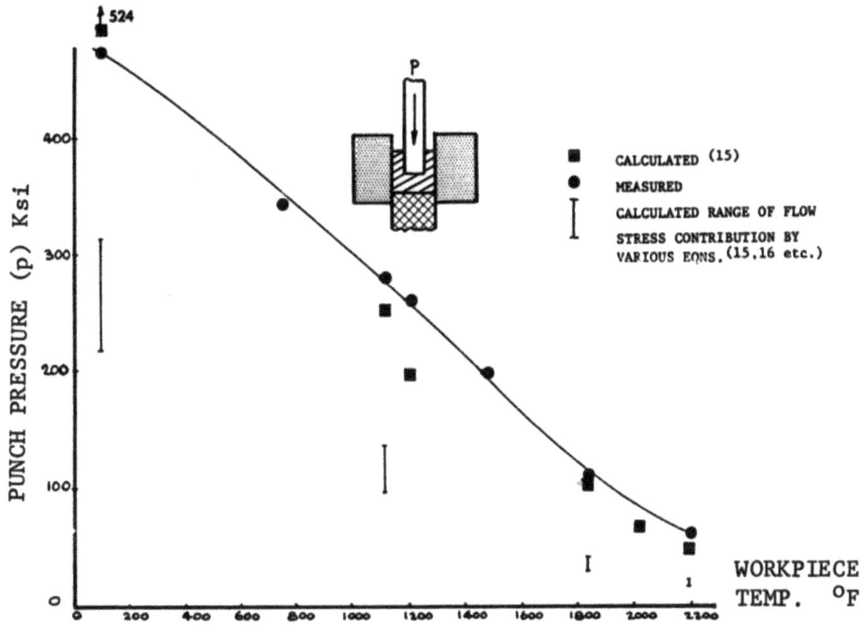

FIGURE 1. BACK EXTRUSION OF 1.0C - 1.2 Cr SPH. ANNEALED CONDITION, 50% REDUCTION.

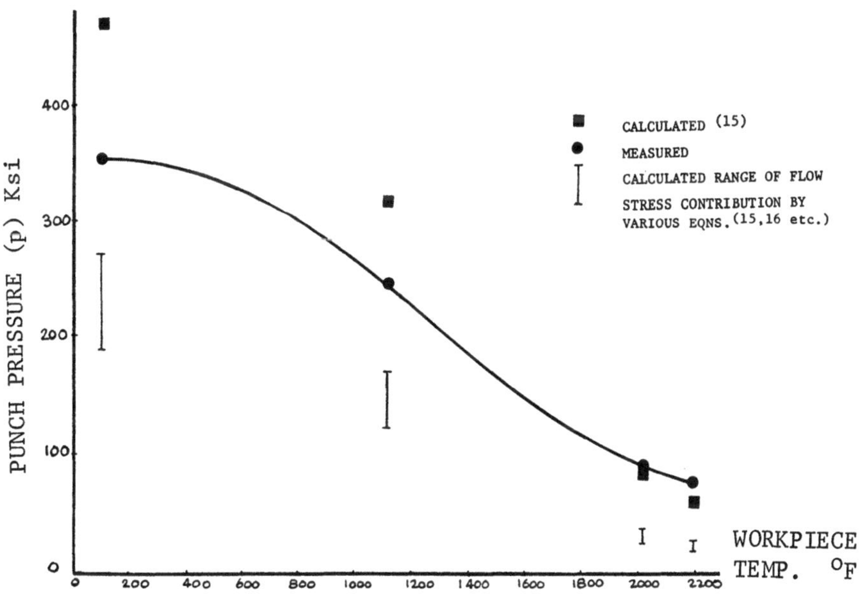

FIGURE 2. BACK EXTRUSION 0.15C - 1.8 Ni - 0.25 Mo 50% REDUCTION

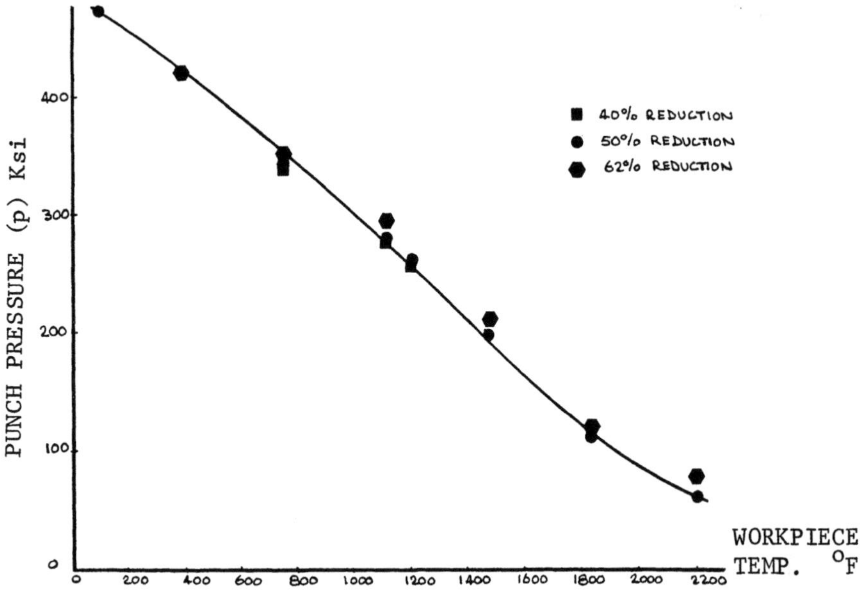

FIGURE 3. BACK EXTRUSION OF 1.0C - 1.2 Cr SPH. ANNEALED CONDITION, 40, 50, and 62% REDUCTION

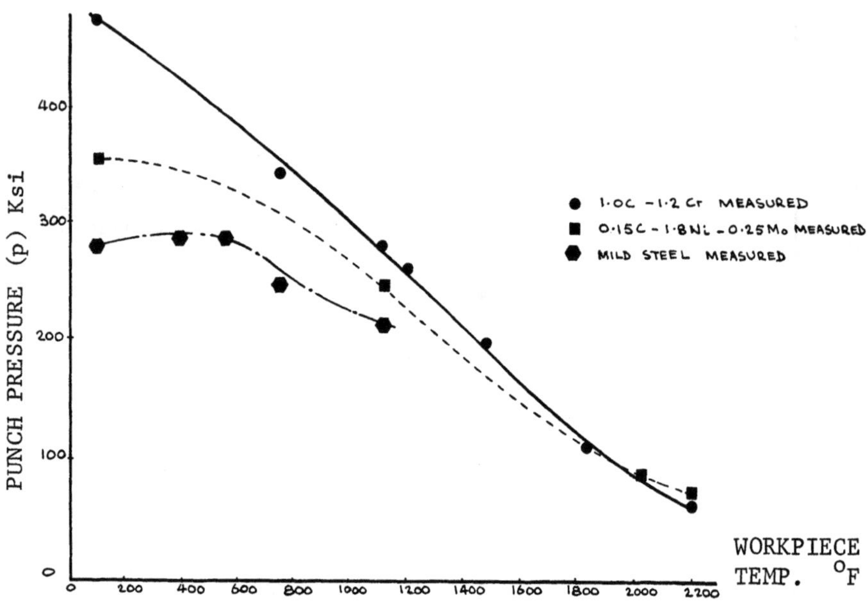

Figure 4. BACK EXTRUSION OF MILD STEEL 0.15 C - 1.8 Ni-0.25 Mo and 1.0C - 1.2 Cr ALLOYS WITH 50% REDUCTION

Certain details of the operation have been published elsewhere[12]. The experiments were again extended over a fairly wide range of temperatures, from room temperature up into the warm forming range (1200 F).

Once more the attractiveness of warm forming operations may be envisaged; lower degree of contamination or oxidation than at hot forming temperatures, but lower forming loads than at room temperatures. It will be recalled that because of nitrogen and oxygen contamination, hot pressings of titanium are currently subjected to an expensive post-form pickling.

A variety of lubricants were surveyed during the investigation. This aspect was given particular attention since preheated tooling was to be used (up to 1000 F for punches, die and blankholder pad, see Figure 5).

At least five pressings were made for each combination of workpiece material, thickness, lubricant and forming temperature. Two blank sizes were utilized 3.5 and 4.5 in. diameter, thicknesses were between 0.030 and 0.070 in. The drawn cup size was 2.0 in. diameter. Experiments were also conducted in a randomized sequence.

Because of the special effects of anisotropy in sheet forming, the tensile properties of the various workpiece materials were sampled in the 0, 45 and 90° directions with respect to rolling. The tensile tests were conducted at a crosshead speed of 0.7 in. per sec. throughout the investigation. The data from the tests were analyzed to provide the factors k and m in the well-known equation

$$\sigma = k\ \varepsilon^{m}$$

The anisotropy coefficient $\bar{R}$ was also obtained from thickness and width strain measurements[12].

The friction coefficient data was generated using plattens of the appropriate die material (AISI H-13) in conjunction with specimens prepared from the above heats of materials. The method utilized was that of Male and Cockroft[13], small scale specimens having similar proportions to those of the original investigation being prepared from the present sheet materials[14]. Some of the results of this aspect of our investigation are included as insets in the Figures 6-9.

It is of interest to note that since the coefficients of friction for the subject lubricants had either negative or positive temperature characteristics[14], an opportunity arose to enable us to examine any effect on frictional work which might follow or

FIGURE 5. VIEW OF TOOLING AND DYNAMOMETRY FOR WARM CUP DRAWING EXPERIMENTS

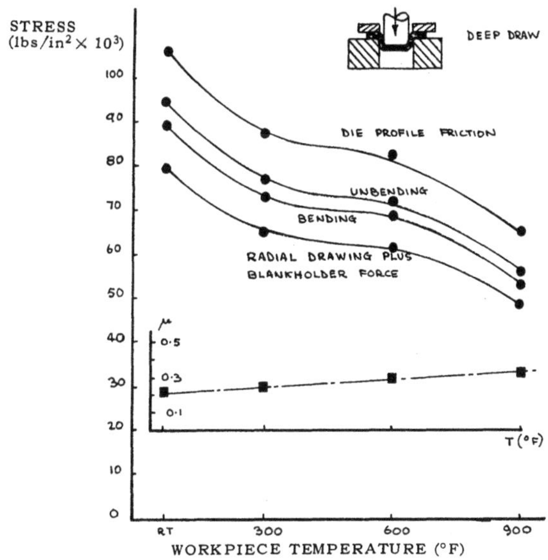

FIGURE 6. STRESS AGAINST WORKPIECE TEMP. FOR 304 STAINLESS WITH PbO AS A LUBRICANT. INSET FRICTION COEF. VS TEMP.

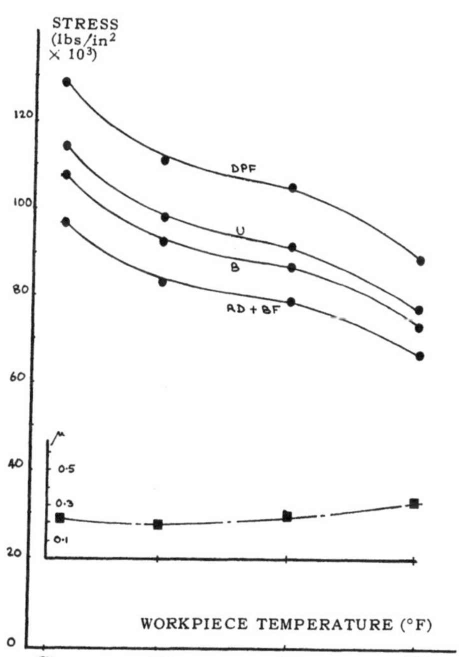

FIGURE 7. STRESS AGAINST WORKPIECE TEMP. FOR Ti6Al4V WITH GRAPHITIC LUBRICANT. INSET FRICTION COEF. VS TEMP.

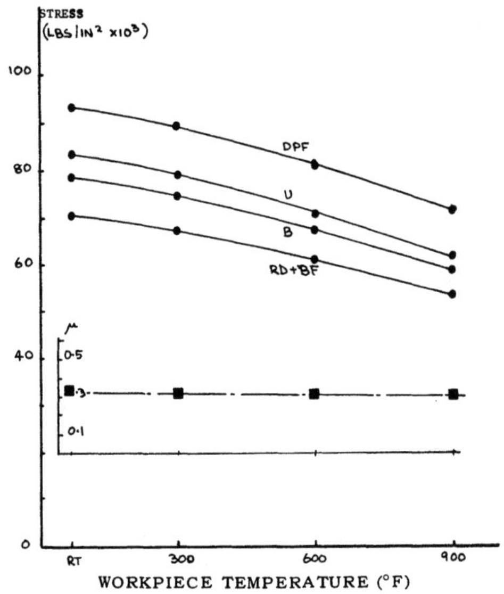

FIGURE 8. STRESS AGAINST WORKPIECE TEMP. FOR INCONEL X-750 WITH BN LUBRICANT. INSET FRICTION COEFFICIENT VS. TEMPERATURE.

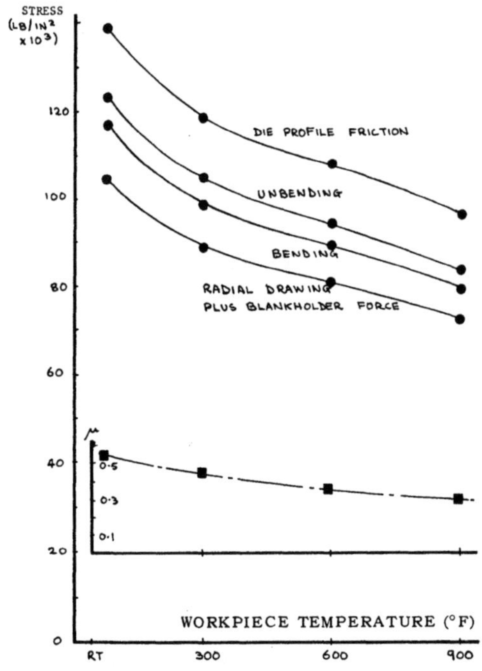

FIGURE 9. STRESS AGAINST WORKPIECE TEMP. FOR Ti6Al4V WITH MoO_2 LUBRICANT. INSET FRICTION COEF. VS. TEMP.

oppose that of plastic work with respect to temperature. The latter, of course, decreases with increasing temperature for a particular operation for the present materials in the solution treated or annealed condition. Figure 10 compares calculated and measured forming loads for a variety of conditions of workpiece temperature and lubricant.

III. METHODS OF CALCULATION

1. Warm backward extrusion

A variety of analyses of backward extrusion have appeared in the metal working literature over the last twenty years. One of the earliest of these was that of Dipper[15] which appeared in 1949 and has since been quoted extensively by Feldman[16] and others. Independently, the Russian workers Kalachev and Machanek[17] performed a similar analysis, specifically oriented towards hot forming. In this country Backofen[6] and Thomsen and Altan[18] have more recently analyzed this problem, the former using the slipline field approach.

The Dipper and the Kalachev and Machanek analyses have become fairly well established in Europe and have the attraction of being fairly simple to handle.

The extrusion punch force F_p according to the former[15] is given by:

$$F_p = a_p \left[\sigma_y \left(1 + 1/3\, \mu \frac{d_n}{h}\right) + \sigma_y' \left(1 + \frac{h}{s} \left[0.25 + \frac{\mu}{2} \right]\right) \right]$$

(The factor 0.25 arises because of an approximation for sticking friction in the author's treatment).

According to the Russian workers[17] the punch force will be given by:

$$F_p = a_p\, \sigma_{y_m} \left[\left(1 + 1/3\, \mu \frac{d_n}{h}\right) + \ln \frac{D_m}{d_n} + \frac{4\mu h' D_m}{(D_m - d_n) d_n} \right]$$

(The last term in the above may be rewritten $2\mu \left(\frac{h'}{s}\right) \frac{D_m}{d_n}$)

Apart from the two different terms to account for extrusion portion of the operation, the second equation specifies a mean flow stress value (σ_{y_m}) as opposed to an initial value (σ_y) and one which allows for the workhardening which Dipper envisions, occurs in the upsetting stage prior to the extruding action (σ_y').

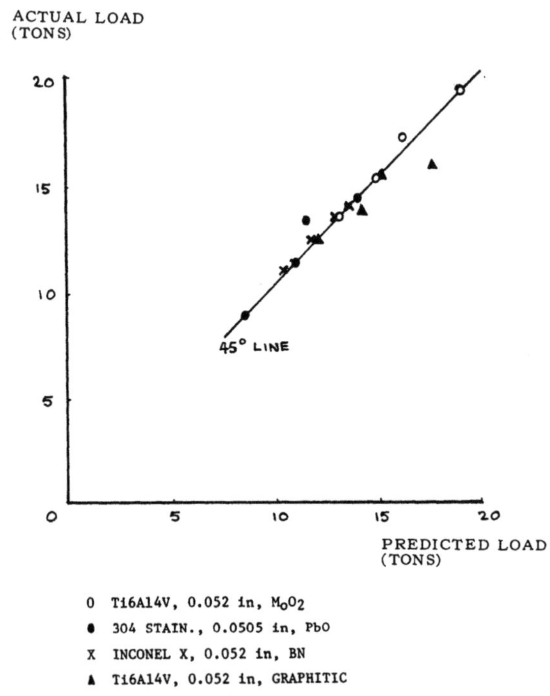

FIGURE 10. DEEP DRAWING TESTS - AGREEMENT BETWEEN ACTUAL AND PREDICTED LOADS.

The above flow stress values are selected to be appropriate to the conditions of strain rate, strain and temperature with the particular workpiece involved.

2. Warm cup drawing

There are an unusually large number of analyses of the case of sample cup drawing. These range from the solution given by Chung and Swift[19] for a non-workhardening, isotropic material through to the more complex picture of an anisotropic work-hardening blank as studied by Chang and Kobayashi[20] and by Ray and Berry[21]. Much of the work prior to 1960 has been reviewed by Alexander[22].

Recently in an investigation of punch load requirements, the present authors[12] have published details of an analysis[23] which considers not only the above material characteristics, but also accounts for redundant work and die-profile friction aspects.

The analysis, which is lengthy and which will not be restated here, provides the manufacturing engineer with a means of determining punch force F_D:

$$F_D = 2\ \pi r_c\ t_c \sin \phi_e\ (\sigma_1 + \sigma_f + \sigma_b + \sigma_u)$$

where the stress terms account respectively for the true radial drawing stress, the stress due to die profile friction and the stresses due to bending and unbending. The hold-down friction is included in the analysis of the first stress. It is interesting to note that a numerical integration is performed to obtain this same stress σ_1.

Involved in the various stress components are a variety of materials and friction related parameters [12,23].

IV. SELECTION OF APPROPRIATE MATERIALS AND FRICTION DATA

1. Warm back extrusion

In the particular investigation cited, the authors had no access to a cam-plastometer or similar device which might be utilized to obtain appropriate flow stress information. However, some limited low strain rate upsetting test results were available[24], which were pressed into use for the lower temperature operations. (It has generally been assumed in estimations of this type that strain rate effects will not be overly important at or near room temperature for the materials concerned).

At the truly 'hot' end of the temperature scale, one can call upon the excellent data of Cook[25] which embraces strain rates between 1.5 and 10^2 sec^{-1} for a variety of steels. The strain rate of 8 sec^{-1} was taken as being appropriate to the present work. Although Cook's data covers two of the steels involved, one (the Ni-Mo steel) was approximated by one of a somewhat higher alloy and carbon content:

0.3 C - 2.5 Ni - 0.6 Cr - 0.5 Mo

Use of this data would tend to provide one with a high estimate, of course.

The friction coefficient values assumed were perhaps less certain:

Room temperature	$\mu = 0.1$
Warm forming	$= 0.2$
Hot forming	$= 0.4$

The first is easily justified from the literature[16], the last is frequently assumed in practical estimates of this sort in industry[17]. The warm forming value, however, is something which might be regarded a convenient compromise (although it might be interpolated in looking through the non-metalworking literature[26]).

The estimated values of punch pressure are included on the Figures 1 and 2, along with the experimental data.

2. Warm cup drawing

As opposed to the back extrusion investigation, in the case of the cup drawing, the authors were able to document all aspects of both materials flow stress-strain data[23] and friction coefficient information[13,23]. The flow stress data, which accounted for the effects of temperature, workhardening and anisotropy as previously mentioned, was accumulated for all the heats of the subject materials in all sheet thickness. The friction coefficient data generated also covered the above range of materials associated with some ten candidate lubricants, as previously stated.

Figure 10 shows a comparison of measured and predicted punch data for four typical experimental conditions.

V. DISCUSSION

Upon analyzing the data in Figures 1 and 2, one is immediately struck by the wholesale dependance we currently have on materials and friction parameter data being readily available.

The Cook data enables us to obtain reasonable agreement between experiment and theory, in spite of our friction approximation for the high temperature back extrusion. We are much less successful at 1200 F and at room temperature. This is particularly true of the lower carbon steel, where frictionless compressive flow stress values from upsetting tests on the actual material were used for estimates of the force requirements at these temperatures.

The standard assumption of strain rate not being of profound importance at room temperature with this type of material is probably not appropriate here. (Recent work by Campbell et al[27] would certainly seem to bear this out particularly well.) Beyond this, we have the possibility that our assumed friction coefficient values contribute to the above disagreement as will be discussed later.

The agreement between theory and experiment in the cup drawing test is seen to be excellent at all the temperatures examined and clearly attests to the desirability of having, at hand, thoroughly appropriate friction and flow stress data.

The form of the cup drawing analysis enables us to look specifically at the part played by friction in each of the operations. In the extrusion investigation an estimate may be had either by allowing the friction parameter to tend to zero in the Dipper equation or by applying the well known expression,

$$F_p = a_p \sigma_y \phi_d$$

which assumes a frictionless mode of working, to each of the two stages of the operation (upsetting and annular extrusion). If the former is done, and if we assume the flow stress value reasonably accurate, friction would account for 25-30% of the force requirement at room temperature with both steels analyzed.

Using similar assumptions at 1200 F, the contribution is up to 50% in the case of the 1.0C 1.2 Cr steel (somewhat less for the low carbon steel, using the rather unreliable flow stress data). At 2200 F, however, where both friction and flow stress data are good, friction accounts for between 60-65% of the force requirement for extruding both steels. This once more underscores the importance of having at hand suitable friction coefficient data for elevated temperature operations of this type.

Looking at the cup drawing results, die profile friction accounts for somewhere between 10 and 14% of total load requirement almost irrespective of temperature and materials combination. If one assumes that blankholder friction is of the order of 10% of the pure drawing stress (it is certainly not greater than this[22]), the total contribution of friction would not seem to be ever greater than 20% of the total load or force requirement. However, since repeatable distinguishable results could be obtained with different lubricants at a particular temperature, the above assumption applied blankholder friction must be viewed only as an upper limit approximation. Furthermore, since the contribution of die profile friction changes only slightly, irrespective of whether the lubricant concerned has a negative or positive temperature characteristic, it might also be assumed that blankholder friction is sometimes very much less than the 10% maximum estimated above.

In the cup drawing, redundant work contribution amounts to about 11 to 17% of the total load according to the material and temperature combination. As would be expected, however, this contribution decreases slightly with increasing temperature (in the case of the 304 stainless it falls from 17 to 12% between RT and 900 F for the 0.0505 in. material in 3.5 in. dia. blank form).

Overall, although friction is of significant importance in all of the cup drawing cases examined, it is certainly not as important there as in warm or hot back extrusion or similar bulk forming operations, where it amounts to major proportions.

VI. CONCLUSIONS

1. The comparison between the theoretical and actual results for warm back extrusion demonstrates the need for meaningful flow stress data for the warm working range.

2. The investigation into warm cup drawing shows the importance of using friction data obtained with the correct workpiece-tool combination, at the correct temperature and under conditions of bulk plastic deformation.

3. Finally, the two investigations have shown the feasibility of utilizing the warm working range with a variety of material-lubricant combinations.

VII. ACKNOWLEDGEMENTS

The authors wish to acknowledge the help provided by many colleagues, past and present, in collecting some of the experimental data discussed in the paper. They particularly wish to thank the Connecticut Research Commission who supported the cup-forming work at the University of Bridgeport.

VIII. NOMENCLATURE

a_p = extrusion punch area
d_n = extrusion punch diameter
D_m = extrusion die diameter
F_d = deep drawing punch force
F_p = extrusion punch force
h - extrusion billet height
h' = length of contact between extrusion punch and extrusion
k = Ludwik constant
m = strain hardening exponent
p = punch pressure
r_c = punch radius
$\bar{R}$ = anisotropy coefficient
s = extrusion wall thickness
t_c = sheet thickness after unbending
ε = true strain
μ = friction coefficient
σ = true stress
σ_y = initial flow stress
σ'_y = flow stress in extrusion phase
σ_{y_m} = mean flow stress
σ_1 = radial drawing stress
σ_f = stress due to die profile friction
σ_b = stress due to bending
σ_u = stress due to unbending
ϕ_d = logarithmetic deformation in extrusion or upsetting
ϕ_e = angle of embrace

IX REFERENCES

1. E. Siebel, Stahl and Eisen (23) 1923 1295

2. G. Sachs, Spanlose Formung der Metalle, Springer (Berlin) 1934

3. P. Heuer, VDI Forschungsheft (493) 1962

4. K. Brill, Modellwerkstoffe fur die Massivumformung von Metallen, Disstn., (Hanover) 1963

5. T. Altan, H. Henning and A. Sabroff, TrASME Jrl of Eng for Ind (92) May 1970 444

6. W. Backofen, Jnl of Metals (13) March 1961 206

7. E. Thomsen, TrASME (77) 1955 515

8. C. Adams and H. Taylor, TrAFS (6) 1953 686

9. J. Berry, G. Martin and V. Kondic, TrAFS (67) 1959 449

10. R. Pehlke and R. Marrone, TrAFS (78) 1970

11. R. Pehlke and R. Marrone, TrAFS (78) 1970

12. M. Pope and J. Berry, Paper to be presented at ASME 1970 Winter Annual Meeting

13. A. Male and M. Cockroft, JIM (93) 1964 38

14. M. Pope, L. Robins and J. Berry, TrASLE (13) 1970 148

15. M. Dipper, Arch. fur das Eisenhüttenwesen, Sept/Oct 1949 275

16. H. Feldman, Cold Forging of Steel, Hutchinson (London) 1961

17. M. Kalachev and M. Machanek, Mashinostroitel' Belorussi 1959 No. 6 78

18. E. Thomsen and T. Altan, Proc. CIRP Conf. 1964 (Italy)

19. S. Chung and H. Swift, Proc. Inst. Mech. Eng. (165) 1951 199

20. D. Chiang and S. Kobayashi, TrASME, Jnl of Eng for Ind (88) 1966 443

21. R. Ray and J.T. Berry, TrASME, Jnl of Eng for Ind (92) May 1970 412

22. J. Alexander, Met. Reviews (5) 1960 19

23. M. Pope, M.S. Thesis, University of Bridgeport, Conn. 1969

24. J. T. Berry, Unpublished work

25. P. Cook, Proc. Conf. Props. of Materials at High Rates of Strain, Inst. Mech. Eng., (London) 1957, 86

26. E. Bisson and W. Anderson, Nonconventional Lubricants in Advanced Bearing Technology, NASA SP 38 (Washington, D.C.) 1964

27. A. Dowling, J. Harding and J. Campbell, JIM (98) 1970 215

THEORIES AND EXPERIMENTS ON FRICTION, DEFORMATION, AND FRACTURE IN PLASTIC DEFORMATION PROCESSES

Shiro Kobayashi

Department of Mechanical Engineering
Division of Mechanical Design
University of California at Berkeley

INTRODUCTION

Various methods of analysis, based on different assumptions and degrees of approximation, have been developed for the solution of plastic deformation problems. The choice of a specific method depends upon the information sought. In this paper, the problems of friction, deformation characteristics, and fracturing involved in plastic deformation processes are discussed, using three different methods of analysis: the upper-bound, the slip-line, and the finite element methods.

One of the decisive factors that influence the deformation pattern is friction. In deformation processes, such as ring compression and extrusion-forging, the overall geometry of the specimens is determined by the state of friction at the tool-workpiece interface. A simple approach to establishing the relationship between friction and geometrical change during these processes is the use of the upper-bound method.

For analyzing plane-strain problems, the well-developed slip-line theory is applicable. A slip-line solution to a specific problem predicts not only the stress and strain distributions in the deforming material, but also the possibly localized deformation zone and perhaps the pattern of fracturing. The slip-line solutions of indentation and compression for rigid, perfectly plastic materials are used here for analyzing the side-pressing of cylindrical rods with machined flats of various widths. Although the slip-line theory has been successfully applied to a number of problems, the effects of elastic behavior and work-hardening on solutions found for real materials are not known; also lacking is a detailed solution for the deformation that takes place during continued loading beyond the yield point.

In many forging operations, the geometrical configuration of a deforming body changes continuously during the process, and these changes in shape can be determined only by following the path of deformation. Furthermore, the occurrence of defects in plastic deformation processes can be predicted only if the local states of stress and strain in the real material have been accurately determined. Complete solutions to problems of this type can be found by using the finite element method. The finite element solution of the simple upsetting of cylinders under conditions of complete sticking is given here as an example.

AXISYMMETRIC EXTRUSION-FORGING

The process of extrusion-forging is illustrated in Fig. 1.

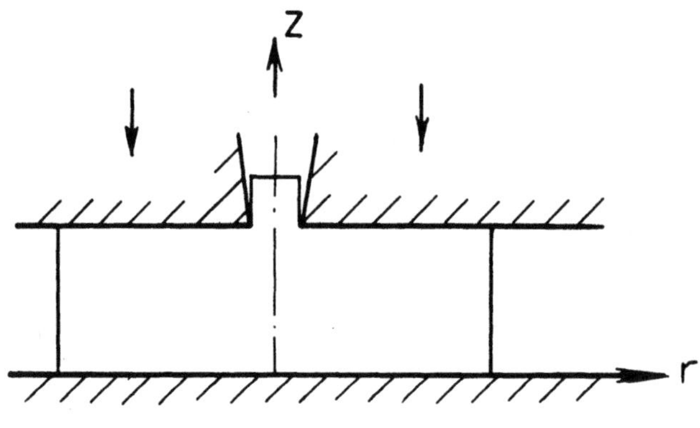

Fig. 1. Schematics of the extrusion-forging process.

A cylindrical billet is forged in its axial direction between two flat parallel dies, one of which contains a circular hole, thus permitting a simultaneous vertical and lateral flow of the material. When the frictional constraint at the tool-workpiece interface is small, extrusion through the central hole in the die does not take place, and the height of the central projection decreases with a reduction in the height of the main forging. With increasing frictional constraint, extrusion of the material through the hole begins. Thus the height of the central projection depends on the die friction. The metal flow in extrusion forging is influenced also by the dimensions of the specimen. The upper-bound approach is applied to the problem of establishing the relationship between the geometrical change and the friction at the interface.

Two modes of deformation are possible in the extrusion-forging process, as shown in Fig. 2. A theoretical analysis, using an assumed velocity field, was developed for extrusion-forging problems. It is assumed that the proposed velocity field approximates

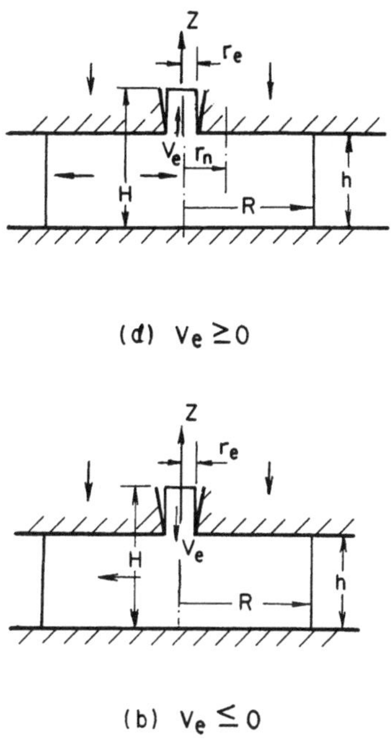

Fig. 2. Two modes of deformation in extrusion-forging.

the actual flow behavior of the material, although the field contains velocity discontinuities. Two types of velocity fields, shown in Fig. 3, are considered. In both types the radial and axial velocity components in zone ① are defined by

$$u_I = \frac{1}{2h}\left[r - (1 + V_e)\frac{r_e^2}{r}\right] = u_{II} ,$$
$$v_I = -\frac{1}{h} z = v_{II} , \tag{1}$$

for a unit downward velocity of the upper die.

In zone ② the velocity components for type I are

$$u_I = -\frac{r_e}{2h} V_e , \qquad v_I = \frac{r_e}{2h} V_e\left(\frac{z}{r}\right) , \tag{2}$$

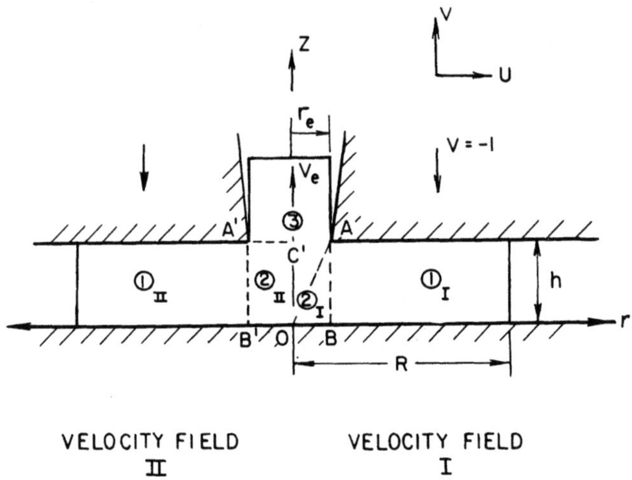

Fig. 3. Two types of velocity fields assumed in extrusion-forging.

while for the type II velocity field, they are

$$u_{II} = -\frac{V_e}{2h} r , \qquad v_{II} = \frac{V_e}{h} z . \tag{3}$$

Zone ③ moves as a rigid body with the axial component V_e. The boundaries between the zones, namely, AB and AO for the type I velocity field and $A'B'$ and $A'C'$ for the type II field, shown in Fig. 3, are the surfaces across which the tangential velocity components are discontinuous. In determining the velocity field, the incompressibility condition and the continuity of the material across the velocity discontinuity surfaces must be satisfied. For the deformation mode given in Fig. 2(a) for $V_e \geq 0$, the neutral surface, where the radial velocity component u is zero, appears in zone ①. The neutral radius r_n, in this case, is shown to be

$$r_n = \sqrt{1 + V_e} \cdot r_e . \tag{4}$$

The total energy dissipation rate can be obtained by adding the deformation energy rate in zones ① and ②, the frictional energy dissipation rate along the tool-workpiece interface, and the energy rate due to velocity discontinuities. The unknown parameter V_e can be determined so that the total energy dissipation rate becomes minimum. Once this parameter is known, the velocity field can be defined completely at any configuration during the operation. The overall geometrical change of the specimen is then calculated by integrating incremental displacements. In order to select the proper type of velocity field at various stages during the process, a choice must be made at each stage for the velocity field which yields a smaller total energy dissipation rate under identical conditions of geometry and friction. The theory is explained in greater detail elsewhere [1, 2].

Experimental investigations of extrusion-forging have been reported in the work of Jain and Bramley [3, 4]. They studied the metal flow and the energy and load requirements for a wide range of process variables, using billets cut from cold drawn En8 bars of appropriate diameter and preheated to 1120°C in an electric resistance furnace. Results obtained from these experiments are compared with theoretical calculations in Figs. 4, 5 and 6; special reference is made to the variation of extrusion height as functions of friction and specimen dimensions. Figure 4 shows the effect of friction at the interface on the total height variation. In the theoretical curves of Fig. 4(a), m is the ratio of the frictional stress to the shear strength of the deforming material, and it is assumed to be the same at two interfaces and constant across the interface surface. The datum conditions for the experiments shown in Fig. 4(b) are as follows: $H_0 = 0.75$ in.; $D_0 = 2R_0 = 1.0$ in.;

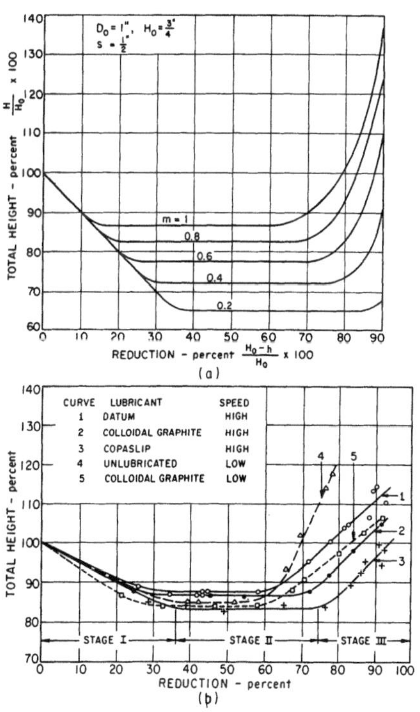

Fig. 4. Effect of friction on total height of extrusion.

$d = 2r_e = 0.50$ in.; high speed; unlubricated. Variations in forging speed were obtained on a hydraulic press with a constant closing speed of 0.07 ft/sec. An MK II Petro-Forge high-energy-rate forging machine [5] was used to obtain high forging speeds in the range of 15 to 50 ft/sec. As seen in Fig. 4(a), the theory indicates the presence of three stages of deformation. The theory shows also that lubrication reduces the height of extrusion and delays the transition between the three stages of flow. This agrees with the observed effect of lubrication at both high and low speeds, as shown in Fig. 4(b). The effect of billet size on the metal flow is given in Fig. 5. It is seen that the theory is again in good agreement with the experiments.

Theoretical and experimental results on the effect of hole size are given in Figs. 6(a) and 6(b), respectively. Figure 6(a) shows that the range for stage I becomes narrower with increasing hole size, and at a 3/4-in. hole diameter stage I vanishes. The transition from stage II to stage III is delayed with increasing hole size, thus giving a wider range of stage II for the larger hole diameter. Also, the curves for stage III cross each other. These characteristics, deduced from the theory, are observed in the experimental results given in Fig. 6(b).

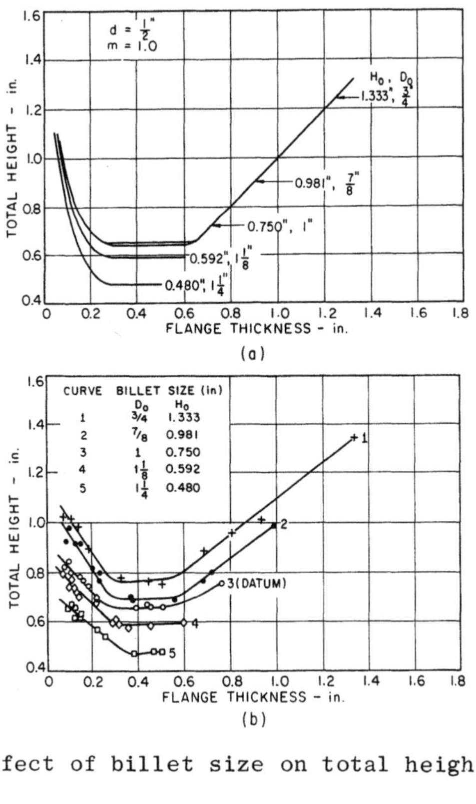

Fig. 5. Effect of billet size on total height of extrusion.

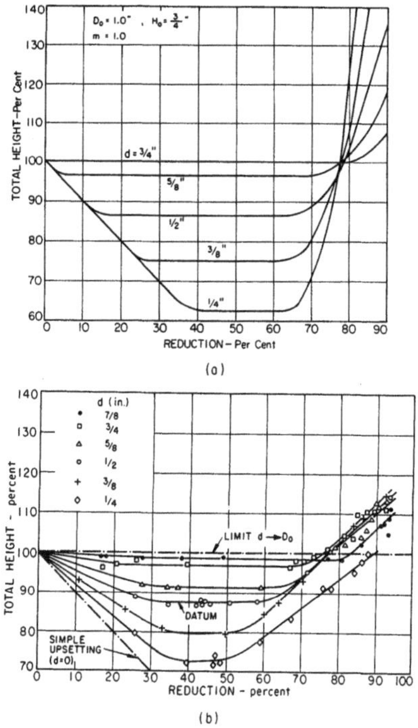

Fig. 6. Effect of hole size on total height of extrusion.

The theory assumes velocity fields that contain velocity discontinuities. The theory also assumes a constant flow stress and neglects the effects of temperature and strain rate; the experiments, however, were conducted with heated billets and at both high and low speeds. Furthermore, the friction conditions at the upper and lower interface were considered identical and were represented by a constant frictional stress. In spite of these assumptions, it was seen that the dimensional changes predicted theoretically were very close to those observed in the experiments. It must be noted, however, that the degree to which the assumptions are approximated should be carefully examined if the quantitative accuracy of the predictions is of primary interest [1].

PLANE-STRAIN-SIDE-PRESSING

The well-developed slip-line theory is a useful tool for the analysis of plane-strain problems. Slip-line solutions for the indentation and compression of rigid, perfectly plastic materials [6, 7, 8] are used in this analysis of the side-pressing of cylindrical rods with machined flats of various widths prepared from aluminum alloy 7075-T6 [9]. In the problem of side-pressing, three types of deformation must be distinguished. When the height-width ratio of the specimen is sufficiently large, deformation occurs only in the material adjacent to the dies (type I deformation). When the height-width ratio is reduced beyond a critical value, a plastic zone develops throughout the material between the dies, and the two ends of the specimen move apart as rigid masses (type II deformation). For a range of height-width ratio less than unity, the deformation extends to the sides of the specimen (type II deformation). In the side-pressing of cylindrical rods with machined flats, type II and type III deformation modes are significant. The slip-line field for the type II deformation is shown in Fig. 7. This field is valid for $H/W \geq 1$. With the notation given in Fig. 7, the mean die pressure is given by

$$\frac{\bar{p}}{2k} = \frac{1}{2}\left(1 + \frac{W}{H}\right) + 2\psi_0 - \frac{\pi}{4} - \frac{2}{H}\int \psi \sin\psi \, ds , \qquad (5)$$

where the integral is to be performed along the slip-line OBA. In this field the velocity discontinuities are present along the slip-lines shown in Fig. 7 by the heavy lines. It must be noted that the slip-line fields--therefore the mean die pressures--are independent of the friction conditions at the die-workpiece interface for a range of height-width ratio equal to or larger than unity.

In order to apply the slip-line solutions not only for estimating the yield point load, but also for obtaining instantaneous

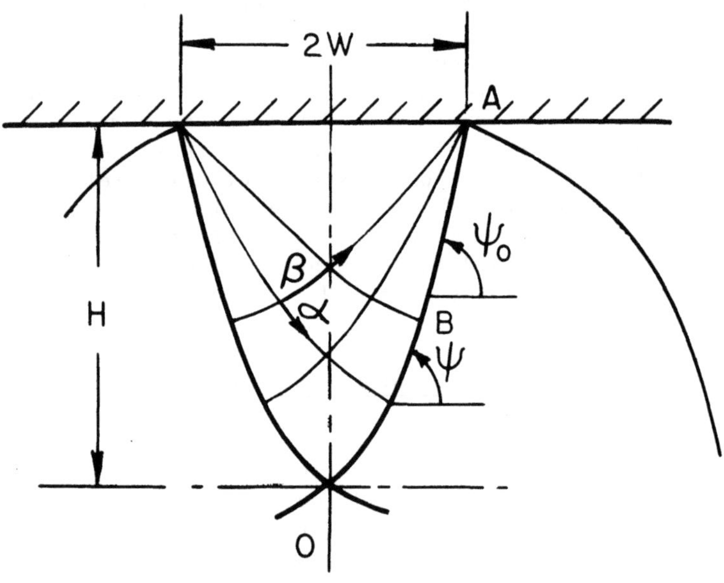

Fig. 7. Slip-line field for type II deformation.

configurations during the continued deformation, the variation of the height-width ratio in the course of side-pressing must be known. For predicting this variation as a function of reduction in height, the simple first approximation is obviously to assume that the radius of curvature of the free surface remains unchanged, because both sides of the specimen move as rigid masses in the type II and type III deformation modes. Than an instantaneous width of contact can be calculated, according to

$$W = \frac{1}{4H}\left\{A_0 - R_0^2\,(\theta - \sin\theta)\right\}, \qquad \theta = 2\sin^{-1}(H/R_0), \tag{6}$$

where A_0 is the initial cross-sectional area and R_0 is the initial radius of curvature of the free surface of the workpiece.

When the variation of the height-width ratio during side-pressing is known, the load displacement curves can be constructed theoretically by applying slip-line solutions to the instantaneous configurations. Figure 8 shows experimental and calculated load-

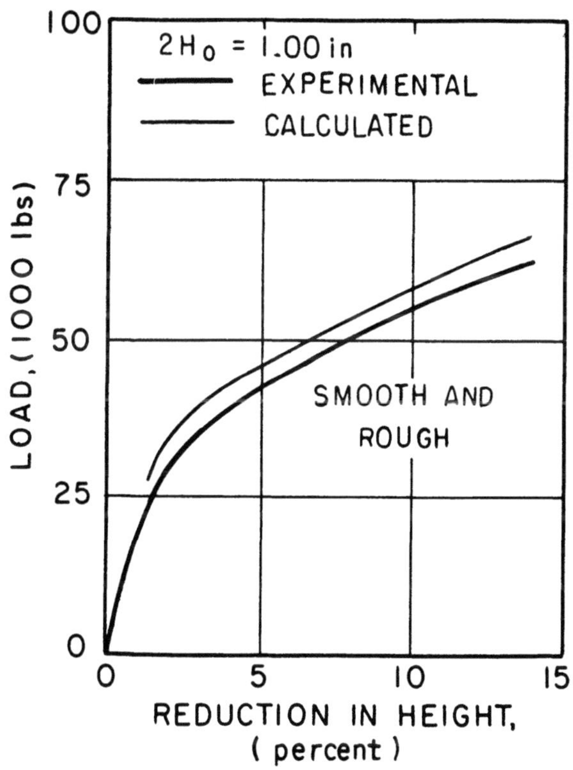

Fig. 8. Comparison of experimental and calculated load-displacement curves for specimens of $2H_0 \mp 1.00$ in.

displacement curves for a circular cylindrical specimen. As the slip-line solutions predicted, the experimental load-displacement curves were identical for both smooth and rough dies.

Specimens of various geometries fractured in the range of type II or type III deformation. Figure 9 shows the fracture for a type II deformation. This type of fracturing was observed for all the specimens when the height-width ratios at fracture were equal to or larger than unity. The localized deformation zone revealed by etching suggests that fracturing occurred along the velocity discontinuity line characterized by the angle ψ_0 in the slip-line solution given in Fig. 7. This was further confirmed in Fig. 10, where the angles between the contact and fracture surfaces were measured and compared with the angle ψ_0, calculated from the slip-line solution.

Slip-line solutions for the type III deformation become dependent on the friction conditions at the tool-workpiece interface. Load-displacement relationships for specimens with machined flats were determined by the same calculation as that in Fig. 8, and the results are given in Fig. 11. Experiments show that the load-displacement curves were identical for smooth and rough dies for a specimen of $2H_0 = 0.85$ in., but the effect of interface friction appeared for a specimen of $2H_0 = 0.70$ in. These observations again confirmed the predictions of the slip-line theory. For smooth dies the mean die pressure oscillates, with increasing W/H being equal to 2k for integer values of W/H. Slip-lines are straight lines for integer values of W/H, as shown in Fig. 12(a), but they should be modified according to Green [7] for other values of W/H in such a way that the velocity discontinuities terminate at the die corner (Fig. 13(a)). For rough dies the slip-line fields given in Fig. 14(a) and (b) are valid, and the mean die

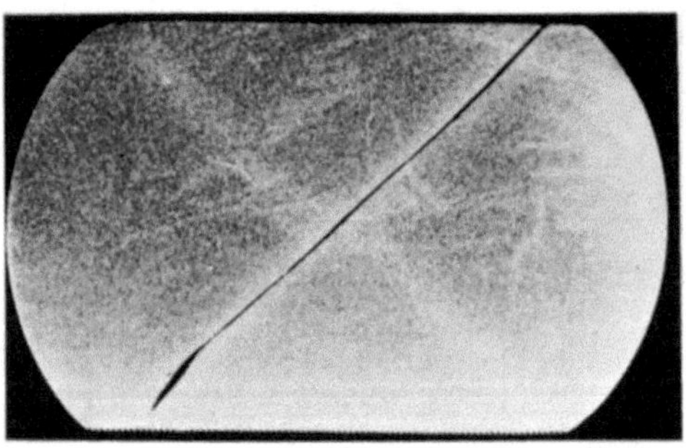

$2H_0 = 0.80$ in: $(W/H)_f = 1.05$
ROUGH DIES

Fig. 9. Fracture and localized deformation zone for type II deformation.

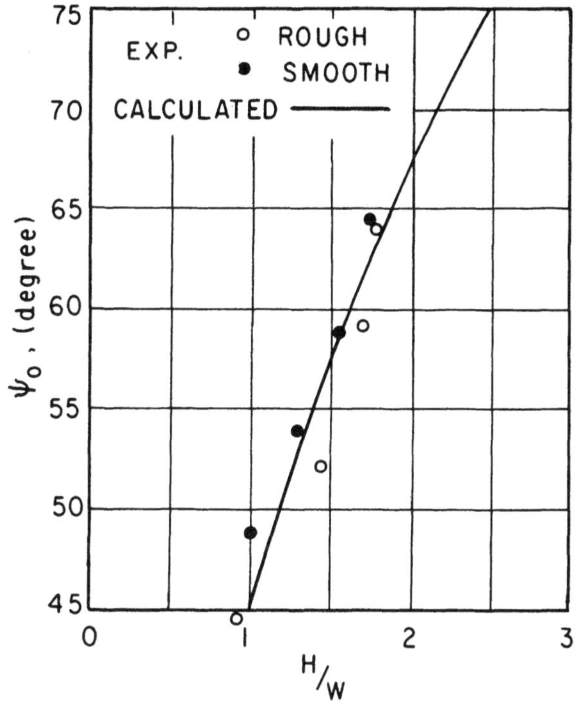

Fig. 10. Angles characterizing the centered fan slip-line fields as functions of the height-width ratio.

pressure corresponding to these fields is very nearly approximated by the equation

$$\frac{\bar{p}}{2k} \cong \frac{3}{4} + \frac{W}{4H} . \qquad (7)$$

The slip-lines along which the velocity discontinuities exist are again shown by heavy lines. When the height-width ratio at fracture is less than unity, the fracture patterns must be compared with these fields. As the value of W/H approaches integer, the slip-lines become straight with smooth dies, as shown in Fig. 12(a). This was the case for Fig. 12(b), and the slip-line that makes an angle of 45° with the contact surface at the corner of the specimen along which fracturing occurred is a velocity discontinuity line. Figure 13(b) corresponds to the case given in Fig. 13(a). In this case the fracturing occurred not along the velocity discontinuity line but along the boundary slip-line. According to Green [7], this boundary slip-line makes an angle of 55° with the contact surface at the corner of the specimen for $W/H \approx 1.30$. The measured angle was 51° in Fig. 13(b). Figure 14(c) shows a case

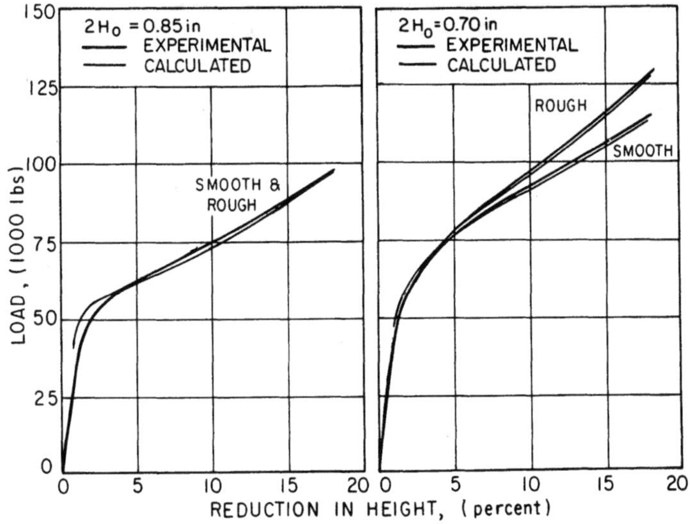

Fig. 11. Comparison of experimental and calculated load-displacement curves for specimens of $2H_0$ = 0.85 in. and 0.70 in.

where the slip-line solution given in Fig. 14(a) applies. Once again, agreement between the velocity discontinuity line in the slip-line field and the localized deformation zone shown by etching is striking. In this case, too, the fracturing took place along a boundary slip-line making a 45° angle with the contact surface, where no velocity discontinuity exists in the solution. It should be emphasized that the localized deformation zones were revealed in the configurations at the time of fracture and that fracturing took place along the characteristics. These observations confirmed the findings by Spretnak and his coworkers [10, 11].

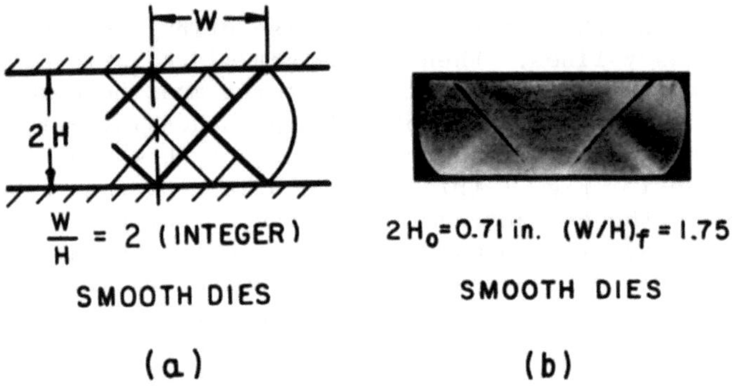

Fig. 12. Slip-line field and fracture for smooth dies with W/H ≅ 2.

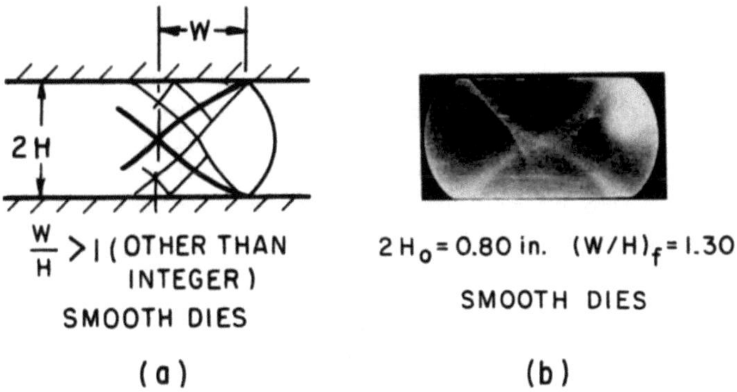

Fig. 13. Slip-line field and fracture for smooth dies with W/H > 1 (other than integer).

Despite the fact that the slip-line theory used here is based on the assumption of rigid, perfectly plastic materials, the theory agrees very well quantitatively with the observations of the mean die pressure at the yield point, the load-displacement relationship, the effect of friction at the interface, and the effect of specimen geometry on the deformation characteristics. However, for the determination of critical conditions of fracture initiation and propagation, accurate information on the local stress and strain distributions must be available, taking into account the work-hardening property of the material.

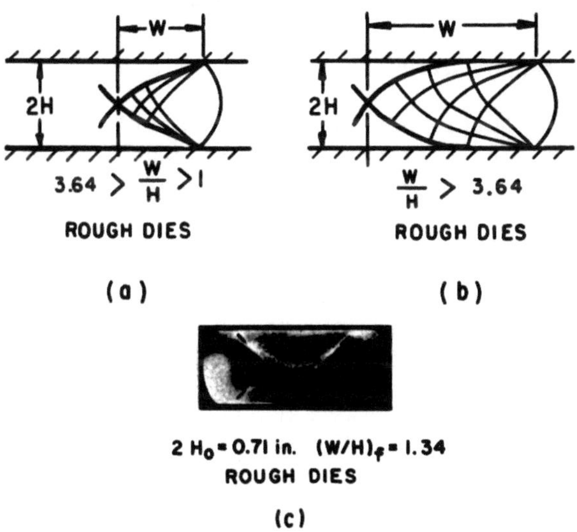

Fig. 14. Slip-line field and fracture for rough dies with W/H > 1.

AXISYMMETRIC UPSETTING OF SOLID CYLINDERS

In many forging operations, the geometrical configuration of a deforming body continuously changes during the process. The body may contain an elastic region as well as a plastic region, and one region may experience unloading while another undergoes continued loading. One of the problems in this category is the axisymmetric upsetting of solid cylinders. An attempt was made to obtain by the finite element method [12], a complete theoretical solution for the axisymmetric upsetting of solid cylinders, under conditions of sticking at the tool-workpiece interface.

The finite element method is based on variational principles. A variational formulation of elastic-plastic behavior of materials in the absence of body forces assumes a stationary value of the functional

$$\Phi = \frac{1}{2}\int_V (\dot{\underset{\sim}{\sigma}}^T \dot{\underset{\sim}{\varepsilon}})dV - \int_S (\dot{\underset{\sim}{U}}^T \dot{\underset{\sim}{F}})dS , \qquad (8)$$

where $\dot{\underset{\sim}{\sigma}}$ is the stress-rate vector; $\dot{\underset{\sim}{\varepsilon}}$, the strain-rate vector; $\dot{\underset{\sim}{U}}$, the velocity vector; and $\dot{\underset{\sim}{F}}$, the distribution of the given external stress-rate vector. The superscript T denotes the transpose. The distribution of the stress-rate vector, $\dot{\underset{\sim}{\sigma}}$, is associated with the strain-rate vector, $\dot{\underset{\sim}{\varepsilon}}$, which is derivable from a continuous velocity vector, $\dot{\underset{\sim}{U}}$, differing infinitesimally from the actual distribution, and satisfying the boundary conditions. The surface integral of the second term in Eq. (8) exists only over the part of the surface which is subjected to the stress-rate, $\dot{\underset{\sim}{F}}$, while the first integral is taken through the volume of mass.

A continuum of elastic-plastic body is divided into elements interconnected at a finite number of nodal points. The functional is then approximated by a function of nodal point values. In the displacement method, these nodal point values are the displacement (or velocity) components.

The discretization of the variational problem is performed on the elemental level by approximating the functional with respect to the m-th element by a function of the m-th element nodal point values. This approximation is accomplished by replacing the actual distribution with an approximating velocity distribution in each element. When an appropriate operator is applied to the approximated velocities, the strain-rate components in the element are derivable and the stress-rate components can be expressed in terms of strain-rate components. For isotropic materials the stress-rate and strain-rate components are associated by Hooke's law in the elastic region and by the Prandtl-Reuss equations during continued loading in the plastic region. The functional (8), summing the functionals of subregions, is then approximated by the function of nodal point variables. Applying the variational principle to this approximating function results in simultaneous equations for the unknown velocity components at the nodes of the region, in the form

$$\dot{\underset{\sim}{R}} \approx \underline{K}\,\dot{\underset{\sim}{u}} \tag{9}$$

where $\dot{\underset{\sim}{u}}$ is the nodal point velocity vector, $\underline{K}$ is the stiffness matrix, and $\dot{\underset{\sim}{R}}$ is the equivalent nodal point force-rate vector. The solution of the simultaneous equations for the nodal point velocities and the elemental distributions then provide the approximate solution to the actual velocity distribution.

When we consider mixed boundary conditions where the velocity vectors over a part of the surface are described, the matrix equation for the unknown nodal point vectors is obtained by using the submatrices of the global stiffness matrix, $\underline{K}$. The detailed formulation of the theory and the sequence of the computer program are given elsewhere [12, 13].

A cylindrical specimen with a height-diameter ratio of 0.8 was compressed in its axial direction between two flat parallel dies. Since the dies were assumed to be rigid and rough, no relative displacement was permitted at the tool-workpiece interface. The material properties used for the analysis were E (Young's modulus) = 10×10^6 psi, ν (Poisson's ratio) = 0.33, H' (the slope of the effective stress and plastic strain curve) = 20,000 psi, and Y_0 (initial yield stress) = 13,000 psi. The finite element analysis was carried out until a 33 percent reduction in height was achieved.

Although our primary concern is the solution in the range of plastic compression, it may be of interest to find how the plastic zone develops from the initial stress-free state. The shapes of plastic zones at some selected reductions are given in Fig. 15. As shown in Fig. 15(a), two plastic zones were nucleated, first along the edge of the tool-workpiece interface and then at the center of the specimen. The average die pressure increased linearly with a steep slope. When the plastic zone spread and two zones were connected, as shown in Fig. 15(b), the pressure-displacement curve started to bend. The slope of the pressure-displacement curve kept decreasing during the formation of the elastic region of a truncated cone shape beneath the die (Fig. 15(c)). Then, the sharp bend of the pressure-displacement curve was completed (yield point) and the slope of the curve became approximately constant at the stage shown in Fig. 15(d).

When comparing actual values with some of the computed results, it must be noted that the material properties used in each case should have similar patterns so that a quantitative comparison

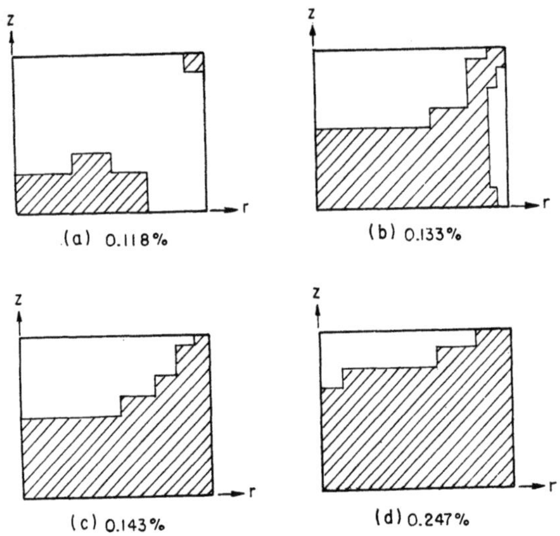

Fig. 15. The plastic zones in axisymmetric upsetting at various reductions in height.

can be made. In order to produce similar flows for two strain-hardening metals,

$$(E/Y)_1 = (E/Y)_2 \ , \tag{10}$$

and $\bar{\sigma}/Y$ must be the same function of $\bar{\epsilon}^p$ where E is the modulus of elasticity; Y, the initial yield stress; $\bar{\sigma}$, the flow stress; and $\bar{\epsilon}^p$, the plastic effective strain [14]. When a large plastic deformation is involved, the effect of the elastic property can be negligible. The solutions computed with a specific material property are then applicable to a group of materials with similar plastic properties. Because of its plastic behavior, aluminum 1100-F was found to be one of the materials to which the computed solution is quantitatively applicable.

The theoretical and experimental load-displacemtn curves are compared in Fig. 16. Agreement is excellent, with the small deviation attributed to the fact that some differences exist between the theory and experiment with regard to the materials' behavior and the constraint conditions at the tool-workpiece interface.

Figure 17 shows the bulge profiles at various reductions in height. For small reductions the theory well predicts the geometrical change of the free surface. The predictions, however, deviate more from the observations as the reduction in height

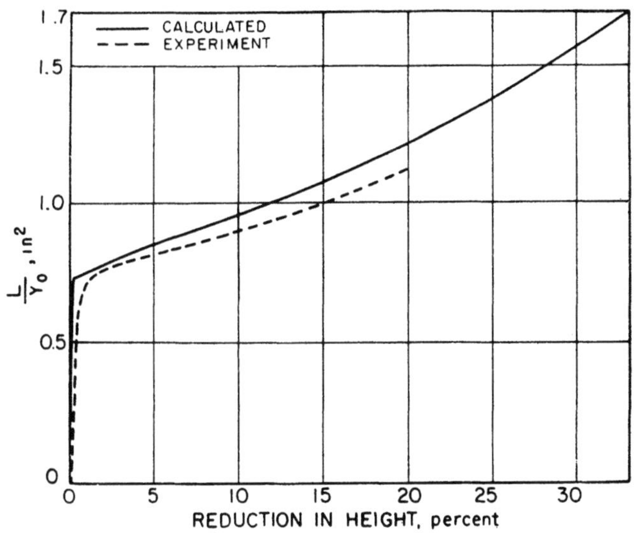

Fig. 16. Computed and experimental load-displacement curves.

increases. It is a well-known fact, as measurements in Fig. 17 indicate, that originally free surface comes into contact with the die at some stage of compression. This phenomenon is more pronounced with increasing frictional constraint at the interface. The results of the finite element analysis did not produce this fact even at a reduction of 33 percent. When originally free surface comes into contact with the die, severe distortion occurs locally, and it appears that much finer elements are required to produce this severe distortion by the finite element method.

The effective strain is an indication of the degree of deformation, and can be calculated by following the deformation at any point incrementally. The computed distribution of the effective strain at a 20 percent reduction in height is shown in Fig. 18. Experimental hardness distribution across the contact surface at the same reduction is also shown for comparison.

In simple upsetting, ductile fracture occurs at the barreled surface and the condition of fracture depends on the complete history of the plastic flow. The strain path of the small element on the equatorial free surface was plotted in Fig. 19. Considerable deviation of the path from that for homogeneous deformation was apparent since the radial strain increases first and then begins to decrease as the reduction in height increases. These results are in good agreement with the observation made by Kobayashi [15].

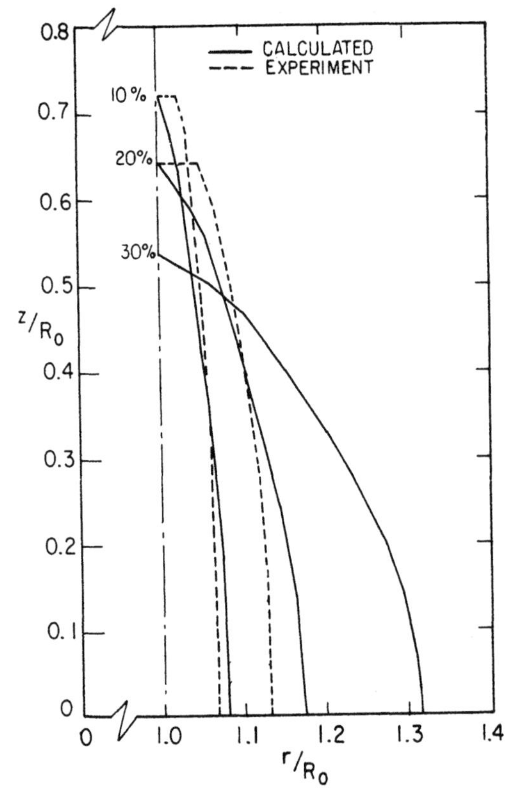

Fig. 17. Theoretical and experimental bulge profiles at various reductions in height.

Figure 20 gives the distributions of normal and shear stresses along the elements in contact with the die at various reductions in height. The contact pressure distribution is more or less uniform over the central area, but increases near the edge of the contact surface. A trend showing a slight increase in pressure toward the center appears at a 33 percent reduction. This result indeed coincides with Takahashi's [16, 17] findings for the compression of copper specimens. In his experiments, Takahashi measured the contact pressure distribution for specimens of various height-diameter ratios by using the pressure-senstive pin method. He found that in specimens with height-diameter ratios larger than unity, the distribution of contact pressure was highest at the edge of the specimen and lowest at the center. He also found that the pressure distribution for specimens with height-diameter ratios less than 0.5 showed a friction hill.

The shear stresses are almost zero at the center, but they increase first rather sharply and then gradually toward the edge

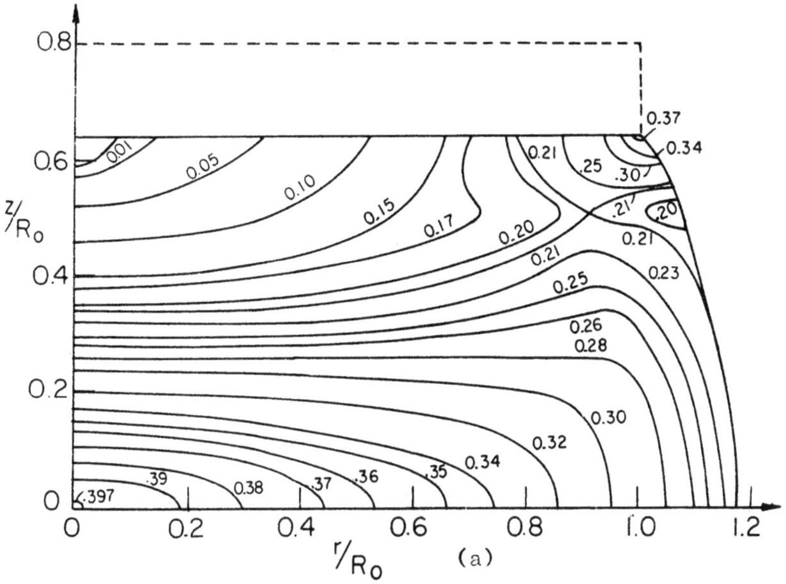

Fig. 18(a) Computed effective strain distribution at 20 percent reduction in height.

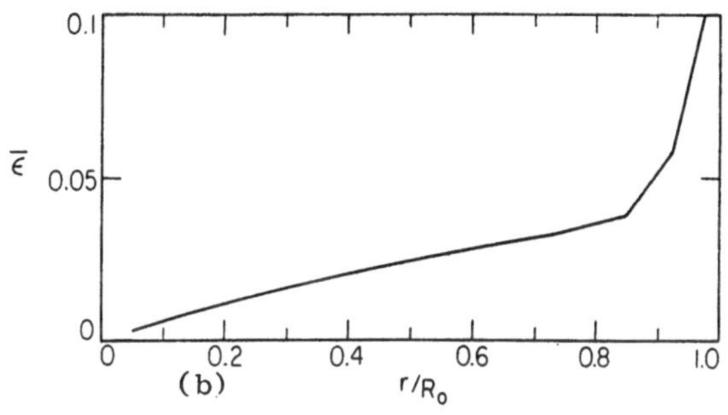

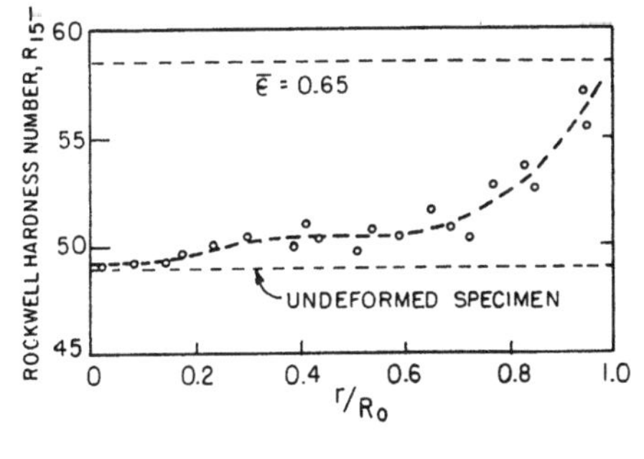

(c)

Fig. 18(b) Computed strain distribution across the contact surface; Fig. 18(c) Hardness distribution across the contact surface. Hardness number, R_{15T} (1/16" ball indenter with a 3 Kg minor load and a 15 Kg major load).

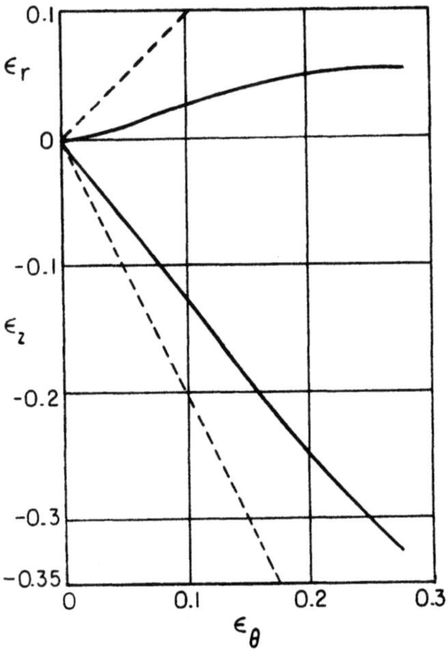

Fig. 19. Strain variations in the element at the equatorial free surface.

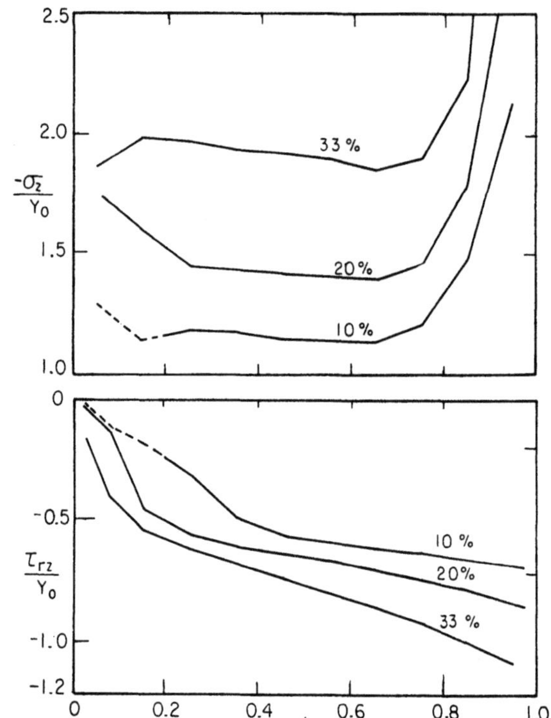

Fig. 20. Distributions of contact pressure and frictional stress at various reductions in height.

of the specimen. Although the condition of complete sticking used in the computation is an extreme boundary condition, and may never be achieved in reality, the measurements of the interface shear stresses in plastic compression of aluminum disks by Backofen and coworkers [18, 19] confirmed the trend of the computed distribution.

SUMMARY

The use of a specific method of analysis depends on the information sought. In a process like extrusion-forging, simple calculations based on an assumed velocity field can be used to determine the overall geometrical change of the workpiece caused by various friction conditions and specimen geometry. In the plane-strain side-pressing process, slip-line solutions for rigid, perfectly plastic materials well predict the load-displacement relationships and the effect of die friction on the deformation characteristic; the solutions also give good correlation with localized deformation zones and the pattern of fracturing.

The usefulness of the finite element method for detailed studies of deformation characteristics in plastic deformation problems was demonstrated. The plastic zone development, load-displacement curves, geometrical changes of the free surface, and stress and strain distributions were computed and shown to predict well the actual behavior of the material. Evidently, the finite element method is a powerful tool for the analysis. Questions remain, however, with regard to the accuracy of the solutions and the efficiency of the computation.

ACKNOWLEDGMENTS

The author wishes to thank the National Science Foundation for its grant GK-14946 and the Air Force Materials Laboratory, Wright-Patterson Air Force Base, for its contract AF33615-68-C-1314 under which this paper was prepared.

REFERENCES

1. Y. Saida, C. H. Lee, and Shiro Kobayashi, "Some Aspects of Friction in Forging Problems," Proceedings of the II Inter-American Conference on Materials Technology, Mexico City, Aug. 24-27, 1970.

2. S. C. Jain, A. N. Bramley, C. H. Lee, and Shiro Kobayashi, "Theory and Experiment in Extrusion Forging," presented at the 11th International Machine Tool Design and Research Conference, Birmingham, England, Sept. 14-18, 1970.

3. S. C. Jain and A. N. Bramley, Proc. I. Mech. E. 182, 783 (1967-68).

4. S. C. Jain and A. N. Bramley, Proc. 9th Mach. Tool Des. Res. Conf., Pergamon Press, 1968.

5. F. Bakhtar, L. T. Chan, R. Davies, A. J. Organ, B. W. Rooks, and S. A. Tobias, "Recent Developments in High Energy Rate Forming With Petro-Forge," Proceedings of the 2nd Inter. Conf. of The Centre For High Energy Forming, Estes Park, Colorado, June 23-27, 1969.

6. R. Hill, The Mathematical Theory of Plasticity, Oxford at the Clarendon Press, 1950.

7. A. P. Green, "A Theoretical Investigation of the Compression of a Ductile Material Between Smooth Flat Dies," Philosophical Magazine, vol. 42, 1951, p. 900.

8. J. F. W. Biship, "On the Effect of Friction on Compression and Indentation Between Flat Dies," Journal of the Mechanics and Physics of Solids, vol. 6, 1958, p. 132.

9. S. C. Jain and Shiro Kobayashi, "Deformation and Fracture of an Aluminum Alloy in Plane-Strain Side-Pressing," presented at the 11th International Machine Tool Design and Research Conference, Birmingham, England, Sept. 14-18, 1970.

10. R. H. Ernst and J. W. Spretnak, "Effect of Shear Strain Rate on the Susceptibility of Selected Engineering Alloys to Plastic Instability in Torsion," Trans. Iron and Steel Institute of Japan, vol. 9, 1969, p. 361.

11. C. A. Griffis and J. W. Spretnak, "An Analysis of the Energy Absorbed and the Fracture Process in Conventional Notched Bars of AISI 4340 Steel as a Function of Tensile Strength," Trans. Iron and Steel Institute of Japan, vol. 9, 1969, p. 372.

12. C. H. Lee and Shiro Kobayashi, "Analysis of Axisymmetric Upsetting and Plane-Strain Side-Pressing of Solid Cylinders by the Finite Element Method," to be presented at the ASME Winter Annual Meeting, New York, Nov. 1970.

13. C. H. Lee and Shiro Kobayashi, "Elastoplastic Analysis of Plane-Strain and Axisymmetric Flat Punch Indentation by the Finite Element Method," to be published in the International Journal of Mechanical Sciences, 1970.

14. A. H. Shabaik and Shiro Kobayashi, "Investigation of the Application of Visioplasticity Methods of Analysis to Metal Deformation Processing," Final Report, prepared under Navy, Bureau of Naval Weapons, contract NOw 65-0374-d, February 1966.

15. Shiro Kobayashi, "Deformation Characteristics and Ductile Fracture of Steel SAE 1040 in Simple Upsetting of Solid Cylinders and Rings," presented at the ASME Winter Annual Meeting, Los Angeles, Nov. 1969.

16. Soji Takahashi, "The Distribution of Contact Pressures in Compressing Cylindrical Specimens," Journal of Japan Society for Technology of Plasticity, vol. 6, No. 52, 1965, p. 271.

17. Soji Takahashi, "Some Experiments on the Contact Pressure Distribution in the Compression of Copper Cylindrical Specimens," Journal of Japan Society for Technology of Plasticity, vol. 9, No. 84, 1968, p. 30.

18. G. T. Van Rooyen and W. A. Backofen, "A Study of Interface Friction in Plastic Compression," International Journal of Mechanical Sciences, vol. 1, 1960, p. 1.

19. G. W. Pearsall and W. A. Backofen, "Frictional Boundary Conditions in Plastic Compression," Trans. ASME, Journal of Engineering for Industry, vol. 85, 1963, p. 68.

THE USE OF WORKABILITY TEST RESULTS TO

PREDICT PROCESSING LIMITS

A. L. Hoffmanner

TRW Inc.

Cleveland, Ohio

ABSTRACT

The concept of workability testing was developed from accurate determinations of the states of stress under which a material is deformed during testing. These results provided a general relationship between stress and fracture strain from a variety of tests and materials over a wide range of stress, strain rate and temperature. It was found that the major stress parameter determining the fracture strain $\bar{\varepsilon}_f$ was the average value of the normal stress σ_T perpendicular to the direction of alignment of grain boundaries and second phase particles. This definition of stress provided the ductile fracture criterion $\ln\bar{\varepsilon}_f = A + B\ \sigma_T/\bar{\sigma}$ in which $\bar{\sigma}$ is the flow stress and A and B are material parameters related to the mechanical behavior of the solid solution and mechanical texturing respectively. The workability concept subsequently was amplified by applying the test results to make accurate predictions of working limits during forging, extrusion and rolling. Good agreement was found between predictions and observations when the mechanics of the processes were defined by analytical and experimental-analytical procedures. These results will be extended to support qualitative procedures for process improvements by isolating the factors which are most prominent in determining the limiting reductions during deformation processing.

INTRODUCTION

The value of workability tests for process development has been well appreciated. However, quantitative uses of these test results have been attempted only in the last few years. This recent impetus has grown from the need for improved processing procedures which, in turn, led to applications of plasticity theory. The major advances in workability have resulted through the development and application of plasticity theory to better define the ductile fracture phenomenon and to determine the conditions imposed by the process on the work material. However, considerable empiricism still exists due to the lack of a mathematical theory for ductile fracture; the complexity of metalworking processes which does not easily permit exact or complete analytical solutions; and a great lack of understanding of the effects of test specimen design on stress and/or strain, hence, on test results. An attempt will be made to clarify these points with specific examples at room and elevated temperatures.

Implicit in the use of plasticity theory is that the deforming material can be treated as a continuum. Although metals and alloys cannot be strictly considered as continua, they do exhibit this macrosopic behavior when their basic structural features are much less in extent than the dimensions of the deforming specimens. In general, the behavior of the material must be properly considered in constructing the mechanics of the deformation process. Description of this behavior requires accurate tests to insure that size effects are averted and that flow stress anisotropy is properly described[1]. Furthermore, test specimens must be carefully scrutinized to insure that the stress and strain rate distributions correlate with the appropriate plastic stress-strain rate relations and that the specimen design does not lead to an unknown state of stress or strain concentrations. Because workability tests are concerned with the response to stress of the average effect of the microstructural features (grains, second phase particles, voids, etc.) of the alloy, a deformation induced phenomenon, particularly propagating cracks, larger than the microstructural features will dominate the deformation behavior of the specimen. Therefore, in the following macroscopic treatments of ductile fracture, the only apparent valid operational definition of fracture is a crack which has dimensions significantly larger than the major microstructural features or exists with a concentration sufficient to affect the average deformation behavior of the deforming piece. This definition requires that the presence of a crack or cracks is not a sufficient condition for fracture. Fracture occurs when the crack size or distribution is sufficient to disturb the deformation behavior. It appears that fracture is ductile crack propagation which occurs when the deformation is perturbed over a region about the crack extending greater than the dimensions of the crack. Damage [2] appears to be associated with ductile crack multiplication and initiation. These fine cracks and voids form along planes of maximum shear strain and are rotated in accordance with linear displacement fields characteristic of continua. Although these fine or microcracks proliferate as the deformation progresses, the deformation across them appears continuous. This continuity of the deformation in the presence of microcracks has been established. Once a discontinuity develops, which usually occurs by growth and linkage of microcracks, fracture is assumed to have occurred and the ensuing deformation is overwhelmingly influenced by the presence of the propagating fracture toward complete separation. Although fracture strain has been conventionally defined from the specimen geometry at complete separation, such a description can result in a 0 to 500 percent error in the fracture strain determination when compared with the definition of fracture as the first crack causing a discontinuity in the deformation field.

Any mechanical description of the ductile fracture process must include the effects of second phase particles and stress. McClintock[3] has presented a model based on void growth describing the effects of stress on ductile fracture. An empirical relation based on McClintock's model was found to describe the fracture strain dependence on stress and anisotropy for several engineering alloys with characteristic mechanical texturing. At a particular temperature and strain rate, this relation is of the form $\ln\bar{\varepsilon}_f = A + B(\sigma_T/\bar{\sigma})_{Avg.}$, where $\bar{\varepsilon}_f$ is the fracture strain, σ_T is the average of the maximum values of the normal stresses perpendicular to the direction of texturing, $\bar{\sigma}$ is the average value of the flow stress over the range of strain $\bar{\varepsilon}_f$, and A and B are material constants. The quantity A is most sensitive to the condition of the matrix or solid solution and, therefore, is dependent on temperature and strain rate; whereas B is dependent mainly on mechanical texturing and, therefore, responds strongly to solutioning and recrystallization heat treatments.

The following discussion will deal with material characterization within the framework of the analyses; use of the analyses to describe the fracture strain dependence on stress and anisotropy determined from test results; and application of these results with additional plasticity analyses to determine limiting reductions during deformation processing. Workability will be viewed as consisting of two parts: 1) the material contribution which is the material's fracture strain dependence on stress and anisotropy; and 2) the process contribution, which is the stress and strain requirements of the process which must not exceed the capability of the material for a successful deformation process. Other phenomena of scientific and technological significance were observed during these investigations and will also be discussed. The most significant observations were microcracking at very small strains, the total effect of temperature on workability, and the significance of proper specimen and experimental designs.

EXPERIMENTAL PROCEDURE

A. Material

The specimens for room temperature testing were obtained from either three-inch diameter bar stock of the aluminum alloy 2024-T351 or from 1.5-inch diameter bar of spheroidized annealed and cold drawn AISI 1095 steel (drill rod). The 2024 aluminum alloy was tested in the as-received T351 condition and in the annealed (0) condition. Typical microstructures of the 2024-T0 material exhibited numerous spheroidal precipitates of 2×10^{-4} inch average diameter within the grains; clear evidence of heavy, but discontinuous grain boundary precipitation; and a denuded region adjacent to the grain boundaries. The grains have average diameters of 6×10^{-3} inch in the direction of working and 2×10^{-3} inch in the transverse direction.

Elevated temperature testing was performed on the titanium alloy Ti-6Al-4V and the Ni-base superalloy U-700. The Ti-6Al-4V material was hot rolled (at approximately 1700°F) from 24 inch diameter ingots to approximately 2.25 inch diameter bar stock and mill annealed for two hours at 1100°F. This material meets AMS Spec. No. 4928C and is typical of forging stock. The nickel-base alloy U-700 was obtained from 9 inch diameter vacuum induction melted and vacuum cast ingots forged to 3.5 inch round cornered square bar at 1950°F to 2025°F. This bar stock has the following chemistry in weight percent: 0.069C, 0.004S, 0.04Si, 0.04Mn, 15.10Cr, 5.24Mo, 0.14Fe, 3.32Ti, 4.26Al, 17.8Co, 0.02Cu, 0.02Zr, 0.031B, Balance Ni.

B. Workability Tests

1. Tensile Testing of Cylindrical Specimens

Room temperature tensile testing was performed on two types of cylindrical specimens: 1) originally smooth (0.375 inch diameter, 0.250 inch diameter, and 0.125 inch diameter with 4:1 gauge length-to-diameter ratios); and 2) radiused notched specimens (0.250 inch minimum gauge section diameter with either 0.125 inch or 0.250 inch radiused notches). Specimens were accurately sectioned at angular increments of 15° or greater to the working direction and at various distances from the rod centers to ascertain the stress and fracture strain dependence on orientation and position. Testing was performed at room temperature on an Instron testing machine at a nominal strain rate of 0.1 in./in./min. which was determined to be near the average strain rate for the overall program. Measurements of extension were performed continuously until fracture with an extensometer exhibiting a strain sensitivity of at least 0.0005 in./in. Measurements of the minimum cross-sectional radius (a) and neck radius of curvature (R) were performed in two directions at 90° to each other either continuously by photographing the specimen or discontinuously by removing the specimen from the test fixtures and performing these measurements on an optical comparator.

Elevated temperature tensile testing of 0.125 inch diameter specimens originally smooth and with radiused notches was performed on a Model 510 Gleeble manufactured by Duffers Associates. All specimens were heated at a rate of 350°F/sec. to within 90% of the test temperature and then heated at a decreasing rate from 20°F/sec. to the test temperature in approximately 25 seconds. The specimens were held at temperature for at least 1-1/2 minutes. Holding times from 1-1/2 to 5 minutes at the test temperature did not affect the subsequent test results. In addition to holding and testing at the same temperature, a specific thermal cycle was also used in which the specimen was heated to a prescribed temperature; held 1-1/2 minutes or longer; cooled at a rate of approximately 20°F/sec. to a new temperature; and held at this new temperature for a prescribed time which was varied between 0 and 4.5 minutes before testing.

2. <u>Groove-Plate Plane Stress Specimens</u>

The grooved plane stress specimen proposed by Hill[(4)] was used for this test following the experimental procedure used by Ellington[(5)]. This test and procedure appears to be one of the most accurate for determining the anisotropy and the flow stress-strain behavior of materials under very well defined conditions of stress and strain. These plane stress tests were run continuously at a nominal strain rate of 0.10/min. An extensometer was placed across one groove on each specimen and a minimum of ten photographs were taken of both groove dimensions and scribe line displacements during testing. In addition, measurements of through-thickness strain were also taken past maximum load.

3. <u>Torsion Testing</u>

The torsion test specimens used in this study are shown in Figure 1. Initial tests were performed on specimens with a cylindrical gauge section with a 0.030 inch radius where the shoulders and gauge section meet. This type of specimen was found unsatisfactory because deformation became concentrated at the base of the shoulders, due to work hardening through the gauge section, and resulted in fracture on a conical surface at the gauge section extremities. To avert this problem of strain concentration, the design was subsequently changed to a 0.150 inch radiused section for which fracture was found to occur on a plane perpendicular to the specimen axis.

After the torsion test specimens had been machined,fine, short circumferential scribe lines (0.0004 inch wide and 0.070 inch length) and 0.0004 inch wide axial scribe lines were placed on the specimen surface with a 60° included angle diamond point. These lines permitted the accurate measurement of the axial and shear strain components of the surface strain. It was found that short and shallow circumferential scribe lines were necessary to

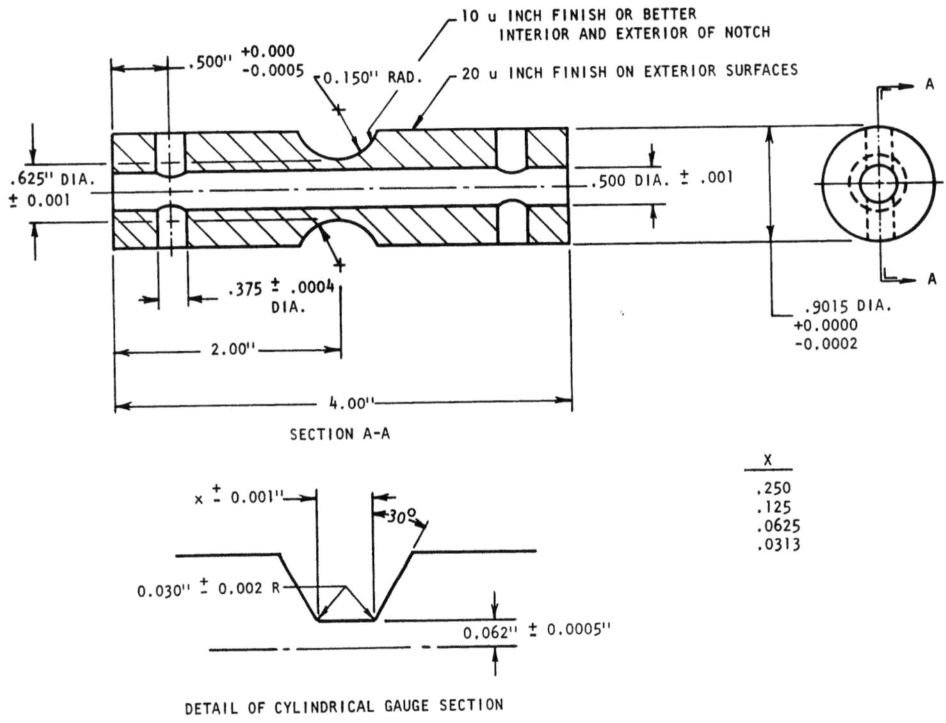

Figure 1. Torsion Test Specimen Designs

avoid shear strain concentration at the base of the line. Axial scribe lines were placed at 120° separations about the inner and outer diameters of the gauge section and were continued on the specimen shoulders and shank to provide a zero strain base for alignment. All measurements were made on a Gaertner toolmaker's microscope (No. 366AP) with a specimen centering fixture, which included a goniometer to determine angular displacement of the axial scribe lines. Measurements with this apparatus can be made to ±0.00005 inch and ±0.5 minute of an angle.

The torsion test apparatus consisted of a lathe with precisely aligned grips on the tailstock and chuck sections. The tail stock was modified to accept a 2.5000 ±0.0001 inch shaft containing an integrally machined load cell for torque and axial load measurements. The shaft and load cell were connected to a pneumatic cylinder through a universal joint. The entire assembly, shaft, with integral load cell, and grip, was finish machined as a unit to insure axial alignment. The axial load applied to the specimen was present at the beginning of the test and the extension or contraction of the specimen was followed by a linear variable differential transformer. A timing disk, which activated a photocell at every 15 degrees of angular rotation was connected to the headstock grip and used for measurements at very large strains. Also, a direct mechanical drive was available for measuring precise rotation with an angular transducer. Axial load, torque, angular rotation, axial displacement, and temperature were recorded directly on a Visicorder.

4. Compression Testing

Compression testing was investigated by using simple upsetting of cylinders and the side pressing of initially cylindrical shafts, disks, and disks with machined flats. The purpose of these different specimen geometries was to modify the stress history during deformation. All flat surfaces were machined parallel to within ±0.0002 inch in 1.0 inch and all specimens were prepared with a 10 microinch rms surface finish. In upsetting, the effects of length-to-diameter ratio were investigated for unlubricated and lubricated specimens (0.002 inch thick teflon film). The variables in the side pressing studies were also lubrication and specimen geometry (circular disks, disks or cylinders with machined flats, and length-to-diameter ratio). After machining, an orthogonal array of Vickers diamond pyramid impressions was accurately placed at 0.050 inch separations about the exact center (±0.0001-inch) of the specimens along subsequent directions of principal normal stress. These directions corresponded to the direction of loading and the direction normal to it through the center of the specimen. Eight impressions were placed in each of the two directions. Depending on the specimen material, the impressions were initially between 0.0025 and 0.005 inch wide. Although surface microcracks could be observed about some impressions after deformation, fracture appeared to occur in a random manner which could not be attributed to the presence of the impression. It was concluded, therefore, that the presence of the impression did not affect our results. It was found necessary, under certain conditions in side pressing, to extrapolate the strain to zero point separation at the specimen center. This necessity arises principally from the specimen curvature. This extrapolation was necessary in some cases during upsetting because the gradient of vertical point separation (or strain) can be large; however, the circumferential surface strain during upsetting does not exhibit a gradient in any horizontal plane because of axial symmetry.

C. Processing

1. Rolling

Unlubricated rolling was performed on a two-high Farrel ("mint" mill) with rolls ten inches in length and diameter. The roll speed was 20 ft/min. with rolls "superfinished" to produce a 2 $\pm$1 microinch rms finish. Specimens were machined to produce 0.250 inch thick strip approximately 6.0 inches long of 2.50 inch and 1.25 inch widths. Edge cracking was investigated in strips of three different edge contours: square, radiused (0.125 inch radius), and chamfered (60° included angle with a 0.015 inch radius at the chamfer apex). The specimens were finish machined to produce a 8 to 12 microinch finish except at the edges which were 4 to 6. Subsequently, either 1.0 or 0.5 inch long scribe lines 0.0004 inch wide were placed on the specimen edges. After measurement of the initial line separations and length on a Gaertner toolmaker's microscope, the specimens were rolled incrementally as follows: square edge specimen, 10, 20, or 40 percent per pass except near fracture; and radiused- and chamfered-edge specimens, 10 percent per pass until fracture. To avoid exceeding the fracture strain, the reduction at fracture was usually determined by the first of duplicate specimens to fracture or, in the case of the square-edge strip, extra specimens were run without scribe lines. Maximum reductions of approximately ten percent were taken near fracture on subsequent specimens.

2. Extrusion

Extrusion was performed at a ram speed of 12 in./min. through dies of 60°, 90°, and 120° included angles with anhydrous lanolin as a lubricant. Experimental visioplasticity results (6) were used to determine the relation between maximum centerline strain rate (which corresponds to nearly two times the average rate) and ram speed as a function of reduction ratio $(R_o/R_f)^2$, where $2R_o$ and $2R_f$ are the initial and final diameters of the extrusion. Both the billets and the container were lubricated prior to extrusion and load-time curves were recorded on an oscillograph. The billets were machined with one end contoured to match the die and approximately 0.005 inch under the container diameter. Initial billet diameters of 1.780, 1.412, and 0.812 in.

were used. The two larger diameters were used in extrusion through 90° dies to study single- and multiple-pass extrusion and the 0.812 diameter billets were used in single-pass reductions through all three die angles. Because the basis for the workability criterion was the centerburst phenomenon, an internal defect, all solid extrusions were radiographed.

EXPERIMENTAL RESULTS

A. Tensile Testing of Cylindrical Specimens

The results in Figure 2 demonstrate the fracture strain anisotropy exhibited by the alloys tested at room temperature during this study. These data correspond to measured effective strain based on the final diameters of the completely separated specimens. As indicated by the results of Larson and Nunes(7), this value of the fracture strain could be in error by a 0 to 10 percent overestimate. Such an error corresponds to the increment of strain between a discontinuity in the load-extension curve for a stiff machine and final specimen separation, and appears to be associated with a rapidly propagating ductile crack originating at the center of the necked area. Similarly occurring discontinuities in the incrementally measured a/R ratios were also observed. The vlaue of a/R at separation always was significantly larger than extrapolated results from larger values of a or specimen diameter. Because the discontinuity appeared to be associated most strongly with R and no unique criterion could be established for defining fracture for tensile specimens, the specimen dimensions at complete separation were used to determine the effective strain, but the extrapolated values of a/R were used to determine the associated stress.

The three curves in Figure 2 exhibit a marked similarity of form. However, significant features for understanding the effects of specimen geometry on fracture strain are the longitudinal fracture strain ($\bar{\varepsilon}_L$) and the ratio of the longitudinal-to-transverse fracture strain ($\bar{\varepsilon}_L/\bar{\varepsilon}_T$) which are as follows:

	$\bar{\varepsilon}_L$	$\bar{\varepsilon}_L/\bar{\varepsilon}_T$
Al-2024T0	0.55	1.64
Al-2024T351	0.37	2.85
Drill Rod	0.82	1.60

The significance of the fracture strain anisotropy will be discussed in a later section on torsion testing.

The tensile test results for flow stress $\bar{\sigma}$ and axial centerline stress σ_{3max}, based on the Bridgman analysis (8), are shown in Figure 3 for initially smooth and radiused-notched specimens of both longitudinal and transverse orientations. Within the limits of the Bridgman analysis, the flow stress of these alloys

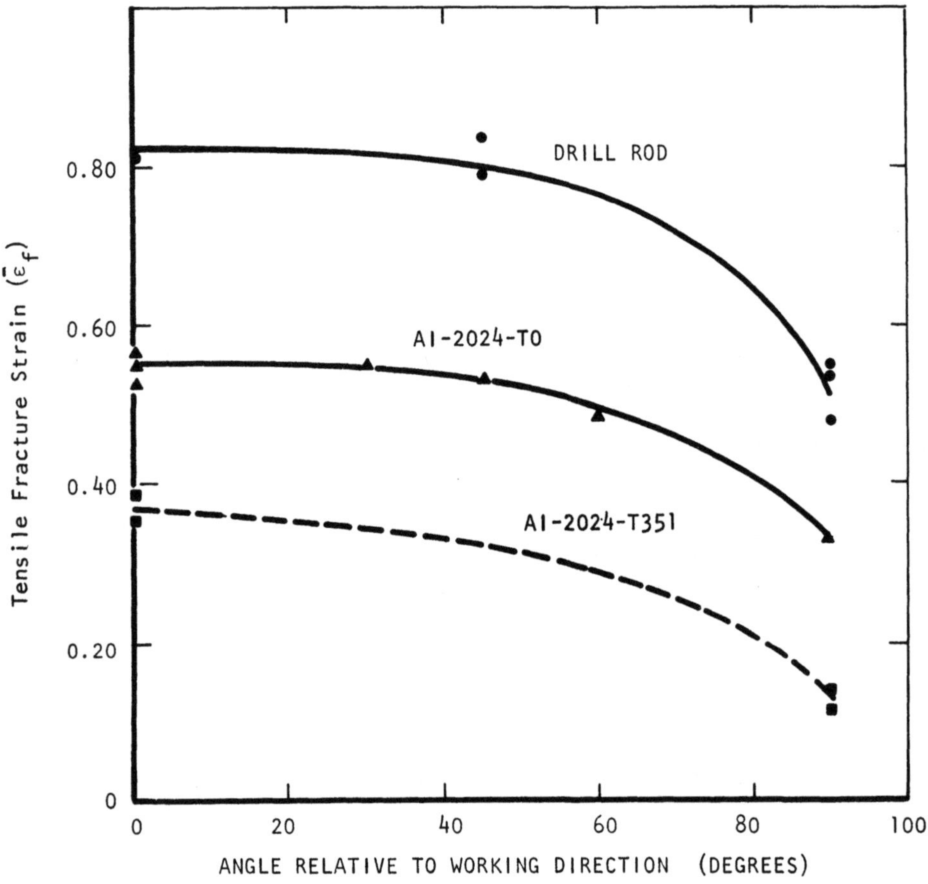

Figure 2. Tensile Fracture Strain as a Function of Orientation Relative to the Working Direction of the Bar Stock

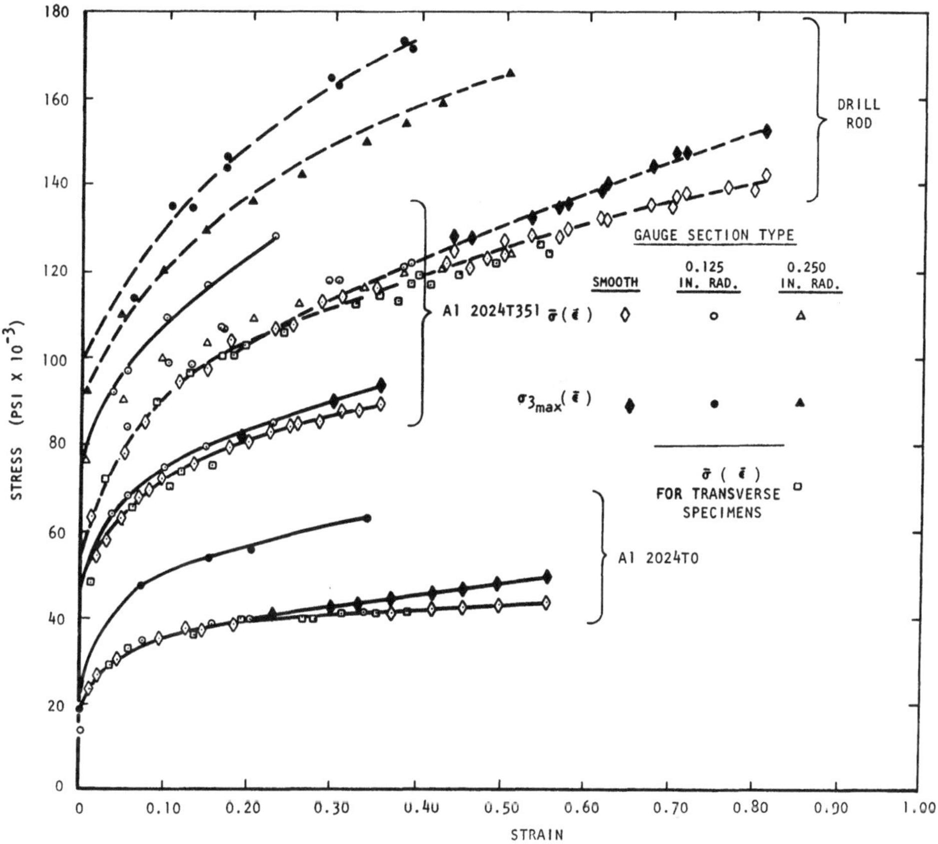

Figure 3. Flow Stress- and Maximum Tensile Stress-Strain Behavior for Longitudinal and Transverse Tensile Specimens

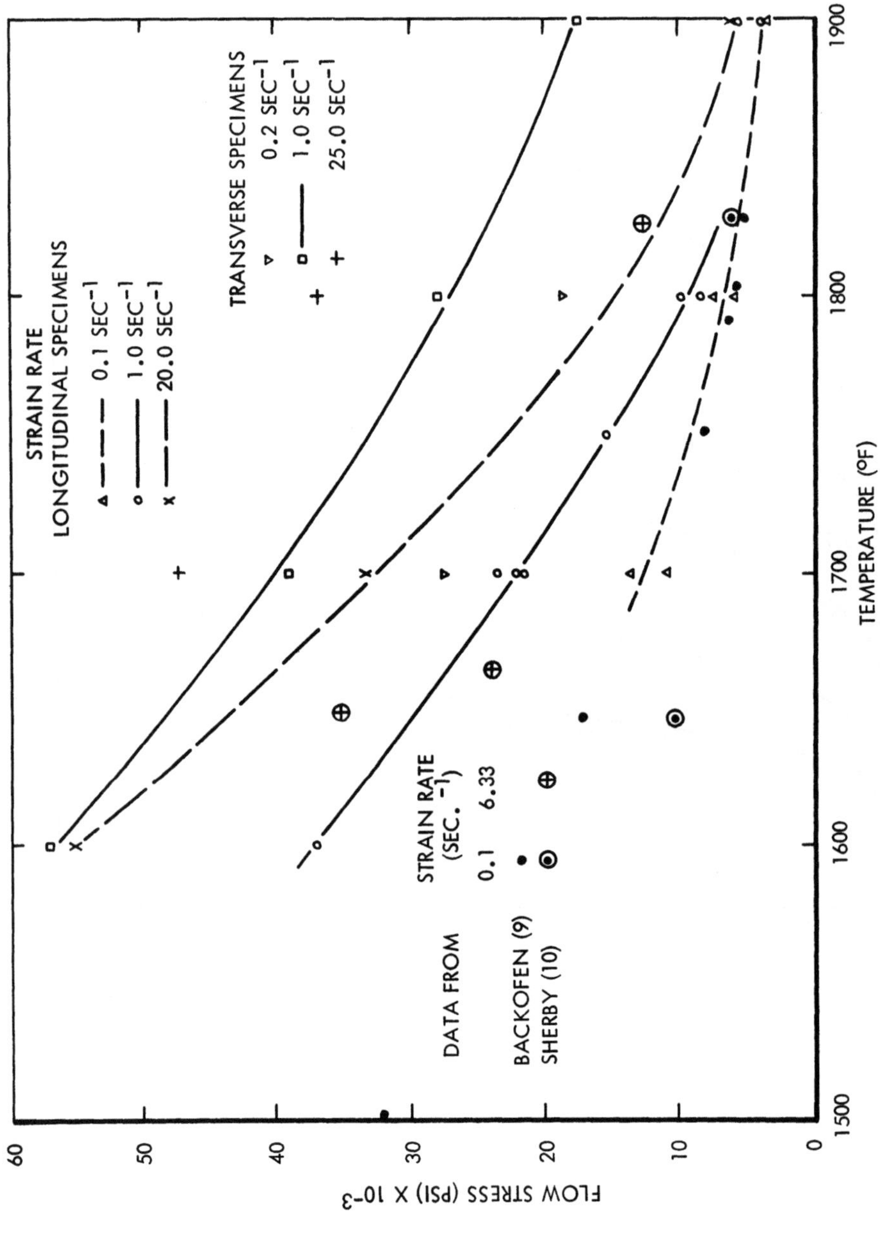

Figure 4. Flow Stress Dependence on Temperature for Longitudinal and Transverse Ti-6Al-4V Tensile and Torsion Specimens

is isotropic as indicated by the coincidence of the flow stress-strain curves for specimens oriented parallel and perpendicular to the original working direction of the bar stock. Additional support for the flow stress isotropy was given by measurements of the incremental diameter changes which showed that no ellipticity developed as a result of tensile testing. However, striated fracture surfaces clearly indicated the direction of prior working for transverse specimens of all three alloys. Al-2024TO specimens sectioned at angles of 30°, 45°, 60°, and 90° to the longitudinal direction also exhibited striated fracture surfaces which were inclined at 35° ± 3° to the tensile axis. The radiused notched specimens exhibited fracture surfaces at 90° to the tensile axis.

Elevated temperature flow stress data from tensile tests on the titanium alloy Ti-6Al-4V are presented in Figure 4. These data obtained on both longitudinal and transverse specimens demonstrate the strong flow stress anisotropy exhibited by this alloy even at temperatures above the α-β transus (1825°F). Similar results were obtained by Lee and Backofen(9) with Zircaloy-4. Data from other investigations for Ti-6Al-4V tensile test(9) and solid torsion test specimens (10) are also presented in this figure for comparison. The fracture strains for the results in Figure 4 are large (greater than 2) and the longitudinal specimens exhibit sharp maxima at a particular temperature for each strain rate. The temperatures at which the maxima occurred increased with increasing strain rate and corresponded to the maximum strain rate exponent "m" (9) at each strain rate. However, only a gradual increase of the fracture strain with temperature was observed for the transverse results at all strain rates. The fracture strain exhibited almost no dependence on strain rate over the range investigated for the transverse specimens.

B. Plane Stress Tests

The plane stress tests were performed with grooved specimens of Al-2024TO. Following Ellington's(5) procedure, fine scribe lines were accurately placed at 1° intervals about both sides of the line calculated for the direction of pure extension(4,5) for flow stress isotropy because the tensile test results indicated that this condition should prevail. Therefore, the actual line undergoing zero rotation, i.e., pure extension, should correspond to the relative displacement vector of the two sides of the groove. The von Mises yield criterion and associated flow rules provide accurate relations among the groove angle $\emptyset$, the velocity displacement vector, load, specimen dimensions, stress, strain and strain rate(4). Additional measurements of the deforming specimens were made to provide independent measurements of strain and strain rate to evaluate the consistency of the results. If one scribe line could be found to exhibit only pure extension during the entire test, then the test could be continued until fracture under constant stress ratios (i.e., constant ratios of the principal stresses to the flow stress). This behavior was observed within the limits of experimental error of $\pm 0.5°$. The fracture strains and stress ratios associated with the particular groove angles $\emptyset$ are shown in Table 1. In this table, $\bar{\sigma}$ is the flow stress, σ_1 is the principal stress component which was nearly parallel to the loading direction in most of the specimens. L is the axial load, h the groove thickness, w the groove width, and P is the average pressure and equals $-(\sigma_1+\sigma_2+\sigma_3)/3$.

C. Torsion Test Results

For the majority of torsion tests, two gauge section designs were used as described in Figure 1. However, the effects of gauge length on stress and fracture strain were investigated using cylindrical gauge sections of 0.0315, 0.0625, 0.125, and 0.250 inch length. With few exceptions, the inner diameter was 0.500 inch and the outer diameter was 0.640 inch.

It is almost a convention to define shear strain during pure torsion as $\gamma = \bar{r}\theta/\ell_o$, in which $\bar{r}$ is the average gauge section radius, θ the total relative angular displacement across the gauge section in radians, and ℓ_o the gauge length. It was found, as reported by other investigators (11), that the shear strain at fracture based on the preceding definition of γ increased as the gauge section decreased. Such results imply that fracture strain is not a unique function of stress or that the state of stress changes with gauge length. It was felt that this observation must be erroneous or that the condition of pure shear could not exist during pure torsion for the gauge section dimensions which were selected. Hence, at the very least, an analysis of the torsion test was in order due to the great attention it has received for workability testing (12) and the relatively small amount of information that is available on the precise definition of stress and strain. To perform this analysis, axial scribe lines were placed at 120° on the inner and outer surfaces of the torsion specimens. The angular displacements of these lines along the axial direction z were measured incrementally and after fracture. The shear strain at fracture γ should equal

$$r\theta/\ell_o = r\Delta\theta/\Delta z = rd\theta/dz,$$

(where r is the radial distance from the centerline to where θ is measured, and usually corresponds to the outer radius) if the shear strain were uniform across the gauge section. Figure 5 shows the dependence of the angular displacements of the scribe lines on z across the gauge section for gauge lengths of 0.125, 0.060, and 0.052 inch. Although the shear strain is very uniform across the central portion of the gauge section and independent of the gauge length, the shear strain increases very rapidly at the gauge section extremities. In fact, the shear strain in the region of fracture is nearly a constant for all three specimens

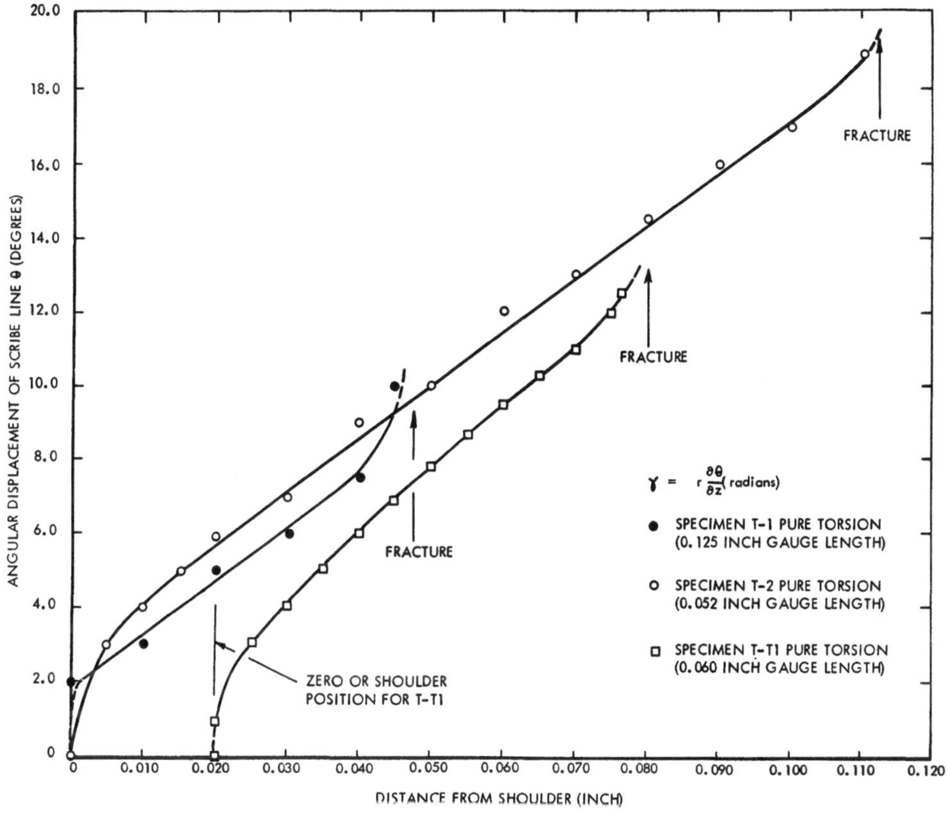

Figure 5. Torsion Test Results Showing Inhomogeneity of the Shear Strain Across the Specimen Gauge Section

Table 1

Effect of Groove Angle Ø on the State of Stress for the Grooved Strip Tensile Specimen

Angle Ø	σ_2/σ_1	$\sigma_1/\bar{\sigma}$	σ_1(hw/L)	$-P/\bar{\sigma}$*	Measured Fracture Strain (in/in)
54°44'	0	1.00	1.00	0.33	0.607
60°	0.088	1.043	1.034	0.38	0.610 ± 0.04
70°	0.26	1.113	1.052	0.47	0.574 ± 0.012
80°	0.42	1.15	1.024	0.54	0.508 ± 0.012
90°	0.50	1.155	1.00	0.57	0.489 ± 0.011

* Average pressure $P = -\left(\frac{\sigma_1+\sigma_2+\sigma_3}{3}\right)$

independent of gauge length. These observations indicate that even for pure torsion, fracture strain cannot be determined from the gross rotation of the specimen. For this reason, all of the subsequent specimens had axial scribe lines and specimens for tests with axial loading had short circumferential scribe lines. Furthermore, fully radiused (0.150 inch) specimens were subsequently used to insure that fracture occurred on a plane perpendicular to the axial direction.

Another benefit of the axial scribe lines was the exact definition of fracture strain. Macroscopic fracture during torsion testing initiates by the formation of short discontinuous circumferential segments of fine surface or microcracks which propagate by growth and linkage. Once a surface crack runs around the circumference, the surface has now fractured and the surface scribe line displacements cease. The continuous surface crack now propagates radially inward by localizing deformation to its base and the deformation is extremely localized. The discontinuity of this displacement (about 2 to 3° of total rotation) can be observed near the shoulder region in the data in Figure 5.

The data in Table II provide a review of the effect of gauge length on fracture strain at room temperature for pure torsion and torsion with a nominal tensile stress of 27,000 psi. These results show that for pure torsion fracture strain is independent of gauge length if this strain is determined by $d(r\theta)/d\ell$ at fracture, and for torsion with nominal axial tensile stress of 27,000 psi, the fracture strain decreases with an increase of the stress ratio σ_θ/σ_z, which can be related to an increase of overall average tensile stress. The approximated value of the radial strain $\sim\varepsilon_r$ shown in Table II is the strain calculated on the assumption of homogeneous deformation ($\sim\varepsilon_r = \ln[r_o-r_i]/(r_o-r_i)_o]$. The value of ε_r

Table II

Strain Measurements on Torsion Test Specimens
Tested at Room Temperature

Nominal Axial Stress of 27,000 psi

Al-2024-T0 (gauge length)	$\tilde{\varepsilon}_r$	ε_z	ε_θ	ε_r	Stress Ratio σ_θ/σ_z	γ	$\bar{\varepsilon}$
TB2 (0.0342 in.)	-0.31	.30	-.014	.28	.46	0.36	0.39
TB8 (0.1265 in.)	-0.42	.42	-.042	-.38	.42	0.37	0.51
TB14 (0.2503 in.)	-0.36	.40	-.065	-.33	.37	0.38	0.55
TB21 (0.0730 in.)	-.37	.37	-.027	-.34	.43	0.23	0.43

Pure Torsion

Al-2024-T0 (gauge length)							
TB3 (0.075 in.)	0	0	0	0	0	1.15	0.66
TB5 (0.137 in.)	0	0	0	0	0	1.06	0.61
TB20 (0.250 in.)	0	0	0	0	0	1.21	0.63
TB22 (0.031 in.)	0	0	0	0	0	1.06	0.61
Al-2024-T351							
TTHN1 (0.150 in.Rad.)	0	0	0	0	0	-	.320
TTHN2 (0.162 in.)	0	0	0	0	0	-	.316
Drill Rod							
RITTIX (0.125 in.)	0	0	0	0	0	0	1.026
R3TTG2 (0.125 in.)	0	0	0	0	0	0	1.12
R6TTS1 (0.150 in.Rad.)	0	0	0	0	0	0	1.21

in this table corresponds to the value obtained from the incompressibility relation using the measured values of ε_θ and ε_z (additional definition of these quantities and the description of the analysis are provided in Ref.11).

The torsion test results in Figure 6 for Ti-6Al-4V hollow specimens, which were obtained at a strain rate of 2.5/sec. with hollow specimens, exhibited a behavior similar to the longitudinal tensile test results. The arrows on the solid curve for the fracture strain dependence on temperature indicate the measured temperature rise during each test. This temperature rise should not be equated to the total energy evolved during deformation because this energy is liberated over a period time during which conduction to the grips and radiation and convection to the specimen environment appear to account for approximately one-half of the deformation energy. The fracture strain presented on this curve is the natural or effective strain $\bar{\varepsilon}$ related to $\bar{\gamma} = r\bar{\theta}/l_o$ by $\bar{\varepsilon} = \bar{\gamma}/\sqrt{3}$. This strain, which was determined from the total angular displacement $\bar{\theta}$ across the gauge section, agreed with the measured value from surface marks to within ± 5 percent. The procedure used for accurately determining fracture strain for the room temperature tests was necessary due to the workhardening at low temperatures which concentrated shear strain to the gauge section extremities. Because workhardening is negligible during hot working, fracture usually occurred through the center portion of the gauge section. Although $\bar{\gamma} = r\bar{\theta}/l_o$ could be related to surface shear strain for the torsion test specimens, the fracture strain, particularly for the titanium alloy specimens, was difficult to define. It was found during testing that shear stress would either rapidly achieve a maximum and subsequently continuously decrease for tests below 1800°F or would exhibit a maximum very close to the point of initial yielding and then continuously decrease with further strain for tests above 1800°F. This continuous decrease of the flow stress after a maximum at very small strains could have developed from one or a combination of the following: 1) the temperature rise during testing, 2) rotation of the microstructure or principal directions of anisotropy, and 3) a slow rate of ductile crack growth.

The effect of temperature rise on the flow stress can be determined from the temperature ranges for the test results in Figure 6 indicated by the arrows in this figure. The solid curve on which these arrows are drawn was obtained from the measured values of the total strain at complete separation of the specimens (zero torque) at temperatures corresponding to the initial test temperature. The dashed curve in Figure 6 corresponds to the strain where the measured flow stress had dropped to 90 percent of the value corresponding to the instantaneous test temperature. This value of strain has been defined as the first indication of fracture, and for all of the test results in Figure 6 correspond-

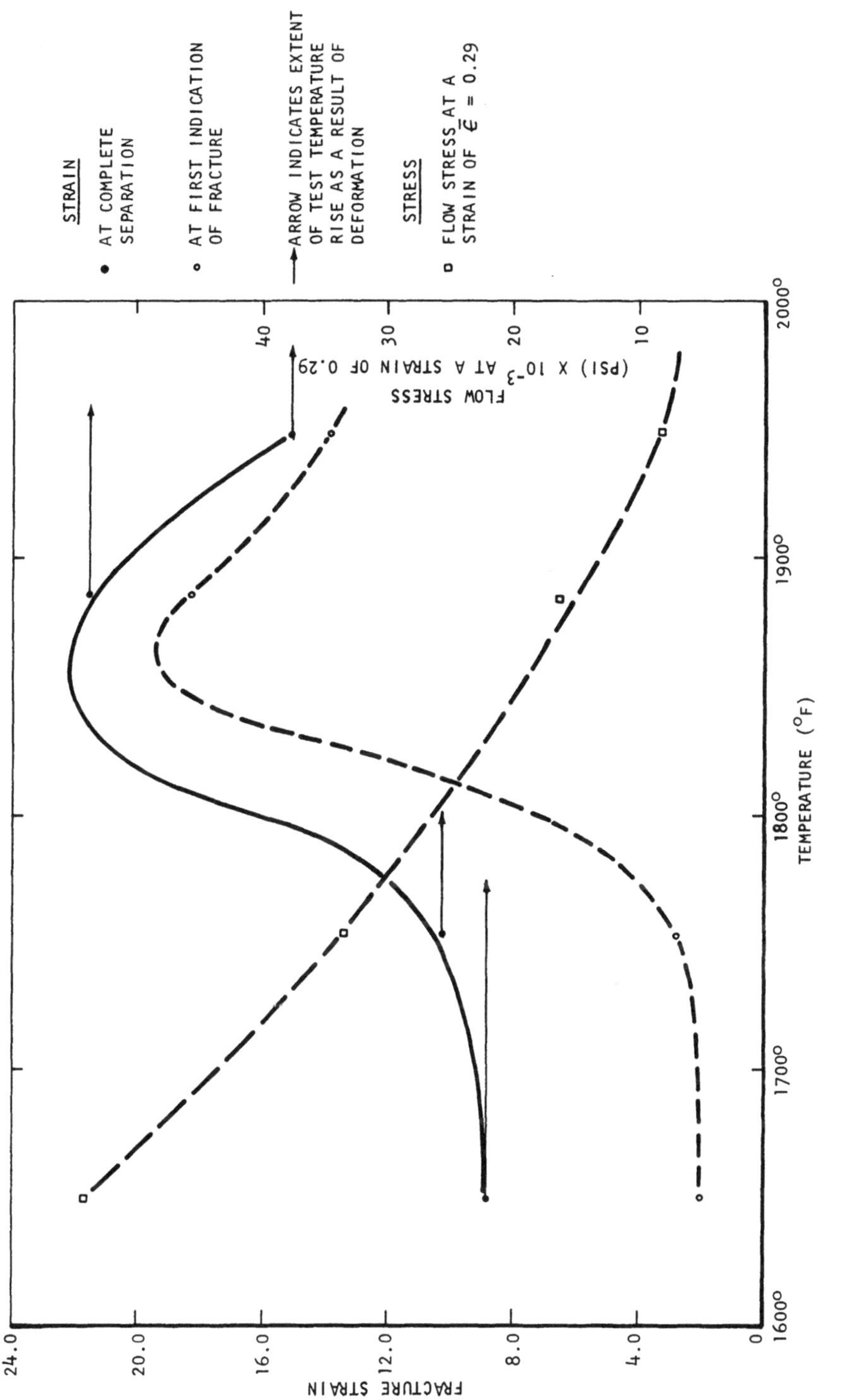

Figure 6. Ti-6Al-4V Torsion Test Results for Hollow Specimens

ed to a rotation of the originally longitudinal structure or direction of prior work to between 2 to 15° of the transverse plane of the torsion test specimen. Therefore, it must be concluded that the temperature rise produced only a part of the flow stress decrease during torsion testing. Because of the rotation of the microstructure during torsion testing and the strong fracture strain anisotropy below 1800°F, a portion of this flow stress decrease may be associated with ductile crack propagation due to the rotation of the principal directions of fracture strain anisotropy.

The test results in Table III on solid and hollow specimens were obtained to determine the effect of gauge section thickness on the apparent fracture strain and flow stress. The significant disparities between the fracture strains for hollow and solid specimens are associated with ductile crack propagation as shown by the photographs in Figure 7.

D. Compression Testing

Compression testing was performed in a similar manner for both upsetting and side pressing of cylinders, although fracture occurred on the barrelled cylindrical surface during upsetting and on the original circular ends of the side pressed specimens. For both designs, the fracture originated at the center of these surfaces. The procedure for analysis of the grid deformations is described in Ref. 11 where the stress ratio

$$\alpha = d\varepsilon_2/d\varepsilon_3 = (2\sigma_2-\sigma_3)/(2\sigma_3-\sigma_2)$$

was defined using the Levy-Mises strain increment relations. The principal normal stresses were determined and defined by the following: σ_1, the component normal to the deforming surface or the radial component σ_r for upsetting; σ_2, the horizontal component or the circumferential component σ_θ for upsetting; and σ_3, the vertical component. These stress components have the following values for a particular value of α :

State of Stress

α	
-0.2	$\sigma_1 = 0$, $\sigma_2 = -0.378\bar{\sigma}$ and $\sigma_3 = 1.134\bar{\sigma}$
-0.5 uniaxial compression	$\sigma_1 = 0 = \sigma_2$, and $\sigma_3 = -\bar{\sigma}$
-1.0 pure shear	$\sigma_1 = 0$, $\sigma_2 = -\sigma_3 = \bar{\sigma}/\sqrt{3}$
-2.0 uniaxial tension	$\sigma_1 = \sigma_3 = 0, \sigma_2 = \bar{\sigma}$
-5.0	$\sigma_1 = 0$, $\sigma_2 = 1.134\bar{\sigma}$, and $\sigma_3 = 0.378\bar{\sigma}$

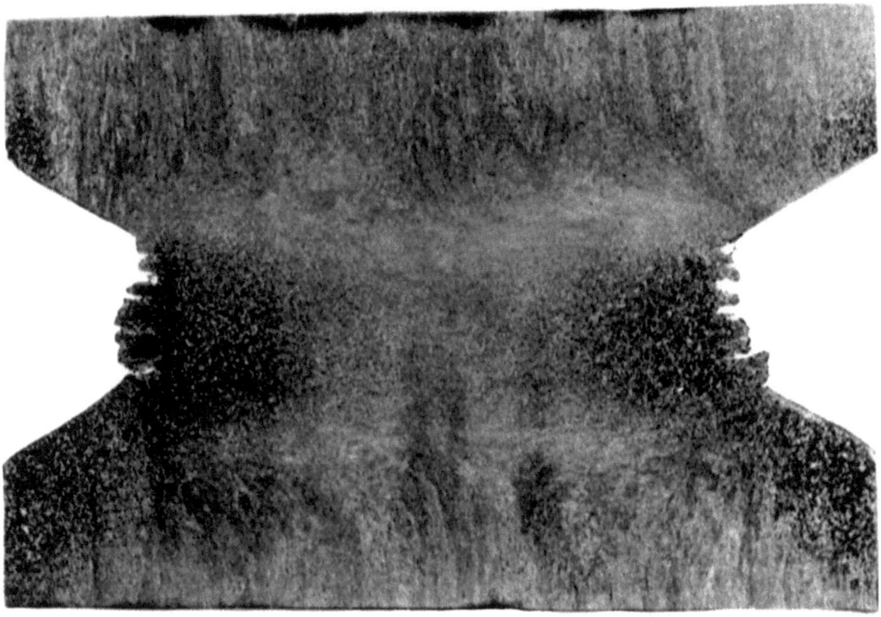

Solid Ti-6Al-4V Torsion Test Specimen
TiTorS-5 Tested at 1806°F
Strain Rate of 2.5/second
Specimen Shown After a Strain of $\bar{\varepsilon}$ = 44.6
6.25X Magnification

Hollow Ti-6Al-4V Torsion Test Specimen
TiTor-9 Tested at 1885°F
Strain Rate of 2.5/second
Specimen Shown After Fracture at a Strain of $\bar{\varepsilon}$ = 21.45
6.25X Magnification

Figure 7. Photographs of Solid and Hollow Ti-6Al-4V Torsion Test Specimens

Table III

Ti-6Al-4V Torsion Test Results

Specimen No.	Initial Temp. (°F)	Strain Rate ($sec.^{-1}$) at r	Initial Flow Stress (psi) $\bar{\sigma}$	Fracture Strain $\bar{\varepsilon}$	Remarks
Hollow Specimens					
Ti-Tor-3	1753	2.88	19,200	10.23	
Ti-Tor-8	1648	2.61	35,300	8.87	
Ti-Tor-9	1947	2.73	10,800	15.10	Y.P.*
Ti-Tor-10	1885	2.57	12,800	21.45	Y.P.
Solid Specimens					
Ti-TorS-1	1610	1.59	17,700	42.9	**
Ti-TorS-2	1620	1.56	13,700	24.1	
Ti-TorS-4	1628	1.58	13,800	28.0	
To-TorS-5	1806	1.59	5,060	***	
Ti-TorS-6	1783	1.60	5,310	47.9	
Ti-TorS-7	1860	1.60	4,210	***	Y.P.

* Y.P. indicates a maximum torque near yielding.
** Test stopped before fracture, held 30 seconds at temperature, and run to fracture.
***Test stopped before fracture.

The values of α observed during this investigation involving 27 upset and side pressed specimens ranged from -0.46 to -3.44. In general, the side pressed specimens could be deformed at more compressive states of stress than for upsetting. The range of stress capable during side pressing is much larger than upsetting. For all of the results which will be discussed the prior working direction of the original bar stock was parallel to the axes of the cylindrical specimens. Therefore, for α = -2.0, the compression test was equivalent to pure uniaxial testing of a transversely sectioned tensile specimen. An example of these test results are shown in Figure 8 in which the α and ln h_o/h (h_o is the initial and h the final height of the steel upset specimen) dependence on ε_3 are shown. The state of stress at fracture for specimen US-1 corresponds to α =-1.42 for which $\sigma_3 = -0.27\bar{\sigma}$ and $\sigma_2 = 0.84\bar{\sigma}$. Therefore, since $\sigma_1 = 0$, the plane of maximum shear strain or stress is inclined at 45° to the vertical and horizontal directions and isparallel to the radial direction. These planes are the fracture planes for specimen US-1 shown in Figure 9. The 45° cracks occur for values of α greater than -2.0 for which the planes of maximum shear strain are at 45° to the horizontal and vertical directions. For values of α less than -2.0 the planes of maximum shear strain are parallel to the loading direction and are inclined to the surface at 45° as shown for specimen U-1 in

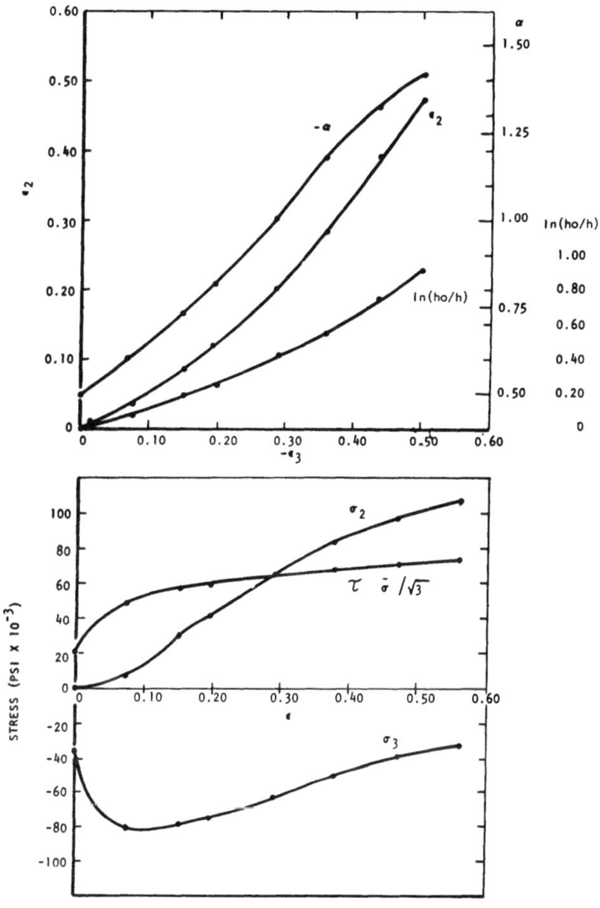

Figure 8. Experimental Data and Resulting Stress and Strain Determinations for Upset Specimen US-1

Figure 9.

Microcracks, fine, shallow cracks, along directions of maximum shear strain appear very early during cold deformation, but require relatively high magnification to be observed. An example of this phenomenon is shown in Figure 10a for which the microcracks were observed at a strain 0.092 for a specimen which exhibited fracture at a strain of 0.927 as shown in Figure 10b. Fracture was defined as the first indication of a discontinuous motion of grid marks usually corresponding to a misalignment.

E. Processing

1. Rolling

The rolling tests were performed similar to the upsetting tests to determine the state of stress leading to fracture. A typical example of fracture during rolling, i.e., edge cracking, is shown in Figure 11 for square edge strip. The 45° cracks are typical of edge cracking; however, the center plane failure resulted from the strip-end curvature and the intersection of 45° cracks initiating from each edge of the strip on perpendicular planes. For all of the strip specimens, the direction of mechanical texturing was parallel to the rolling direction.

The results in Figure 12 demonstrate the relation between the thickness reduction during rolling of square-, radiused-, and chamfered-edge strip and the tensile strip fracture strain. It was found[11] that increasing the severity of the curvature resulted in an increasing tensile character of the average state of stress at the strip edge. This fact can be anticipated from the definition of $\alpha = d\varepsilon_2/d\varepsilon_3 = (2\sigma_2-\sigma_3)/(2\sigma_3-\sigma_2)$, where ε_2 is the strain in the rolling direction and ε_3 is the strain in the through-thickness direction. The through-thickness strain ε_3, at the strip edge, which is always negative, will decrease in absolute value relative to ε_2 as the undeformed or free length of the edge is increased. This free length is zero for square edge strip and increases as the strip edge curvature and/or taper is increased. Due to continuity, ε_2 must conform approximately to the average axial strain in the rolling direction of the strip. These conditions require that α becomes more negative as this free length increases. Therefore, in this investigation α is most negative or the state of stress is most severe for the chamfered-edge strip, and, consequently, σ_2, which is always tensile, and σ_3 are most tensile. Therefore, edge curvature introduced artificially or by inhomogeneous deformation can result in a more tensile state of stress at the edge of rolled strip. This behavior is demonstrated in the experimental and analytical results in Figure 13 for rolling of drill rod strip with originally square edges. The experimental strain distribution, ε_2 and ε_3, across the strip are shown for various stages of the reduction. As the deformation becomes large a significant strain gradient, corresponding to

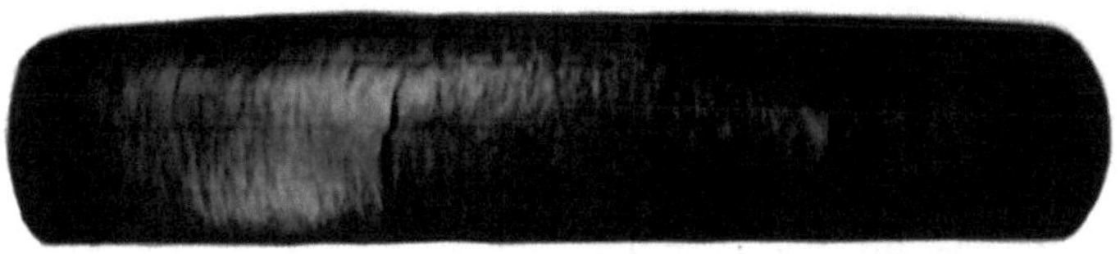

Specimen Type U-1

Specimen Type US-1

Figure 9. Ductile Fracture in Upset Specimens US-1 (45° Inclined Cracks $\alpha = -1.42$ and U-1 (Vertical Crack, $\alpha = -2.21$)

Figure 10a. $\bar{\varepsilon} = 0.092$ 500X

Figure 10b. $\bar{\varepsilon} = 0.927$ 500X

Figure 10. Microcracks Along Directions of Maximum Shear Strain at a Strain of 0.092 on a Specimen (SS-2) which Exhibited Macroscopic Fracture at a Strain of 0.982

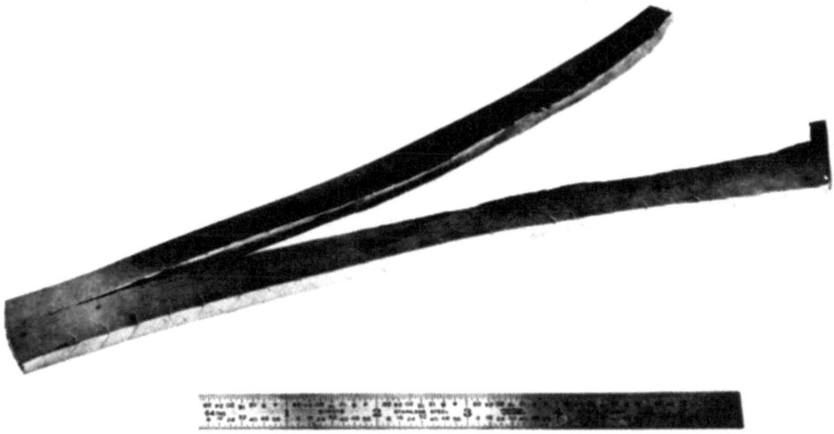

Figure 11. Edge Cracking in Al-2024T351. Propagation of the Edge Cracks to the Center Resulted in Splitting. (Note: Dulling spray was used to decrease reflectivity.)

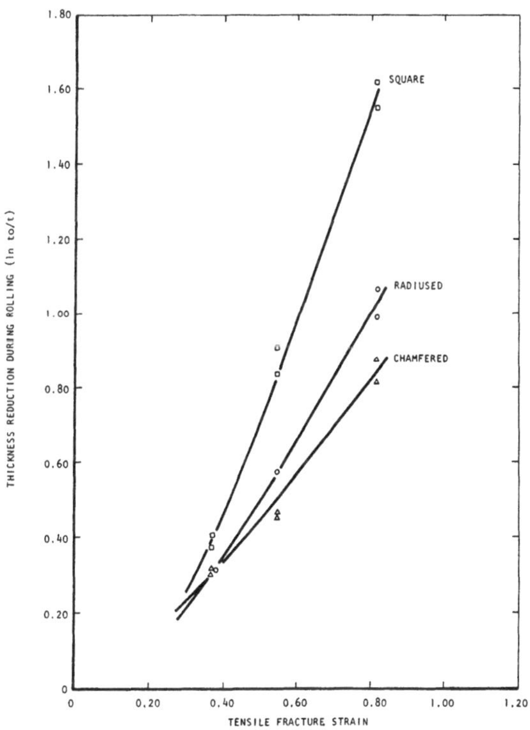

Figure 12. The Relation Between the Logarithm of the Thickness Reduction in Rolling and the Tensile Fracture Strain

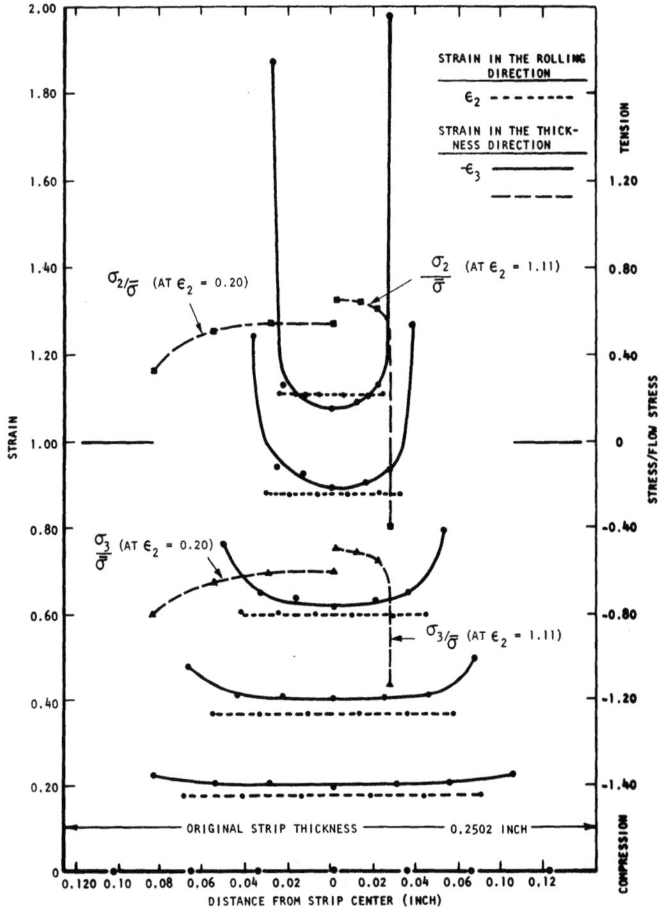

Figure 13. The Incremental Strain and Stress Distributions Across a Square-Edge Strip (Specimen PSB2-1)

edge curvature, becomes apparent. As this strain gradient increases both the gradient of stress and the tensile character of both stress components increase. However, even over relatively large reductions (e.g., $\ln(t_o/t) = 1.6$ for drill rod), the average state of stress for square edge strip corresponds to pure shear $\sigma_2 = -\sigma_3$. Stress determinations similar to the results in Figure 13 were obtained for square edge, radiused edge and chamfered edge strip of all three alloys. These result in terms of fracture strain and the α variations are presented in Table IV.

TABLE IV

COMPARISON OF TESTING AND ROLLING RESULTS

Material	Edge Contour	Observed Fracture Strain	α Initial	α Average	α Final	Predicted Fracture Strain Based on the Average Value of α
Drill Rod	Square	1.48	-0.81	-1.02	-1.10	1.45
Al-2024T0	Square	0.95	-0.96	-1.02	-1.10	0.92
Al-2024T351	Square	0.48	-0.96	-1.04	-1.08	0.45
Drill Rod	Radiused	0.92	-2.25	-2.11	-1.87	0.91
Drill Rod	Chamfered	0.77	-3.80	-2.10	-1.52	0.91
Al-2024T0	Chamfered	0.39	-3.81	-2.44	-1.85	0.56
Al-2024T351	Chamfered	0.26	-3.14	-2.62	-2.18	0.31

2. Extrusion

The extrusion results in Figure 14 are presented in terms of an energy quantity (plastic strain energy, divided by $\bar{\sigma}$) at various initial R_o to final R_f ratios of the radii of the extrusions. The reduction ratio equals $(R_o/R_f)^2$. The circled points in this figure correspond to the observance of centerburst defects either by X-ray examination or by visual examination due to fracture of the extrusion. Most of the data in this figure was obtained with Al-2024T351 because multiple passes were required to fracture the TO alloy and drill rod. The numbers on the photographs in this figure are also shown adjacent to the corresponding data points on the graph. All of the photographed specimens were extruded through dies of 90° included angle with identical conditions of lubrication and ram speed. The number 1 photograph was obtained from the TO alloy and shows both the radiographed and external appearance of this specimen after three consecutive reductions of $R_o/R_f = 1.18$. The radiograph illustrates the chevron or centerburst defect which was observed after the third pass. This particular defect is relatively large and has affected the external surface of the extrusion as indicated by the two annular bonds. In general, visual observation of extrusions is not sufficient for detecting this internal defect. The photographs numbered 2 and 3 were obtained from the T351 alloy extrusions which fractured completely upon the occurrence of the centerburst defect.

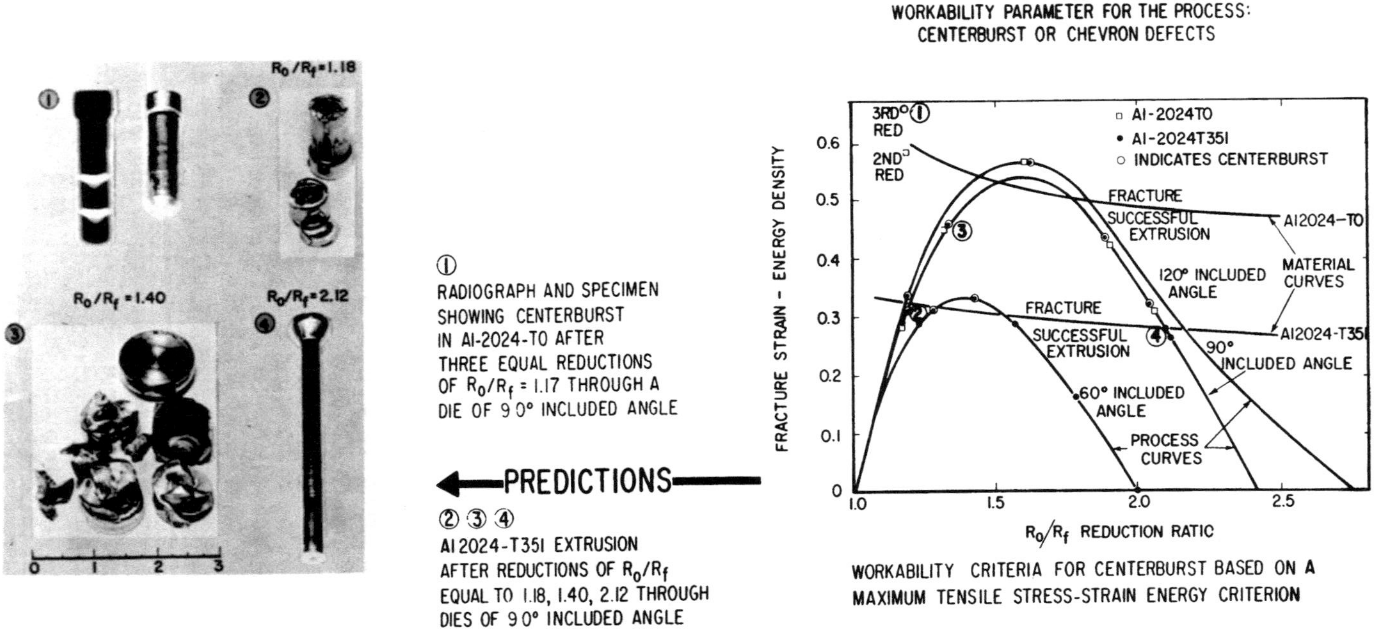

Figure 14. Workability Parameter for the Process: Centerburst or Chevron Defects

The largest reduction experienced by the four photographed specimens was achieved with number 4 which was not defective. These data show that the centerburst defect is favored by large die angles, small reductions per pass and low fracture strain of the work material.

DISCUSSION OF RESULTS

The application of plasticity theory requires a formalism, similar to other fields of science, in which the state or condition of the system and subsystems being analyzed are properly described. This formalism was followed by first defining the flow stress behavior of the materials used in this study. Since it is well known that stress and structure interact to determine the fracture strain, a significant portion of this work was devoted to determining this interaction and in describing the fracture strain anisotropy. The results demonstrated that the alloys behave macroscopically as a continua. However, the effects of structure can confound the results by providing fracture along surfaces which cannot produce separation of the specimens. This observation during torsion testing resulted from the combined effects of structure and specimen design whereby fracture occurred first along directions of microstructural alignment inclined at an acute angle to the gauge section. The torsion test specimens were sectioned from wrought bar with the specimen axis parallel to the rolling direction; hence, parallel to the direction of microstructural alignment. The planes of maximum shear strain during torsion testing are perpendicular to the gauge section and lie perpendicular and parallel to the axial direction. Therefore, the initial state of stress is pure shear parallel and perpendicular to the principal direction of mechanical texturing, the axial direction. This direction will rotate during testing, thereby complicating interpretation of the results. However, materials exhibiting fracture strain anisotropy can undergo a significant amount of deformation with a large shear component along the direction of microstructural alignment. This condition can produce failure along the directions of microstructural alignment inclined to the axis of the specimen(11). This observation indicates that the transverse fracture strain, i.e., the strain normal to the direction of particle alignment might control or, at least, affect fracture during torsion testing. Although approximate proportionalities between the tensile longitudinal fracture strain $\bar{\varepsilon}_L$ and the fracture strain in pure trosion $\bar{\varepsilon}_{TOR}$ have been found to have the form $\bar{\varepsilon}_{TOR} = (1.2 \pm 0.2)\bar{\varepsilon}_L$ (10,12), the results of this study indicate that the transverse tensile fracture strain $\bar{\varepsilon}_T$ is more appropriate for mechanically textured alloys of medium ductility. The validity of this relation is demonstrated by the following results:

$\bar{\varepsilon}_L$	$\bar{\varepsilon}_T$	$\bar{\varepsilon}_L/\bar{\varepsilon}_T$	$\bar{\varepsilon}_{TOR}$	$\bar{\varepsilon}_{TOR}/\bar{\varepsilon}_L$	$\bar{\varepsilon}_{TOR}/\bar{\varepsilon}_T$
AL-2024T0					
0.55	0.31	1.77	0.65	1.18	2.11
AL-2024T351					
0.37	0.146	2.53	0.32	0.86	2.20
Drill Rod					
0.82	0.51	1.60	1.11	1.35	2.17

These results show that the fracture strain in torsion for the three mechanically textured alloys tested at room temperature correlates best with the transverse fracture strain in tension. This deduction is confirmed by results in Figure 15 and Ref. 11 showing initial fracture and crack linkage along the rotated direction of texture alignment.

The effect of rotation of the microstructure on fracture strain during torsion is one of several factors which complicates interpretation of these test results. Other significant factors are gauge section geometry, work hardening and crystallographic texturing. The results in Figure 5 demonstrated the combined effects of gauge section geometry and workhardening which produced a shear strain concentration at the extremeties of the gauge section. This behavior has resulted in erroneous measurements when the gross specimen rotation was used to determine strain. However, for alloys which do not workharden, a condition typical of the hot working temperature range for most alloys, fracture occurs randomly throughout the cylindrical gauge section. Therefore, workhardening is a significant consideration in experimental design for the torsion test. Even in the absence of workhardening, the specimen design, e.g., gauge section thickness, can affect the flow stress and fracture strain measurements as shown by the results in Figures 4, 6, 7 and Table III. The results for the hollow and solid Ti-6Al-4V specimens differ significantly. The hollow specimens exhibited higher average flow stresses and lower fracture strains and, in general, appear to follow the behavior of the transverse tensile specimens. Because of the large fracture strains for the elevated temperature tests, strain during torsion testing of the hollow and solid torsion specimens was determined by the gross rotation across the gauge section, not by grid line displacements as was performed in the room temperature tests. Therefore, fracture in these tests was determined by complete separation. However, complete separation requires more rotations as the gauge section thickness increases as shown by the photo-

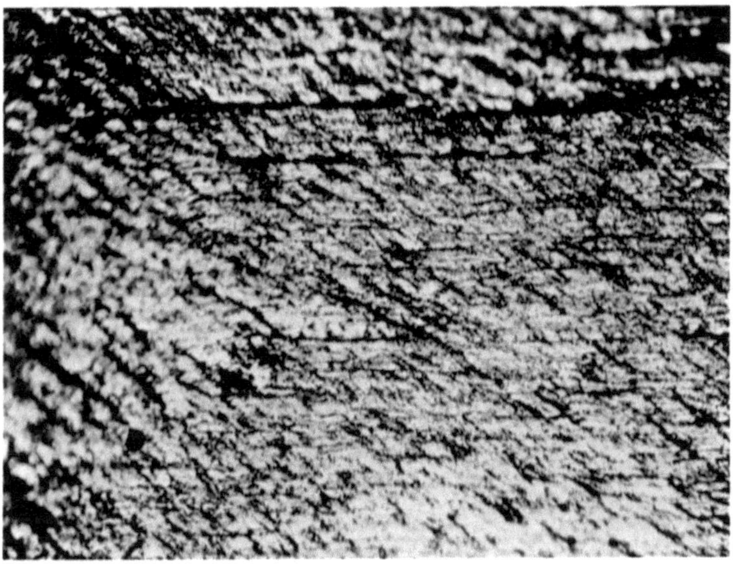

Figure 15. Center of Region of the Radiused Gauge Section of Drill Rod Specimen R6TTG1 Showing Fracture Along Periphery of the Plane of Minimum Section Thickness (Horizontal) and Crack Linkage Along Deformed Direction of Particle Alignment (Diagonally Oriented from lower right to upper left.) 75X

graphs in Figure 7. Therefore, the validity of the test results must be established by determining the gauge section thicknesses for which the fracture strain is independent of thickness.

The flow stress differences between the results for the two gauge section thicknesses is not understood. It may be associated with the strain rate sensitivity of this alloy which is superplastic in the hot working range and the spread of the actual strain rates about the average. For the solid specimen this spread is very large which could result in the outer portion of the gauge section sustaining most of the torque, whereas the inner portion would sustain little torque due to a superplastic response.

The torsion test is unique because the structure is rotated relative to the principal stress directions. For most other tests and during deformation processing, the structure rotation is not as severe as in torsion. However, the magnitudes of the principal stress components in other tests may vary during deformation as was shown for compression testing in Figure 8. McClintock [3] postulated that the stress components relative to the structure control the initiation of fracture. The simplest demonstration of the interaction of stress and structure should be obtained under conditions where both these factors could be set at various prescribed levels and maintained constant during a test. This procedure was followed utilizing concepts first proposed by McClintock which required definition of the average normal stress component σ_T perpendicular to the direction of maximum extension or mechanical texturing. A relation of the form $\ln\bar{\varepsilon}_f = A + B(\sigma_T/\bar{\sigma})_{avg.}$ was found for longitudinal and transverse tensile specimens for all gauge section geometries and for the plane stress tests. This relation is demonstrated by the results in Figure 16 in which the range of σ_T associated with the respective specimen geometries are schematically shown. Although the range of $\sigma_T/\bar{\sigma}$ for torsion and compression tests are shown on this figure these data have been omitted. The problems with the torsion test results were previously discussed. The difficulty with analyzing the compression test results arose from the large variation of σ_T during a test. The fracture strain was either larger or smaller than anticipated from the average value of σ_T depending upon whether σ_T became more compressive or more tensile, respectively, with increasing deformation. In other words, an increasing compressive state of stress appeared to have a healing effect on fracture initiation. The results in Figure 16 demonstrated that the longitudinal and transverse fracture strains of a particular alloy are not independent quantities. In general, the quantity A describes the effects of strain rate, temperature and heat treatments which do not significantly change the microstructure. Therefore, A determines the position of the line in Figure 16, whereas B describes the effects of mechanical texturing and does not change unless this structure is modified by solution or recrystallization heat treatments.

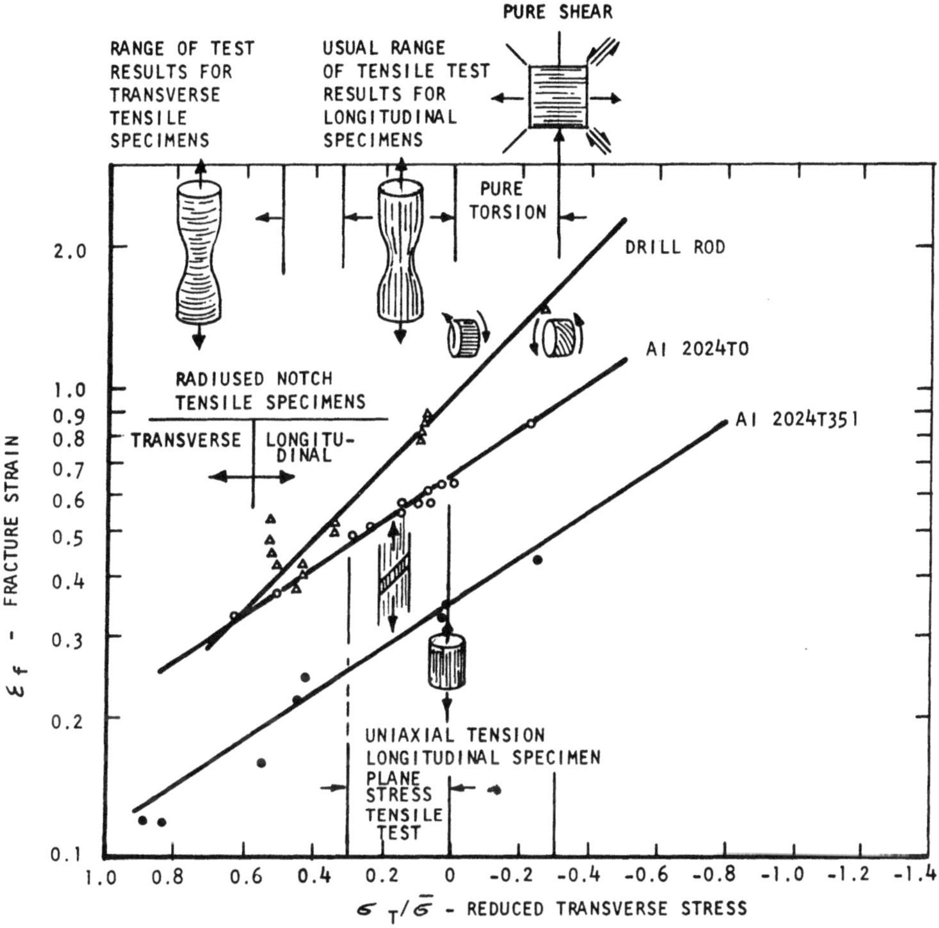

Figure 16. Fracture Strain Dependence on the Reduced Stress $\sigma_T/\bar{\sigma}$ with Superposition of the Ranges of $\sigma_T/\bar{\sigma}$ Corresponding to Each Test

The workability test results in Figure 16 were used to predict fracture or limiting reductions during rolling and extrusion. The rolling results in Figures 12 and 13 were used to determine the stress and strain behavior caused by the different specimen geometries and to use this information to predict edge cracking. For all three edge geometries, a relation of the form $\varepsilon_2(-\varepsilon_3)^{-n} = K$ described the strain behavior along the center of the edge. The quantities n and K in this relation were dependent primarily on the edge contour, not the material, for tests performed at room temperature. For square edge strip n and K were nearly 1, for the radius edge strip n = 0.94 and K = 1.94 and for the chamfered edge strip n = 0.72 and K = 1.65. In general, n is proportional to the radius of curvature of the strip edge.

The general relation between ε_2 and ε_3 for a particular edge configuration permitted graphical determination of the relation between $\bar{\varepsilon}$ and α and, therefore, between $\bar{\varepsilon}$ and $\sigma_T/\bar{\sigma}$. The predicted fracture strain $\bar{\varepsilon}_f$ was determined as the value $\bar{\varepsilon}_f$ for the appropriate alloy where

$$(\bar{\varepsilon}_f)^{-1} \int_0^{\bar{\varepsilon}_f} (\sigma_T/\bar{\sigma})\, d\bar{\varepsilon}$$

or the average value of $\sigma_T/\bar{\sigma}$ over the strain $\bar{\varepsilon}_f$ first exceeded the plot of $\ln\bar{\varepsilon}_f$ versus $\sigma_T/\bar{\sigma}$ in Figure 16. The agreement of the fracture strain predictions based on workability test results with the actual observations is shown in Table IV. The agreement is excellent for square- and radiused-edge strip but becomes worse for chamfered-edge strip due to the variations of α during rolling of strip with this edge configuration. Because the state of stress becomes less tensile with reduction due to blunting of the chamfered edge, the predicted values of fracture strain for the chamfered-edge strip are larger than the observed.

The use of the workability test results for predicting center-burst defects during extrusion was based on the empirical procedure demonstrated by Cockcroft and Latham (13). This procedure involves a maximum tensile stress σ^* strain energy criterion in which it is proposed that an alloy fractures when it has achieved a strain energy density equal to the integral

$$\int_0^{\varepsilon_f} \sigma^* d\bar{\varepsilon} = C_i,$$

where C_i is a material constant for a particular strain rate and temperature. Therefore, C_i can be determined from independent test data such as the tensile test using Bridgman's analysis (8). The value for the strain energy density in Figure 14 is $C_i/\bar{\sigma}_{avg.}$, where $\bar{\sigma}_{avg.}$ is the average flow stress over the strain interval to fracture during a test or the reduction during extrusion. Since the maximum tensile stress during extrusion occurs along the centerline where the deformation is homogeneous, the strain during extrusion $\bar{\varepsilon} = 2\ln(R_o/R_f)$. Cockcroft and Latham assumed this fracture criterion and the possibility that plane strain slip-line field results for extrusion would describe the state of stress for axisymmetric extrusion through dies of 120° included angle when the two were compared on the basis of equal cross sections (i.e., if the initial and final plate thicknesses equaled the respective diameters). The feasibility for such a correlation was first demonstrated by Thomsen (14) based on comparisons of experimentally determined flow fields for axisymmetric extrusions and the plane strain predictions. Because of the few analytical and experimental results for stress and strain during axisymmetric extrusion, the slip-line field results provided a means of at least acquiring approximate values for the centerline stress.

The results in Figure 14 present an extension of the Cockcroft and Latham procedure to dies of 60° and 90° included angles, as well as results for 120° dies. The purpose of these experiments was to determine if the hypotheses of Cockcroft and Latham could be extended to die angles other than 120°. The results in Figure 14 definitely support their hypotheses. The parabolic curves in this figure are the calculated strain energy densities divided by the flow stress at strains corresponding to the frictionless extrusion reductions for 60°, 90° and 120° dies. Therefore, these curves correspond to the workability requirements of the process. The assumption of frictionless extrusion was made for ease of analysis. In reality, lubricated extrusion was performed with coefficients of friction in the range of 0.04 to 0.07 (friction factors in the range of 0.1 to 0.2) as determined by the ring test (15). The frictionless assumption is not expected to differ significantly from the experimental conditions. The nearly horizontal curves are the strain energy densities divided by the flow stress for the aluminum alloy 2024 in the T0 and T351 conditions and, therefore, represent the material's strain capability or workability. Results for drill rod are not shown because this alloy required multiple passes for defect formation as predicted by the analysis. These curves were determined by first obtaining the strain dependence of the maximum tensile stress and the flow stress during tensile testing by using the Bridgman analysis (8). The results for the maximum tensile stress, the centerline stress during tensile testing, provided the quantity C_i, the strain energy density. The strain dependence of the flow stress $\bar{\sigma}$ was used to normalize the strain energy density, i.e., $C_i/\bar{\sigma}$, to present the results in a more compact form. The slight curvature of the material curves is due to the strain dependence of $C_i/\bar{\sigma}$ arising from $\bar{\sigma}$. Since the process and material curves were both constructed on the basis of the strain-energy density divided by the flow stress, centerbursting should occur when the process curves exceed or require a higher strain energy than the material curves. The open squares and solid circles represent data points for T0 and T351 specimens respectively. The circled points correspond to the observation of centerburst. The maximum departure of the predictions from the experimental results in Figure 14 was obtained for point number 2 and corresponds to 7 percent, which is approximately the magnitude of the variation about the average value of the strain energy density for these specimens. The results in Figure 14 demonstrate the general observation that centerburst is favored by large die angles and light reductions particularly in multiple pass processes.

The success of the Cockcroft and Latham procedure for predicting centerburst is clearly demonstrated by the results in Figure 14. Attempts to use the workability results in Figure 16 were only successful at small reduction ratios R_o/R_f, less than about 1.4. At the larger reductions, the predicted occurrence of centerbursts was more prevalent than the experimental observations. These observations were surprising in view of the good correlations of predicted and observed results for test specimens and rolling with the relation $\ln\bar{\varepsilon}_f = A+B(\sigma_T/\bar{\sigma})$ where accurate stress determinations could be performed. Because these calculations required few assumptions with the exception of the validity of the slip-line results, the approximation procedure of Cockcroft and Latham was investigated to determine if any discrepancies existed in their procedure. These investigations included evaluation of their maximum tensile stress-strain energy criterion, and the assumption of the maximum centerline tensile stress and resulting strain occurring simultaneously and instantly as predicted from the slip-line field results for an ideally plastic material (i.e., no strain hardening or strain rate sensitivity).

The fracture criterion proposed by Cockcroft and Latham is questionable because anisotropy, which in general does exist, is not implied in the analysis. Although many of the results in this paper confirm the findings of Cockcroft and Latham, when their proposed criterion is exposed to states of stress other than uniaxial tension and pure torsion, it is no longer obeyed. For example, the following energy densities were obtained for longitudinally sectioned specimens of drill rod:

	C_i
Smooth-round tensile specimen	99,700 psi
Radiused notch (a/R = 0.5) tensile spec.	71,000 psi
Radiused notch (a/R = 1.0) tensile spec.	56,600 psi
Tubular torsion spec. (pure torsion)	99,000 psi

In general, the energy density C_i decreases as the state of stress becomes more tensile. Therefore, the success which Cockcroft and Latham observed with their fracture criterion appears fortuitous and is related to the specimen geometries which they investigated and not the general applicability of their fracture criterion.

The slip line field analyses, which are reviewed in Ref. 11, were performed for an ideally plastic material which, by definition, does not exhibit strain hardening or strain rate sensitivity. With this assumption, the slip-line field touches the centerline of the extrusion at a point which requires that material along this centerline is instantly deformed at the strain $\bar{\varepsilon} = 2\ln(R_o/R_f)$ under the calculated stress. This condition requires an infinite strain rate which is impossible for a real material. In reality, the stress and strain would be spread over a finite volume. Visioplasticity results(6) for axisymmetric extrusion of lead through a die of 90° included angle were reviewed to determine the following approximate percentages of the total centerline strain which occurred at the peak stress:

R_o/R_f	Reduction Ratio	Percent Strain at Peak Stress
1.41	2/1	80
1.73	3/1	60
2.0	4/1	60

The remaining fractions of the total strain did not occur at the peak stress, but under a state of axial stress equal to approximately zero with the radial and circumferential components nearly equal to the flow stress in pure compression. Therefore, the best agreement, in terms of stress-strain behavior, between the plane strain and axisymmetric results appears to occur at low reductions. These observations are in agreement with the calculations based on the results in Figure 16.

The effect of friction in causing the disparity between the predictions based on the Cockcroft and Latham criterion and the results in Figure 16 appears comparatively insignificant in view of the validity of their criterion and the assumption of instantaneous occurrence of the centerline strain and peak stress. The results in Figure 17 are presented to demonstrate the range of die angles and reductions (R_o/R_f) necessary for centerburst formation based on the Cockcroft and Latham analysis and Avitzur's(16) upper bound analysis based on the assumption of a spherical velocity field for extrusion. Avitzur's results demonstrate the effect of friction on the conditions necessary for defect formation. The slip-line field results correspond to zero axial centerline stress and the upper bound curves correspond to the reduction-die angle-shear factor (m) combinations where the completely spherical flow field requires the same deformation power as the spherical field with a cylindrical cavity. Processing conditions corresponding to reductions and die angles in region below

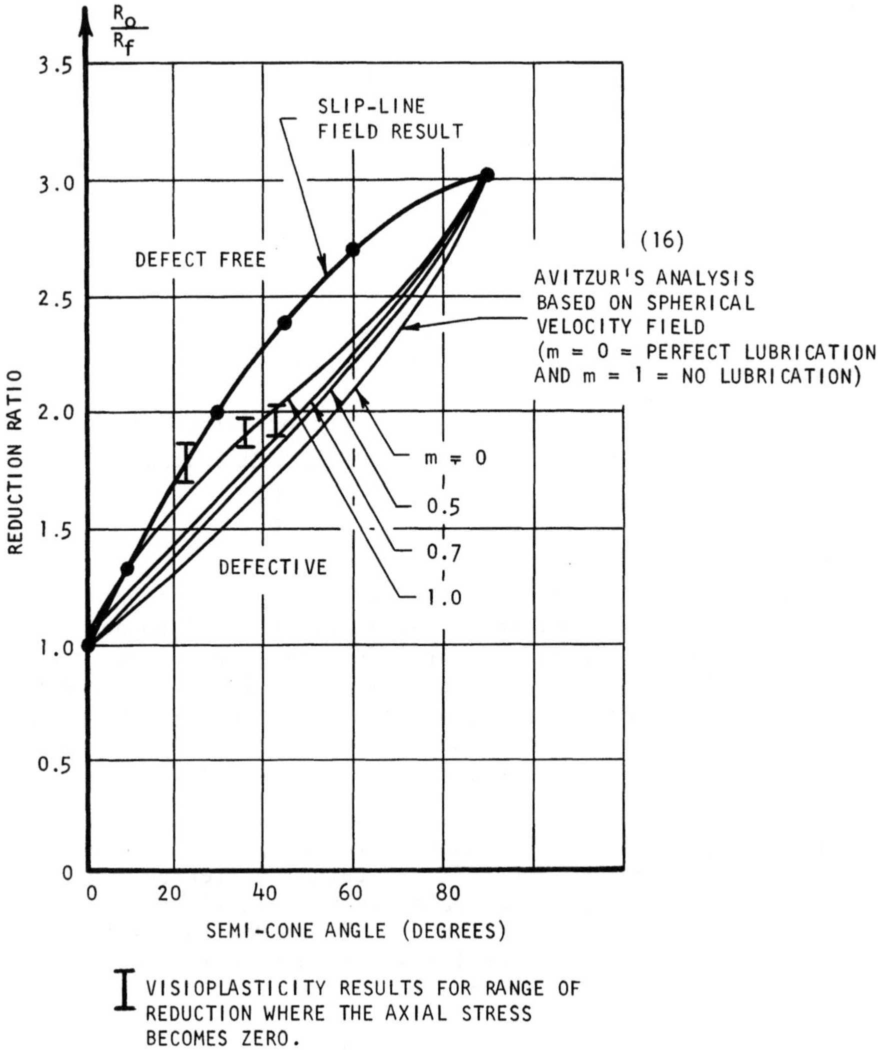

Figure 17. Reduction Ratio (Initial/Final Radius Ratio) as a Function of Die Semi-Angle Defining Range for Centerburst or Chevron Formation.

the curves are predicted to produce centerbursts. However, the results in this figure cannot be used to specify the exact number of passes to produce this defect. Visioplasticity (6) calculations are superimposed on these curves to show the range of reductions where the axial centerline stress is nearly zero for dies of 50°, 70°, and 90° included angle. The visioplasticity results which were performed with m equal to approximately 0.1, show good agreement with the slip-line field results at low die angles and small reductions. This behavior would be anticipated based on the previous observations that the fraction of the centerline strain occurring at the peak stress increases as the reduction decreases. Furthermore, at light reductions the centerline stress approaches uniaxial tension, the basis for the energy calculation in the Cockcroft and Latham analysis. At larger reductions, where the state of stress is less tensile, the slip-line analyses provide a greater range of conditions for defective extrusions than anticipated from either the visioplasticity or upper bound results. However, as the state of stress becomes less tensile,the true energy density for failure increases as was demonstrated by the drill rod results. Therefore, the slip-line analysis should provide very conservative predictions for the region of defect formation. This behavior would be anticipated from the results in Figure 17. However, an Al-2024T-351 specimen (see Figure 14) extruded through dies of 90° included angle at a 2.06 reduction (R_o/R_f) was predicted and extruded defective. This extrusion should have been defect free based on the visioplasticity and the upper bound results (for all m values).

It must be concluded that in view of the presently available data the Cockcroft and Latham analysis provides an accurate working procedure for defect prediction during extrusion. Although the method is empirical, it is significant that accurate predictions can be made based on laboratory test data. However, it must be recognized that the reason why this procedure does work is not understood.

CONCLUSIONS

The results presented in this paper contain one of a very few attempts to quantitatively relate basic material performance or workability data and observations during deformation processing. The relation between theory and practice was demonstrated to depend strongly on knowing strain, stress and the orientation of stress to the major features of the microstructure. Complications in analyzing torsion test data were shown to arise from the rotation of the microstructure relative to the principal stress directions, and the interaction of specimen geometry and work hardening which could produce unanticipated strain gradients and erroneous fracture strain measurements if conventional test procedures were used. For tests in which the orientation of the principal stress directions relative to the major microstructural features of the alloy remained nearly constant, a fracture criterion of the form $\ln\bar{\varepsilon}_f = A+B(\sigma_T/\bar{\sigma})$ was found to be obeyed. The material constant A was found to describe the effects of temperature and strain rate on the fracture strain independent of the alignment of microstructural features, whereas B was related to the fracture strain dependence on stress and isotropy through the mechanical texturing of the microstructure. The quantity B was most strongly affected by heat treatments or deformation conditions which altered the degree of mechanical texturing or microstructural alignment.

The use of workability test results with analyses of the processing conditions were demonstrated for rolling and extrusion. Although accurate predictions were made in both cases, the procedures for the centerburst calculations during extrusion are not well understood.

The results of this study demonstrated that accurate predictions of processing performance can be made if the process mechanics can be defined. Inhomogeneous structures such as alloys exhibit continuum behavior up to the point of macroscopic fracture, which usually determines the limiting reduction in most metalworking processes. Deformation damage which appears to be a prelude to fracture, does affect subsequent service performance of a wrought product. Quantitative prediction of the inception and extent of damage has not been attempted, although it appears to be proportional to the factors affecting the fracture strain.

ACKNOWLEDGEMENTS

This work was sponsored by the Materials Processing Branch, Manufacturing Technology Division of the Air Force Materials Laboratory. This sponsorship and the permission for publication are greatfully acknowledged. Helpful review and direction of this program was provided by Mr. V. DePierre and Mr. W. T. O'Hara of the Air Force Materials Laboratory, Mr. C. S. Cook of the Westinghouse Research Laboratories, formerly of AFML, Professor S. Kobayashi of the University of California and Professor G. E. Dieter of Drexel Institute of Technology.

REFERENCES

1. R. Hill, "A Theory of Yielding and Plastic Flow of Anisotropic Metals," Proc. Roy. Soc. London, Ser.A, 193, (1948), 281.

2. H. C. Rogers, "The Effect of Material Variables on Ductility," Ductility, ASM, (1968), 31.

3. F. A. McClintock, "On the Mechanics of Fracture From Inclusions," Ductility, ASM, (1968), 255.

4. R. Hill, "New Method for Determining the Yield Criterion and Plastic Potential of Ductile Metals," J. Mech. and Phys. of Solids, 1, (1953),271.

5. J. P. Ellington, "An Investigation of Plastic Stress-Strain Relationships Using Grooved Tensile Specimens," J. Mech. Phys. Solids, 6, (1958), 276.

6. A. H. Shabiak and E. G. Thomsen, "Investigation of the Application of Visioplasticity Methods of Analysis to Metal Deformation Processes," Final Report - Part II prepared on Contract No. N000019-67-C-0509, University of California, (January 1968).

7. F. R. Larson and J. Nunes, "The Low Temperature Plastic Flow and Fracture Tension Properties of Heat Treated SAE 4340 Steel," TASM, 53(1961)663.

8. P. W. Bridgman, Studies in Large Plastic Flow and Fracture, McGraw-Hill, (1952).

9. D. Lee and W. A. Backofen, "Superplasticity in Some Titanium and Zirconium Alloys," Trans.AIME, 239, (1967)1034.

10. C. M. Young and O. D. Sherby, "Simulation of Hot Forming Operations by Means of Torsion Testing," Technical Report AFML-TR-69-294 (Feb. 1970).

11. A. L. Hoffmanner, "Workability Testing Techniques," Final Engineering Report on Contract No. F33615-67-C-1466, (June 1969).

12. J. L. Robbins, H. Wagenaar, O. C. Shepard, and O. D. Sherby, "Torsion Testing as a Means of Assessing Ductility at High Temperatures," J. of Materials,(June 1967), 271.

13. M. G. Cockcroft and D. J. Latham, "Ductility and Workability of Metals," J. Institute of Metals, 96, (1968),33.

14. E. G. Thomsen, "Comparison of Slip-Line Solutions with Experiment," TASME, J. Appl.Mech., 23(June 1956)225.

15. A. T. Male and M. G. Cockcroft, "A Method for the Determination of the Coefficient of Friction of Metals under Conditions of Bulk Plastic Deformation," J. Inst. of Metals, 93, (1964-65) 38.

16. B. Avitzur, "Analysis of Central Bursting Defects in Extrusion and Wire Drawing," ASME Paper No. 67-Proc.-5,(1967).

HOT WORKABILITY TESTING TECHNIQUES

H.J. McQueen and J.J. Jonas

Mechanical Engineering, Sir George Williams University
and Metallurgical Engineering, McGill University
Montreal, Canada

ABSTRACT

Processes for the hot forming of metals are of great industrial importance because of the low flow stresses and high ductilities exhibited by most metals at homologous temperatures greater than one half. High rates of working which are desirable for reasons of economy have, however, the effect of increasing the flow stress and, in some circumstances, of decreasing the ductility. A critical comparison is made of the usefulness of a variety of high temperature mechanical tests to determine suitable conditions for working. Examples are presented of successful correlations of laboratory testing and industrial production. The important testing parameters and the tolerances in their control are reviewed.

The variations in workability from alloy to alloy and from heat to heat are the result of differences in the mechanisms of deformation, softening and fracture. The examination techniques which disclose the complex microstructural changes associated with these mechanisms are surveyed, and the hot work tests are compared with regard to their ability to provide suitable samples for metallography.

The microstructures produced by hot working and subsequent cooling determine the properties and, therefore, the suitability of the material for cold forming or for service. Since the possibility of determining these properties depends largely on whether the hot workability specimens are in suitable form for further mechanical testing, the capabilities in this respect of the different experimental methods are assessed.

INTRODUCTION

Hot working enters significantly into the manufacture of more than eighty-five percent of all metal products. In addition to changing the shape, it plays an important role in improving the structure for further forming operations and for service. In consideration of its critical role, it is essential that it be carried out as economically and effectively as possible. The ease of working, called the workability, is measured in terms of the power consumed and the rate and size of the possible reductions. These in turn are related to the flow parameters (temperature, strain rate, flow stress, and strain), to the initial macrostructure, and to the developed microstructure (grain size and shape, precipitate distribution, substructure and preferred orientation)[1-4]. The workability is also judged by the quality and properties of the product, which are dependent on the final microstructure.

Workability can be determined directly and reliably by deforming the material on standard production equipment. This is not usually economical, however, and several laboratory methods[5-14] have been developed which permit the simulation of industrial processes and the selection of suitable working conditions. The present paper is concerned with the capabilities of various hot working tests for a) measuring the flow parameters, b) determining the ductility, c) studying the microstructural changes taking place during and after deformation, and d) providing for the determination of room temperature properties. The following modes of laboratory testing will be analyzed: tension [15-35], compression [6 7 14 36-74], torsion [1 22 24 35 75-121], rolling [40 122-33] and extrusion [40 124 133-44]. Before the different tests are compared, the various measurements of interest will be discussed.

WORKING LOAD and FLOW STRESS

The design of new equipment, or the extension of existing equipment to new operations, requires calculation of the forces generated and the power consumed. Such calculations can be made by means of the mathematical techniques of plasticity theory, for which an estimate of the mean values of the strain, strain rate and temperature is required [35 140 143-7]. Minimization of the inaccuracies inherent in the calculations is possible only if precise information about the dependence of the flow stress on the mean flow parameters is available from hot working tests. It should be noted that plasticity calculations generally employ the true stress, that is the total force divided by the instantaneous area [146]. The necessity for this measure can be seen in simple upsetting, where the load rises rapidly as the section area and flow stress both increase.

At constant temperature and strain rate, the flow stress

increases with strain up to strains of 0.2 - 1.0 (higher values for lower temperatures or higher strain rates) and then remains approximately constant as the strain is further increased (Fig.1)[1 3 22 96 106 108]. Thus, in hot working tests, a steady state regime is attained, in which the strain rate, temperature and flow stress are constant and independent of strain. In some materials, the steady state is preceded by a maximum in the flow stress, which is usually designated as the peak flow stress[1-4 22 38 106 108 118].

In most industrial forming operations, the strains are less than the steady state strains determined in laboratory tests. It is therefore important to establish the full stress-strain curve if at all possible. Once a set of curves is determined, an algebraic equation can be fitted to them, most conveniently by the use of a computer. Some examples of equations quoted in the literature are:

$$\sigma = A\varepsilon^m \qquad (1)$$ [60 64]

$$\sigma = \sigma_0 + B\varepsilon^m \qquad (2)$$ [60 95]

$$\sigma = \sigma_0 + C[1 - \exp(-D\varepsilon)]^m \qquad (3)$$ [110]

where σ is the flow stress, σ_0 the yield stress, ε the strain and A,B,C,D and m are experimental constants. In comparison with Eq. 1, Eq. 2 allows for a yield stress. Eq. 3 allows for steady state deformation at high strains and it reduces to Eq. 2 at low strains. Once the constants of the equations have been determined at a series of temperatures and strain rates, it is possible to estimate values for intermediate temperatures or strain rates and to generate the desired flow curve on the computer[4]. Whenever possible the deformation should be carried into the steady state region, for then, in cases where the industrial deformation is non-uniform in strain rate, the steady state or peak flow stress for the appropriate mean strain rate can be used as the upper limit of stress.

For strains greater than 0.2, logarithmic or true strain

$$\varepsilon = \int_{\ell_0}^{\ell} (1/\ell)\,d\ell = \ln(\ell_f/\ell_0) \qquad (4)$$

is used in plasticity calculations rather than engineering strain

$$e = 100\ (\ell - \ell_0)/\ell_0 \qquad (\%) \qquad (5)$$

because true strain has the same value for equivalent deformations in extension or compression[1 4 7-8] (here ℓ is the instantaneous length and ℓ_0 is the initial length). A true strain of 2.3 may be either a compression to 90% reduction or an extension to

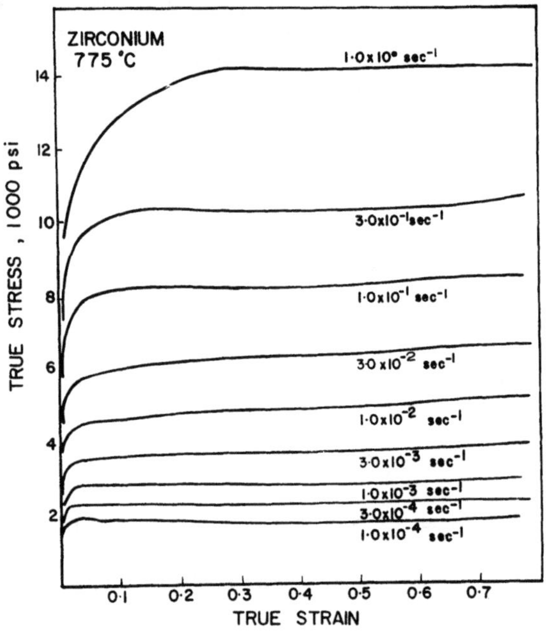

Fig. 1a Influence of strain rate on the stress-strain curves derived from hot compression data for sponge zirconium at 775°C. [73]

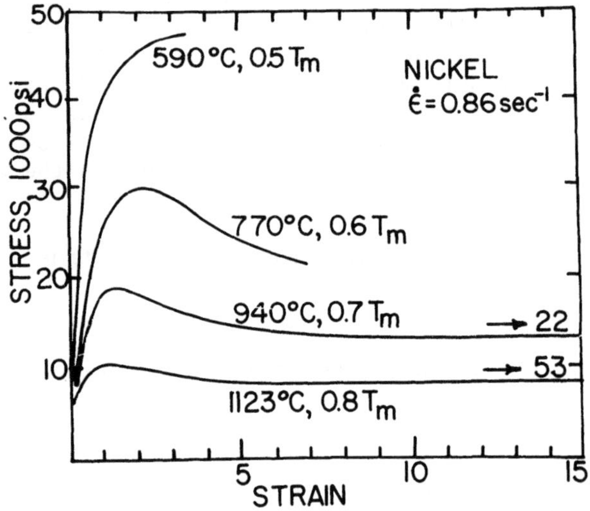

Fig. 1b Influence of temperature on the stress-strain curves derived from hot torsion data for nickel of commercial purity at a strain rate of 0.86 sec^{-1} [105].

1,000% elongation; in either case, the increase in flow stress from strain hardening and the work expended in deformation is the same. Confusion can be avoided by reporting engineering strain in percent and logarithmic strain without units. In this paper, strain and strain rate refer to true or logarithmic values.

Initial Structure

The concentration and segregation of dissolved impurities and the distribution, size and strength of second phase particles have a strong influence on the flow stress, regardless of the strain [7-9 46 53 69 98-101]. The grain size and shape and previous deformation affect the high temperature yield strength considerably, but they usually have less effect as the strain is increased toward the steady state value [25 92 104 116]. For satisfactory analysis, the microstructure actually present at the start of a test should be determined. Because of the size difference, it is not usually possible to simulate in test specimens the macrostructures found in production workpieces, in particular those produced by casting.

In multipass operations, the structure at the start of each stage subsequent to the first is the result of the previous deformation and of the structural modifications that have taken place in the interval between the deformations[4 8 37 67 91]. The temperature and elapsed time between passes determine the extent of recovery, recrystallization, precipitation or homogenization which take place [12 67 116 132]. Multipass deformations can be studied in the laboratory by means of programmed tests in which temperature, strain rate, and delay intervals can be controlled and varied [18 36 67 118].

Strain Rate Control

The flow stress at a given strain or in steady state deformation varies greatly with the strain rate and the temperature of deformation [1-4]. This behaviour is in strong contrast to the weak dependence observed under conditions of cold working [148]. It is therefore very important to provide, throughout a test, a constant true strain rate

$$\dot{\varepsilon} = (1/\ell)(d\ell/dt) \qquad (\mathrm{sec}^{-1}) \qquad (6)$$

which requires a crosshead speed which varies as follows:

$$d\ell/dt = \mathrm{constant}\cdot\ell \qquad (7)$$

For a given material at a given temperature and true strain rate, a steady state of deformation prevails in tension, compression and torsion with approximately the same flow stress [22 84]. At constant engineering strain rate, on the other hand, the flow stress drops in tension (true strain rate decreasing), rises in compression (true strain rate increasing), and remains steady in torsion (true and engineering strain rates equal). Moreover, the use of engineering strain rate

$$\dot{e} = (100/\ell_0)(\Delta\ell/\Delta t) \quad (\%/sec) \qquad (8)$$

leads to the anomaly that, during the simultaneous compression between the same anvils of two blocks, of which one was originally one inch high and the other a half inch high, the engineering strain rate in the former is half that in the latter.

In most hot-working operations, the strain rates of interest extend from 10^{-2} sec^{-1} to $10^3 sec^{-1}$. In order to measure the flow stress at strain rates above 10^{-1} sec^{-1}, it is necessary to use a high speed galvanometric recorder, an oscilloscope or a magnetic-tape data storage system [73]. Furthermore, as the strain rate increases, it becomes more difficult to maintain the actual strain rate equal to the programmed strain rate. During the initial part of a test, the strain rate is likely to be too low because of the difficulty of accelerating the specimen and the loading train [149]. This can be overcome to some extent by the use of a high inertia loading system running at the correct speed and a suitable engagement device. A further problem is that, while the specimen is work hardening, the frame is also deforming elastically, thus lowering the strain rate. Once the maximum force is passed, the strain rate may be higher than expected, due to elastic unloading of the frame [149].

In most working operations, the strain rate is not constant but increases to a maximum and then diminishes to zero. A profile of strain rate against strain or time can be constructed for each operation (Fig. 2) [86 143-6]. At the strain for maximum strain rate, the flow stress is less than that measured in a test at a constant strain rate equal to the maximum. On the other hand, during the decreasing strain rate portion, the flow stress is higher than for the applicable constant strain rate test. These differences arise because of the deformation structures which are inherited from the immediately preceding strains at lower or higher strain rates, respectively [4 54 72 116]. Because of these factors, the true mean flow stress and power may not equal the flow stress which is calculated from constant strain rate test data. The ideal way to determine the power and the maximum flow stress is to use a test programmed to simulate the strain rate profile of the working process [86].

Temperature Control

The provision of a constant temperature is hindered by two factors which tend to counteract each other. On the one hand, the work of deformation is transformed into heat which, unless it can

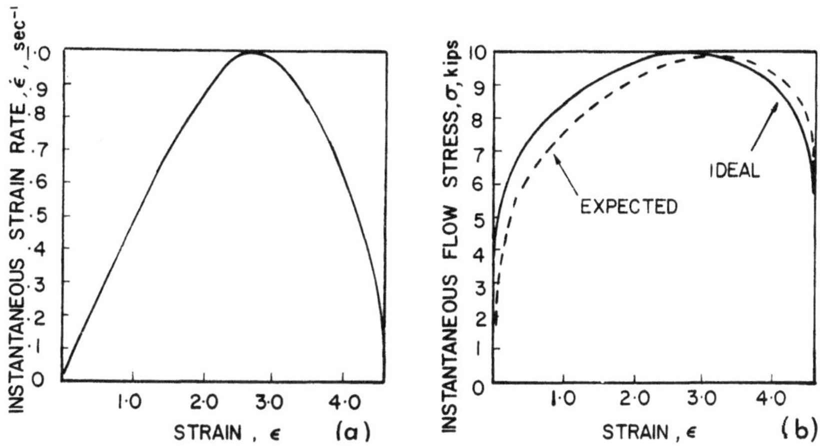

Fig. 2 a) Strain rate profile as a function of strain for a 100:1 extrusion die and a ram speed of 0.36 ipm (143). b) Flow stress profiles for material flowing through the strain rate profile of a). The full line is the ideal profile based on constant strain rate, steady state data for α-Zr at 850°C. (73). The broken line is the expected profile after the deformation history is taken into account. The area under the broken curve is the actual work done, and is greater or less than the ideal area depending on whether the peak strain rate is attained early or late along the profile.

be dissipated, raises the temperature of the work piece[54 78 95 141]. This effect becomes appreciable at strain rates above 1 sec^{-1} and can result in an increase of about 35°C. in steel deformed at 10^2 sec^{-1} and 1155°C., to a strain of 0.5 [47]. Temperature rise due to adiabatic heating is promoted by large specimens and by deformation within a furnace. On the other hand, contact with relatively cool loading tools lowers the temperature, especially for small specimens [35 130]. The temperature should always be measured throughout the test, at least in trial runs, preferably by embedded thermocouples [132]; unfortunately in many investigations, only the temperature at the start of the test is reported. It should be added that, as many metals exhibit roughly similar behaviour at the same fraction of their absolute melting temperatures, it is frequently convenient to employ the homologous temperature scale,i.e., fractions of the absolute melting temperature. Alloys are generally referred to the melting point of the base metal. The hot working temperature range lies above approximately half the melting temperature.

In a real working operation in which the temperature diminishes with strain or time, the flow stress will clearly be higher at a given strain than if the temperature had remained constant. However, during cooling, when a given strain is reached, the flow stress is less than that measured in an isothermal test at the lower temperature because of the softer deformation structure inherited from the previously higher temperatures [4]. The best way to obtain precise data for such operations is by programmed tests [18].

Mathematical Correlation of Flow Parameters

In most metals, for either the steady state or a selected strain, use has been made of several mathematical expressions relating the stress, strain rate and temperature over wide ranges of the latter variables [1-4]. The most generally applicable equation has been found to be [107]

$$\dot{\varepsilon} = A[\sinh\ (\alpha\sigma)\ ]^n\ \exp(-Q/RT) \qquad (9)$$

which can be written in the form

$$Z = \dot{\varepsilon}\ \exp\ (Q/RT) = A[\sinh(\alpha\sigma)]^n = f(\sigma) \qquad (10)$$

Here R is the gas constant and A, α, n and Q are material constants determined from the data. The most satisfactory manner for finding them is to use a computer program which finds the best fit to the test results (Fig. 3) [65 140-1]. The temperature-corrected strain rate Z is constant in a hot working test since both $\dot{\varepsilon}$ and T are held constant [2]. The plot of log Z against sinh $(\alpha\sigma)$ in Fig. 3 permits the flow stress to be found for any temperature and strain rate once Z is calculated [2-4 107]. Analysis of the interdependence between the flow parameters gives considerable insight into the mechanism of deformation; e.g. the exponential temperature relationship indicates a thermally activated mechanism and the value of Q indicates the type of mechanism [2-4 22 80 139 148].

FORGEABILITY, MALLEABILITY, AND DUCTILITY

Working a material in a temperature range of high ductility is economically desirable because it permits greater reductions per pass and reduces the number of failures. One of the important uses of hot working tests is to determine how ductility varies with composition [87-8 109], grain structure, phase distribution [9 21 31 46 77-8 98-101], temperature, and strain rate, and thus to define the optimum working conditions [5 8 9 12 46 75-6 119]. However, the ductility varies with the test method, since it depends on the magnitude of the hydrostatic compression relative to the maximum tensile component. In the selection of a test which correlates with the process in question, it is therefore important to consider the stress states imposed and the conditions of friction, inhomogeneous deformation and non-uniform cooling [24 35].

A frequent cause of failure in hot forming processes is the cast structure, with its inhomogeneity, inclusion distribution,

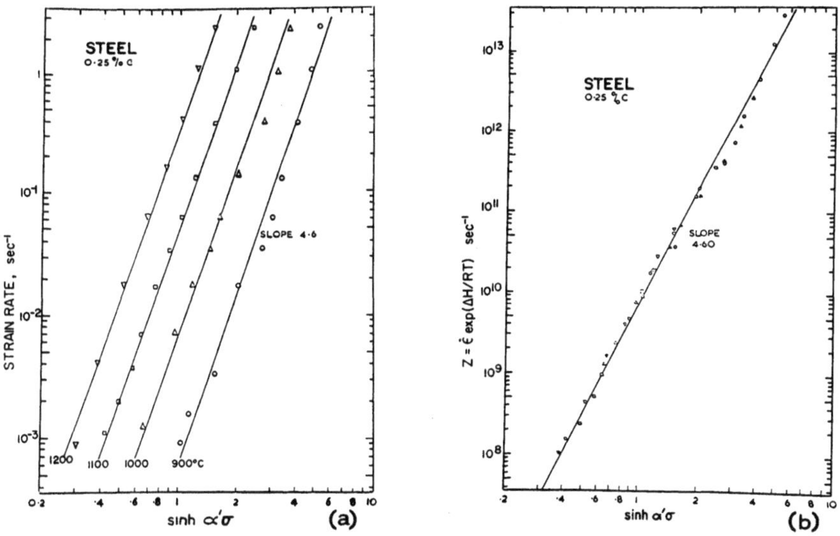

Fig. 3 a) The power relationship (Eq.9) between strain rate and a hyperbolic sine function of the flow stress observed in hot torsion experiments on medium carbon steel [22 107]. b) The data in a) replotted in terms of the temperature-corrected strain rate Z (Eq.10).

columnar grains and planes of weakness [9-14 24 30-1 147]. Unfortunately, it is very difficult either to simulate such a macrostructure in small test samples or to duplicate precisely the interaction of the stress field and the macrostructure found in the forming process [6]. Because all stages of working subsequent to breakdown are on wrought material, laboratory tests on such materials are very useful. Even in wrought material, it is important to know the structure at the start of the deformation, because changes in the grain size and the precipitate morphology during preheat considerably alter the ductility [7 10 17 24 116]. Multipass processes must be simulated by discontinuous tests, since the structural changes during the delay intervals usually cause an increase in ductility [10 12 114 132].

In certain tests the limit of malleability is defined by the appearance of the first cracks and in others by complete fracture [6 11 12]. The former is more rigorous and is therefore preferable for specifying the limits of industrial forming operations in which the goal is a completely sound product. Specimens should be subjected to visual, non-destructive and microscopic inspection. Metallographic examination of etched sections also gives useful information regarding the cause and mechanism of cracking [31 49 109 114 118 119].

HOT-WORK MICROSTRUCTURES

The microstructure resulting from hot working depends mainly on the composition, the temperature and the strain rate. In aluminum the inverse subgrain size and in copper the inverse recrystallized grain size increase linearly with log Z, the temperature corrected strain rate [3 12 50 61 70 107 120]. Examination at a series of different strains shows that the microstructure gradually changes during initial straining, but remains stable throughout the steady state [3 22 72 91]. Such observations of the microstructural evolution give valuable insight into the mechanisms of deformation and have led to the conclusion that the microstructure developed determines the flow stress at each strain [2 3].

Since structural changes such as recovery or recrystallization occur rapidly at high test temperatures, the delay between the end of deformation and quenching should be made as short as possible if the as-worked structure is to be examined [2 3 22 120 122 132-3]. However, it is also interesting to study the effect on structure of various delay times and cooling rates, especially in relation to industrial practice [22 32 37 41 70 74 91 112 125]. For research purposes an initial structure of large recrystallized, grains (1-5 mm) facilitates the examination of the deformed structure (especially in electron microscopy) and the establishment of the mechanisms of deformation [70 120 132 137-9].

Optical microscopy (Fig. 4) is useful for studying the size and shape of grains, the presence of substructure, the extent of recrystallization, and the distribution, size and shape of second phases [1-4]. It also serves as a check on the uniformity of deformation and as a guide to specimen selection for electron microscopy [61 70 120 132 137]. Transmission electron microscopy (Fig.4) is an essential tool for studying the dislocation substructure and the extent of recovery, for determining subgrain size and sub-boundary thickness [6 28 61 70 96 132 137-9], and in recrystallized grains, for distinguishing between those formed during deformation and those nucleated subsequent to deformation but prior to cooling [96 108 120 132]. Transmission electron diffraction supplements microscopy in many ways. With a small aperture it is possible to determine the misorientation across a sub-boundary and to characterize the dislocations in the sub-boundary [61 70 96 137-9]. With a larger aperture it is useful in estimating the degree of polygonization, in locating grain boundaries which cannot be seen easily in the microscopic image, and in ascertaining the presence of dynamically recrystallized nuclei [61 120 132].

X-ray diffraction can be used to measure the crystallite size, the internal distortion and the preferred orienta-

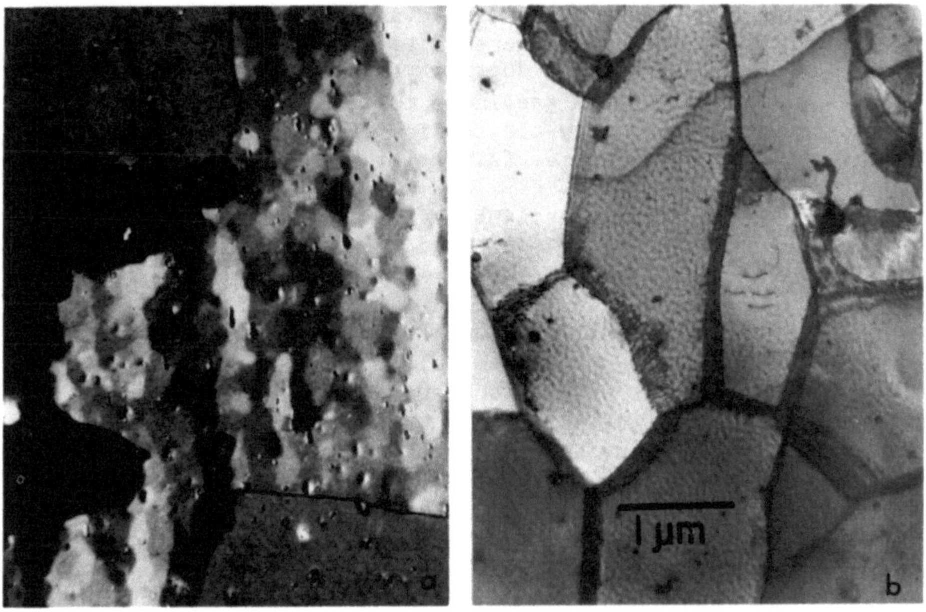

Fig. 4 Microstructures in hot worked commercial purity aluminum: a) worked grains with dynamically recovered substructure and statically recrystallized grains in an extrusion produced at a ram speed of 1.5 ipm, 450°C, and an extrusion ratio of 40:1. X 350 polarized light [138]; b) polygonized dislocation substructure resulting from compression at 400°C and 220 sec^{-1} to a strain of 0.7 [61].

tion [1 3 28 51-2 96 107 111 115 124 126 134-5]. The latter property can be a valuable criterion for differentiating restoration mechanisms, e.g., the texture of a heavily worked metal which recrystallized during cooling will differ from that of a metal which recrystallized repeatedly in the course of deformation [1 96 115]. Since X-ray diffraction averages the structure in a volume of crystal, it may not be satisfactory for the analysis of microscopically inhomogeneous specimens and should be combined with optical and electron microscopic examination.

PROPERTIES OF HOT WORKED METALS

The mechanical properties at the end of a hot working operation or test are directly related to the microstructure present, which depends in turn upon the deformation conditions and on the rate of cooling [28 40 62 122 124 129 132 137 150-2]. The properties are of interest in some cases with respect to service applications and in others with a view to further forming operations. An important consideration in selecting an experimental technique is its ability to produce samples which have dimensions suitable for further testing. Hardness tests are advantageous in that they can be performed on small or inhomogeneous specimens, but they give no measure of the yield stress or ductility. Compression, which can be carried out on very small specimens to determine the yield, does not indicate ductility. Only the tension test gives yield stress, UTS and ductility, but requires fairly large specimens.

The properties after annealing, ageing or other heat treatment are strongly dependent on the microstructure at the end of hot working [42-3 128-9 131]. The rate of recrystallization and the size of the grains depends on the temperature and rate of deformation [40 44 112 125]. The strength and fracture toughness of tempered steels depends partly on the austenite grain size which was produced by the last hot working operation prior to transformation [92-3 153]. The strength increase produced by ausforming depends on the dislocation substructure generated, the carbide precipitation induced, and their effect on the transformation to martensite [93 154-7]. In a Ni-base superalloy, hot working promoted formation of a grain boundary precipitate of Ni_3Cb which remained stable during aging and which did not alter the precipitation of γ' [26].

TENSION

Tension testing at or below room temperature has given very useful results with respect to applications at these temperatures. With the necessary equipment widely available and the techniques well known, it is unfortunate that tension testing is not entirely suitable for hot working studies [15-35]. This is because a) the strain rates produced are usually too low and decrease during a test and b) necking and fracture prevent sufficiently large strains from being attained.

Necking and Ductility

The plastic instability known as necking is peculiar to tensile deformation and occurs when the incremental increase in strength due to strain hardening in a local region is insufficient to offset the incremental decrease in cross-section [148]. Necking limits the uniform deformation to strains of less than 0.3, which is much less than that usually attained in hot-forming operations and is frequently less than that required for the inception of the steady-state regime [13 33-4]. In the neck, the strain rate increases and becomes non-uniform and the temperature may also rise because of the work of deformation [34 148]. Thus the flow stress and strain rate calculated on the basis of uniform deformation are no longer valid [13]. Moreover, it is usually not possible to make corrections based on the instantaneous neck radius, as it cannot be measured in a typical test due to the enclosing furnace and the rapidity of the test [16].

The strain to neck formation is lower when the rate of strain hardening becomes less; thus, omitting from consideration changes in strain rate at the neck, one would reason that the uniform strain is reduced as the temperature is increased or as the strain rate is lowered [23 33-4 148]. In reality, because of the increased strain rate sensitivity at elevated temperatures, the increased strain rate at the constriction raises the flow stress there, and brings about a transfer of strain towards the edges of the constriction, thus lengthening the necked region. Because the change in cross-section is more gradual than at low temperatures, the reduction in area proceeds to a high degree before the final fracture is induced by hydrostatic tension [15 23 31 33-4 148]. Furthermore, tests at constant true strain rate exhibit greater ductility than those conducted with a crosshead velocity constant at the initial value, in which the true strain rate is diminishing [23]. The growth of the neck is also delayed when the tensile test is conducted in small increments of strain with intervening delays (simulating a multi-pass operation), if in these unloaded periods either recrystallisation or recovery restores the rate of strain hardening when the deformation is resumed [23]. The superposition of hydrostatic compression on a tension specimen increases the reduction in area without altering the uniform deformation.[148-9 158] However, the complexity of this technique at elevated temperatures has prevented anyone from attempting it.

In ductile materials, the elongation is a poor measure of the deformation possible in processing because the instability arising from simple tensile loading is not present in hot working processes [11 13]. The reduction in area gives a better measure of the true ductility because it is directly related to the strain in the fracture region [10-1 24-5]. However, the final tensile fracture, which occurs by the coalescence of pores and fissures under the hydrostatic tensile stress, is considerably different from fracture in forming processes, where hydrostatic tension is usually absent. The net effect of the above factors is that, while there is no exact correlation between tensile and process ductility, the variations of tensile ductility with temperature are similar to the variations in processing formability [29]. Notched tensile specimens have been used to check the workability of steel for rotary hot piercing[31]. The notch induces a hydrostatic tensile stress, which opens up pores at inclusions and defects, as does the piercing operation. In low ductility materials, the tensile test gives a satisfactory indication of brittleness because there is no prior necking.

Strain Rate

The normal tensile tester has a maximum strain rate of about 10^{-1} sec^{-1} ($\sim$ 10% per sec) which is about one thousandth of the rate of common working processes. With constant rate of crosshead motion, the strain rate decreases as the specimen elongates (Eq. 6) [13 16 33].

However, in the neck, the strain rate actually increases to more than 20 times nominal as the constriction becomes more severe [16 33]. Thus, even in tests conducted at constant nominal strain rate, the strain rate in the neck continuously rises, causing the flow stress to increase. This augmentation is over and above the increase in flow stress due to the increasing hydrostatic stress component [33 148 158]. Rossard has constructed a special tensile machine in which the rate of pulling is adjusted with respect to the necking behaviour of the alloy under study so that the strain rate at the neck remains constant [22]. On this machine, strains as great as 2.0 were attained and the stress-strain curve was similar to that for torsion of the same alloy.

Other special tensile machines which produce sufficiently high rates of deformation have been constructed [17]. The "Gleeble" uses a simple hydraulic cylinder which gives maximum elongation rates of 10 sec^{-1} at a constant crosshead velocity [18 25-6 32]. The strain rate in the Nemlab pneumatic machine is controlled by adjusting the pressure of the compressed air supply [27 29]. In the impact tension test, the movable end of the specimen is propelled by a dropping weight [9 13 24]. In impact tension, the strain rate decreases rapidly as the impact energy is absorbed by the deformation. Even if the load and elongation are measured as functions of time for a series of impact energies, it is not possible to calculate a set of true stress-true strain curves. This is essentially because different microstructures are developed under different energy inputs [13]. Impact tension does permit the determination of a relationship between energy and elongation, but one must take into account the reduction in cross-section by using true strain.

Temperature Control

The specimen is usually deformed within a furnace, the constant temperature zone of which should be at least as long as the gage length at maximum uniform elongation. At high strain rates, the deformation is increasingly adiabatic and the temperature reaches a maximum at the neck. With resistance furnaces, cooling is delayed until the furnace is swung away; induction or focussed radiant heating offer greater quenching speed. The "Gleeble" instrument, which heats by passing a current through the specimen, provides the capability of rapid heating (1,650°C/sec) and cooling (120°C/sec, without quenchant) [18 25-6]. However, the temperature is not constant along the specimen length, initially as a result of cooling by the water cooled grips, and later upon necking, due to the concentration of current in the neck. The temperature in the zone to which the thermocouple is attached is held constant.

Microstructure and Properties

The necked region is small and is traversed by a gradient of strain and strain rate. Thus only the microstructure immediately adjacent to the neck can be considered to be that associated with the flow stress. The strain gradient is frequently so severe and the neck so small that even hardness testing cannot be used with assurance.

COMPRESSION TESTING

Hot compression, which can be performed on the same equipment as tension tests, is more suitable for hot working studies since the stress system is closer to those found in deformation processing, and no basic instabilities such as necking arise [13 148]. Compression tests may be very simple or highly refined. For example, the amount of upset which induces edge cracking, or the complete true stress-true strain curve can be determined. Since the area increase is inversely proportional to the decrease in height, the force for deformation increases rapidly, requiring a strong testing machine [13]. The friction between specimen and anvils, unless it is closely controlled, has detrimental effects on the results.

Barreling

In axisymmetric compression, friction between the platens and the specimen leads to greater deformation in the midsection of the specimen than in the constrained ends (Fig. 5a)[14 146]. In severe cases, there are cones of negligible flow at the ends and the bulging of the specimen causes the cylindrical surface to roll over into contact with the platens (Fig. 5b) [158]. The friction can be greatly reduced by using suitable lubricants [54 57], such as Teflon [64] and liquid glass [55-6 58 64 71-3]. Circular grooves or dimples in the end faces retain the lubricant, improving its efficiency [54 58-60 68 71-3]. The friction is also decreased by use of hard, polished platens, which should be protected from oxidation by testing in a controlled atmosphere [58 72-3]. With such techniques it is possible to attain a strain of 0.7 without barreling and of about 1.2 with slight barreling [13 56 58 71]. The constraints due to friction increase the flow stress and formulas for correction have been derived [48 56-7 71]. This effect diminishes as the diameter decreases relative to the height, thus the real flow stress can be found from the flow stresses of specimens of different diameters by extrapolating to zero diameter[13 48 146 159]. When the specimen height exceeds twice the diameter, the experimental problem shifts from barreling to collapse by buckling.

In plane strain deformation, a narrow band across a wide strip is compressed by means of narrow platens which are wider than the strip [13 57 146]. Under these conditions, uniform strains of 2.3

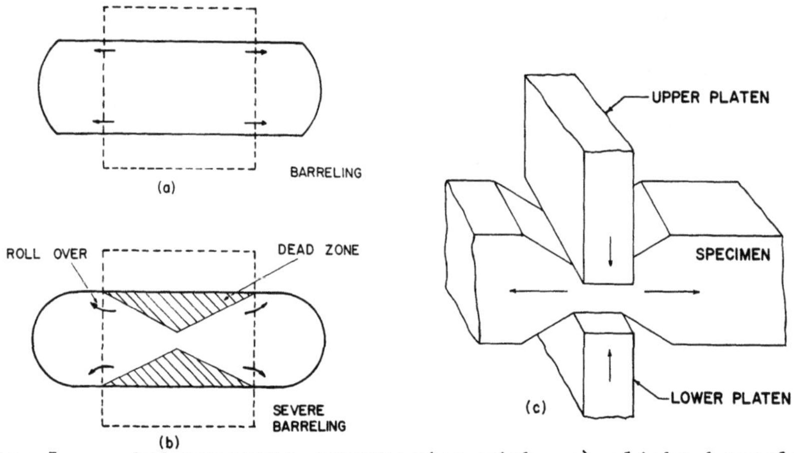

Fig. 5 Axisymmetric compression with a) slight barreling as a result of light friction which hinders sliding and b) severe barreling as a result of high friction which prevents slipping. c) Plane strain compression.

have been attained [57]. The constraints of the undeformed shoulders on both sides of the platens prevent extension parallel to the long dimension of the platens (Fig. 5c). The result is the formation of a groove and extension normal to the long dimension of the platen, much as in rolling, which is also a form of plane strain compression. As the area under the platens is constant, the total force does not rise as rapidly as in axisymmetric compression [13 57]. If the lubrication is insufficient, a dead zone will form in the specimen next to the face of each platen. As the reduction proceeds, the effect of friction increases [13].

Universal Testing Machines

In a compression test at a constant anvil speed, the true strain rate continuously increases and by the time a reduction of 90% ($\varepsilon = 2.3$) has been reached, it has increased by a factor of 10. In order to avoid this difficulty, the anvil speed must be continuously reduced during the test [71]. At strain rates below 1 sec^{-1}, a servo-controlled universal tester can be adapted to maintain a constant strain rate [72-4]. For such capability, the testing machine must be equipped with a generator of a suitable control signal and a screw operated machine must have a variable speed motor. It should be noted that a given machine can maintain higher true strain rates in compression than in tension, because in the former the crosshead speed is reduced during the test and shorter specimens are generally used.

With a testing machine so equipped, it is possible to interrupt a test, hold the specimen at temperature for a certain time to permit recovery or recrystallization, and then to measure the subsequent yield stress in order to determine the extent to which the mechanism has progressed [32 74]. It is also possible to change strain rates in the middle of a test to see how the flow stress varies as the deformation structure gradually changes to that characteristic of the new strain rate [72]. Furthermore, with suitable function generators, it should be possible to simulate the successive stages of a hot forming operation such as slow pressing where the strain rate is low. The flow stress measured in compression has been shown to satisfactorily correlate with the stress for hot shearing of billets [38].

Cam Plastometer

For strain rates between 1 sec^{-1} and $10^3 sec^{-1}$, the only instrument capable of giving a constant rate of strain is the cam plastometer [54-67]. On the cam plastometer (Fig. 6) the platens are driven together by a cam whose radius increases with the angle of rotation, so that the strain rate is constant [58 63]:

$$r = r_0 + \ell_0 \left[1 - \exp(\dot{\varepsilon}_c \, \alpha/\omega_c)\right] \tag{11}$$

Here r is the radius of the cam lobe, r_0 the base radius, $\dot{\varepsilon}_c$ the design, constant, compressive strain rate, α the angle subtended by the lobe and ω_c the design angular velocity. With the specimen inserted between the anvils, the test is carried out by inserting a block, called the cam follower, between the movable anvil and the rotating cam at the instant that the minimum radius of the cam is opposite the anvil. Since the cam is rotating rapidly, the cam follower must be slid in quickly by a pneumatic piston automatically actuated by a signal from the cam [58-9 67]. The cam follower must be removed after the compression to prevent repeated blows on the specimen and interference with its removal. To facilitate insertion of the cam follower, the arc of minimum radius should not be less than 90°. The angular velocity of the cam must be maintained constant as it compresses the sample; for this, a strong motor and heavy flywheel are required. Different strain rates are obtained by changing the speed of rotation of the cam.

Specimens with different reductions can be obtained by using specimens shorter than the one for which the cam was designed. It should be noted that, if tests are conducted in which the span between the cam axis and the fixed anvil is altered from the designed value, the strain rate will not be constant.

When the alteration in span is Δs, the strain rate for angular velocity ω at an instantaneous specimen height, ℓ is:

$$\dot{\varepsilon}_\ell = \omega \, (\dot{\varepsilon}_c/\omega_c)\left[\ell/(\ell+\Delta s)\right] \tag{12}$$

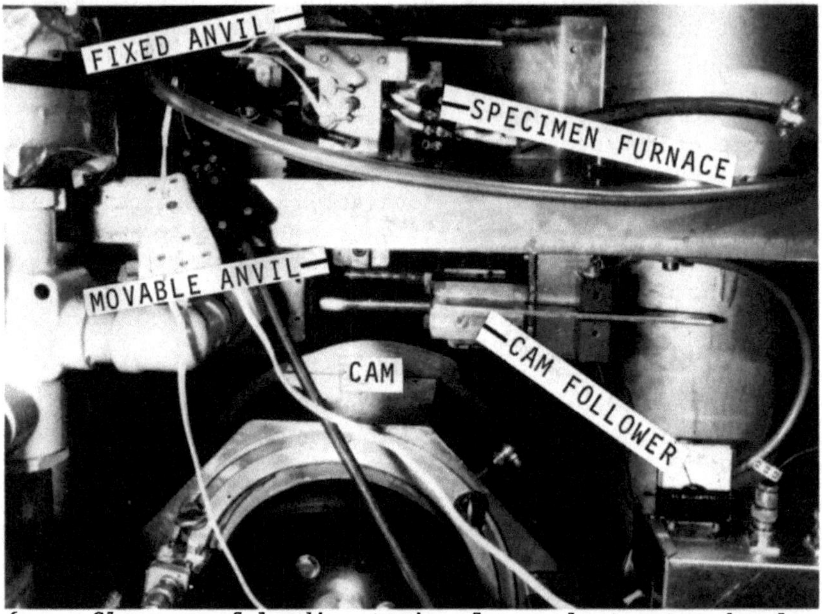

Fig.6 Close-up of loading train of cam plastometer developed by Hockett [58-60] at the Los Alamos Scientific Laboratory. The cam is driven, via a series of gearboxes, by a 50 HP d.c. motor and produces strain rates from 10^{-1} to 2.3×10^{2}sec.$^{-1}$. Here the cam follower is withdrawn; in order to transfer the lift of the cam, it is moved leftward into position below the movable anvil.

This means that constant strain rate reductions greater than the design reduction cannot be produced by a given cam and that specimens higher than the design height cannot be compressed at constant strain rate.

The cam plastometer produces the deformation defined by the cam in a single operation. The strain rate profile of a particular process can be reproduced with the proper cam contour [53-4].

Furthermore, various schedules of deformation, with intermittent anneals or deformations at different strain rates, could in principle be produced by cutting complicated cams. A series of deformations can also be produced by several cams which are mounted on the same drive shaft and which can be slid into position successively [67]. The superposition of supplementary stress fields appears to be very difficult, with the exception of plane strain.

The stress can be measured by a load cell on the fixed anvil. The strain is defined by the cam, with allowance for elastic distortion. The strain, strain rate and uniformity of strain have been observed in room temperature experiments by the high-speed cinematography of specimens on which a grid had been ruled [60].

Upset Tests

Simple upsetting tests [36-49] have been carried out in drop, pendulum [45], or pneumatic hammers which produce a compression rate that decreases as the specimen spreads and work hardens. The initial strain rate can be calculated, but its variation with strain is generally not known, as it varies from material to material depending on the work hardening characteristics[13]. Recently, Samanta has shown, for certain conditions of tup mass and velocity, that the strain rate is almost constant up to strains of 0.3, but diminishes to half that value by a strain of 0.7 [47-8]. The mean strain rate can be increased by increasing the velocity and mass of the tup. If differences in strain are produced in different samples by limiting the reduction mechanically, the average strain rates are not the same [13]. Because the conditions in this type of test are much like the normal forging operation, [147], it gives an adequate measure of force and energy [9 36].

Swaging

Hot swaging can be used to produce controlled reductions, but it is not a suitable workability test since it involves multiple blows, and the magnitude of the following important parameters cannot be ascertained: flow stress, strain rate, temperature and quenching time. However, when the specimen is advanced a fixed amount between successive blows, it is possible to make some estimate of the mean strain rate, the reduction per blow and the interval between reductions [152].

Temperature Control

Compression specimens may be deformed within a furnace, but in many investigations they are simply transferred from the furnace to the testing machine. To minimize cooling, the furnace should be close to the press, and transfer to the precise deforming position should be mechanically simple and rapid. In some cases, disposable insulating containers have been used to prevent both changes in temperature and oxidation of the specimens [55]. Heating by passing a current through the specimen is usually precluded by the presence of non-conducting lubricants. Because of the compact shape of the specimen, deformation in a furnace is almost adiabatic; outside a furnace, the self-heating is counteracted by the cooling effect of the anvils [35]. Because of the pancake shape of the specimen, it is very difficult to quench it in position between the platens; a fast quench can be achieved by separating the platens and ejecting the specimen into a quench bath. If the deformation is within a furnace, this can be accomplished by knocking the specimen into a tube passing through the lower anvil to the quenchant [71-3].

Malleability

The limit of malleability is usually defined by the appearance of edge cracks. These are the result of circumferential tension stresses resulting from the barrelling which cannot be eliminated at high reductions [6-7 49]. The correlation between compression tests and forging operations is generally good, but may be diminished by differences in lubrication and other factors [35]. The resistance to edge cracking can be studied more critically by cutting notches in the surface parallel to the compression axis [9]. Materials which are comparatively brittle may crumble or fail along a plane at 45° to the axis [148].

Upset tests give a rapid and inexpensive measure of the malleability exhibited in industrial forging and rolling [7 49 147]. This test has been used to determine the temperature ranges for edge checking in different heats of Ni-base alloys[6]. Whenever checking appeared in the test billets, it also appeared on edges of strip rolled under the same forming conditions.

Metallography and Mechanical Properties

Compression specimens are suitable for optical microscopy and for the preparation of thin foils for electron microscopy [61 72]. The pattern of non-uniformity of deformation can be examined on etched cross sections [61]. Hardness can be measured on the cross-section and can be used to check uniformity [150]. Small specimens for determination of the flow stress by compression can easily be cut from the samples [62 151]. Tensile specimens cannot usually be made, except from rods prepared by swaging.

Metallographic study of upset specimens is not amenable to fundamental interpretation because of the varying strain rate during the test. However, these specimens are generally useful for investigating the microstructures, the mechanical properties or the heat treatment response of forgings [40-3 51]. An entire series of experiments of this type can be replaced by the compression of a single wedge-shaped sample [9 37]. The specimen can be sliced to give sections with a range of strains and strain rates from zero to maximum. The effects of the percent hot reduction on aging can be determined by microscopy and hardness testing. The wedge specimen also shows the strain at which the first cracks appear.

TORSION

The hot torsion test (Fig. 7),[1 22 24 35 75-121], which consists of twisting a specimen with a heated gage section, is capable of producing strains of the order of 20. Since the dimensions remain constant, the true strain rate and the engineering strain rate are equal and constant. The difficulties with the torsion test arise from the variation in the axis to surface strain and strain rate, and the influence on the ductility of a shear-to-normal stress ratio of unity, which is much higher than that commonly found in forming operations.

Center to Surface Variation of Strain and Strain Rate

When a solid cylinder is twisted, the strain and strain rate vary from zero at the axis to a maximum at the surface [22 91 100-1 105 107 148]. It is the surface values which are commonly reported. This variation gives rise to problems of interpretation, because the surface work hardens more than the core and the mechanisms of deformation may be different [22 101 148]. Nevertheless, most torsion experiments have employed solid specimens and usually a correction has been made in the calculation of the flow stress from the torque,[22 101 148]. The use of tubular specimens largely avoids these difficulties, but gives rise to others [35 84 118 21]. The tubular specimen is usually larger than the solid one, and its dimensions must be proportioned so that the walls are sufficiently thick to resist flattening [120]. Thinner walls can be used as the gage length is shortened.

The strain normally quoted is the surface strain, which is calculated from the total angle of twist and the length and diameter of the test section [22 86 100-1 105 111 118-9 148]. The angular rotation may be measured by a rotary variable differential transformer, a helical potentiometer, or a photoelectric device. The gage length is usually the distance between the fillets leading to the heavier grip sections. Since the highest strain and stress occur in the surface layer, the surface must be carefully finished and protected from oxidation.

In hot twist tests of rods without reduced sections, the length of the deformation zone varies during the test as a result of differences in temperature and in strain hardening along the length of the rod [9 75-8 82]. Sometimes, increased deformation and hardening in the hotter regions cause the deforming zone to lengthen, leading to high ductility. In other tests, the deformation heats up the deforming region so that it becomes softer and the strain and strain rate become localized, leading to low ductility. These two developments can occur in the same alloy at different temperatures or strain rates [82].

When the shear stresses and strains are converted to equivalent tensile stresses and strains according to the von Mises criterion, torsion flow curves are almost identical to tension or compression flow curves for the same strain rate and temperature [22 84 148]. The interdependence of surface strain rate, effective stress and temperature is the same as that for compression and extrusion tests.

Fig. 7 Overall view of torsion tester developed by Rossard [22] at IRSID in France and manufactured by SETARAM in Lyon. From left to right can be seen a 5 HP, 1500 rpm, electric motor, a 30:1 continuously variable hydraulic speed reducer, 30:1 mechanical speed reducer, an electro-pneumatic engagement and braking device, a photoelectric rotation transducer, a resistance furnace and a fixed specimen support with strain gage bridge.

Simulation of Processes Involving Multiple Deformations

In torsion testing, the strain rate is altered simply by changing the rate of rotation and it is possible to achieve rates from 10^{-5} to $10^{3}sec^{-1}$ [22 84 105 118]. The strain rate may be changed in the course of a test to study the interrelation of deformation structure, strain rate and flow stress [22 86 91]. Deformation can also be stopped and restarted to study the effects of annealing on the structure [22 91 93 115-6]. These techniques have been used to simulate multi-pass processes, such as slab rolling [117] and continuous strip rolling, in order to study the effects of initial temperature, reduction per pass, and final cooling rate [89 90].

Furthermore, since axial stresses can be applied during testing [78 118-9], it is possible to simulate to some degree the complex stresses present in such processes as extrusion or tube piercing. Compressive axial loading requires very careful alignment of the specimen and grips to prevent buckling.

Self-Induced Longitudinal Strain or Stress

In the course of a torsion test, if the stationary end is free to move in an axial direction, the specimen may change length, giving rise to changes in torsional strain and strain rate. The dilatation may either be extension or contraction and, for certain materials and conditions, may reverse as the test proceeds [4 94 101 106 118]. In a given material, the effect usually varies with temperature and strain rate, passing through a maximum. The cause of these changes is not yet known, but it may be related to grain boundary sliding and pore formation, to misalignment of the grips, or to the formation of a preferred orientation. In tubular specimens, longitudinal contraction is accompanied by radial bulging of the gage section [120]. If the stationary end of the sample is not free to move axially, induced tensile or compressive stresses arise and alter the stress state [79 101 106].

Certain materials have been observed to develop intrusions and extrusions at the surface and also on the inner surface of tubular specimens [85 101 118 120]. Their characteristics vary with strain rate and temperature; again their origin is not known.

Temperature Control

Torsion specimens are usually deformed in a furnace which has a constant temperature zone longer than the gage length. The deformation is not entirely adiabatic at low rates, but becomes increasingly so at high rates. The heating may be produced by a resistance furnace, passage of an axial current [24], focussed radiation [118], or induction [120]; the last three have the advantage of permitting more rapid cooling. However, even with a resistance furnace, a cooling time of 0.1 second has been attained by pneumatically actuating the fixed grip to withdraw the specimen into a cooling spray [108-11], or by mounting the quenchant jets between the windings and quenching the specimen within the furnace[96]. Tubular specimens offer the possibility of quenching by injecting the coolant through the axial hole [120]. The large size of tubular specimens usually leads to long quenching times (3 sec.)[120]. Because of the excellent control of strain rate, strain, temperature and quenching rate, the torsion test is suitable for simulating thermomechanical treatments [92-3].

Ductility

The surface strain at fracture is usually several times that attained in any forming operation, even extrusion. The high ductility results from a ratio of shear stress to normal stress of unity, which is much higher than in any other mode of deformation. Fracture commences with the formation of pores by grain boundary sliding, which is enhanced by the high shear strain [1 79 99 109]

[113-4 118-9]. Propagation by pore coalescence and tearing is promoted by the equality of the perpendicular components of tensile and compressive stress. This mechanism of fracture does not appear to play such a significant role in a typical forming process, because there the shear stress is relatively lower than in torsion and the normal stress is largely compressive. In torsion testing an applied or self-induced tensile stress appears to hasten fracture, but a compressive stress slows propagation [101 106 114 118]. If a torsion test is halted prior to fracture, it is often observed that an annular zone contains a high density of cracks, which indicates that the strain rates in that zone are more conducive to fracture than higher or lower ones [114].

Despite the differences between the stress states in the torsion test and those in hot working processes, the torsional ductility appears to be a good measure for practical purposes [24 75-9 82-3 87-8]. For example, Waspalloy, which exhibits three times as many twists at fracture as does Udimet 700, has about three times as much formability as the latter in forging [9]. It has been used to select alloys and to determine optimum conditions for extrusion [97 121] and hot piercing [78-80].

Microstructural Studies

Because of the radial variation in strain and strain rate, only a thin layer at the outer surface is representative of the nominal flow conditions [22 91 100 108-11]. Examination should therefore be concentrated on sections parallel to the axis and as close as possible to the surface. Specimens for determination of the preferred orientation at the surface have been prepared by boring out the center to leave a thin shell, which is then slit and flattened [96]. Thin foils for electron microscopy can be prepared by cutting small, dished discs from the chemically polished cylindrical shell by jet machining. Optical examination of transverse sections or of tangential sections at different radii, will reveal the effect of the variation in strain and strain rate [12 22 91 105 108-11]. Specimens with different amounts of surface strain can be obtained by stopping the test at the desired values [22].

Product Mechanical Properties

The room temperature mechanical properties can be measured satisfactorily by torsion testing [93], however, it may be necessary to refinish the surface to remove the irregularities resulting from the high temperature deformation. Specimens for tensile testing could conceivably be obtained by boring out the center to leave material of uniform properties; either the entire cylinder or strips from the shell could be used. Longitudinal tensile tests are, of course, transverse to the fiber direction [160].

HOT ROLLING

Laboratory hot rolling [40 122-33] is a technique which is widely used for studying hot working because the equipment is readily available and produces samples suitable for further testing. A further factor favoring its use is that industrial hot rolling produces larger tonnages than all other processes combined.

Strain and Strain-Rate

In hot rolled material, the deformation is fairly uniform, although a section normal to the rolling direction does become curved, with the surfaces leading as it passes through the rolls[124 146]. However, the uniformity decreases as the strain in a single pass increases. Strains as high as 2.3 (90% reduction) can be achieved in a single pass [132]. For such tests, the leading edge of the specimen must be reduced sufficiently for the rolls to grip it and draw in the remainder of the specimen. The transition from the nose to the full thickness may be in the form of a wedge or of steps; the latter are easier to machine.

As a section passes through the rolls, the strain rate increases gradually from its initial value to a maximum and then drops rapidly [86 146]. Thus, when the effects of different reductions were studied by producing them at constant roll speed [123], the result was a different strain rate for each reduction. Moreover, even if the roll speed is adjusted to give the same average strain rate, the strain rate profile as a section passes through the rolls is different for different reductions.

This technique lends itself to multiple pass tests, but it is difficult to maintain the rolling temperature of a small sample unless suitable furnaces are provided to restore the temperature[132]. Moreover, it is difficult to provide the high strain rates and short rest intervals of the final stands of a mill [13 126 152].

Flow Stress

Rolling theory provides formulas for calculating the flow stress of the material from the roll separating force [124 146]. This is an average flow stress and cannot be directly associated with the dislocation structure in a particular section at any stage of its deformation. The average flow stress as a function of strain at a given temperature has been determined by rolling a long, wedge-shaped specimen [40 123]. This is not acceptable, because both the average strain rate and the degree of quenching by the rolls continuously increase as the strain increases.

Temperature Control

The standard practice is to heat the specimen to the rolling temperature (sometimes slightly higher, to allow for cooling during transfer to the rolls) and then to roll it on cold rolls [124 128]. With small specimens (up to 1" thick), the heat of deformation is insufficient to compensate for cooling by the rolls; at greater reductions, the cooling is greater because there is a larger contact surface-to-volume ratio [130 132]. Since the softening point of the rolls limits the temperature to which the rolls may be heated, the desired average or finishing temperature is attained by controlling the preheat temperature. In order to be certain of the conditions, the specimen temperature should be measured continuously during rolling with embedded thermocouples which pass through the rolls [132]. Optical and radiation pyrometers are not accurate enough because of oxidation and cooling of the surface. With this procedure, the temperature still changes between entry and exit from the rolls and from surface to center of the workpiece [132]. Rolling the specimen with an insulating coating is not feasible at high reductions. The only completely satisfactory solution is to use rolls maintained at the desired temperature, or to use a much larger specimen and rolling mill. The specimen can be rapidly transferred from furnace to rolls by using a furnace in line with the mill and moving the specimen into the mill either with a rod embedded in the leading edge or with a pusher.

Since the specimen is free of the rolls right after it is deformed, it can be quenched rapidly. If it is quenched by dropping into a bath, there is a gradient of time at exit temperature along its length which permits the study of the effect of brief annealing on the worked structure [123 128 132]. If this is not wanted, sprays must be used for quenching.

Malleability

In this test, the limit of malleability appears as edge cracking and is a satisfactory criterion for industrial rolling operations [146]. The malleability limit observed in rolling tests on a given material is greater than that in upset tests because the edge barrelling in rolling is not nearly as severe as barrelling in unlubricated upsetting.

Microstructural Examination and Property Measurement

The final specimen shape puts no limitations whatsoever on the removal of samples for metallography or X-ray studies of texture [40 122-4 133]. It is wise, nevertheless, to examine the central layer since the surfaces may have been deformed at temperatures considerably lower than the center.

Hardness measurements can easily be performed on the surface or on mounted sections [123]. Specimens for tensile tests can be obtained in the longitudinal and the transverse directions [122, 123, 133].

EXTRUSION

The extrusion test is capable of producing large strains (up to 5) and has been used to determine relationships between the mean strain rate, average stress and temperature which compare well with steady-state results from other techniques [137-40]. Extrusion experiments can be carried out on any compression machine by using billets a few inches in diameter and a small portable container [40, 137-45].

Nonuniformity of Deformation

The deformation is extremely non-uniform because of the container and die friction [134-6, 145]. The deformation is highest near the surface and is least at the center. The homogeneity of deformation can be improved by the use of suitable lubricants or of backward extrusion, in which the billet does not move relative to the chamber but is extruded through a die mounted on a hollow ram. Hydrostatic extrusion leads to much more uniform deformation, but requires considerably more complex equipment.

With constant ram speed, the strain rate is not uniform across the transverse section and in a given region varies by as much as three orders of magnitude as that region approaches and passes through the die [143-5]. The transverse variation in strain rate is an important source of inhomogeneity in the finished extrusion.

Flow Stress and Flow Relationships

The flow stress can be calculated from the extrusion pressure by means of formulas which make allowance for the friction and the work expended on redundant deformation within the container; this is only an average flow stress, as in hot rolling. To obtain the average flow stress for different strains, different ratios of die opening to billet diameter must be used. The relationship between flow stress, strain rate and temperature can be determined from a series of tests at constant extrusion ratio, billet size and condition of lubrication.

With small billets, the chamber is heated to the extrusion temperature [137-41]. The use of hot work tool steels for the extrusion tooling limits this procedure to the deformation of aluminum and lower melting point metals; however, the use of refractory metals or of suitable superalloys would permit its application to carbon steels. The temperature rise during extrusion at low speeds or low extrusion ratios can be quite small [140-1]. However, at high speeds and ratios, it can be greater than that due to adiabatic deformation to the nominal strain alone because of the work of friction and redundant deformation.

Ductility

The presence of induced hydrostatic compression in the deformation zone means that high strains can be produced without failure. This test is, therefore not a useful measure of ductility.

TABLE 1 SUMMARY OF TEST PERFORMANCE

Test Method	Constant Strain Rate	Maximum Strain Rate	Temperature Control	Flow Stress Measurement	Max. Uniform Deformation
Tension					
Universal Tester	yes	10^{-1}	good	good	0.3
Gleeble	no	10^{2}	fair	fair	0.3
Impact Tension	no	high	poor	poor	0.3
Compression					
Universal tester	yes	1	good	good	1
Cam,axisymmetric	yes	10^{3}	fair	good	1
Cam,plane strain	yes	10^{3}	fair	good	2
Drop weight	no	high	poor	poor	0.2
Forging hammer	no	high	poor	no	0.2
Torsion					
Solid specimen	yes*	10^{3}*	good	fair	20*
Tubular specimen	yes	10^{3}	good	good	20
Rolling	no	10^{2}	poor	fair	-
Extrusion	no	10^{3}	fair	fair	-

* Does not vary with time, but varies radially.

Under certain conditions, cracks can be produced in laboratory extrusions; because of problems of geometrical scaling, however, this can only be used as a guide to commercial extrusion practice. Studies of hot shortness have been made by extruding ice in transparent dies [142].

Microstructure

Because of the inhomogeneity of strain and strain rate, specimens should always be taken from the same position in different extrusions [40 124 133 136-9]. The nose and tail of the extrusion

TABLE 1 SUMMARY OF TEST PERFORMANCE (continued)

Max. Strain In one pass	Limiting Instabilities	Suitability for Micro-Structural Examination	Specimens for Subsequent Mechanical Testing	Ease of Multi-Pass Simulation	Ease of Quenching
2	necking	poor	poor	good	fair
2	necking	poor	poor	fair	good
2	necking	poor	poor	poor	poor
2	barreling	good	good	good	fair
2	barreling	good	good	fair	fair
3	friction	good	good	fair	fair
2	barreling	fair	fair	poor	poor
2	barreling	fair	fair	poor	poor
100	(change in axial length)	fair	fair	good	good
100	(change in axial length)	good	good	good	good
2.3	edge-crack	good	good	fair	good
5	cracking	good	good	poor	good

should not be used because stable extrusion conditions have not yet been established or have started to break down. Although the calculated mean flow stress may not be precisely related to the dislocation structure observed, it is probably a good approximation [137-9].

Because of the variation in structure across the section, mechanical properties are roughly a volume weighted average of those of the different structures. Surface hardness measurements are generally much higher than those from the interior. Longitudinal tensile properties can readily be measured [124]; transverse properties may be determined in compression tests.

CONCLUSIONS

Hot working experiments can provide information which falls into four categories. These are:

1) Information concerning the relationship of the flow stress to the deformation parameters, which is necessary for calculation of working loads;

2) The dependence of the ductility on the forming conditions, which makes known the limits for successful production;

3) Analysis of the microstructural changes taking place during working, which discloses the mechanisms of deformation and suggests directions of control for enhanced workability; and

4) Data concerning the mechanical properties of the product, which indicate possible process alterations to surpass specifications.

The important hot working test methods have been described and their relative utilities assessed. The comparison is summarized in Table 1. There are those that are quick and simple enough to perform on the forging floor, those which satisfactorily determine the flow stress as a function of temperature and strain rate, and those which can simulate multistand rolling mills. The test methods which are recommended for their diverse capabilities are constant strain rate compression and torsion with tubular specimens. The former is closely related to forging and rolling and the deformation is quite uniform; however, the latter is capable of producing extremely high strains for research purposes and is easily programmed for the simulation of multipass industrial processes.

REFERENCES

1. H.P. STÜWE and B. DRUBE: Z.Metallk.,1967, 58, 499-506.
2. H.J. McQUEEN: J.Metals, 1968, 20, (4), 31-38.
3. J.J. JONAS, C.M. SELLARS and W.J.McG. TEGART: Met.Rev.,1969, 14, 1-24.
4. C.M. SELLARS and W.J.McG. TEGART: Met.Rev.,1971, 16, in press.
5. A. PORTEVIN, E. PRÉTET and J. LACOMBE: Rev.Mét.,1936, 33, 114-32.
6. L.H. MARTIN and L.O. BIEBER: Non-Ferrous Rolling Practice, AIME, N.Y.,1948, 15-31.
7. G.P. CONTRACTOR and W.A. MORGAN: Metal Treat.Drop Forging, 1959, 26, 65, 107.
8. A. GUEUSSIER and R. CASTRO: Rev. Mét.,1958, 55, 1023-40.
9. H.J. HENNING and F.W. BOULGER: Mechanical Working of Steel I, Gordon and Breach, N.Y.,1964, 107-125.
10. A. NICHOLSON: Iron Steel, 1964, 37, 290-94, 363-67.
11. P. BLAIN: Rev. Mét.,1965, 62, 881-92.
12. W.J.McG. TEGART: Ductility, ASM, Metals Park, Ohio, 1968, 133-77.
13. P. MOORE: Deformation Under Hot Working Conditions, Iron and Steel Institute, London, 1968, 103-06.
14. E. WINKLER: Can. Met. Q.,1968, 1, 49-54.
15. M. MANJOINE and A. NADAI: Proc. ASTM, 1940, 40, 822-39.
16. C.W. MacGREGOR and J.C. FISHER: J. Appl. Mech. 1945, 12, A217-27.
17. J.R. KATTUS: Short Time High Temperature Testing. ASM, Cleveland, 1958, 92-113.
18. S.E. BRAMER, K.G. KAHMAN and R. TITUS: ibid.,114-34.
19. R. TAMHANKAR, J. PLATEAU and C. CRUSSARD: Rev. Mét.,1958, 55, 383-400.
20. C. CRUSSARD and R. TAMHANKAR: Trans AIME, 1958, 212, 718-30.
21. M PELABON: Rev. Mét.,1961, 58, 599.
22. C. ROSSARD: Métaux Corros. Inds. 1960, 35, 102-115, 140-153, 190-205.
23. C. ROSSARD: Rev. Mét.,1966, 63, 225-35.
24. P.A. PORTEVIN: ibid.,1962, 59, 915, 1043.
25. R.S. CREMISIO: Trans. Vac. Met. Conf. 1965, 75-97.
26. R.S. CREMISIO, H.M. BUTLER and J.F. RADAVICH: J. Metals, 1969, 21, [11], 55-61.

27. N.J. GRANT: D.M.I.C. Reports 1965, (218), 19; 1965,(221),18; 1966, (225), 16.

28. F. BOLLENRATH, V. HAUK and W. WIEDEMANN: Z. Metallk, 1967, 58, 643-49.

29. A.I. KEMPPINEN: Deformation Under Hot Working Conditions. Iron and Steel Inst.,London, 1968, 117-21.

30. J.H. DECROIX, A.M. NEVEU and R.J. CASTRO: ibid.,135-144.

31. D.M. COTTINGHAM: ibid.,145-56.

32. G.A. WILBER, J.R. BELL, J.H. BUCHER and W.J. CHILDS: Trans. TMS-AIME, 1968, 242, 2305-08.

33. P.J. WRAY and O. RICHMOND: J. Appl. Phys.,1968, 39, 5754-61.

34. P.J. WRAY: J. Appl. Phys.,1969, 40, 4018-29.

35. A.L. HOFFMANNER: Sym. Thermal Mechanical Treatment of Metals, London, 1970, in press.

36. O.W. ELLIS: Engineering J.,1941, 24, 466-75.

37. H. KORNFELD and G. HARTLIEF: Archiv. Eisenhuttewes.,1944, 17, 221-26.

38. W. LUEG and H.G. MULLER: ibid.,1957, 28, 505-16.

39. K. FINK, W. LUEG and G. BURGER: ibid.,1955, 26, 665-88.

40. J. SCHEY: Acta Techn. Acad. Sc. Hung.,1957, 16, 131-52.

41. S. GORCZYCA: Mém. Scient. Rev. Mét.,1960, 57, 153-58.

42. P. RAVIZZA: ibid.,1962, 59, 93.

43. P. RAVIZZA: Rev. Mét.,1964, 61, 577, 586.

44. A.T. ENGLISH and W.A. BACKOFEN: Trans. TMS-AIME,1964,230,396-407.

45. I. GOKYU and J. KIHARA: Trans. Iron Steel Inst. Japan, 1966, 6, 196-203.

46. A.J. De RIDDER and R.J. NOEL: SAE Publication Paper 690101, 1969.

47. S.K. SAMANTA: Int. J. Mech Sci.,1968, 10, 614-36.

48. S.K. SAMANTA: Deformation Under Hot Working Conditions, Iron and Steel Inst.,London, 1968, 122-30.

49. K.A. REYNOLDS: ibid.,107-16.

50. Ye.N. SOKOLKOV and Yu.P. SURKOV: Phys. Metals Metallog.,1963, 16, (6), 107-09.

51. A.P. GULYAYEV and A.S. SHIGAREV: ibid.,1964, 18, (2), 73-79.

52. Ye.N. SOKOLKOV, Yu.P. SURKOV and D.I. GURFEL: ibid.,1965, 20, (4), 83-88.

53. R.R. ARNOLD and R.J. PARKER: J.Inst.Metals, 1959-60, 88, 255-59.

54. N. LOIZOU and R.B. SIMS: J. Mech. Phys. Solids, 1953, 1, 234-43.

55. J.F. ALDER and V.A. PHILLIPS: J. Inst. Metals, 1954-55,83,80-86.

56. P.M. COOK: Proc. Conf. Properties of Mat. at High Rates of Strain, Inst. Mech. Engin.,London, 1957, 86-97.

57. J.A. BAILEY and A.R.E. SINGER: J. Inst. Metals, 1963-64, 92, 288-9, 378, 404-08.

58. J.E. HOCKETT: Proc. ASTM, 1959, 59, 1309-19.

59. J.E. HOCKETT; High Speed Testing, Vol VI; The Rheology of Solids, Interscience, N.Y.,1967, 205-26.

60. J.E. HOCKETT: Trans. TMS-AIME, 1967, 239, 969-76.

61. H.J. McQUEEN and J.E. HOCKETT: Met. Trans.,1970, 1, in press.

62. J.E. HOCKETT and H.J. McQUEEN: Proc. 2nd Int. Conf. Strength of Metals and Alloys (Asilomar, 1970), ASM, in press.

63. O. KIENZLE and H. BÜHLER: Z. Metallk.,1964, 55,668-73.

64. H. BÜHLER and H.W. WAGENER: ibid.,1967, 58, 136-44.

65. J.J. JONAS and J.P. IMMARIGEON: ibid.,1969, 60, 227-31.

66. J.M. JAQUERIE and L. HABRAKEN: Cobalt, 1968, 38, 13-19.

67. G.R. DUNSTAN and R.W. EVANS: Metallurgia, 1969, 79, 96-99.

68. D.V. WILSON and G.W. ROWE: J. Inst. Metals, 1967, 95, 25-26.

69. M.J. HARRIGAN: Stanford University,DMS Report 67-23, 1967.

70. J.L. UVIRA: Ph.D Thesis, McGill University, 1968.

71. J.L. UVIRA and J.J. JONAS: Trans. TMS-AIME, 1968, 242, 1619-26.

72. J.P. IMMARIGEON: Master's Thesis, McGill University, 1970.

73. M.J. LUTON and J.J. JONAS: Proc. 2nd Int. Conf. Strength of Metals and Alloys (Asilomar, 1970), ASM, in press.

74. R. DJAIC: Master's Thesis, McGill University, 1971.

75. C.L. CLARK and J.J. RUSS: Trans. AIME, 1946, 167, 736-48.

76. H.K. IHRIG: ibid.,1946, 167, 749-90.

77. F.K. BLOOM, N.C. CLARKE and P.A. JENNINGS: Metal Progress, 1951, 59, 250-6.

78. J. DAUVERGNE, M. PELABON and J. IVERNEL: Rev. Mét.,1954,51,254-64.

79. D.E.R. HUGHES: J. Iron Steel Inst.,1952, 170, 214-20.

80. T. MORISHIMA: Tetsu to Hagane, 1958, 44, 552-59; 660-68.

81. K. TAJIMA and K. KUGAI: ibid.,1956, 42, 980-85.

82. C.S. TOUT and L.H. BANNING: U.S. Bureau of Mines Report 5928,1962.

83. J.B. BARBER: Report, H.Wiggin Co.,Hereford, U.K.

84. F.A. HODIERNE: J. Inst. Metals, 1962-63, 91, 267-73.

85. C. ROSSARD and P. BLAIN: Rev. Mét.,1958, 55, 573-94.

86. C. ROSSARD and P. BLAIN: ibid.,1958, 55, 595-98.

87. C. ROSSARD and P. BLAIN: ibid.,1959, 56, 175-80.

88. F.E. WHITE and C. ROSSARD: ibid.,1962, 59, 237-41.

89. C. ROSSARD and P.BLAIN: ibid.,1964, 61, 949-61.

90. C. ROSSARD and P.BLAIN: ibid.,1962, 59, 233-36.

91. C. ROSSARD and P.BLAIN: Mém. Scient. Rev. Mét.,1960, 57, 173-78.

92. C. ROSSARD: Rev. Mét.,1968, 65, 181-95.

93. C. ROSSARD and A. LeBON: Harterei Tech. Mitt.,1967, 24, 221-225.

94. H.P. STÜWE and H. TURCK: Z. Metallk.,1964, 55, 699-703.

95. H.P. STÜWE: ibid.,1965, 56, 633-42.

96. B. DRUBE and H.P. STÜWE: ibid.,1967, 58, 799-804.

97. R. AKERET and A. KUNZLI: ibid.,1966, 57, 789-92.

98. R.A. REYNOLDS and W.J.McG. TEGART: J. Iron Steel Inst.,1962, 200, 1044-59.

99. J.L. ROBBINS, O.C. SHEPARD and O.D. SHERBY: ibid.,1961, 199, 175-80.

100. J.L. ROBBINS, O.C. SHEPARD and O.D. SHERBY: Trans. ASM, 1967, 60, 205-16.

101. J.L. ROBBINS, H. WAGENAAR, O.C. SHEPARD and O.D. SHERBY: J. Mat., 1967, 2, 271-99.

102. H. ORMEROD and W.J.McG. TEGART: J. Inst. Metals,1960-61,89,94-96.

103. H. ORMEROD and W.J.McG. TEGART: ibid.,1963-64, 92, 297-99.

104. G.P. LEWIS and W.J.McG. TEGART: ibid.,1963-64, 92, 249.

105. D. HARDWICK and W.J.McG. TEGART: ibid.,1961-62, 90, 17-20.

106. D. HARDWICK and W.J.McG. TEGART: Mém. Scient. Rev. Mét.,1961, 58, 869-80.

107. C.M. SELLARS and W.J.McG. TEGART: ibid.,1966, 63, 731-46.

108. M.J. LUTON and C.M. SELLARS: Acta Met.,1969, 17, 1033-43.

109. M.J. LUTON and W.J.McG. TEGART: Met. Sci. J.,1969, 3, 142-46.

110. J.P. SAH, G.J. RICHARDSON and C.M. SELLARS: J. Aust. Inst. Metals, 1969, 14, 292-97.

111. G. GLOVER and C.M. SELLARS: BHP Melbourne Res. Lab. Rep. MRL 5/2, 1970.

112. G. GLOVER and C.M. SELLARS: BHP Melbourne Res. Lab. Rep. MRL 5/3, 1970.

113. F.E. WHITE: Rev. Mét.,1966, 63, 991-98.

114. F.E. WHITE and C. ROSSARD: Deformation Under Hot Working Conditions, Iron and Steel Inst.,London, 1968, 14-20.

115. H.P. STÜWE: ibid.,1-6.

116. M.M. FARAG, C.M. SELLARS and W.J.McG. TEGART: ibid.,60-67.

117. T.B. VAUGHAN: ibid.,68-77.

118. G.E. DIETER, J.V. MULLIN and E. SHAPIRO: ibid.,7-13.

119. E. SHAPIRO and G.E. DIETER: Met. Trans.,1970, 1, 1711-19.

120. H.J. McQUEEN and S. BERGERSON: Met.Sci. J. (submitted).

121. A.G. WYSIEKIERSKI: Report, Canadian Westinghouse, Hamilton, Ont.

122. D. McLEAN and A.E.L. TATE: Rev. Mét.,1951, 48, 765-75.

123. E. BELLA: Acta Techn. Acad. Sc..Hung.,1959, 25, 393-425.

124. R. LEGUET, D. WHITWHAM and J. HERENGUEL: Mém. Scient. Rev. Mét., 1962, 59, 649-71.

125. R. WUSATOWSKI: J. Iron Steel Inst.,1966, 204, 727-36.

126. P.C. HANCOCK and W.T. ROBERTS: ibid.,1967, 205, 547-50.

127. Ya.B. GUREVICH: The Uses of Vacuum in Metallurgy, Oliver and Boyd, London, 1964, 315-20.

128. R.S. SHKLYAR, et al: Phys. Metals Metallog.,1966, 21,(1), 47-52.

129. V.D. SADOVSKY, et al: ibid.,1964, 17,(6), 47-53.

130. S.I. BULAT and L.V. MIRONOVI: ibid.,1966, 21,(5), 97-102.

131. H. OHMORI: Trans. Japan Inst. Metals, 1966, 7, 153-57.

132. H.J. McQUEEN: ibid.,1968, 9 sup., 170-77.

133. D. ALTENPOHL: Z. Metallk.,1967, 58, 434-39.

134. K.V. GOW and R.W. CAHN: Acta Met.,1953, 1, 238.

135. E. HIROSAWA: J. Inst. Metals, 1963-64, 92, 78-81.

136. Yu.M. VAYNBLATT, E.P. BELOVA and T.B. SAGALOVA: Phys. Metals Metallog.,1967, 22,(2), 44-49.

137. W.A. WONG, H.J. McQUEEN and J.J. JONAS: J. Inst. Metals, 1967, 95, 129-37.

138. J.J. JONAS, H.J. McQUEEN and W.A. WONG: Deformation Under Hot Working Conditions, Iron and Steel Inst.,London, 1968, 49-59.

139. H.J. McQUEEN, W.A. WONG and J.J. JONAS: Can. J. Phys.,1967, 45, 1225-35.

140. W.A. WONG and J.J. JONAS: Trans. TMS-AIME, 1968, 242, 2271-80.

141. G. GAGNON and J.J. JONAS: ibid.,1969, 245, 2581-89.

142. J.J. JONAS and F. MÜLLER: Can. J. Earth Sc.,1969, 6, 963-68.

143. T. CHANDRA and J.J. JONAS: Met. Trans.,1970, 1, 2079-82.

144. T. CHANDRA and J.J. JONAS: Met. Trans.,1971, 2, in press.

145. C.E. PEARSON and R.N. PARKINS: The Extrusion of Metals, John Wiley & Sons, N.Y.,1960, 142-222.

146. J.M. ALEXANDER and R.C. BREWER: Manufacturing Properties of Materials, Van Nostrand, Princeton, N.J.,1963.

147. A.M. SABROFF, F.W. BOULGER and H.J. HENNING: Forging Materials and Practice, Reinhold, N.Y.,1968.

148. W.J.McG. TEGART: Elements of Mechanical Metallurgy, MacMillan, N.Y.,1966.

149. D. HUTCHISON: private communication.

150. D.J. ABSON and J.J. JONAS: Met. Sci. J.,1970, 4, 24-28.

151. O. KOSIK: Master's Thesis, McGill University, 1971.

152. J.L. UVIRA: Report, Stelco Research Centre, Burlington, Ont.

153. R.L. CRAIK, M.J. MAY and D.J. LATHAM: Met. Eng. Q.,1969, 9, (3), 12-21.

154. E.B. KULA and S.V. RADCLIFFE: J. Metals, 1963, 15, 755-62.

155. W.E. DUCKWORTH, P.R. TAYLOR and D.A. LEAK: J. Iron Steel Inst., 1964, 202, 135-142.

156. O. JOHARI and G. THOMAS: Trans. ASM, 1965, 58, 563-78.

157. D.J. SCHMATZ: Met. Eng. Q.,1966, 6, (2), 20-24.

158. P.W. BRIDGMAN: Studies in Large Plastic Flow and Fracture, McGraw-Hill, N.Y.,1952.

159. M. COOK and E.C. LARKE: J. Inst. Metals, 1945, 71, 371-90.

160. W.A. BACKOFEN, A.J. SHALER and B.B. HUNDY: Trans. ASM, 1954, 46, 655-80.

SIMULATION OF EXTRUSION STRUCTURES BY MEANS OF TORSION TESTING FOR A HIGH STRENGTH NICKEL-BASE ALLOY, UDIMET 700

Conrad M. Young, Graduate Research Assistant, and Oleg D. Sherby, Professor, Department of Materials Science, Stanford University, Stanford, California

I. INTRODUCTION

The simulation of hot forming operations by means of torsion testing has been performed on Udimet 700, a nickel base alloy. The purpose of this research was threefold: 1) to evaluate the hot working (T > .7 Tm) properties of the U -700 alloy, (tests were conducted as low as .45 Tm), 2) to attempt to simulate the extrusion of this alloy by means of torsion testing under comparable conditions of temperature, strain rate and strain, and 3) to evaluate the torsion test as a laboratory technique for developing hot worked microstructures under controlled conditions of temperature, strain, and strain rate.

The first objective required the determination of the alloy's resistance to plastic flow (strength) and ductility as a function of temperature and strain rate over the range of temperature and strain rate commonly used in forming this alloy. The ductility was determined as the true strain at fracture in the torsion test and the strength was continuously recorded as a function of the strain. Tests were also interrupted at predetermined amounts of strain to duplicate the extrusion strains and these were examined using microhardness and optical microscopy to characterize the hot worked microstructure. The advantages of a hot torsion test are shown in the successful development of microstructures similar to those developed in extrusion, the accurate determination of the stress, strain and temperature with control of the atmosphere not possible in hot working operations, the large deformations possible, and the ability to perform tests at a laboratory scale which produce samples massive enough so that subsequent evaluations of mechanical properties could be done. The goal of developing a simple laboratory test which can easily measure the hot workability or ductility of new alloys by means of a few tests has not been reached. Since the ductility is a strong function of the stress state, several different types of tests must be performed such as tension, compression and torsion. However, the ability of laboratory testing to predict hot working parameters should make the introduction of new alloys to practical forming techniques much more economical and also allow a greater control over the final properties of the product material.

II. EXPERIMENTAL PROCEDURE

The experimental work was carried out on the torsion test machine shown in Figure 1.

Figure 1. View of the Stanford torsion test machine showing the drive section at the left and testing section to the right.

The machine consists of two components, the drive train and the test section. A 10 h.p. electric motor drives two four-speed truck transmissions and two chain sprocket assemblies in series. The highest strain rate is $10.45 sec^{-1}$ while the lowest strain rate is 3.16×10^{-5} sec^{-1} with 32 different strain rates available over these five orders of magnitude. Large fly-wheels maintain a constant rate of rotation and the drive section is isolated from the testing section to minimize vibrations. A heavy magnetic multiple disk clutch couples the drive train to the test section. The test section input shaft is held by an electric band brake. At the moment the clutch is engaged the brake releases and the sample is deformed essentially at the speed of the rotating machine. At the highest strain rate the sample is deformed at a uniform rate by 0.2 twists as determined from the strain rate indications. The testing section, as shown in Figure 2 consists of two shafts: a rotating input shaft and a stationary shaft.

Figure 2. Testing section showing input shaft on the left joined by the sample to the stationary shaft on the right. The clam shell elliptical oven is open showing the quartz atmosphere tube. The water quench enters through the small side arm on the quartz tube.

The specimen connects the two shafts and is placed in slotted grips which align the sample with the axis of rotation. One end of the specimen is pinned while the other "free" end has clearance to move axially during the deformation. The torsional strain is measured by means of contacts mounted on a disk which can give an event mark for each 30° of input shaft rotation on the Sanborn two-channel recorder. The torsional moment is measured by a lever arm attached to the stationary shaft. The torsional force pulls on a 5 KG Instron load cell and the output from this strain-gage Wheatstone bridge is amplified for the recorder. The samples used for this investigation were primarily of the "standard" type shown in Figure 3. The torsion machine can deform this type of sample at strain rates up to 10.45 sec^{-1}. For higher strain rates a shorter gage length and a larger diameter sample can be used. One type of high strain rate sample is shown in Figure 3 and this specimen can be deformed at up to 41.8 sec^{-1}. A dual elliptical radiant heating furnace encloses the test section. The furnace is water jacketed and capable of temperatures up to 1200°C with very rapid heating rates. A quartz tube runs through the furnace enclosing the shafts and sample. High purity (99.999%) argon is circulated along the entire length

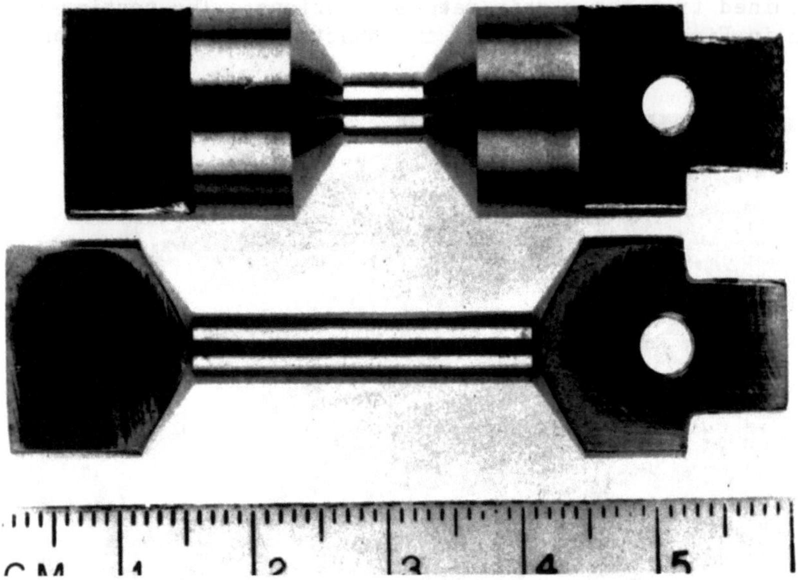

Figure 3. Torsion samples presently used. Bottom: "standard" sample, 1" x 0.20" dia. gage length, top; "high strain rate" sample, 1/4" x 0.20" dia. gage length.

of the shafting during heating and testing to prevent oxidation. Distilled water is used to quench in place inside the quartz tube at the end of a test.

To simulate the extrusions and interrupt the test at a predetermined strain at strain rates as high as 14 sec^{-1} it was necessary to build a timer circuit to program the clutch and brake. At the start of a test, the brake is turned off by the timer and the clutch engages. After the pre-set time of deformation, the clutch releases and the brake stops the deformation. Power to the furnace is immediately switched off, the water quench solenoid is actuated, and the sample is quenched in situ.

Torsional strain is defined as the distance moved, under an applied torque, by a reference point on the surface of the specimen divided by the original gage length. Thus the torsional strain, γ, is equal to $\frac{\pi D}{L_o}$ x N where N = number of twists, D = gage diameter,

N = number of twists, D = gage diameter, and L_o = the original gage length of the specimen.

Torque can be converted to torsional stress, τ, by using an analysis described by Nadai(1) and further developed by Fields and Backofen (2) to include the strain rate sensitive materials. Using the maximum shear stress formula, surface torsional stress is given by $\tau = \frac{M\ (3 + n + m)}{2\ \pi\ r3}$ where M is the applied torque, r is the radius of the gage section, n is the slope of log M versus log γ at constant strain rate, and m is the slope of log M versus log of the shear strain rate, $\dot{\gamma}$, at constant strain. At temperatures above .5 Tm, n can be generally taken as zero for metals (zero strain hardening rate) and m is typically about 0.2 (3). Substitution of these values into the above equation yields $\tau = \frac{0.51\ M}{r3}$. This equation can be used to calculate the maximum torsional stress values. However, it should be realized that m varies with temperature for each of the materials tested. The variation of m with temperature has only a minor influence on the calculated effective stress because it must be added to the much larger factor in the equation (values of 0.2 compared to 3).

The distortion-energy theory for yielding (4) is used to calculate effective stress, $\bar{\sigma}$, which is equal to the normal stress in tension and compression and $\sqrt{3}\ \tau$ in a torsion test. The effective strain, $\bar{\varepsilon}$, is equivalent to the true strain in tension or compression and equal to $\gamma / \sqrt{3}$ in torsion. Hodierne (5) has shown that there is a very small divergence between experimental results and the distortion-energy theory.

The torsion test samples were first polished then thoroughly cleaned to remove all machining oil. The specimen was then placed in the stationary grip and pinned to maintain the axial alignment. The thermocouple which was used to monitor the sample temperature was then wired to the sample with nickel wire to maintain thermal contact. The temperature was read at the center of the gage section but probing along the gage length at various temperatures gave only $\pm$ 2°C variation. The shaft with the thermocouples and sample in place was then pushed through the pillow block bearings, engaging the other grip. High purity argon was used to flush air out of the quartz tube around the sample prior to heating. The sample was then rapidly brought up to temperature and held at temperature for 20 minutes prior to testing. To begin a test the machine was first set at the desired strain rate then brought up to speed, the clutch being disengaged and the brake engaged. To start the test the clutch is engaged and the brake disengaged at the same time causing the sample to deform. The clutch remains engaged for only a predetermined time if the test is to be interrupted at a set strain and the brake is engaged at the end of this time as the clutch disengages. The recorder registers the torque and the strain events continuously. The sample is quenched in place immediately at the end of deformation by distilled water from a pressure tank controlled by a solenoid valve and removed when cooled to be examined.

The billets to be extruded are first lightly sandblasted to remove surface oxides and grease. This is followed by heating at 180°F and coating with a glass slip which acts as a lubricant and controls oxidation. The billet is then placed in a heating oven with an argon atmosphere for at least 1 hour to bring it up to the extrusion temperature. The billet is removed from the furnace and placed in the heated extrusion liner (500°F) which has been coated with a graphite lubricant. The transfer time is approximately 15 sec. The billet is backed-up with a graphite follower and the ram brought up close to the follower. The press is a 700 ton Lombard with specially instrumented stem and die so that the die, liner, and stem loads can be determined. The ram continues to move forward until the high pressure switch is actuated and the billet is upset and extruded at a speed which has been approximately pre-set by the high pressure valve. The ram actually has an initial velocity before extrusion and the velocity is discontinuous during extrusion but an approximate running speed can be determined. The billet is extruded through the die and is often coated with the glass lubricant into a run-out tube with a de-accelerator. The specimens used for comparison to torsion were removed from the run-out tube and water quenched.

The effective strains and strain rates for the experimental extrusions were based on relationships developed by Avitzur. (6) Using the assumption that all calculations are surface strains and strain rates and that the extrusion die half-angle (α) is 60°, then

$$\bar{\epsilon} = 2.24 \,\ell n \frac{D_o}{D_f} = 1.12 \,\ell n \, R$$

where $\bar{\epsilon}$ = effective strain at surface

D_o = initial upset diameter

D_f = final product diameter

R = reduction of area ratio

$$\dot{\epsilon} = \frac{\mathrm{Vram}\left(\frac{R_o}{R_f}\right)^2 R_f^{\,2}\, f(\alpha)\; 3\,(1 + \cos\alpha)\,\sin\alpha\left(\ln\frac{R_o}{R_f}\right)}{\left(R_o^{\,3} - R_f^{\,3}\right)}$$

substitution for f (α) = 1.034 = f (60°)

$$\dot{\epsilon} = \frac{4.04}{D_f}\;\mathrm{Vram}\left(\frac{D_o}{D_f}\right)^2 \frac{1}{\left(\frac{D_o}{D_f}\right)^3 - 1}\;\ln\left(\frac{D_o}{D_f}\right)^2$$

where $\dot{\epsilon}$ = average effective strain rate over the entire deformed volume in sec^{-1}

Strain rates and effective strains calculated from these relationships give higher values than the simple relations previously used by the authors to compare the torsion and extrusion deformations. The gradient of strain in an extrusion from the surface to the center is much lower than that achieved in torsion so the extrusion relationships are considered to be reasonable for comparison to torsional strains and strain rates. Micro-structures in both the torsion and extrusion must be examined fairly close to the surface to have any correlation based on comparable hot working conditions.

III. MATERIAL

Udimet 700 (U-700) alloy bars, hot rolled, centerless ground, 1/2" diameter (heat No. 3213) and 5/8" diameter (heat No. 3571) were used to characterize the hot working properties of the wrought alloy. This material was purchased from Special Metals Corporation, New Hartford, N.Y. and the analysis is given below.

U-700 Alloy Analysis (Wt % by Vendor)

Heat No.	C	Mn	Si	Cr	Ni	Co	Fe	Mo	Ti	Al
3213	.06	<.10	<.10	15.2	bal	19.0	.13	4.95	3.43	4.40
3571	.06	<.10	<.10	15.3	bal	17.6	.20	4.85	3.43	4.50
3138A	.135	<.10	<.10	14.7	bal	19.2	.21	5.05	3.20	4.45

For the extrusion simulations 3" diameter by 6" long billets were obtained from Allvac Metals Company, Monroe, North Carolina, designated as heat No. C-6299, Allvac No. 3138A, heat treated 4 hours at 1080°C, air cooled. These billets were from the same lot as the extrusions which were done on this alloy. The torsion samples were machined from these billets with the axis of the torsion specimen aligned with the billet axis. Photomicrographs of the as-received billet material are shown in Figures 4 and 5. These microstructures show the very fine gamma prime (γ') or Ni_3 (Ti, Al) precipitate which has formed as the result of the heat treatment together with well distributed large carbide (MC or $M_{23}C_6$) and boride precipitates (M_3B_2) also plainly evident in the gamma (γ) matrix. It was not possible to determine the γ grain size in the billet materials. The phases present in wrought U-700 are very dependent upon the thermal history of the alloy as are the mechanical properties so the simulation of the hot working process requires close control of heating rates, time held at temperature, and cooling rates. The significant difference in carbon content between the billets (heat No. 3138A) and the wrought bar stock would seem to account for the observed lower ductility and higher flow stress of the billet material due to the greater influence of the carbides on the mechanical properties (11).

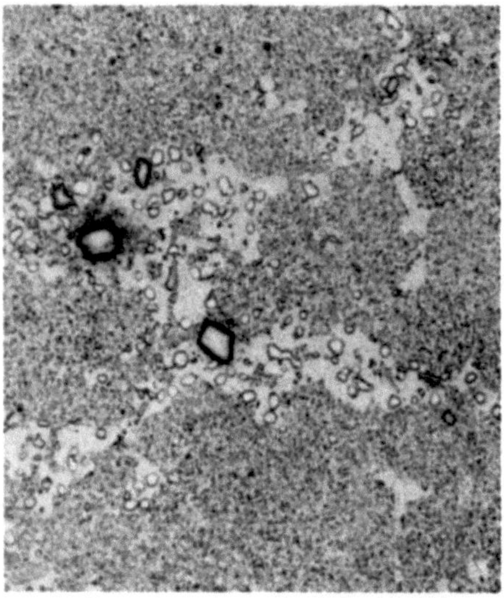

Fig. 4 U-700 billet material as-received, longitudinal section (500X).

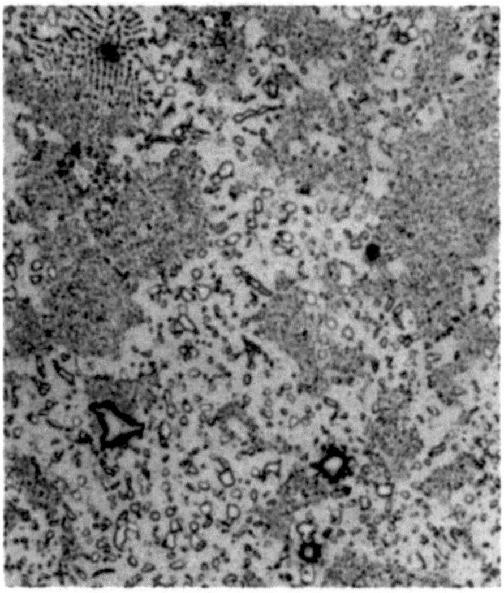

Fig 5 U-700 billet material as received, transverse section (500X).

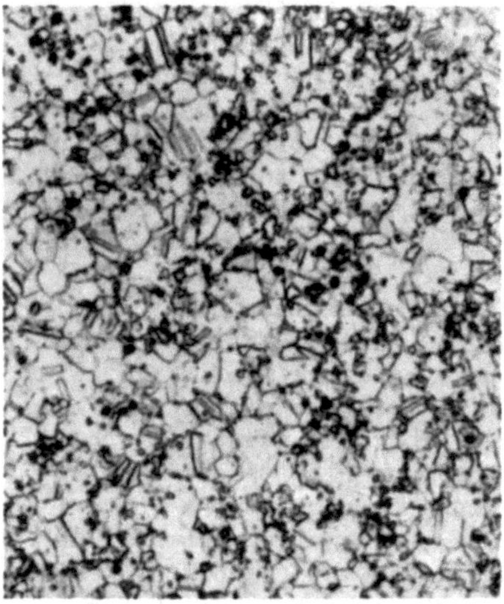

Fig. 6 Extrusion #2459, 1062°C, $\dot{\varepsilon} = 7.95\ sec^{-1}$, $\varepsilon = 1.95$, transverse section (500X).

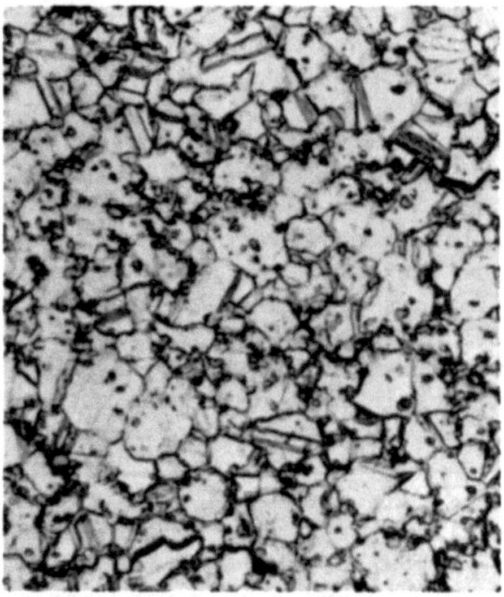

Fig 7 Extrusion #2459, longitudinal section (500X).

IV. RESULTS AND DISCUSSION

The microstructures of U-700 developed in extrusion can be compared to microstructures developed in torsion at the same temperature, strain rate and strain. The microstructure of an extrusion done at 1062°C, $\dot{\epsilon}$ = 7.95 sec^{-1}, ϵ = 1.95 is shown in Figures 6 and 7. Grain boundaries are now evident and a fine grain size has been developed by the deformation. The carbide and boride precipitates now appear mainly in the grain boundaries. This should be compared to the undeformed material shown in Figure 8 which has experienced nearly the same thermal histroy yet retains a structure similar to the original billet material shown in Figures 4 and 5. A torsion specimen which developed a microstructure very similar to the extrusion is shown in Figure 9. The strain is somewhat lower than that experienced by the extrusion, but the grain size and precipitate size and distribution are very similar. There are more annealing twins evident in the extrusion, possibly indicating a slower cooling rate for the extrusions. Simulation of larger extrusion reduction ratios are shown in Figures 10 and 11. The grain boundaries can no longer be etched and the precipitates appear spheroidized. The former longitudinal texture still evident in Figure 9 has disappeared and a radial plane texture is clearly developing at a strain of 2.76.

The extrusion microstructure developed at 1100°C, $\dot{\epsilon}$=8.95 sec^{-1}, ϵ = 2.39 is shown in Figures 12 and 13. The transverse section shows a fairly uniform distribution of precipitates both in the grain boundaries and within the grains which are nearly equiaxed. Undeformed material which has experienced nearly the same thermal history as both the extrusion and the torsion specimens is shown in Figure 14. This structure strongly resembles that of the extrusion except that the precipitates are present mainly in the grain boundaries and there is inhomogeneous banding of precipitates and a large amount of very fine structure, possible γ', still present. Photomicrographs of torsion samples deformed to simulate the extrusion operation are shown in Figures 15-17. The rotation of the texture with strain is again evident. Straining appears to lead to recrystallization which refines the grain size and forms coarsened spheroid precipitates. The torsion and extrusion microstructures appear quite similar. There has been considerable solution of the second phase γ', as well as the other precipitates when compared to the microstructure of the original billet material but grain growth has not been excessive due to the large number of precipitates still present.

Photomicrographs of the U-700 extrusion deformed at 1150°C, $\dot{\epsilon}$ = 4.46 sec^{-1} and ϵ = 2.23 are shown in Figures 18 and 19. These appear quite similar to the U-700 billet material deformed in torsion under similar conditions to three different strains as shown in Figures 21 to 23. The majority of the precipitates are now in solution but a fine grain size is still retained. There appears to be

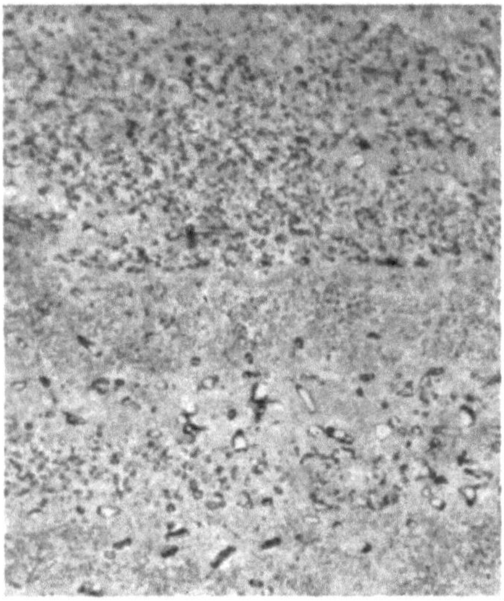

Fig. 8 Undeformed longitudinal grip section of U-700 sample deformed in torsion at 1050°C (500X).

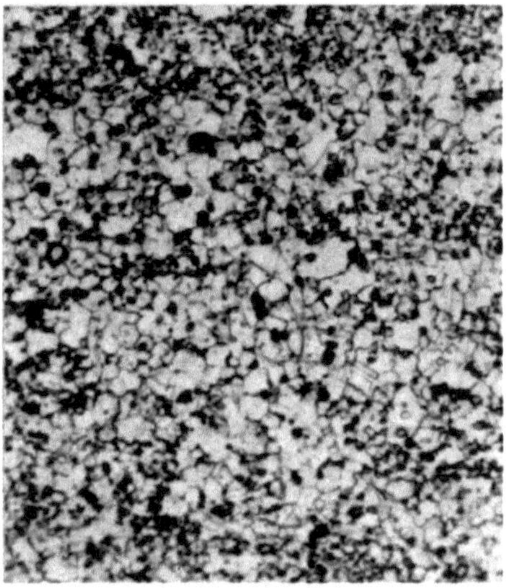

Fig. 9 U-700 torsion specimen, 1058°C, $\dot{\varepsilon}$ = 6.33 sec^{-1}, ε = 1.77 longitudinal section (500X).

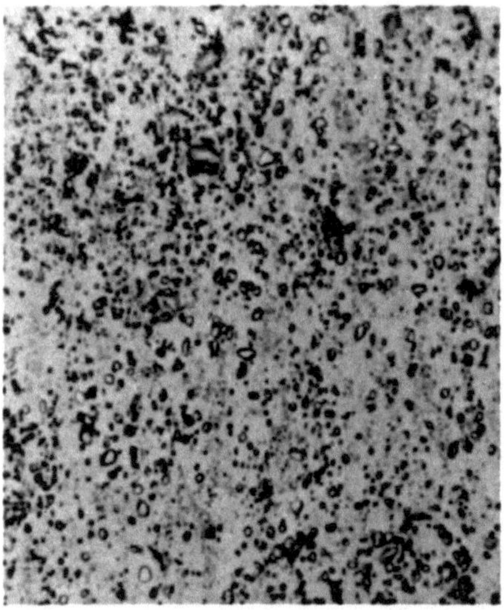

Fig. 10 U-700 torsion specimen, 1050°C, $\dot{\varepsilon}$ = 6.33 sec^{-1}, ε = 2.2, longitudinal section (500X).

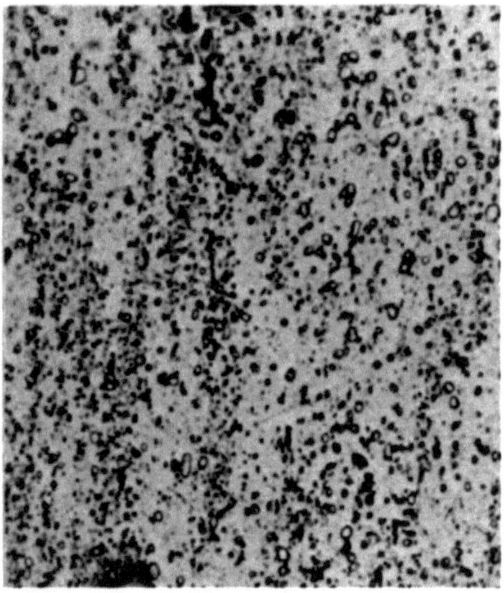

Fig 11 U-700 torsion specimen, 1050°C, $\dot{\varepsilon}$ = 6.33 sec^{-1}, ε = 2.76, longitudinal section (500X)

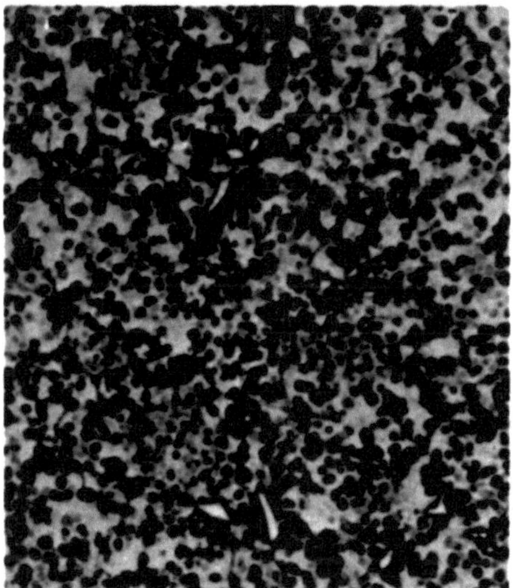

Fig. 12 Extrusion #2449 1100°C, $\dot{\varepsilon}$ = 8.95 sec^{-1}, ε = 2.39, transverse section (500X).

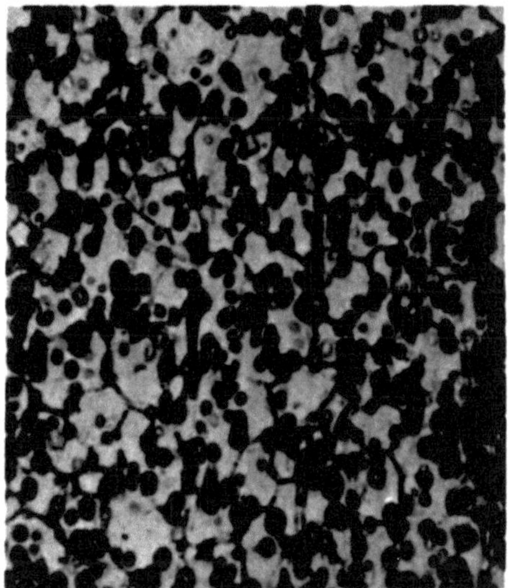

Fig. 13 Extrusion #2449, longitudinal section (500X).

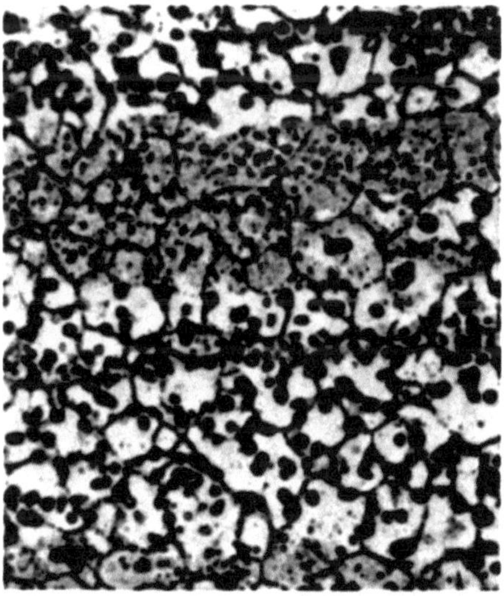

Fig. 14 Undeformed longitudinal grip section of U-700 sample deformed in torsion at 1100°C (500X).

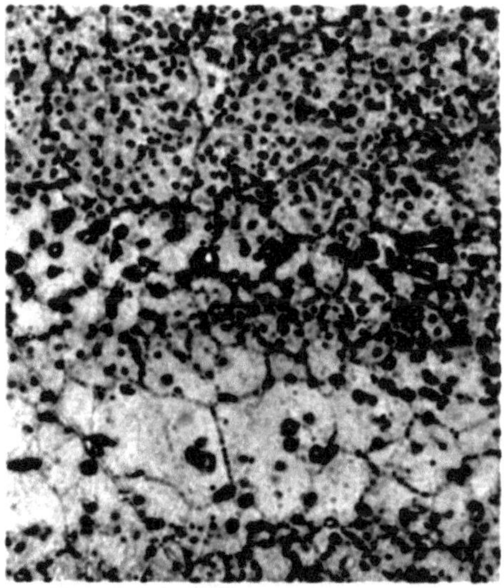

Fig. 15 U-700 torsion specimen, 1100°C, $\dot{\varepsilon}$ = 13.88 sec^{-1}, ε = 1.46, longitudinal section (500X).

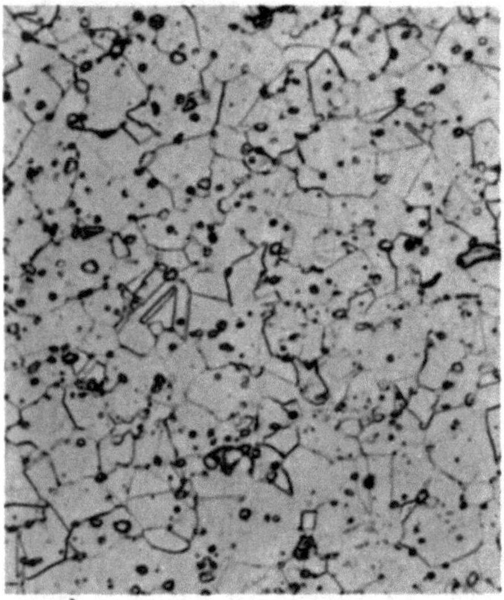

Fig. 16 U-700 torsion specimen, 1098°C, $\dot{\varepsilon}$ = 10.5 sec^{-1}, ε = 2.3, longitudinal section (500X).

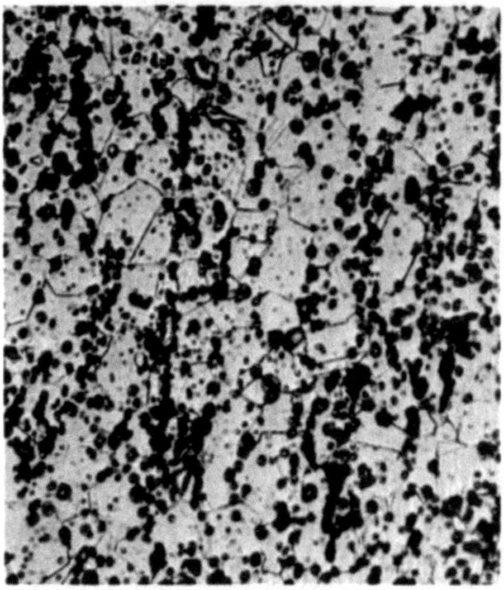

Fig. 17 U-700 torsion specimen, 1100°C, $\dot{\varepsilon}$ = 13.88 sec^{-1}, ε = 2.77 longitudinal section (500X).

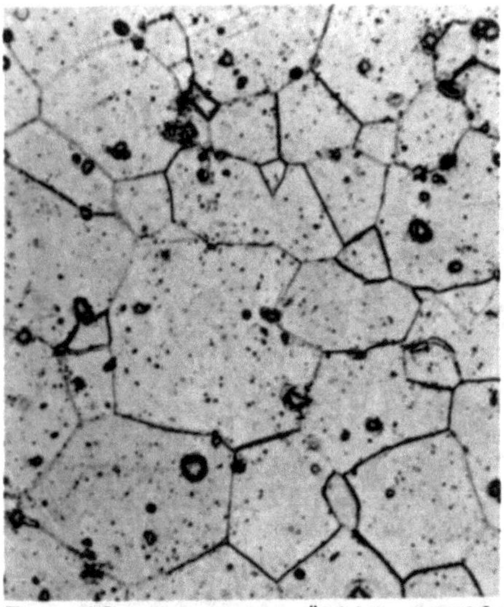

Fig. 18 Extrusion #2442, 1150°C, $\dot{\varepsilon}$ = 4.46 sec^{-1}, ε = 2.23, transverse section (500X).

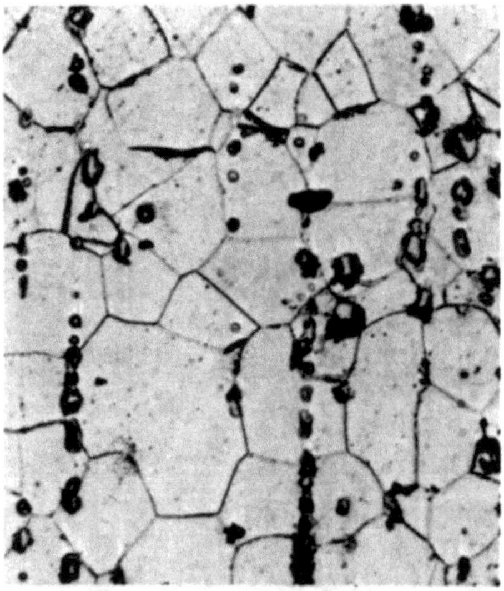

Fig. 19 Extrusion #2442, longitudinal section (500X).

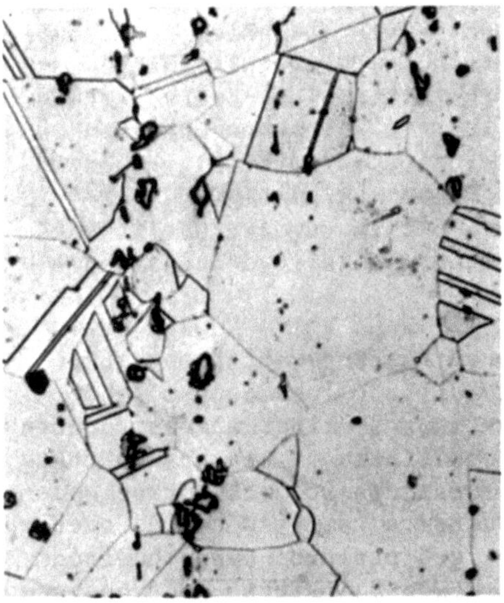

Fig. 20 Undeformed longitudinal grip section of U-700 sample deformed in torsion at 1152°C (250X).

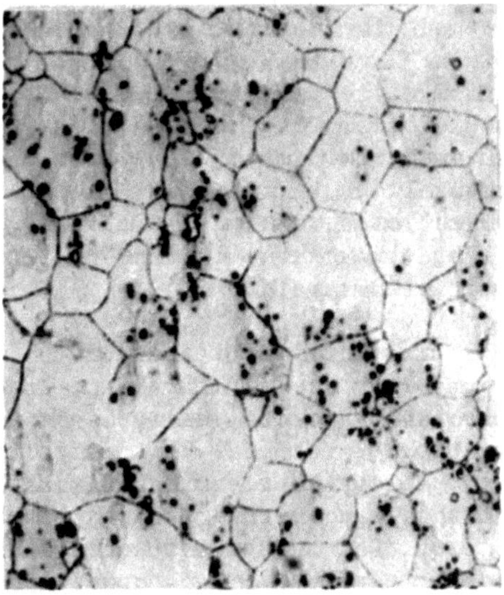

Fig. 21 U-700 torsion specimen 1152°C, $\dot{\varepsilon}$ = 3.78 sec^{-1}, ε = .82, longitudinal section (250X).

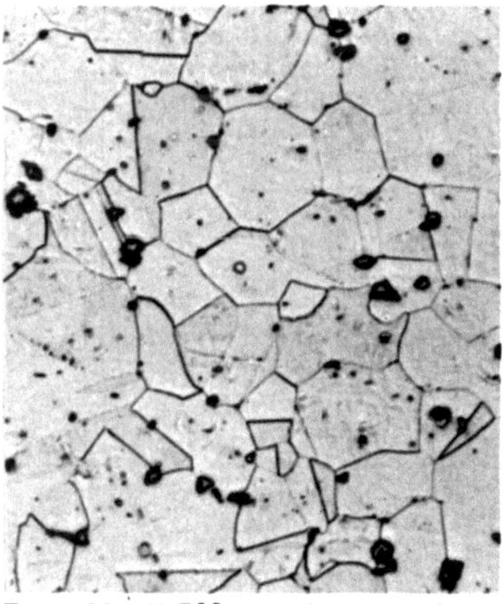

Fig. 22 U-700 torsion specimen 1142°C, $\dot{\varepsilon}$ = 3.78 sec^{-1}, ε = 2.1, longitudinal section (250X).

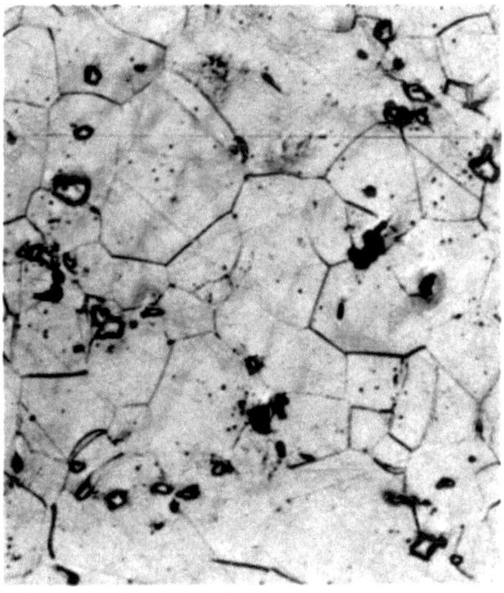

Fig. 23 U-700 torsion specimen, 1146°C, $\dot{\varepsilon}$ = 3.8 sec^{-1}, ε = 4.36, longitudinal section (250X).

more grain growth in the undeformed material as shown in Figure 20 and annealing twins are very evident. The difference in grain size between the deformed and the undeformed sample (compare, for example, Figure 22 with Figure 20) lends support to a recrystallization mechanism of hot working. Continuous recrystallization during working would be expected to produce a smaller grain size than observed in an undeformed material even though the deformed material is heated to a higher temperature than the undeformed (grip) region. The effect of strain on the microstructure at this temperature is not clear, though the quenched-in grain size appears to decrease somewhat with strain.

Extrusion simulation of U-700 over the range of temperature normal for hot working this alloy (1050-1150°C) by means of the laboratory torsion test has shown that very similar microstructures could be developed. The controlled conditions of the torsion test can be used to determine this alloy's response to hot working conditions, and further testing of the deformed samples, could develop information on the influence of the microstructure upon the room temperature mechanical properties. This data in turn could be used as a guide to control the hot working parameters.

Typical examples of the stress-strain curves obtained in torsion for U-700 alloys are shown in Figure 24.

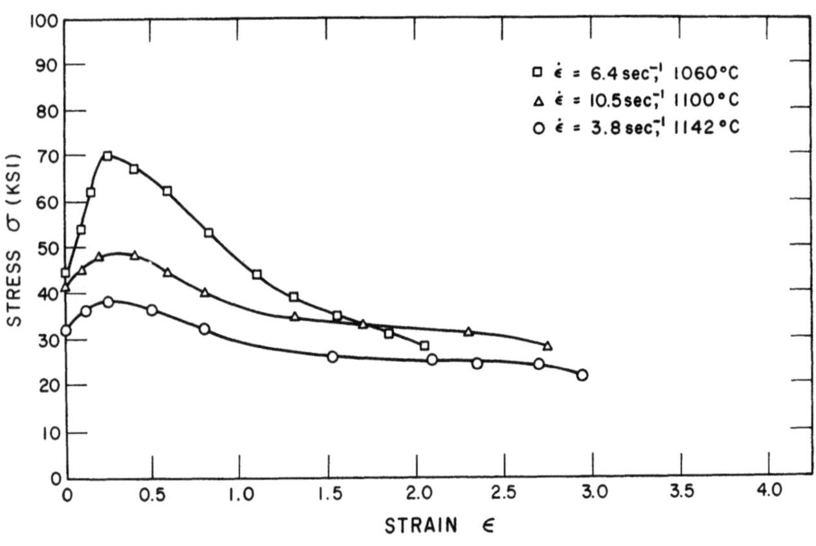

Fig. 24 Interrupted test stress-strain curves for U-700 billet material which simulates the extrusions done on this alloy.

Each curve represents one of the torsion tests done to simulate an extrusion on the identical material. It was observed that increasing the strain rate or decreasing the temperature increased the peak flow stress over the narrow range of hot working temperatures and strain rates studied. It is believed that the strain softening which occurs after the peak flow stress is due to both recrystallization and the increase in temperature of the sample from the deformation. After the strain softening a plateau flow stress is evident where a balance has been reached between hardening processes and the temperature increase-recrystallization softening processes.

There has been a large amount of speculation as to what the drop in stress with strain in torsion means. Dieter and Shapiro (7) have found in high purity nickel and Udimet 600 that microcracks form just beyond the peak flow stress. These cracks are prevented from propagating by the formation of new recrystallized grains nucleated at triple points in the grain boundary. It is believed by Dieter that the large strains which are possible in hot working result from the migration of the new recrystallized grain boundaries which blunt the cracks. There has been no evidence for the formation of cracks in the U-700 alloy just after the peak stress and examination in the uniform flow stress plateau region has not shown any intergranular cracking.

Recently there has been evidence presented to support both recrystallization and dynamic recovery by glide and climb of dislocations to explain the strain softening mechanism in hot working. Stüwe (8) has proposed that aluminum, copper, nickel and 18/8 stainless steel all deform by the mechanism of dynamic recovery where a stable substructure is formed within the original grains and is maintained by climb and glide of dislocations during deformation. Rapid recrystallization can occur immediately after the start of deformation leading to a decrease in the resistance to plastic flow. Tegart (9) has presented evidence that recrystallization occurs during deformation in hot torsion and is the true mechanism while the material is being strained. The actual shape of the torsional stress-strain curve may be calculated using recrystallization kinetics of nucleation and growth. This recrystallization during deformation occurs in metals with a low stacking fault energy where climb would be more difficult because of the widely spaced partial dislocations. A different activation energy is calculated for the hot working of those materials which recrystallize during deformation than would be expected from the dynamic recovery model operating in creep. Tegart proposed the following relationship based on the work of Garofalo (10).

$$\dot{\epsilon} = A(\sinh \alpha \sigma)^{n'} \exp(-Q/RT)$$

where $\dot{\epsilon}$ is the effective strain rate, A, α, n' are constants independent of temperature, σ is the effective stress, Q is the activation energy, R is the universal gas constant, T is the absolute temperature. From this relationship an activation energy can be calculated from hot torsion tests which gives an indication of the deformation mechanism. Tegart has found that the activation energy for creep and hot working is nearly that for lattice self-diffusion in metals such as aluminum and α iron. That is, the activation energy is nearly constant over a wide range of strain rates and indicates that the same recovery processes are operative over this entire range of strain rates from creep to hot working. However, for other metals such as 18/8 stainless steel, copper, nickel and nickel-iron alloys the activation energy for hot working is different from that of lattice self-diffusion, typically being much higher and can be associated with a different process which is believed to be recrystallization. Thus when hot working some alloys, and this is believed to be the case for U-700, recrystallization operates to reduce strain hardening and large strains are reached by repeated recrystallization. This explains the more equiaxed structure and decrease in grain size observed after large deformations.

Other factors which may contribute to the peculiar drop in stress with strain observed in torsion may be, 1) heating due to nearly adiabatic conditions during deformation at the high strain rates, 2) localized deformation which subsequently propagates through the sample, 3) impact and inertia effects from the machine, and 4) texturing of the microstructure with straining.

The calculated temperature increase with strain for the U-700 alloy assuming adiabatic conditions has been shown to be significant even at fairly low strains at high strain rates (11). The peak flow stress normally occurs at a strain less than 1.0 and at this strain the temperature rise is 60°C for a starting temperature of 900°C at a strain rate of 10.5 sec^{-1}. The decreasing stress with strain after the peak flow stress can at least partially be attributed to this temperature rise.

The possibility that some form of localized deformation which would strain harden a small region of the gage length and thus propagate along the gage length with strain, was suggested by several tests which were interrupted and showed evidence of localized strain. The stress-strain curves for these samples were nearly identical to those obtained for duplicate tests taken to fracture. The uniform flow stress, however, was slightly below that of the samples tested to separation and subsequent tests showed that a misalignment had led to localized deformation.

The effect of the inertia of the recording system was tested to be certain the stress-strain curve was not a function of the machine. The inertia of the stationary torque arm was doubled and duplicate tests run at the highest strain rates. The peak in the stress strain curve was not changed as might be imagined if acceleration and overshoot of the torque arm was a factor and virtually identical stress-strain curves were obtained.

A further complication arising at large deformations is that, as the deformation increases, the specimen may assume a more and more pronounced deformation texture. This has been proposed as a possible explanation for the steady state flow stress (12) in polycrystalline copper where the hardening-induced increase in the shear stress from cold working is compensated for by a decrease in the shear stress caused by a change in the original grain texture. It seems that this is not the case, however, for the U-700 alloy. For this alloy the deformation mechanism is apparently one of continuous recrystallization. The microstructure as observed after quenching appears to suggest a recrystallized texture.

It is our impression that the main cause of the decrease in flow stress with straining at high strain rates is due primarily to adiabatic heating. Meaningful correlations between strain rate and flow stress at a given initial test temperature should therefore only be made with the maximum flow stress observed (which occurs at small strains).

The influence of temperature on the torsional ductility of U-700 under hot working conditions is shown in Figure 25.

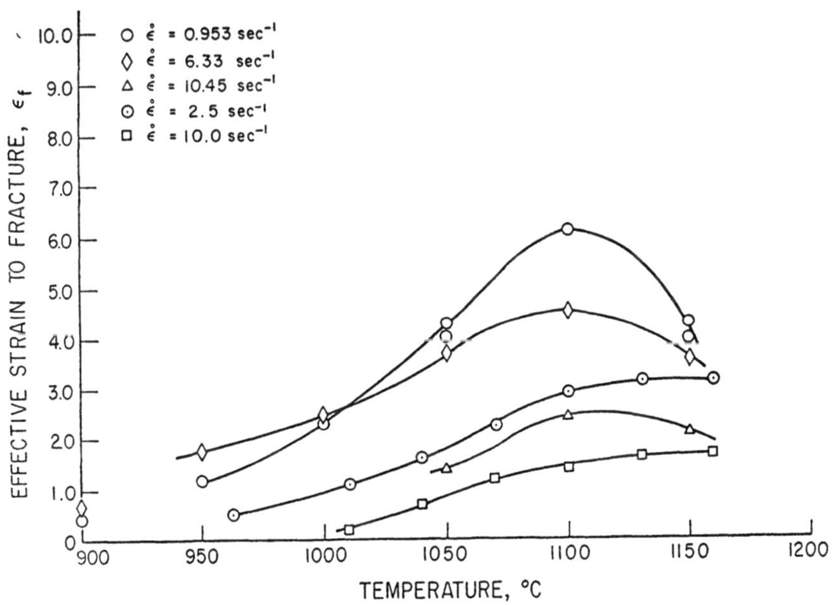

Fig. 25 Influence of temperature on the torsional ductility of U-700.

The normal hot working range for this alloy is 1050-1150°C and the torsion tests indicate that this is the range where the ductility is quite high. The data of Hoffmanner (13) also show the same trend of ductility with temperature. The maximum ductility occurs at 1100°C for strain rates from 0.953 sec^{-1} to 10.45 sec^{-1}. It was found, however, that for a strain rate of 0.0006 sec^{-1} a minimum in the ductility occurred at 1100°C (to be described later). This has not been explained other than the possibility that the deformation mode is different at slow strain rates, that is, the deformation process could be dynamic recovery at slow strain rates and recrystallization at high strain rates.

A comparison was made of the data obtained from our hot torsion tests on Udimet 700 with the results obtained in tension by Grant et al. (14). Analyses of the results obtained revealed three important findings:

(1) A close correlation was obtained between the effective stress calculated from torsion data and the tensile flow stress when compared at the same effective strain rate; (2) the torsional ductility-temperature curves were very similar to the tensile ductility-temperature curves when compared at the same effective strain rate and (3) it would appear that a relation exists between the true strain to fracture in tension and the true strain to fracture in torsion for ductile materials; when the ductility is low (i.e. below $\bar{\varepsilon} \approx 1.0$) the correlation is such that $\bar{\varepsilon}$ torsion = $\bar{\varepsilon}$ tension; but when the strains to fracture are large then $\bar{\varepsilon}$ torsion > $\bar{\varepsilon}$ tension. The findings mentioned above are described in detail in the following paragraphs.

Grant et al. (14) performed very short time stress-rupture tests on cast and wrought Udimet 700. Their data are plotted as initial stress versus strain rate in Figure 26 for a temperature of 1150°C (2100°F).

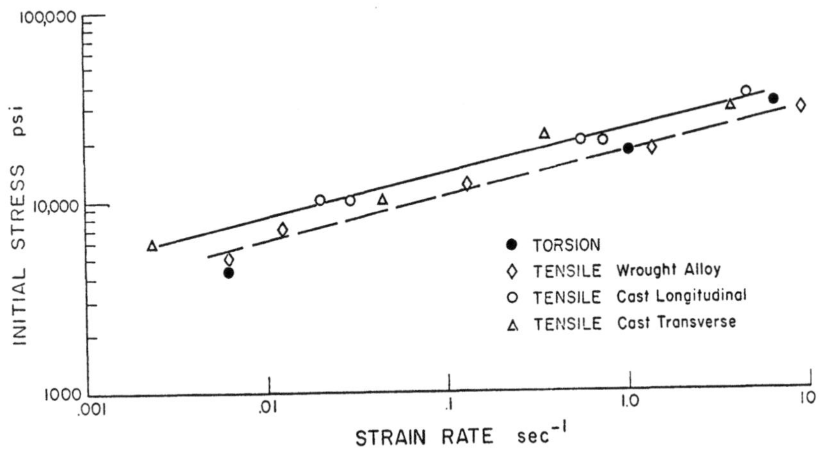

Fig. 26 Relation between strain rate and initial stress for Udimet 700 at 1150°C.

Strain rate was determined by dividing the elongation to fracture by the time to rupture. As can be seen from Figure 26, the cast Udimet 700 is slightly stronger than the wrought product. Our torsion results on wrought Udimet 700 are also shown in Figure 26 by the data given as filled circles; the initial torsional stress during deformation was plotted as effective stress ($\sigma = \tau \sqrt{3}$) and the torsional strain rate as effective strain rate ($\dot{\varepsilon} = \dot{\gamma}/\sqrt{3}$). It can be seen that the torsion data on wrought Udimet 700 fall in very closely with the corresponding tensile data.

Grant et al. reported tensile ductility of Udimet 700 in terms of elongation to fracture and also percent reduction of area. In order to properly compare their data with our torsional ductility data, the reduction of area values were converted to true tensile fracture strain. Thus, it can be readily shown that

$$[\varepsilon \text{ fracture}]_{\text{tension}} = \ln \frac{Ao}{Af} = \ln \left(\frac{1}{1 - (\frac{\% \, RA}{100})} \right)$$

The data of Grant et al. plotted as ε fracture versus strain rate at various temperatures are shown in Figure 27.

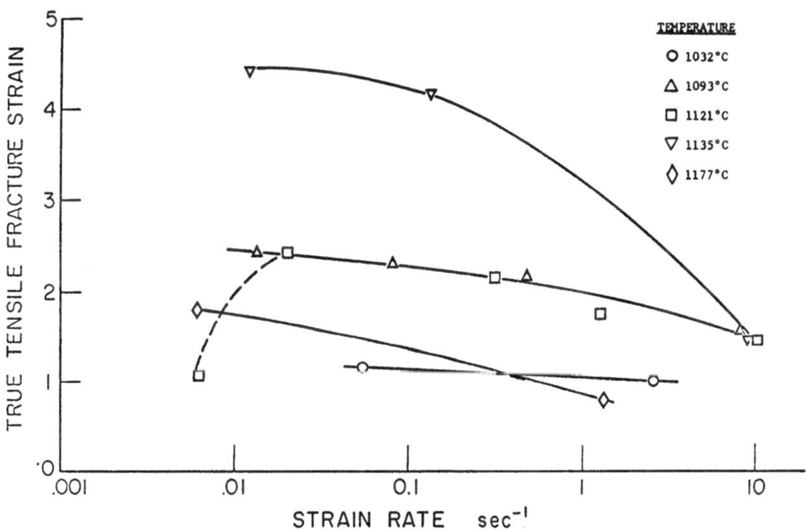

Figure 27 Influence of strain rate and temperature on the true tensile fracture strain for Udimet 700 [derived from data of Grant, et. al (14)].

These results are compared with the torsional ductility results ($\varepsilon_f = \gamma/\sqrt{3}$) at two effective strain rates in Figures 28 and 29.

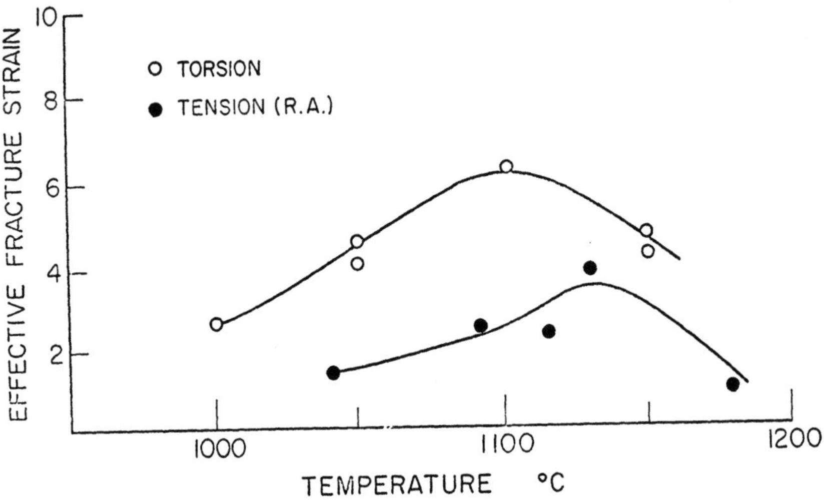

Figure 28. Comparison of torsion and tensile data of effective fracture strain vs. temperature for Udimet 700 ($\dot{\varepsilon}$ = 1 sec^{-1}).

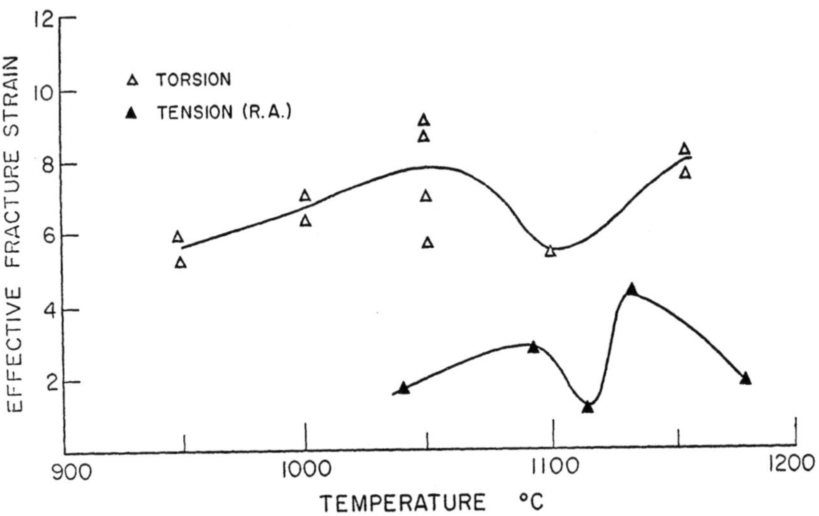

Figure 29. Comparison of torsion and tensile data of effective fracture strain vs. temperature for Udimet 700 $\dot{\varepsilon}$ = .006 sec^{-1}).

At a relatively high strain rate of $\dot{\epsilon}$ = 1 $secs^{-1}$, Figure 28, the torsion and tension data show similar shaped curves revealing a maximum ductility at about 1100°C. At a slow strain rate of $\dot{\epsilon}$ = .006 $secs^{-1}$ the ductility-temperature curves again appear similar for the torsion and tension tests (Figure 29). In this case, however, a ductility minimum is noted at about 1100°C with the possibility of two maxima on each side of the minimum.

In the previous paragraphs evidence was presented showing the close relation between torsion and tension data. A recent study by Latham and Cockroft (15) also revealed close correlation between torsion and tension data in relation to ductile fracture. Their study involved a number of materials tested at room temperature. Halford and Morrow (16) also related tension and torsion ductility data for four metallic alloys at room temperature. All of these data are plotted in Figure 30 together with the elevated temperature data for Udimet 700 described earlier (Figures 28 and 29). The general correlation obtained suggests that $\epsilon_{torsion} = \epsilon_{tension}$ at low strains to fracture, whereas $\epsilon_{torsion} > \epsilon_{tension}$ at large strains to fracture. Further theoretical and experimental studies would appear to be in order to determine the reason for the interesting correlation obtained.

V. CONCLUSIONS

1. Microstructures of U-700 developed in extrusion could be simulated by means of the torsion test in the temperature range of 1050 to 1150°C. At each temperature in the hot working range a distinctive structure was developed which could be characterized by optical microscopy. The microstructures developed in torsion seem to be the result of recrystallization during deformation and after a strain of approximately 2.0 the structure appears to become independent of strain. The original longitudinal texture present in the billet material is sheared and a new texture develops on radial planes at large strains in torsion.

2. Hot working of U-700 appears to involve repeated recrystallization. This is manifest in the shape of the stress-strain curve which shows a strain softening leading to a steady state flow stress and by the difference in grain size between the deformed and non-deformed sections of the same as-quenched sample.

3. The ductility versus temperature curves for U-700 clearly defines the optimum hot working range of temperature and strain rates for this alloy and serve as an example of the usefulness of the torsion test to predetermine working parameters.

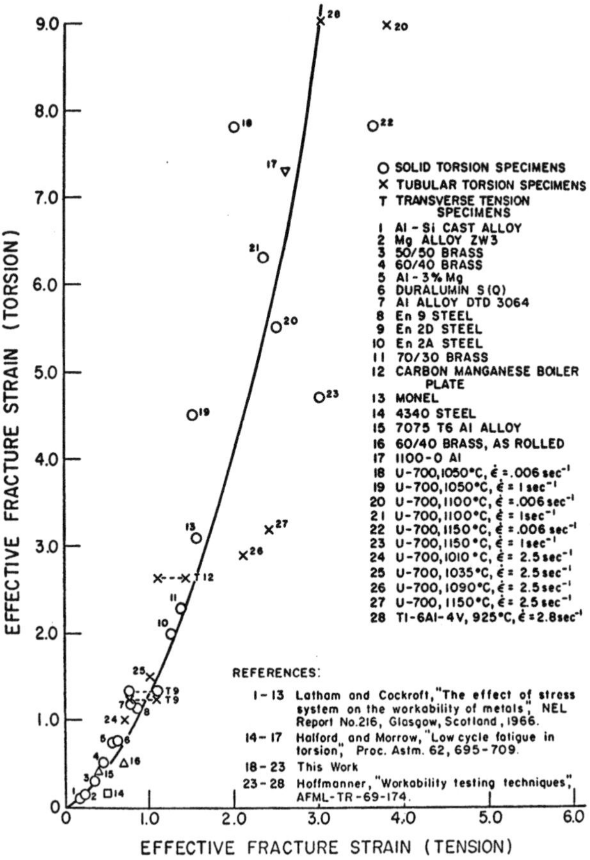

Fig. 30 Relation between torsion and tension values of effective fracture strain for a number of alloys.

ACKNOWLEDGMENTS

The authors are very grateful to the personnel in the Processing and Nondestructive Testing Branch of the Air Force Materials Laboratory for their generous assistance during the conduct of this program. This paper has been based on the work reported in AFML-TR-69-294. Special thanks are extended to Messrs. Shingo Inouye, Vincent dePierre, and Atwell Adair for their cooperation and encouragement, especially in relation to the extrusion experiments performed at their laboratory. The authors would also like to acknowledge the important contribution to the program made by Mr. Nels Wicklund and Mr. Timothy Reiley in the torsion testing program and in the metallographic studies.

VI. REFERENCES

1. A Nadai, Theory of Flow and Fracture of Solids, McGraw-Hill Book Co., Inc., New York, N.Y., (1950)349.

2. D. S. Fields and W. A. Backofen, Proc. ASTM, 57, (1957), 1259-1271.

3. O. D. Sherby, Acta Met., 10, (1962), 135.

4. G. E. Dieter, Jr., Mechanical Metallurgy, McGraw-Hill Book Co., New York, (1961), 60.

5. F. A. Hodierne, J. Inst. Metals, 91, (1962), 267.

6. Betzold Avitzur, Metal Forming: Processes and Analysis, McGraw-Hill Book Co., New York, 1968.

7. G. E. Dieter and E. Shapiro, "Fracture of Metals During Deformation Processing under Conditions of Hot Working," Final Report on Contract No. N00019-67-C-0251, April 30, 1968, Drexel Institute of Technology, Philadelphia, Pennsylvania.

8. H. P. Stüwe, Acta Met., 13, (1965) 1337.

9. W. J. Tegart, ASM, Metals Park, Ohio, (1968), 133.

10. F. Garofalo, Fundamentals of Creep and Creep-Rupture in Metals, McMillan Co., New York, 1965.

11. M. J. Harrigan, A. R. McIlree, C. M. Young, O. D. Sherby and O. C. Shepard, "Simulation of Hot Working Operations by Means of Torsion Testing," AFML-TR-68-225, August 1968.

12. H. G. Grewe and E. Kappler, Physica Status Solidi, 6, (1964), 339.

13. A. L. Hoffmanner, "Workability Testing Techniques, "AFML-TR-69-174, June 1969.

14. N. J. Grant, P. Bridenbaugh, B. Green and R. Kane, "Investigation of Fracture in Connection with Hot Deformation Processing of Metals," Naval Air Systems Command NOw66-0104-d, Final Report December 1966.

15. D. J. Latham and M. G. Cockroft. "Ductility and the Workability of Metals," Journal of the Institute of Metals, 96, (1968), 33.

16. G. R. Halford and J. Morrow, Proc. ASTM 62 (1962) 695.

PREDICTION AND EFFECTS OF MATERIAL DAMAGE DURING DEFORMATION PROCESSING

H. C. Rogers

Department of Metallurgical Engineering

Drexel University, Philadelphia, Pa.

Conventional deformation processes such as rolling, forging, wire drawing, and extrusion are continually being supplemented by newer processes like power spinning. Alloy designers are producing new and tougher materials at an accelerating rate. At the same time, there is a demand for improved product reliability at reduced costs through the use of higher speed continuous processes. The metal fabricator is therefore continually forced to reassess and readjust his processing practices. He can no longer enjoy the historical luxury of a completely empirical approach to the development of a new processing schedule with only minor extrapolations from known practices, but must use guide lines that at least limit the range of process variations to be considered.

The metal fabricator becomes aware that his processing practice could be improved in one of three ways. The most immediate and obvious is an inability to carry out the operation - fracture or visible cracks occur during processing. Only slightly less severe is the failure of the product to pass in-plant quality control tests or its failure to complete a subsequent step in the processing sequence because of the production of large internal fractures such as occur in "cuppy" wire.

The most subtle type of evidence that an improvement in processing procedures is needed is in the form of customer dissatisfaction - the product does not perform well in service, or it cannot be fabricated satisfactorily, or it compares unfavorably with a competitors material. This degradation in expected properties can be correlated with the degree of structural damage introduced into the metal during its deformation processing history. The term "structural damage" refers to the aggregate of voids and cracks found in a material following extensive deformation.

Once the fabricator is aware that a problem exists there are a number of process variables under his control that may be altered in an attempt to improve processing conditions. For example, in wire drawing these might be temperature, drawing speed, lubrication, reduction per pass, die angle or shape, and intermediate thermal treatment, if any. The degree of effectiveness of the remedy may not become apparent, however, at the same rate as that at which the problem is detected. For example, all failures in ductile metals must begin by the generation of structural damage. It is only the degree to which this is propagated to fracture that varies. Hence, a remedy that would satisfactorily prevent complete fracture during the initial processing stage might still leave the product with substantial damage or even internal cracks. It becomes obvious, therefore, that the minimization of structural damage through process control is one of the necessary goals in process design.

It was shown over a decade ago by Puttick(1), Rogers(2) and others that fracture in ductile metals initiates by the development of voids at a number of sites where compatibility of deformation is difficult such as grain boundary triple points, inclusions and second phase particles. Fracture proceeds by the growth and coalescence of these voids to form an internal crack. In a tensile specimen, substantial void growth and coalescence occurs only at the neck which forms after uniform elongations of approximately 30 percent. Not only does the strain concentrate in this region but the state of stress becomes one of increasing triaxial tension as the neck develops. This leads to a central internal crack on the axis of the specimen at the plane of minimum cross section. Further deformation causes this crack to propagate outward leading to a conventional cup-cone failure.

In drawing, on the other hand, a copper wire can undergo at least a 250,000 fold increase in length without failure. Furthermore, the resulting wire retains substantial ductility as well as maintaining or increasing its strength despite the enormous elongation it has undergone. This ability to exploit the plastic flow capability in metals to produce sound and ductile shapes after extensive deformation defines the role of deformation processing.

One distinction between the drawing of a wire, for example, and the extension of a rod in simple tension is the employment of dies or other means to confine or restrict the flow to the local region under the die. This confined deformation acts to prevent free and uncontrolled flow as occurs in a simple tension and which leads to fracture through necking and failure.

A second, and probably more important, distinction between wire drawing and simple tension is the state of stress existing in the metal as it undergoes plastic deformation. As is well known, the stress system at any point in a solid subjected to external forces can be described in terms of a hydrostatic component and three deviator stresses. The former acts to change the volume of a hypothetical element of the solid while the deviator stresses act to alter its shape. Because the negligible compressibility of most solids limits their ability to change volume homogeneously, plastic deformation under states of stress with a substantial component of hydrostatic tension should lead to early nucleation of voids and enhanced void growth just as hydrostatic tension in fluids leads to cavitation.

Suppression of the hydrostatic tensile components of the stress by the addition of external hydrostatic pressure during the deformation has been found to markedly influence the ductile fracture process. The work of Bridgman (3) provides the most comprehensive source of this information. He has shown that, by testing an SAE 1045 steel in tension with the superposed hydrostatic pressures up to 400,000 psi, the reduction in area at fracture increased linearly with increasing pressure. Rogers (4) has discussed the details of the changes in the character of the fracture under pressure. The principal influence of an increasing external hydrostatic pressure on tensile fracture is to suppress void nucleation until the neck is further developed and the accompanying hydrostatic tensions are enhanced. This delayed fracture initiation naturally leads to markedly increased reductions in area.

It was, therefore, proposed (5-10) that in any deformation process the degree of structural damage generated locally in a wrought metal should depend strongly on the hydrostatic component of the stress acting there while the metal is being plastically deformed. In principle, then, the ability to predict the amount of structural damage to a particular metal during processing, and its effect on the mechanical properties of the deformed metal depends on:

1. A knowledge of the <u>local</u> state of stress throughout the metal as it undergoes plastic deformation.
2. A knowledge of the tendency of a particular metal to suffer structural damage under unfavorable states of stress.
3. A knowledge of the sensitivity of the mechanical properties of a particular metal to the degree of structural damage previously generated within it during processing.

A comprehensive analytical and experimental program, described in References 5-10, was undertaken to investigate this concept. The deformation process chosen for study was strip drawing even though it is not a commercial process. It closely approximates commercial wire drawing yet, being a plane strain or two-dimensional analogue of wire drawing, it can be analyzed using the slip line field method.

ANALYSIS

Historically, plasticity solutions developed for deformation processing have had as their prime objective the determination of separating force, work and power for a given set of process parameters (11). A common characteristic of these solutions is that they assume the flow of metal in the process to be homogeneous, i.e., that planes perpendicular to the direction of flow remain as planes at all times. In fact, however, the flow processes in drawing, extrusion, and rolling can be extremely nonhomogeneous (12), such that the detailed stress state produced in the deforming regions can be expected to differ substantially from the predictions of homogeneous solutions.

The complexities of nonhomogeneous solutions in deformation processing are formidable and with the exception of analysis employing the slip-line filed method, little progress has been made towards the development of detailed solutions for the stress states. The slip-line field approach is applicable under the restrictions of plane strain for a plastic-rigid material of constant yield stress. A variety of solutions employing this method of analysis have been developed for strip-drawing,(13) hot-rolling,(14) and extrusion (15). These solutions describe in detail the flow pattern and the stress state, but it is the magnitude and distribution of the hydrostatic compression or tension within the deforming zone that is required for a greater understanding of the development or suppression of structural damage during processing.

The slip line field solution for strip-drawing by Hill and Tupper (13) is complete in form and permits determination of the stresses and velocities in the deforming region as a function of the process parameters, the die angle and reduction per pass. Details of the method for constructing the slip line field for a particular value of die angle and reduction are given by Hill and Tupper. From this, the die pressure and drawing stress were determined. Also available from the solution, but requiring further calculation, are the principal stresses and hydrostatic stress in the deforming region under the die.

Two restrictions in the applicability of the solution exist. One restriction is that, for a given die angle, a minimum reduction exists below which flow occurs ahead of the die (upsetting or bulging) and the solution is no longer applicable. The second restriction relates to the maximum reduction for a given die angle, beyond which the solution approaches the homogeneous solution. Stating this second restriction another way, the solution is applicable only when the field intersects the center line of the strip at a single point. Green and Hill (16) have expressed these limitations as

$$\alpha(0.230 + \frac{\alpha}{9}) < r < 2 \sin\alpha/(1 + 2 \sin\alpha) \qquad [1]$$

where α is the die semi-angle and r the reduction. The expression for the lower limit is empirical, with an error in r of less than 1%. For reductions below the lower limit, a bulge forms ahead of the die. For reductions beyond the upper limit in [1], the plastically deforming zone extends over a finite range along the center line. Green and Hill (16) have extended Hill and Tupper's analyses to this latter case, which approaches the classical homogeneous solution of Sachs.(11) The drawing stress t and the die pressure q are given as

$$\frac{t}{2k} = \ln(\frac{1}{1-r}) \qquad [2]$$

$$\frac{q}{2k} = 1 + \ln (1-r). \qquad [3]$$

The solution defined by Eq. [1] is of greatest interest because the more inhomogeneous the flow process, the greater becomes the possibility for high hydrostatic tension and structural damage.

The slip line field compatible with both stress and velocity considerations in the frictionless case is shown in Fig. 1.

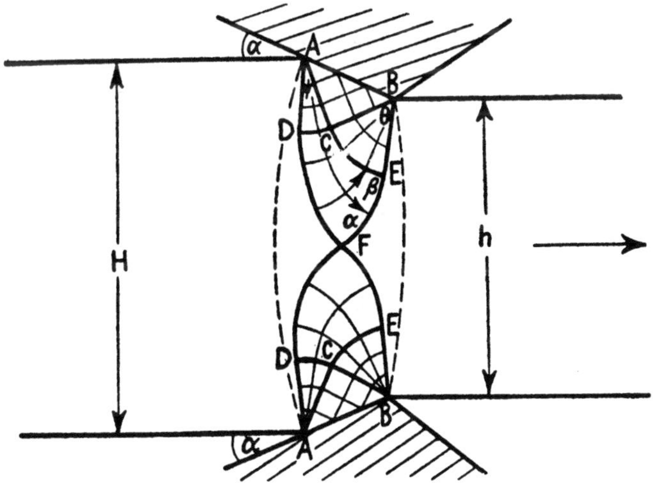

Figure 1. Slip-line field and plastic zone for drawing through frictionless, wedge-shaped dies (after Hill).

The mechanics of the construction of the field are covered in Hill (17). It is necessary, however, to determine the hydrostatic component of the stress in greater detail. The results of Hill's analysis are in the form of the drawing stress, t, and die pressure, q, for specific die angles and reductions. His analysis can be extended, however, using the Hencky relationships (18) and the fundamental properties of the plane strain behavior of a rigid, plastic material. This technique has been presented in some detail previously (5) and will only be outlined here. The Hencky relations are:

$$\begin{aligned} p + 2k\phi &= \text{constant along an } \alpha\text{-line} \\ p - 2k\phi &= \text{constant along an } \beta\text{-line} \end{aligned} \qquad [4]$$

Here p is the mean compressive stress (hydrostatic compressive stress) along a slip-line; 2k is the yield stress in tension; and ϕ the change in angle as one progresses along an α or β slip-line. Thus, depending on how one moves about in a slip-line field, the hydrostatic stress changes by an amount measured by the angular change of the slip-line. Therefore, if the hydrostatic component of the stress can be established at one point in the field, the values of the hydrostatic stresses throughout the field can be determined by the slip-line curvature. The three principal stresses in plane strain deformation can be expressed as p + k, p, and p - k. Thus, the hydrostatic component of the stress in the region directly under the die can be determined from the die pressure, q, using the relation p = q - k. The values of q, as stated earlier, are given in Hill (17) as a function of die angle and reduction. Since the slip lines are straight in the triangular region under the die, the hydrostatic stress is constant and equal to q - k throughout this region, including point C in Fig. 1. This establishes the pressure at one point in the curvilinear field from which the pressures at any other point can be calculated. They can also be calculated from first principles using the fact that the horizontal component of the force along the boundary ADF of the field is zero when there is no back tension (7). Thus, once the field applicable to a specific die angle and reduction is constructed, the hydrostatic component of the stress at any point can be established.

The construction of the specific field for a particular die angle and reduction is based on the general double fan slip line field for which Hill gives the Cartesian coordinates (19) of the nodal points. From geometry (see Fig. 1) it can be shown that the reduction,

$$r = 1 - \frac{h}{H} \qquad [5]$$

and the die semi-angle,

$$\alpha = \theta - \psi. \qquad [6]$$

Thus, for a given die semi-angle α, the values of θ and ψ can be widely varied to accomodate different reductions, the only restriction being that 2. above is satisfied and that both the α and β slip lines intersect the midplane at a point and with a slope of 45° to satisfy symmetry conditions. Appropriate geometrical considerations transform the nodal point values from Hill to locations in the deforming strip.

Using this approach, the hydrostatic pressure at various points within the deforming strip was calculated (5) in terms of the yield stress in tension, 2k, $\alpha = 15°$ and a reduction of 8.5%. This is plotted in Fig. 2. Negative values indicate hydrostatic tension. It should be noted, for this and the slip line fields for other die angles and reductions, that the variation in hydrostatic stress throughout the zone of plastic deformation is quite marked. Under all conditions, however, the greatest hydrostatic compression is in the region directly under the die while at the midplane the hydrostatic stress has its lowest algebraic value.

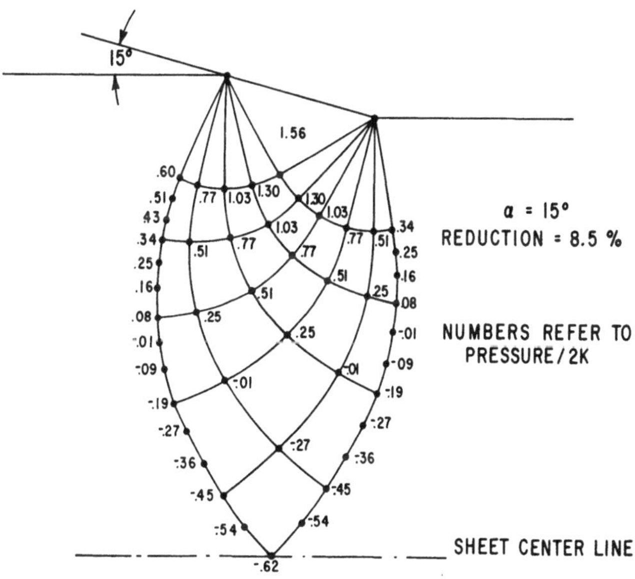

Figure 2. Distribution of hydrostatic pressure in strip for one set of drawing conditions. (2k is tensile yield stress)

Figure 3 presents the calculated values of these two extremes in hydrostatic stress as a function of reduction per pass for six different die semi-angles, α = 5° to α = 30°. It should be noted that for only a few limited reductions at the lower die angles is the hydrostatic stress at the midplane compressive, the other combinations producing various degrees of hydrostatic tension. For each die angle, the highest hydrostatic tension exists for the least reduction permissible in this solution. At these points the hydrostatic tension is well above that which would be produced by the same metal subjected to flow in simple tension, i.e., prior to necking. The magnitude of hydrostatic tension is, at its extreme, equivalent to a fairly severely notched bar. For larger reductions at a given die angle, these tensions decrease to much lower values.

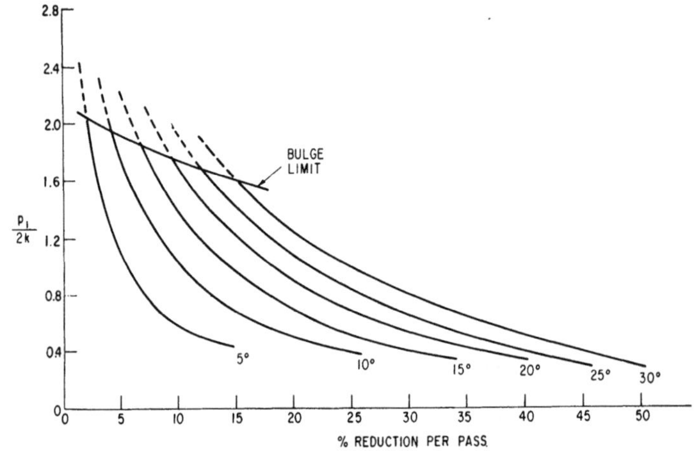

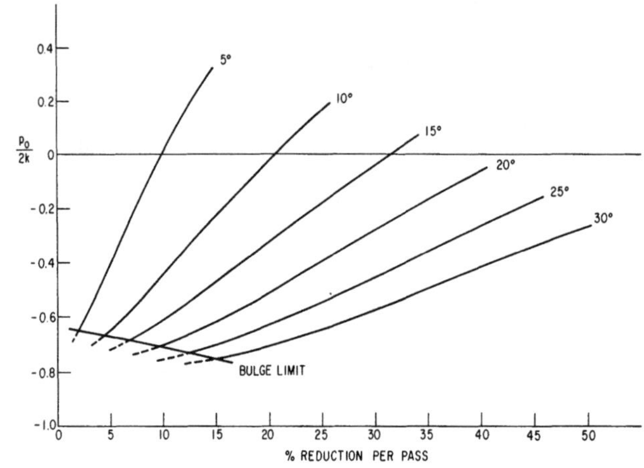

Figure 3. Hydrostatic pressure vs reduction per pass for frictionless dies having various semi-angles. a) Under the die. b) at the midplane. (2k is the tensile yield stress)

EXPERIMENTAL VERIFICATION

Assessment of Structural Damage

Structural damage as defined herein can be determined in a number of ways. Quantitative metallography could be used but is tedious and time consuming. Determination of the density changes that result from deformation combined with standard metallography was found to be entirely adequate (5-10). Density change provides a quantitative measure of the total void or crack volume introduced into the wrought metal while the metallographic observations yield information on the type, size, number, and location of the defects.

The density of the deformed metals was determined to a quite high precision using the fluid displacement technique. The principal elements of the density measuring facility used by Rogers and Coffin consisted of a semimicro balance and a controlled temperature bath positioned directly below it. Weighings were made both in air and in a bath of diethyl phthalate, the temperature of which could be measured to 0.001 C. Completely surrounding this inner bath was a large, temperature-controlled outer bath that could maintain the temperature of the inner bath to 0.003 C for several hours. Further details of the equipment and experimental method have been given elsewhere (7).

The early work of Rogers and Coffin (5) convincingly demonstrated the significance of the magnitude of the hydrostatic component of the stress prevailing locally in the deforming metal. It was shown that in a strip-drawn metal, e.g., electrolytic tough pitch copper, a large gradient in damage exists through the thickness of drawn strip. The damage is extensive at the midplane where the stresses were highly tensile during drawing while under the die the damage is minimal. It was also established that successive draws using the same conditions of die angle and reduction per pass cause the accumulation of damage,frequently leading to fracture during a subsequent drawing operation. Thus any exact analysis of the behavior of a material during deformation processing must take into account its prior processing history.

In a more recent study (10) Rogers has examined the sensitivity of a number of common metals during strip drawing relative to the generation of internal structural damage. The strips were all drawn at 25% reduction per pass using three different die semi-angles, 10°, 20° and 30°. During drawing, these produced at the midplane of the strip, hydrostatic stresses, normalized in terms of the tensile yield stress, 2k, of +0.17, -0.39 and -0.64 respectively (tension is negative). As in other forms of ductile fracture, material characteristics such as cleanliness, rate of strain hardening, prior history, brittleness, etc. played a major role in determining the amount of structural damage produced. A few of the results are

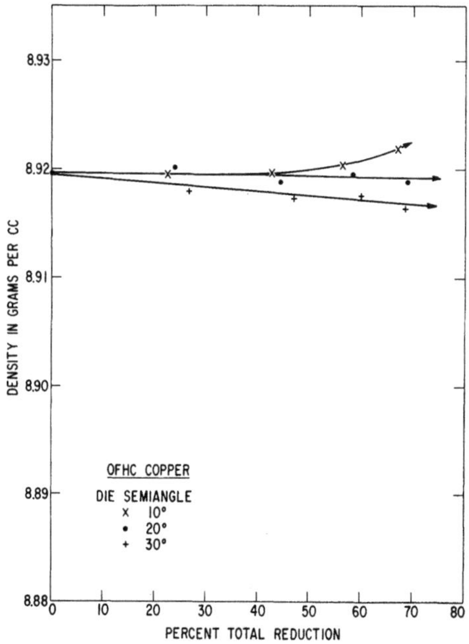

Figure 4. Absolute density of OFHC copper strip as a function of reduction for three different die angles.

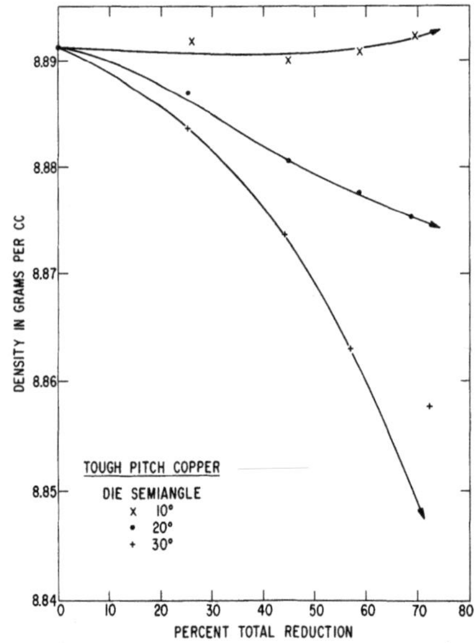

Figure 5. Absolute density of tough pitch copper strip as a function of reduction for three different die angles.

included here. A comparison between "dirty" tough pitch copper and relatively "clean" OFHC copper, Figures 4 and 5, show that the high inclusion content of the tough pitch copper leads to substantial void nucleation and growth. This is also highly sensitive to the stress state during processing. In contrast, OFHC copper is nearly free of structural damage under any of the standardized drawing conditions.

Pearlitic 4340 steel, as shown in Fig. 6, is damaged more easily than any of the materials that were examined. It should be noted that in these figures an arrow at the termination of the curve indicates that drawing could have continued. The absence of such a terminal arrow is indicative of failure during the indicated draw or an inability to start the next draw without fracturing the strip. It was impossible to draw pearlitic 4340 steel with the 30° die, it could only be drawn once with the 20° die, and even with the 10° die it could only be drawn three times. For comparison, the effect of rolling is also plotted. For all the processing variations, the decrease in density is approximately the same at the same total reduction.

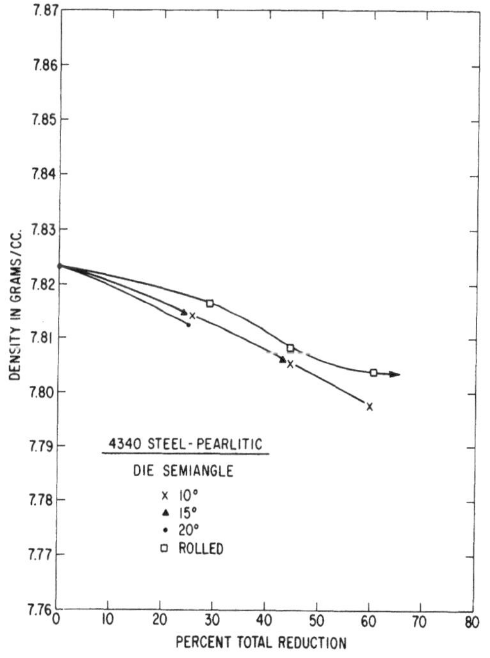

Figure 6. Absolute density of pearlitic AISI 4340 steel strip as a function of reduction for three different die angles.

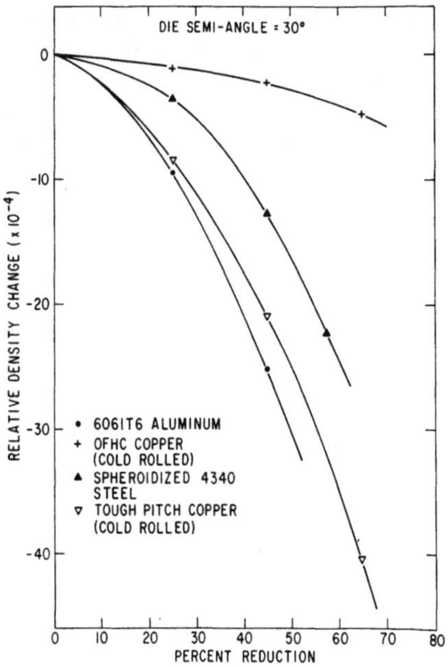

Figure 7. Relative change in density with reduction of several metals with different densities. Drawn through dies having a semi-angle of 30 degrees.

For comparison of the behavior of materials with widely different absolute densities, conversion of the density changes, i.e., $\frac{\Delta\rho}{\rho}$, is required. An example of this is illustrated in Figure 7, which compares the change in relative density with total reduction using dies having a semi-angle of 30° for four materials with considerably different densities. The OFHC copper showed the least damage; the 6061-T6 aluminum alloy, the most.

One of the principal aims of the investigation was to determine whether a "damage sensitivity" parameter could be assigned to the various materials studied which would indicate their propensity to the generation of damage when deformed under equivalent processing conditions. The relative density change for each of the materials resulting from a total drawing reduction of 45% was therefore determined as a function of the hydrostatic component of the stress prevailing during the strip-drawing process. The results are plotted in Figures 8 through 10. To the right of the dotted line in each figure the states of stress are considered favorable, i.e., compressive. Since there is obviously a physical limit beyond which an increase in hydrostatic stresses can no longer cause the strip to densify, all of the curves approach a horizontal asymptote.

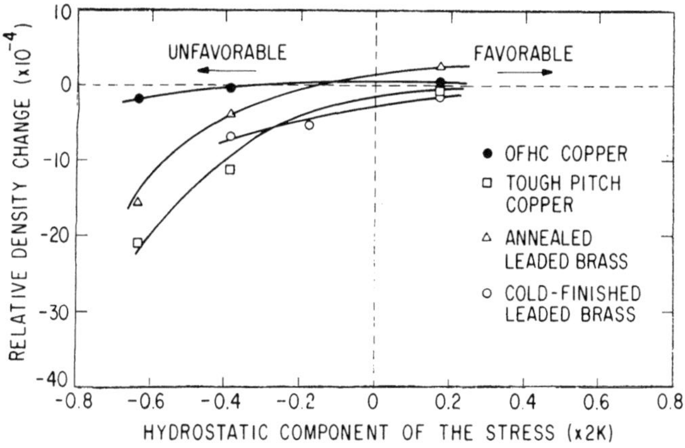

Figure 8. The relative change in density of the drawn copper-base metals as a function of the hydrostatic component of the stress operative during the deformation.

For those materials that have suffered a density decrease when drawn with the mildly favorable hydrostatic stress having a component of +0.17, the value of this asymptote should be zero. For those materials that actually show a small density increase, the level of the asymptote above zero is a measure of the void volume that existed in the starting material prior to strip drawing.

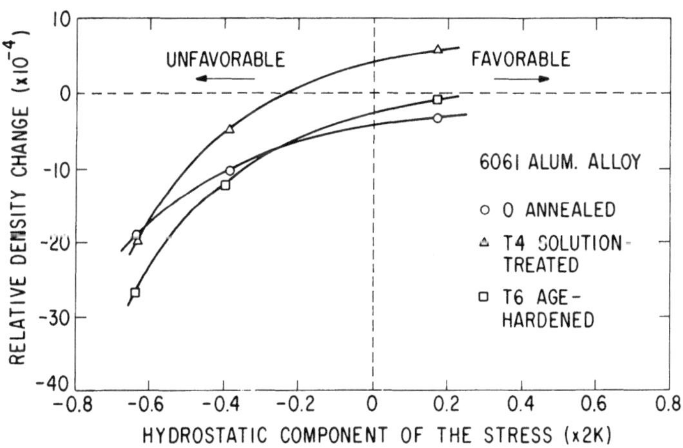

Figure 9. The relative change in density of the 6061 aluminum alloy in three conditions of heat treatment as a function of the hydrostatic component of the stress operative during the deformation.

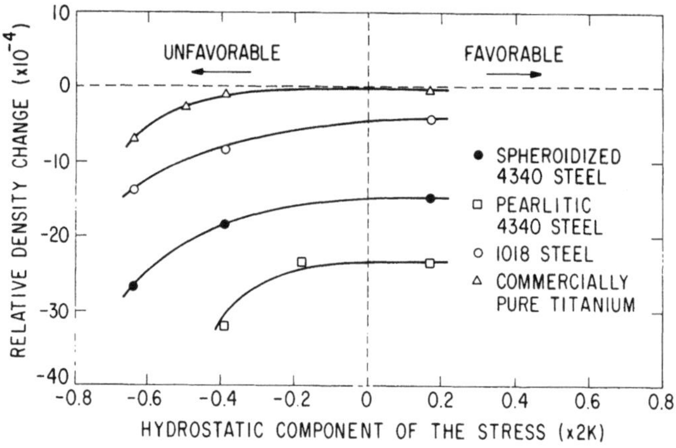

Figure 10. The relative change in density of the three steels and the commercially pure titanium as a function of the hydrostatic component of the stress operative during the deformation.

Except for the very clean materials, the titanium and OFHC copper, all the curves of density change for the nonferrous metals fall in a relatively narrow band with respect to changes in the hydrostatic component of the stress. This band is far from linear, however. The steels, on the other hand, Figure 10, vary considerably in their response to stresses imposed by the deformations used in the present study. The increasing downward displacement of the more or less horizontal portion of damage curves for the mild steel, and the spheroidized and pearlitic 4340 steels,respectively, is a measure of increasing damage response in these materials to deformation alone. The low slope in the region of the favorable stress states indicates that only extremely high pressures will completely prevent damage in a material such as pearlitic 4340. This was corroborated by the comparable drop in density exhibited by this material even on rolling (Figure 6). The nonlinearity and the vertical displacements of these curves make the assignment of a value of "damage sensitivity" to these different materials difficult.

Mechanical Properties of Drawn Strip

Transverse tensile tests were performed on the drawn strips to determine whether or not structural damage hand an effect on the wrought metal. The specimens were small with a square gage section, the size of which depended on the thickness of the drawn strip. In all cases, true strain was used to compare materials. Ductility (E_A) was determined using the relation

$$E_A = \log_e \frac{A_o}{A_f}$$

where A_o is the initial area and A_f is the area at fracture.

In addition to the strip which had been drawn with dies having either 10°, 20° or 30° semi-angles, rolled strips of each material were also tested. Specimens were removed from the strip with the length transverse to the rolling or drawing direction. Since the density measurements indicated that structural damage increases in almost every case as the reduction increases, the strips with the greatest reduction under each condition were compared. With the most ductile materials this meant a 68% total reduction.

The transverse tensile ductility was determined for a variety of metals as a function of the hydrostatic component of the stress operative during the deformation process. These are, to repeat, +0.17, -0.18, -0.39, and -0.64 x 2k for the 10, 15, 20 and 30 degree dies, respectively. To compare the drawn strips with those that had been rolled, a state of hydrostatic stress with components of +0.33 x 2k was ascribed to the rolling process. This in essence equates rolling with simple compression as a first approximation.

The results indicated that for most of the materials tested there was a linear correlation of transverse tensile ductility, with the hydrostatic component of the stresses prevailing during deformation over the range of the stress states employed. There was a significant variation in material response to the change in hydrostatic stress, however.

Examples of the two primary responses are shown in Figures 11

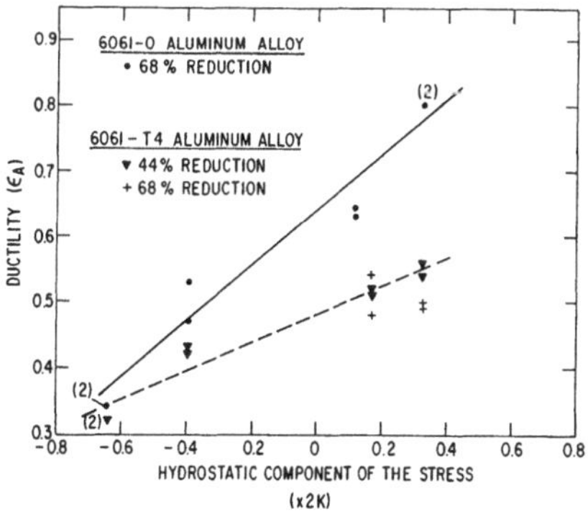

Figure 11. The transverse tensile ductility of rolled and drawn 6061 aluminum alloy strips in both the solution-treated (T4) and annealed (0) conditions as a function of the hydrostatic component of the stress prevailing during the deformation process.

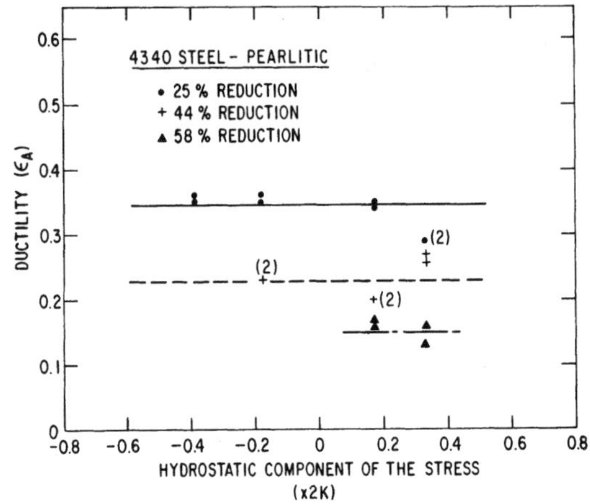

Figure 12. The transverse tensile ductility of rolled and drawn pearlitic AISI 4340 steel strips as a function of the hydrostatic component of the stress prevailing during the deformation process.

and 12. It is primarily the slope of the curve that indicates the response of the metal to the hydrostatic stresses occurring during processing. The f.c.c. metals and alloys for the most part all showed a high sensitivity to the state of hydrostatic stress, although this was reduced by prior cold work. The condition of the aluminum alloy modified its sensitivity to the state of stress to a minor degree. The annealed alloy had the greatest sensitivity while the age hardened and solution treated alloys had a lower slope.

The steels as a group showed the lowest stress sensitivity. In fact, only a relatively soft mild steel showed any stress sensitivity at all. Both pearlitic and spheroidized 4340 steels showed essentially no stress response, even when the hydrostatic stress was substantially compressive as in rolling. For the steel in the pearlitic condition this correlates closely with the observed damage production. The ductility seemed to be more closely related to the amount of strain introduced than to the state of stress prevailing during processing. The probable explanation of this lack of sensitivity is the same one that can be put forth to explain the relatively poor capability of these same materials of withstanding the more severe drawing operations. These materials are sufficiently sensitive to small amounts of structural damage, and the propagation of this damage to complete failure occurs with such ease that all of the degrees of damage studied are so great they are all "super-critical" in these materials, and no change in mechanical properties is observed. Another factor that apparently also plays a role in the tensile ductility of the drawn strip is a possible difference in crystallographic textures produced when different die angles are used in drawing. This is currently under investigation.

The Effects of Friction

The analysis described above was predicated on the condition of zero friction between the die surfaces and the strip being drawn. Experimentally this condition was approximated as closely as possible through the use of tungsten carbide dies and a sprayed on PTFE telomer in a binder as a lubricant. The measured coefficient of friction was normally less than 0.2.

The drawing operation was carried out using a strain gaged tensile weigh-bar to measure the drawing force directly. To obtain die separating force, a specially constructed die holder was developed that deflected elastically under the action of the separating forces. This deflection was also monitored by strain gages. For any given draw, the friction coefficient μ is determined from

$$\mu = \frac{T\cos\alpha - 2S\sin\alpha}{T\sin\alpha + 2S\cos\alpha} \qquad [7]$$

Here T and S are the drawing force and separating force load cell values for any draw experiment and α is the die semi-angle.

All commercial deformation processes must be carried out with friction between the deforming metal and the tools or dies. In general, lubrication will be considerably poorer in a commercial process than in a laboratory investigation of the type considered here. It is therefore imperative that the influence of die friction on the hydrostatic component of the stress at the midplane of the strip also be determined since this component of the stress has been shown to control the degree to which structural damage is generated in strip during drawing, thereby degrading the properties of the drawn product. For this reason Coffin and Rogers undertook a detailed analysis of the effect of friction on the slip line field solutions for strip drawing (20) from which the hydrostatic component of the stress can be calculated.

Friction between the strip and the flat, wedge-shaped dies used in strip-drawing generates shear forces along that interface. No longer do the slip lines intersect the surface at an angle of 45° which made the small triangular region beneath the die isosceles

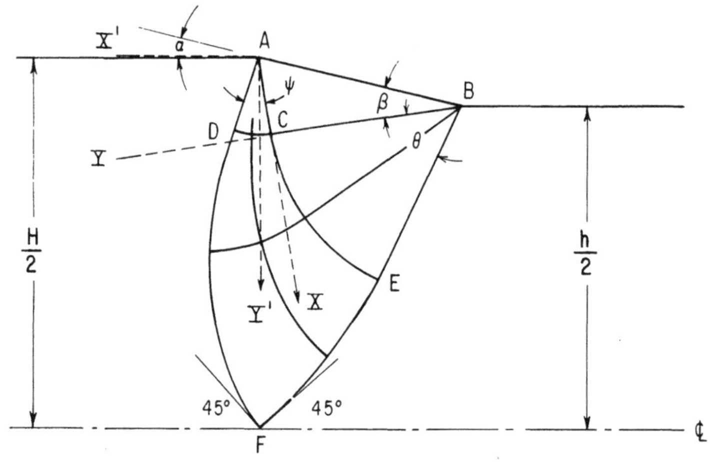

Figure 13. Slip-line field and coordinate system for sheet-drawing with friction.

in the frictionless case, but instead the legs of the triangle are of unequal length as in Figure 13; their ratio equals tan β. Friction reduces angle β, changing the symmetric double-fan field used in the frictionless analysis to one with increasing asymmetry as β decreases. The angular relationship (cf Eq. 6) now becomes

$$\theta - \psi = \alpha + \frac{\pi}{4} - \beta \qquad [8]$$

Downie (21) calculated the coordinates of the asymmetric fields in 5° increments of $\Delta\theta$ and $\Delta\psi$ up to $\theta = \psi = 90°$ and for tan β from 1 to 0 in steps of 0.1. Although Downie's calculations are directly applicable to the case of strip drawing with friction, it was necessary to convert Downie's results, given in increments of tan β, into 5° increments in β itself. Only then could the analysis be applied to a series of dies in which α varied in 5° increments. This required a somewhat extensive computer program to handle the numerical integration and interpolation. The technique is described more completely in Reference 20. Figure 14 is representative of the results of that investigation. The hydrostatic pressure at the midplane of the strip for dies with a 5° semi-angle is shown as a function of reduction and for coefficients of friction, μ, varying between 0 and 0.6. The curve for zero friction is the same as that for 5° dies in Figure 3. The effect of friction then is to make the hydrostatic component of stress more negative than would be expected under frictionless conditions. For a 10% reduction per pass, this amounts to an increase in hydrostatic tension of 0.3 (x 2k) when the friction increases to $\mu = 0.6$ from $\mu = 0$.

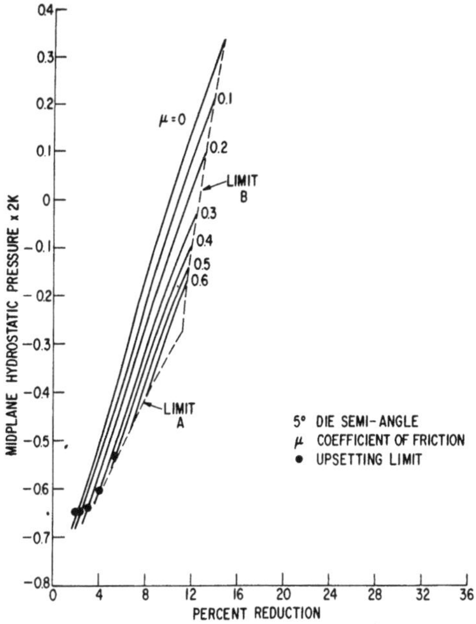

Figure 14. Effect of friction on midplane hydrostatic pressure for dies with 5° semi-angle.

The effect of friction on the hydrostatic component of the stress at the midplane for dies with semi-angles from 5° to 30° is given in Reference 20. Also included are curves for the die pressure, upsetting or bulging pressure, and drawing stress.

In a commercial deformation process such as wire drawing it can also be expected that the increased friction resulting from poor lubrication will change the state of hydrostatic stress in the deforming metal, making it considerably more tensile and thus increasing the rate of generation of structural damage in the product.

Drawing experiments on commercially pure titanium strips (10) provide further confirmation of the analytical predictions (20). As indicated above, the analysis gives quantitatively the change in the midplane hydrostatic pressure with increasing friction coefficient for a given die semi-angle and reduction. In a drawing experiment on titanium lubricated with PTFE using dies having a 20°semi-angle, the strip drew with a low coefficient of friction initially but eventually began to seize. The friction coefficient calculated from the measured forces rose from 0.12 to 0.44. From the calculated curves of Reference 20 this should cause the value of the hydrostatic component of the stress to change by -0.11 (x 2k), i.e., to become more tensile. In addition to the drawing forces, the density changes were also measured for this strip when drawn under conditions of both low and high friction. A definite additional density decrease resulted from drawing under conditions of high friction. The quantitative corroboration of the slip-line field analysis can be seen in Figure 10. When the lower relative density that had been determined for the condition of high friction is plotted as a function of the more tensile hydrostatic stress, it falls precisely on the curve of relative density change as a function of the component of hydrostatic stress determined by drawing titanium under normal low friction conditions!

SUMMARY AND CONCLUSIONS

For a simplified, two-dimensional metalworking process, strip drawing, it has been shown by analysis and experiment that

1. The structural damage generated locally in a drawn strip depends on the nature of the hydrostatic component of the stress prevailing there while it was being plastically deformed during drawing.
2. The magnitude of the structural damage increases as the hydrostatic stresses become increasingly tensile.
3. The nature and magnitude of the hydrostatic stress generated locally in the region where deformation is occurring depends on the process parameters; in the strip drawing experiments described, these are the die angle, 2α, reduction per pass, r, and coefficient of friction, μ. When the die angle is increased or the reduction per pass decreased, everything else being constant, the hydrostatic tension increases with a larger attendant damage in the drawn strip. Whenever friction increases the effect is the same. In addition, increased friction raises the drawing stress substantially.
4. In strip drawing, the hydrostatic stresses are most tensile, and the damage most severe, at the midplane of the strip.
5. Structural damage is cumulative, increasing with successive draws under comparable conditions. When processing conditions are such that large amounts of damage are produced, early fracture frequently results.
6. Material properties play a large role in the damage production process. Materials that have a low inclusion content or with no hard second phase particles to act as void nuclei show minimal damage. Conversely, a pearlitic 4340 steel suffered substantial damage during drawing or rolling with a poor correlation with the prevailing state.
7. The mechanical properties of drawn strip, as measured by transverse tensile tests, in general, show a strong correlation with the stress that prevailed during drawing. The tensile ductility decreases as the hydrostatic stress during drawing becomes more tensile. Again, the 4340 steel showed minimal response to the stress state, the tensile ductility depending

primarily on the total drawing reduction.

The quantitative application of this analytical technique or the experimental results to other types of deformation processes may be difficult. Very few commercial processes produce simple plane strain deformation. Rolling of wide plate and sheet is one such process, however, and should be amenable to similar treatment. Processing of tubular products in some instances is another example.

Although the relationship is difficult to justify theoretically, simple, axisymmetric deformation processes such as wire drawing, extrusion, and swaging closely approximate their plane strain counterparts. Hence, the results of the analyses of their plane strain analogs would be directly applicable. On the other hand, a complex process such as the closed die forging of an intricate shape would be extremely difficult to analyze in its entirety. A piecewise analysis, even though possible, would hardly be justifiable from a practical point of view.

The general features of the results should, however, be applicable to a broad spectrum of metalworking processes. In particular, in metal forming operations, frictional effects are always present. Under conditions of poor or variable lubrication, severe internal damage may be generated in a number of processes. It is also obvious in processes like wire drawing, where the work of deformation is supplied by tensile forces, that excessive friction may cause a tensile failure.

A practical demonstration of the importance of stresses in deformation processing is the beneficial effect of metalworking under high pressure. Here the external application of high hydrostatic pressures counteracts any internally generated hydrostatic tensions, thus minimizing or eliminating damage. These effects have been examined quantitatively in laboratory studies by Rogers and Coffin (7-9).

REFERENCES

1. K. E. Puttick, Phil Mag, 4 (1959) 964.
2. H. C. Rogers, Trans AIME, 218 (1960) 498.
3. P. W. Bridgman, Studies of Large Plastic Flow and Fracture, McGraw-Hill Book Co., Inc., New York (1952).
4. H. C. Rogers, Fundamentals of Deformation Processing, Syracuse Univ. Press, Syracuse, N. Y. (1964), Chap. IX.
5. H. C. Rogers, R. C. Leech, and L. F. Coffin, Jr., "An Investigation of Structural Damage in Metal-Forming Processes," Final Report, Contract NOw-63-0671-c, Bureau of Naval Weapons (July, 1964).

6. H. C. Rogers, R. C. Leech, and L. F. Coffin, Jr., "Investigation of the Nature of Structural Damage in Metal-Forming Processes," Final Report, Contract NOw-65-0097-f, Bureau of Naval Weapons (November 1965).
7. H. C. Rogers, and L. F. Coffin, Jr., "Investigation of the Nature of Structural Damage in Metal Forming Processes," Final Report, Contract NOw-66-0546-d, Bureau of Naval Weapons (June 1967).
8. H. C. Rogers and L. F. Coffin, Jr., "Structural Damage in Metalworking," CIRP Intern. Conf. on Manufacturing Technol., Ann Arbor, Mich., ASTME, Dearborn, Mich. (1967).
9. L. F. Coffin, Jr. and H. C. Rogers, ASM Trans., 60 (1967) 672.
10. H. C. Rogers, "Structural Damage in Metal-Forming Processes," Final Report, Contract NO. 0019-68-c-147, Naval Air Systems Command (March 1969).
11. G. Sachs, Z Angew Math Mech, 7 (1927) 235.
12. L. F. Coffin, Jr., Fundamentals of Deformation Processing, Syracuse Univ. Press, Syracuse, N. Y. (1964), Chap. II.
13. R. Hill and S. J. Tupper, J. Iron Steel Inst. 159 (1948) 353.
14. J. M. Alexander, Proc. Inst. Mech. Engrs., 169, (1955) 1021.
15. W. Johnson, J. Mech. Phys. Solids, 4, 191 (1955).
16. A. P. Green and R. Hill, J. Mech. Phys. Solids, 1, 31 (1952).
17. R. Hill, The Mechanical Theory of Plasticity, Oxford Univ. Press, London (1950), Chap. VI.
18. Ibid., p. 135.
19. Ibid., p. 350
20. H. C. Rogers and L. F. Coffin, Jr., G. E. Research and Development Center Report No. 69-c-344, September 1969. (Accepted for publication in Int. Jnl. of Mech. Sciences).
21. T. M. Downie, Tech. Tept. No. 45, Div. of Appl. Math., Brown Univ. (Nov. 1958).

THE RELATIONSHIP BETWEEN SUPERPLASTICITY AND FORMABILITY

H. W. Hayden, R. C. Gibson, J. H. Brophy

The International Nickel Company, Inc.

Paul D. Merica Research Laboratory

ABSTRACT

Superplasticity is a high temperature deformation phenomenon in which samples exhibit extremely large tensile elongation. The key to obtaining superplasticity is producing grain sizes in the order of ten microns or less, and maintaining these fine structures for reasonable periods of time at temperatures in excess of 50% of the absolute melting point. This is most easily accomplished in two-phase alloys. In most known alloys which exhibit superplasticity, the necessary microstructure is produced by forming operations involving hot or cold working steps. The fine microstructure achieved then contributes to easier hot formability at both high and low strain rates. At high strain rates, where the superplastic effect would not be expected in the tensile test, fine-grained material requires lower working loads than coarse-grained material of identical composition. It has been shown that significant amounts of deformation can be achieved in realistically short periods of time. At low strain rates, where superplasticity is observed in tension tests, the freedom from necking, and the low stresses required for appreciable deformation will probably lead to new forming operations previously impossible in metallic systems. The forming of materials, which will deform superplastically in tension, should not be relegated to the role of a low strain rate laboratory curiosity.

INTRODUCTION

Superplasticity has been observed in many alloy systems and has been described in a variety of terms. Qualitatively, it has been likened to the deformation of taffy, putty, or hot glass(1,2). Semi-quantitatively it is large amounts of apparently neck-free elongation in a tensile test. Quantitatively tensile elongations of 700 to 1000% and over have commonly been observed. Almost universally superplasticity results when the strain-rate sensitivity of flow stress is abnormally high compared to that for an ordinary metal(2). Although the first published reference to superplasticity appeared in 1934(3) the real growth in activity awaited new work published in 1964(2). Since 1962 there have been over 100 technical papers on the subject, and it has been observed in one form or another in more than twenty different alloy systems.

Most of the published literature has concentrated on the basic mechanism of superplasticity, but it has been recognized from the earliest work as a potentially useful process for shaping metals(2). The basic research work has been concerned with the tensile test demonstration of superplasticity in which the phenomenon is most pronounced at relatively low strain rates. In some ways this is unfortunate, because it has caused some observers to relegate the phenomenon to the status of a slow rate laboratory curiosity.

Early work in our laboratory showed attractive reductions in working loads in conventional high strain rate rolling and extrusion of microduplex Ni-Cr-Fe alloys(4,5,6). Moreover, recent work on the application of superplasticity to forming processes has shown that even with low strain rate limitations, useful amounts of deformation can be accomplished in brief periods of time(7-12). The necessary microstructures for superplasticity are easily produced by straightforward heat treatment and working sequences(1). There is increasing evidence of success in applying the structures which deform superplastically to commercially promising shaping operations(13-18). It is the purpose of this paper to summarize this evidence.

PHENOMENON OF SUPERPLASTICITY

In order to apply superplasticity to practical deformation processing, it is important to be aware of effects of processing and materials variables on the strain rates possible in any forming operation. For an alloy to be superplastic, it must be possible to form a fine grain structure (generally 10 microns or smaller) which will be retained for periods of time in the order of minutes to hours at temperatures similar to, or higher than, the recrystallization temperature. Superplasticity has been observed in a few single phase alloys(19,20). However, in the great majority of alloys exhibiting this behavior, high temperature grain size stabilization is accomplished by producing a microstructure composed of an ultrafine distribution of two separate phases. We have used the term "microduplex" as descriptive of such structures.

In materials having a suitably fine microstructure, the relationship between deformation and microstructural variables is of the form:

$$\dot{\varepsilon} = \frac{(K\sigma)^{\frac{1}{m}}}{d^A} \exp(-Q/RT)$$

or

$$\sigma = K\dot{\varepsilon}^m d^{Am} \exp(Q/RT)^m$$

where $\dot{\varepsilon}$ is true strain-rate, σ true stress, d grain size, m and A are exponents, Q an activation energy, R Boltzmann's constant, and T the absolute temperature.

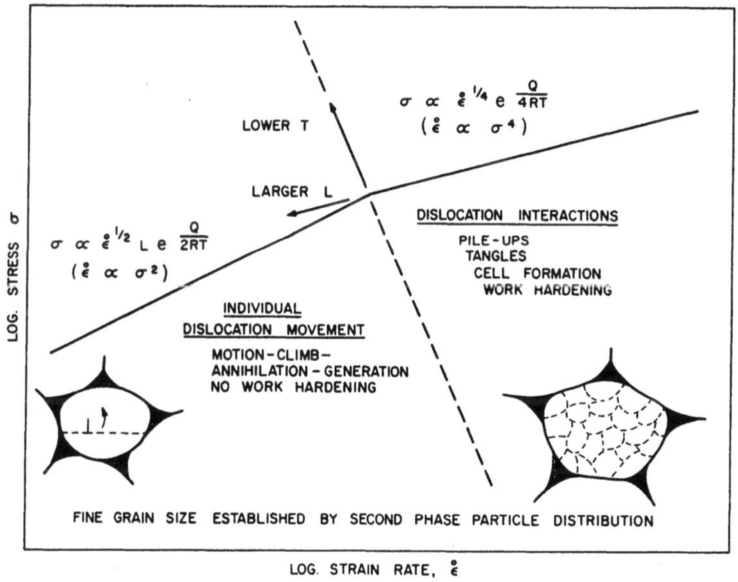

FIGURE 1 - SCHEMATIC STRESS VS. STRAIN RATE RELATION SUMMARIZING EXPERIMENTAL OBSERVATIONS AND THE MECHANISM OF DEFORMATION.

Figure 1 is a schematic representation of the relation between stress and strain-rate typical for many superplastic alloys. It can be seen that there are two domains of behavior.

At high stresses and strain rates, deformation is typical of normal high temperature behavior. In this domain of low strain rate sensitivity (generally $m<1/4$), work hardening effects are observed and tensile elongations are not abnormally high. Although this domain of behavior can hardly be called superplastic, it will be shown that there is an advantage of finer grain sizes in permitting lower load requirements for high speed working operations. There is a transition from low to high strain sensitivity at stress levels where an expected dislocation cell size would be equal to or greater than the grain size. Obviously it is impossible to form cell structures larger than the grain size and hence, in the lower stress domain of superplastic behavior deformation occurs with no work hardening(4,21,22). The exponent of strain rate sensitivity, m, may range from one-third to unity and a value of one-half is typical of several alloys(4,11,12,21,22,23). Work on complex nickel-base alloys has also shown an exponent of 0.5(18).

Within the domain of high strain rate sensitivity, grain size has a pronounced effect in determining the strain-rate produced by an applied force. Several investigations have shown values of 2 or 3 for A, the grain size sensitivity exponent(11,12,14,20,21,23-25). Similarly, temperature is an important variable. Generally, the values of the activation energy Q are similar to those expected for volume diffusion, grain boundary diffusion, or dislocation pipe diffusion. The effects of variations in grain size and temperature on the behavior of a superplastic nickel chromium-iron alloy are shown in Figures 2 and 3.

Knowing the effects of the variables of stress, strain-rate, grain size and temperature on the superplastic behavior of a given alloy, it should then be possible to design forming operations which capitalize on the advantage of great tensile extensibility at low working stress inherent in superplasticity. Similarly, knowledge of the transition stress and strain-rates beyond which deformation behavior becomes normal will establish limits for the maximum possible rates and allowable forces for superplastic processing.

PRODUCTION OF THE MICRODUPLEX STRUCTURE

A microduplex structure is an extremely fine-grained, two-phase microstructure which gives rise to superplasticity

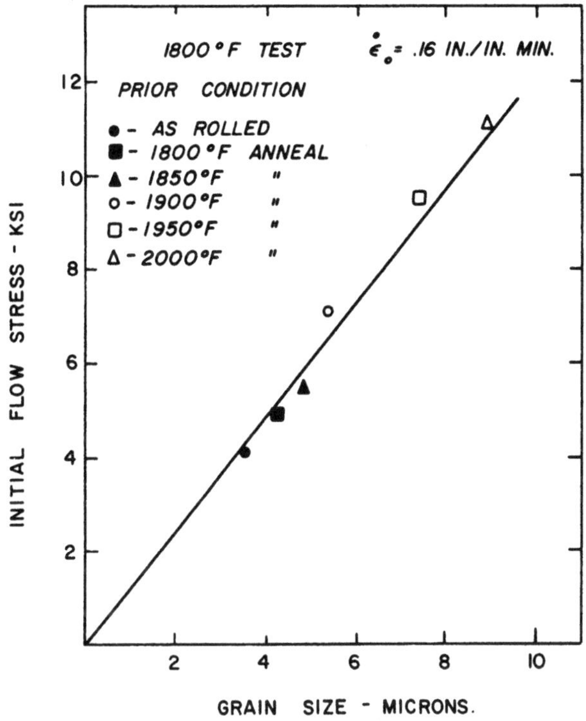

FIGURE 2 - THE EFFECT OF GRAIN SIZE ON THE FLOW STRESS AT 1800°F

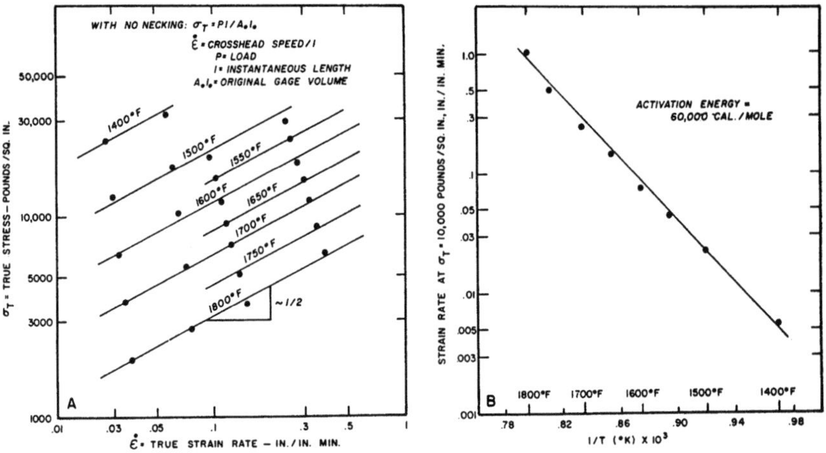

FIGURE 3

(A) Stress vs. Strain Rate and (B) Arrhenius Plot for Samples of a 39%Cr-10%Fe-1.75%Ti-1%Al-Bal.Ni Alloy Pulled First at 1800°F, then at Each Lower Temperature in Order.

at temperatures above about half the absolute melting temperature. At moderately elevated temperatures and at room temperature and below, this structure results in attractive engineering properties as well. An example of this structure in a nickel-base alloy is shown in Figure 4. Chromium-rich alpha prime phase is located primarily at grain boundaries of the face-centered-cubic matrix. Microduplex structures have been produced in a large number of two-phase alloys by a number of relatively simple procedures. Here we shall consider the processes employed for Zn-Al alloys and Fe-Ni-Cr alloys as representative of the types of processing schedules which lead to the desired microstructure.

In zinc-aluminum alloys the effect of solid state phase transformations plays a large role in the production of the microduplex structure. When the 78%Zn and 22%Al is heated to a temperature above 275°C, the eutectoid structure transforms to a single phase. If the alloy is treated at such a temperature, and then quickly cooled to room temperature or a lower temperature such as that of liquid nitrogen, the high temperature single phase will be momentarily retained. The drive toward equilibrium state is so great that the single phase alloy will exothermically decompose into the equilibrium two-phase distribution. The net result of this decomposition reaction is that an initially

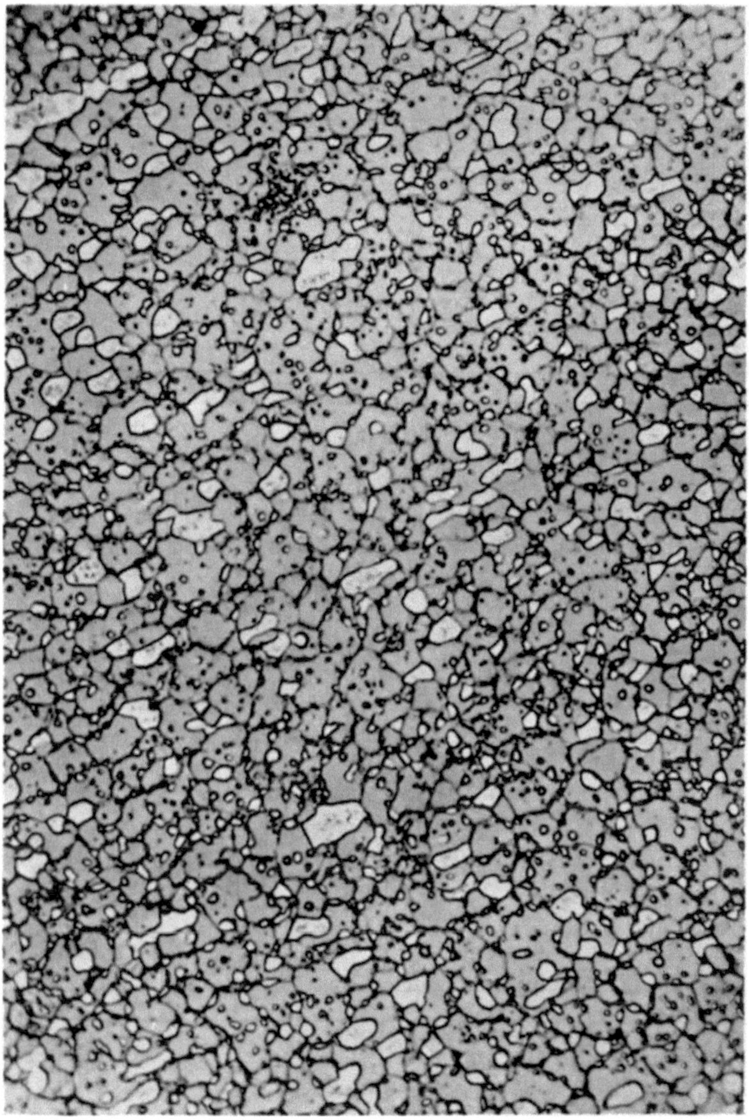

FIGURE 4

Microduplex Structure in
38%Cr-18%Fe-0.6%Ti-Bal.Ni Alloy.
(1000X)

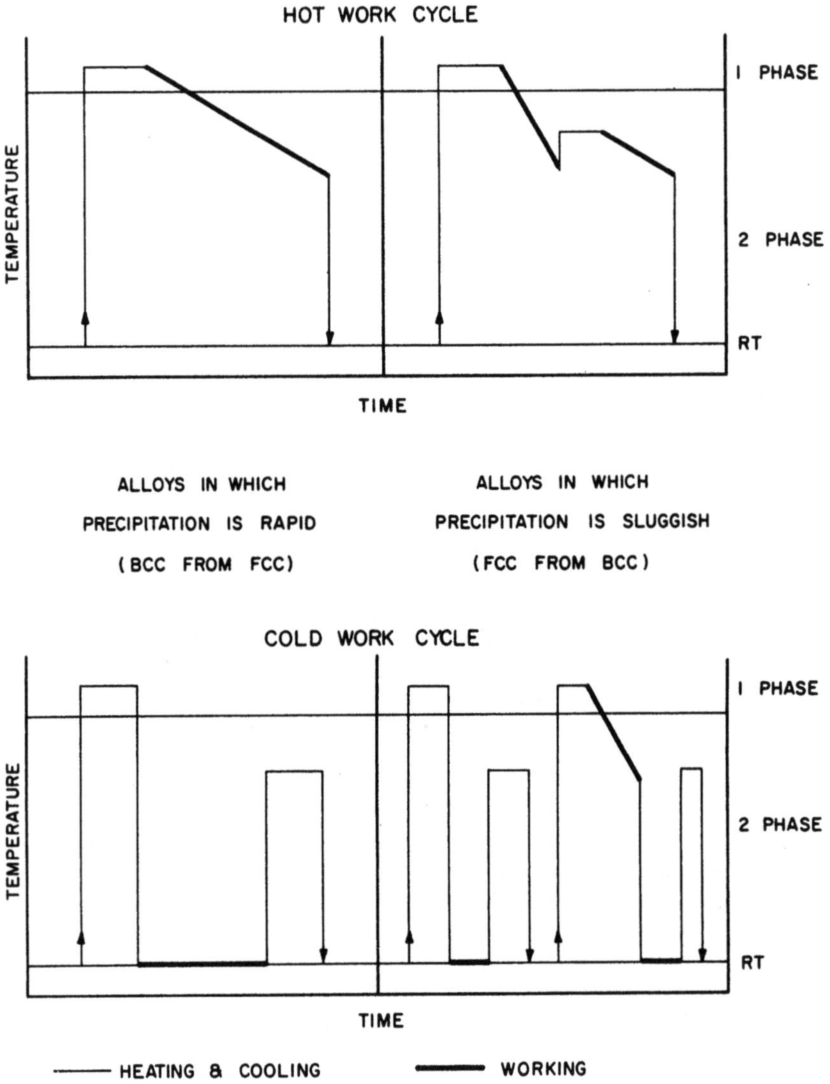

FIGURE 5

Processing Schedules for Producing a Microduplex Structure in Two-phase Ni-Cr-Fe Alloys

coarse-grained high temperature phase decomposes into a very fine-grained mixture of two phases.

If instead of following the quenching procedure, the alloy is slowly cooled from above 275°C to a temperature below the equilibrium transformation temperature, the resultant structure is a coarse-grained aluminum-rich phase with a dispersed zinc-rich precipitate distributed within the coarse grains. Mechanical testing demonstrates that the quenched and transformed material is superplastic, while the slowly cooled material is not.

The thermomechanical treatments necessary to produce the microduplex structure in Fe-Ni-Cr alloys are shown in Figure 5. Two different schedules are necessary, one for alloys in which precipitation of the second phase is rapid (high nickel) and another for alloys in which precipitation is more sluggish (high iron). As can be seen in Figure 5, there is a hot work cycle and a cold work cycle which can be used for each class of alloy. In both hot work cycles, the second phase is first taken into solution in the matrix phase. It then can precipitate and stabilize the grain size of the continuously recrystallizing matrix. This occurs as the working temperature falls. In the cold work cycles the second phase precipitates upon reheating the cold worked single phase matrix, thereby minimizing grain growth. The microduplex structure results only if precipitation accompanies or precedes recrystallization. Essentially the same order of structural size ($<10\mu$) results in both classes of alloys, and the working and heating schedules are simple and easily conducted on conventional equipment.

In Figure 6, it is shown that a two-phase nickel-base alloy

Hot Tensile Bars of a 39%Cr-8%Fe-2%Ti-1%Al-Bal.Ni Alloy in the As-Rolled and Annealed Conditions, Pulled at 1800°F, 0.2 in/min.

could be made to display a whole range of hot tensile behavior from superplastic to conventionally plastic, simply by grain-coarsening anneals(4).

By heat treatment, a coarse-grained, two-phase (macroduplex) structure, shown in Figure 7, can be produced from material originally having a microduplex structure. Both of these structures have useful properties. The properties of a single heat of a 38Cr-18Fe-0.6Ti-Bal.Ni alloy in "microduplex" and "macroduplex" conditions of structure are compared in Table I. In the first row of 1800°F tensile properties, the material was microduplex, having been annealed at 1800°F after hot working. It was superplastic. The material in the second row was converted to the coarse-grained (macroduplex) structure by annealing at 2200°F before testing, and it was not superplastic. The micro-duplex condition shown in the third row of 1800°F properties was then produced by cold working and annealing the coarse macroduplex structure. Superplasticity was "turned on again" after the alloy had been rendered creep resistant in the macroduplex structure. This versatility in properties simply demonstrates the extent to which the structure-property relationship can be controlled.

TABLE I

CYCLIC PRODUCTION OF SUPERPLASTICITY AND CREEP RESISTANCE IN A NICKEL BASE ALLOY(38%Cr-18%Fe-0.6%Ti-Bal. Ni)

	Tensile Test		Stress Rupture		
Structure	UTS psi	El. %	Applied Stress psi	Life, Hour	El. %
	1800° F				
Microduplex	8,700	688	-	-	-
Macroduplex	19,500	39	3,500	712	54
Microduplex	6,400	723	-	-	-
	1600° F				
Microduplex			4,000	0.4	161
Macroduplex			7,000	1,055.6	45

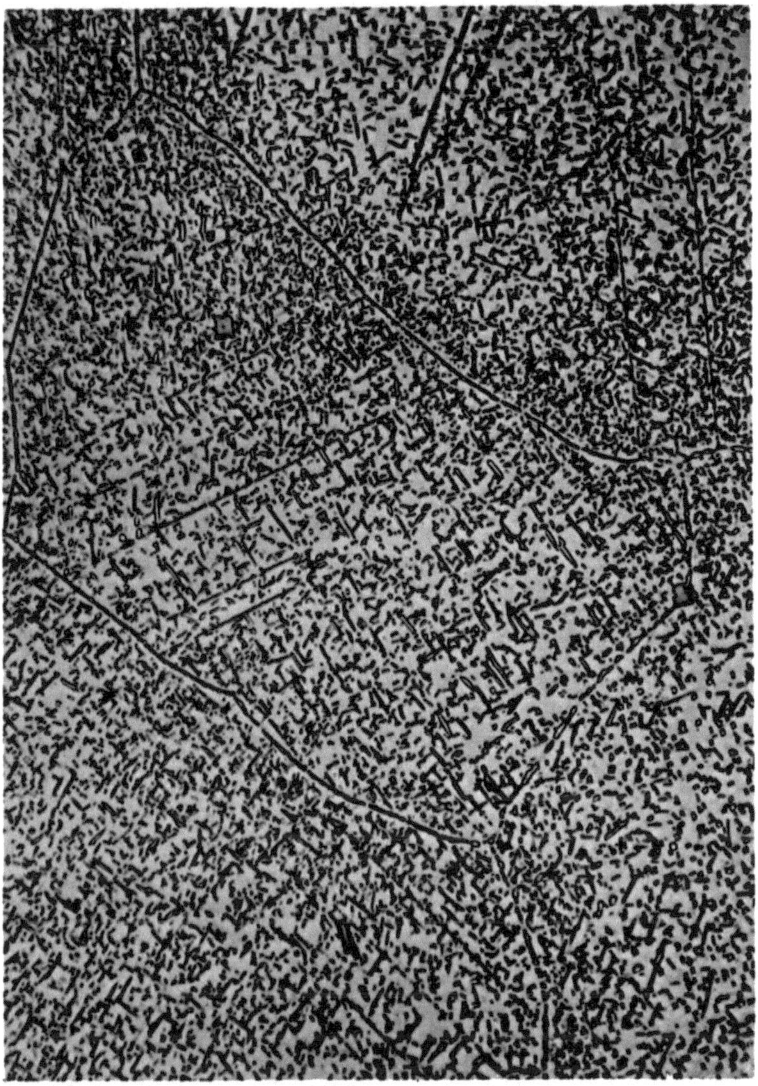

FIGURE 7

Coarse Grained, Two-phase Macroduplex Structure in 38%Cr-18%Fe-0.6%Ti-Bal.Ni Alloy (1000X).

From a practical standpoint, it appears possible to produce an alloy which is capable of superplastic deformation, to form it by an appropriate superplastic process, and then by simple heat treatment to render it creep resistant for elevated temperature service. The effect of the "macroduplex" condition in greatly improving creep resistance is apparent in the 1600°F properties in Table I. In the macroduplex condition the alloy possesses a considerably longer rupture life than the widely used heat resistant alloy, Type 310 stainless steel. Such performance would not normally be expected with conventional processing of such an alloy.

HIGH STRAIN RATE PROCESSING

Although there have been many publications on the effect of microduplex structures on low strain-rate superplastic deformation, there have been comparatively few showing microstructural effects on high speed processing. We have published these for Ni-Cr-Fe alloys(4,5). Although the deformation rates encountered in processes such as forging, hot rolling and extrusion are so high that one would never expect superplasticity in the sense of large tensile extensibility, we nonetheless have found that finer microstructures can lead to processing advantages.

The effects of prior structure and deformation rate on a nickel-chromium-iron alloy have been studied in high strain rate hot rolling experiments. In these tests, strain rates were varied and rolling loads were measured.

Several ingots were forged and rolled to one inch thick plate starting from a temperature of 2200°F. At 2200°F the plate was mostly γ (FCC), but during hot working, α' (BCC) precipitated and caused the retention of a fine-grained two-phase structure. Each plate was then cut in half and one piece was annealed for one hour at 2200°F and water quenched to coarsen the grain size. Both halves were reheated to either 1800, 1900, or 2200°F and immediately rolled in one pass to one-half inch thick. Roll separating force is plotted in Figure 8 versus second rolling temperature for the coarse-grained plates which had been annealed prior to rolling and for those which had been left in the fine-grained as-hot worked condition prior to final rolling. Note that rolling loads for the prior annealed plates were higher than for the plates which had been left in the fine-grained condition prior to final rolling. This shows that in rolling, the fine-grained material offers less resistance to deformation than does the coarse-grained material.

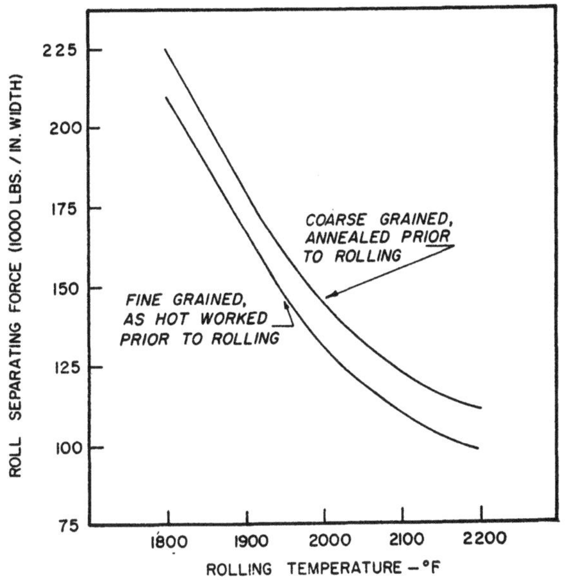

FIGURE 8

Roll Separating Force Versus Rolling Temperature for a 39%Cr-10%Fe-2%Ti-1%Al-Bal.Ni Alloy in Coarse Grained Annealed and Fine Grained Hot Worked Conditions Prior to Rolling.

Comparison of hot working characteristics of a two-phase stainless steel (IN-744) to those of other stainless steels has shown that extrusion pressures (Figure 9) were similar to those for Type 430 stainless steel and considerably less than those for Type 304 stainless steel. The microduplex alloy, which is a mixture of ferritic and austenitic phases, possesses the easy hot workability of a ferritic alloy (Type 430) rather than the characteristics of the austenitic phase or even some intermediate value between ferrite and austenite. The microduplex alloy could be extruded at 1800°F with a flow stress equal to that of Type 304 at 2200°F.

Semi-commercial billets of IN-744 have been extruded to complex shapes and showed excellent die filling ability and low extrusion pressures. Examples of these extruded shapes are shown in Figure 10.

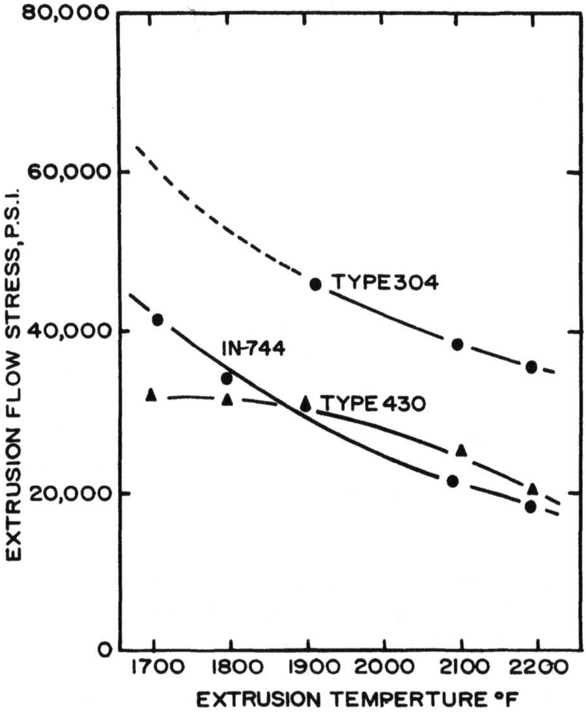

FIGURE 9. Extrusion Flow Stress for Reduction from 3.45 inches to 0.875 inch Round for Several Stainless Steels, Including Type 304, Type 430, and Experimental Composition: 26.4% Cr, 6.2% Ni, 0.36% Ti, 0.035% C, Bal. Fe as Forged from 2200°F

FIGURE 10

Extrusion of IN-744 Showing Excellent Die Filling.

SUPERPLASTIC PROCESSING

In their classic 1964 paper, Backofen, Turner and Avery(2) not only demonstrated unusual tensile extensibility in the superplastic Zn-Al alloy, but also that by simple pressure forming techniques on sheet material superplasticity might easily be applied to practical forming operations. Fields(7) has demonstrated that useful parts of Zn-Al may be fabricated by vacuum forming techniques similar to those common in the forming of polymeric materials. Similar results have recently been shown for the application of a Zn-Al alloy for the production of automotive parts by superplastic forming(8,9). Holt and co-workers have shown that a number of candidate alloys could be shaped superplastically in realistically short times(10,11). Johnson has shown interesting possibilities with dieless drawing processes(12).

At our laboratory, we have been chiefly concerned with experiments on two-phase nickel-base alloys and stainless steels. Similar to other lower melting point superplastic alloys, we have found that these higher temperature alloys can be shaped by rather simple procedures. As in the old chemistry laboratory demonstration, a hollow glass rod when heated to its softening

FIGURE 11

Metallic Capillary Tubes Produced by Superplastic Deformation.

point can be easily drawn to a capillary tube. Figure 11 shows that a superplastic nickel-base alloy can be deformed in a similar manner to produce a metallic capillary tube. In this experiment standard tensile bars were hollowed and then pulled at crosshead rates of 0.2 and 2.0 in/minute at 1800°F. Figure 12 shows dishes of a two-phase stainless steel (IN-744) and a two-phase nickel-base alloy. These were produced by clamping sheet in the rig shown, drawing a partial vacuum, heating to 1700 or 1800°F and holding for a few hours. The times involved seem large. However, the radius of the dish was two inches and the sheet thicknesses were 0.02 inches for the stainless steel and 0.1 inch for the nickel-base alloy. For forming such a dish at one atmosphere pressure differential, the maximum stresses expected for these forming conditions would be 1800 psi for the stainless steel and only 290 psi for the nickel-base alloy.

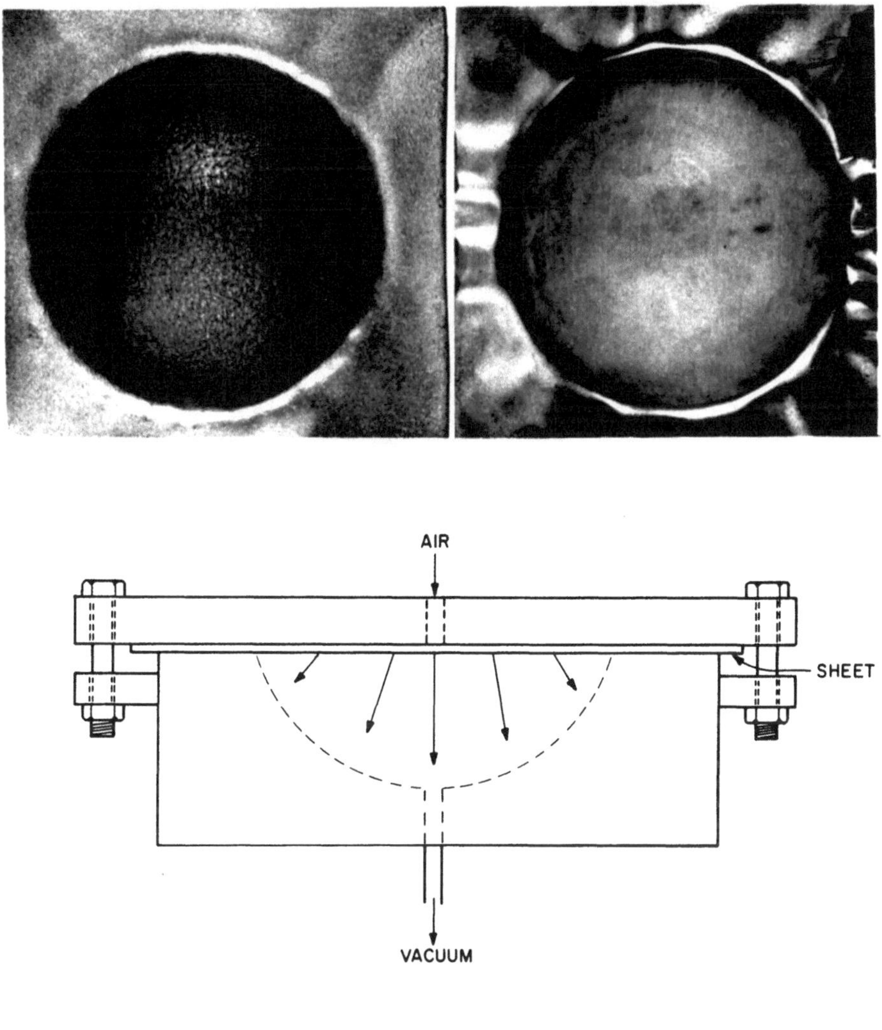

FIGURE 12

Dishes of a Higher Nickel Alloy and a Two-phase Stainless Steel Produced by Vacuum Forming and a Schematic Design of the Vacuum Forming Apparatus

FIGURE 13

Bulged Tubular Part of IN-744 Superplastically Formed by Hot Pressing.

Figure 13 shows a bulged tube geometry of IN-744, having an original wall thickness of 0.04 inches which was made by hot pressing a tube filled with graphite powder in a graphite die. Deformation was accomplished in about 20 seconds at 1700°F with an applied pressure of 3000 psi. The part shows excellent reproduction of the surface of the graphite die. This experiment shows that when working forces are increased in a pressure forming process, useful parts of IN-744 could be made in short periods of time. It also demonstrates that very cheap die materials can be used, inasmuch as the working forces for superplastic forming can be very low.

Several very promising results applying the principles of superplastic forming operations to complex nickel-base superalloys have recently been published. By a process known as GATORIZING* some of the strongest nickel base alloys, normally employed only in cast form, are prepared for convenient close tolerance forging(13). The process involves a working operation somewhat below the normal recrystallization temperature to develop a grain size sufficiently small to require 10,000X magnification for resolution. In this fine-grained condition the alloys can be deformed easily by press forging or superplastically in tension. They are then heat treated to coarsen the grain size and restore creep resistance(14). It has been observed that the putty-like state of the alloy allows close tolerances, complex configurations and lighter forging equipment(15).

Similar observations of superplasticity in superalloys made from consolidated powder have been made(14-18). Again the large tensile deformability resulted from the initial fine grain size of the consolidated product. Heat treatments to coarsen the grain substantially improved the creep-rupture life of the product(16-18).

It now appears well confirmed that the superplasticity observed in the tensile test can be beneficial in conventional metal working operation(4). Even in complex alloys fine grain size can be made to promote superplasticity and then coarsened to "turn off" superplasticity and cause the alloy to show conventional or even creep-resistant properties(1,4,5,26).

PROCESS DESIGN CALCULATIONS

Deformation parameters from tensile tests can be useful information for designing superplastic forming processes. Published data for a superplastic nickel-base alloy having a grain size of 2.8 microns, indicate that at 1800°F for stress up to 15,000 psi and strain rates up to 4 in/in/min, deformation behavior can be represented by the equation(4):

$$\sigma = 7500\ \dot{\varepsilon}^{\frac{1}{2}}$$

From these data one may estimate the minimum times to accomplish a given amount of elongation for several loading or deformation conditions. A value of 200% elongation (or $L_f = 3L_o$ where L is a section length) has been arbitrarily selected for purposes of demonstration. The deformation process might be conducted in several ways: A, constant elongation rate; B, constant applied force, or C, constant true stress. Each of

*Trademark of Pratt and Whitney Aircraft Division, United Aircraft Corporation

these would be accomplished in very short times as follows:

A. Constant Elongation Rate

Such a process might be accomplished with a piston moving at a fixed velocity. For the stress not to exceed 15,000 psi, the engineering strain rate should be no more than 4 in/in/min to ensure superplastic flow. Following the procedures shown by Holt et al.(10,11) the time to accomplish 200% elongation would be:

$$\frac{1}{L_o}\frac{dL}{dt} = \frac{1}{L_o}\frac{\Delta L}{t} = 4$$

$$\Delta L = 2L_o$$

$$t = \tfrac{1}{2} \text{ minute}$$

B. Constant Applied Force

In this process, a sample would be deformed with a constant load. For the true stress not to exceed 15,000 psi at 200% uniform elongation, the initial stress should be 5000 psi. Assuming no necking the instantaneous true stress is:

$$\sigma_i = \frac{PL_i}{A_oL_o}$$

By substitution

$$\frac{PL}{A_oL_o} = 5000\,\frac{L}{L_o} = 7500\,\left(\frac{1}{L}\frac{dL}{dt}\right)^{\frac{1}{2}}$$

$$\left(\frac{5000}{7500L_o}\right)^2 t = \int_{L_o}^{3L_o} \frac{1}{L^3}\,dL$$

$$t = \left(\frac{7500}{5000}\right)^2 \frac{L_o^2}{2}\left(\frac{1}{L_o^2} - \frac{1}{9L_o^2}\right)$$

$$t = 1 \text{ minute}$$

C. Constant True Stress

If a process could be designed where the true stress remained constant during deformation, then the true strain rate would also remain constant. Thus for 200% elongation at a true strain rate of 4 in/in/min, the necessary time would be:

$$t = \frac{1}{\dot{\varepsilon}} \int_{L_o}^{3L_o} \frac{dL}{L} = \frac{\ln 3}{4}$$

$$t = 0.275 \text{ minute}$$

CONCLUSIONS

1. For an alloy to exhibit superplasticity, it must be possible to produce grain sizes in the order of ten microns or less which are stable from coarsening in reasonable times at temperatures in excess of about 50% of the absolute melting point.

2. The production of the necessary microstructure usually requires hot or cold working steps in a thermomechanical processing sequence.

3. Once the proper microstructure is produced, the fine structure contributes to lower load requirements for high strain rate, non-superplastic hot working operations. At low strain rates, this fine structure is an absolute necessity for obtaining the unusual deformation possibilities afforded by superplasticity.

4. Both experience and calculation indicate that forming of a microduplex alloy can be accomplished in realistically short time intervals, on the order of minutes in length.

REFERENCES

1. Hayden, H. W., Gibson, R. C. and Brophy, J. H., "Superplastic Metals," Scientific American, 220 No. 3 (1969) p. 28.

2. Backofen, W. A., Turner, I. R., Avery, D. H., "Superplasticity in and Al-Zn Alloy," Transactions Quarterly ASM, 57 (1964) pp. 980-990.

3. Pearson, C. E., "The Viscous Properties of Extruded Eutectic Alloys of Lead-Tin and Bismuth-Tin," J. Inst. Metals, 54, (1934) p. 111.

4. Hayden, H. W., Gibson, R. C., Merrick, H. F., Brophy, J. H., "Superplasticity in the Ni-Fe-Cr System," Transactions Quarterly ASM, 60 (1967) p. 3.

5. Gibson, R. C., Hayden, H. W., and Brophy, J. H., "Properties of Stainless Steels with a Microduplex Structure," Transactions Quarterly ASM 61 (1968) p. 85.

6. Gibson, R. C., Hayden, H. W., Mihalisin, J. R., Brophy, J. H., U. S. Patent 3,519,419, July 7, 1970.

7. Fields Jr., D. M., Oral Presentation, Metal Science Forum on Superplasticity, ASM Annual Meeting, October 17, 1967, Cleveland, Ohio.

8. "New Material for Car Bodies: Superplastic ZnAl Alloy Sheet," Metallurgia, 78, Dec. 1968 p. 247-248.

9. "Superplastic Metal Alloy," Design Engineering, May 1969, 15 (5) p. 84-85.

10. Holt, D. L. "Superplastic, Fine-Grain Alloys in the Forming of Sheet Metal," Proceedings of the 16th Sagamore Army Materials Conference, 1969, to be published.

11. Thomsen, T. H., Holt, D. L. and Backofen, W. A., "Forming Superplastic Sheet Material in Bulging Dies," Met. Eng. Quart., 10 (1970) p. 1.

12. Johnson, R. J., "Superplastic Metals," Design Engineering March (1969).

13. P&WA Manufacturing Management and Machines, Pratt and Whitney Aircraft Div. of United Aircraft Corp., Dec. 1969, p. 21.

14. Moore, J. B., Athey, R. L., U. S. Patent 3,519,503 July 7, 1970.

15. American Metal Market, December 10, 1969.

16. Freche, J. C., "New Nickel-Base Alloys for Aerospace Applications," Aerospace Structural Materials Conference, NASA Lewis Research Center, Cleveland, Ohio, November 18-19, 1969.

17. Freche, J. C., Waters, W. J., Ashbrooke, R. L., "Evaluation of Two Nickel-Base Alloys, Alloy 13C and NASA TAZ-8A, Produced by Extrusion of Prealloyed Powders," Report NASA TND-5248, NASA Lewis Research Center, Cleveland, Ohio, May 1969.

18. Reichman, S. H. and Smythe, J. W., "Superplasticity in P/M IN-100 Alloy," Int. Journal of Powder Metallurgy, 6 No. 1 (1970) p. 68.

19. Floreen, S., "Superplasticity in Pure Nickel," Scripta Met., 1 (1967) 19.

20. Alden, T. H., "Superplastic Behavior of a Solid Solution Sn-1%Bi Alloy," Trans. AIME, 236, (1966) 1633.

21. Hayden, H. W., and Brophy, J. H., "The Interrelation of Grain Size and Superplastic Deformation in Nickel-Chromium-Iron Alloys," Transactions Quarterly ASM, 61 (1968) p. 542.

22. Alden, T. H., "Strain Hardening and Recovery in Superplastic Pb-5%Cd," Trans. Quart. ASM, 61 (1968) p. 559.

23. Packer, C. M., and Sherby, O. D., "An Interpretation of the Superplasticity Phenomenon in Two-Phase Alloys," Trans. Quart. ASM, 60, (1967) 21.

24. Ball, A., and Hutchison, M., "Superplasticity in the Aluminum-Zinc Eutectoid, Metal Sci. J., 3, (1969) 1.

25. Avery, D. H. and Backofen, W. A., "A Structural Basis for Superplasticity," Trans. Quart. ASM, 58 (1965) p. 551.

26. Gibson, R. C., Brophy, J. H., "Microduplex Nickel-Iron-Chromium Alloys, " Proceedings of the 16th Sagamore Army Materials Conference, 1969, to be published.

INDEX

MIX
Papier aus verantwortungsvollen Quellen
Paper from responsible sources
FSC® C105338

If you have any concerns about our products,
you can contact us on
ProductSafety@springernature.com

In case Publisher is established outside the EU,
the EU authorized representative is:
Springer Nature Customer Service Center GmbH
Europaplatz 3, 69115 Heidelberg, Germany

Printed by Libri Plureos GmbH
in Hamburg, Germany

MIX
Papier aus verantwortungsvollen Quellen
Paper from responsible sources
FSC® C105338

If you have any concerns about our products,
you can contact us on
ProductSafety@springernature.com

In case Publisher is established outside the EU,
the EU authorized representative is:
Springer Nature Customer Service Center GmbH
Europaplatz 3, 69115 Heidelberg, Germany

Printed by Libri Plureos GmbH
in Hamburg, Germany